AF829031

Phytobiomes: Current Insights and Future Vistas

Manoj Kumar Solanki •
Prem Lal Kashyap • Baby Kumari
Editors

Phytobiomes: Current Insights and Future Vistas

Springer

Editors
Manoj Kumar Solanki
Department of Food Quality & Safety
Institute for Post-harvest and Food Sciences,
The Volcani Center, Agricultural Research
Organization
Rishon LeZion, Israel

Prem Lal Kashyap
Division of Crop Protection
Indian Institute of Wheat
and Barley Research
Karnal, Haryana, India

Baby Kumari
Department of Biotechnology
Vinoba Bhave University
Hazaribag, Jharkhand, India

ISBN 978-981-15-3150-7 ISBN 978-981-15-3151-4 (eBook)
https://doi.org/10.1007/978-981-15-3151-4

© Springer Nature Singapore Pte Ltd. 2020
This work is subject to copyright. All rights are reserved by the Publisher, whether the whole or part of the material is concerned, specifically the rights of translation, reprinting, reuse of illustrations, recitation, broadcasting, reproduction on microfilms or in any other physical way, and transmission or information storage and retrieval, electronic adaptation, computer software, or by similar or dissimilar methodology now known or hereafter developed.
The use of general descriptive names, registered names, trademarks, service marks, etc. in this publication does not imply, even in the absence of a specific statement, that such names are exempt from the relevant protective laws and regulations and therefore free for general use.
The publisher, the authors, and the editors are safe to assume that the advice and information in this book are believed to be true and accurate at the date of publication. Neither the publisher nor the authors or the editors give a warranty, expressed or implied, with respect to the material contained herein or for any errors or omissions that may have been made. The publisher remains neutral with regard to jurisdictional claims in published maps and institutional affiliations.

This Springer imprint is published by the registered company Springer Nature Singapore Pte Ltd.
The registered company address is: 152 Beach Road, #21-01/04 Gateway East, Singapore 189721, Singapore

Preface

This book addresses the updates about the plant-associated microbiomes and their contemporary uses. To satisfy the food demands of the global population, advanced technology-based research is needed that can extract the information from the plant metabolism and microbial gene pools up to the complexity. Modern biotechnological tools can unlock the limitations of agricultural practices. However, the application of these tools is not well-equipped. Eco-friendly agriculture using microbial inoculants had positive influences on soil/plant health. Exploring the plant-associated microbial niches, especially endophytes, epiphytes, and soil microbes and how they are benefitting each other, can open new insights to develop sustainable agriculture practices by using consortia of microbes as plant helpers that recover the imbalanced agriculture systems and manage pathogenic diseases.

This book inculcates the gap between soil and plant helper microbiomes and their importance in the intensification of sustainable agriculture. New insights of phytobiome are explored in various chapters on a variety of interrelated aspects of the fascinating areas like plant microbial interaction, integrated pest management, soil fertility intensification, sustainable crop production, and disease management. This book is also entitled to yield a plethora of information about how beneficial plant microbiomes are currently being utilized for smart farming practices. To resolve the global food problem without harming the soil and environment health, this book covers valuable information regarding the significance of microbes in the amelioration of plant and soil health. Application of advanced molecular tools in plant disease diagnosis and pathogen isolation has also been discussed in order to advance the crops and human health. Some chapters have also been written to present the latest information related to phytobiomes and their range as far as their utility is concerned. Information on plant health promoter, endophytes as microbial elicitors, fly ash and plant health, zinc solubilizers, and their role in ecosystem engineering towards sustainable agriculture have also been presented.

This book is intended for everyone who is directly or indirectly involved in agriculture, bioinformatics, and all disciplines related to microbial biotechnology. These include academicians, scientists, and researchers at universities, institutes, industries, and government organizations who want to understand microbial linkages in a shorter time. This book also contains essential information that will help the nonspecialist readers to understand progressive research. We are confident that the current edition will be a milestone as chapters having updated information are anticipated to be published in *Phytobiomes: Current Insights and Future Vistas*. The views conveyed by the expert/authors in their respective chapters are based on their vast experience in the plant-microbe interaction and associated research areas. We thank all the contributors to this book, and we are obliged for their valuable contributions.

Rishon LeZion, Israel — Manoj Kumar Solanki
Karnal, Haryana, India — Prem Lal Kashyap
Hazaribag, Jharkhand, India — Baby Kumari

Contents

About the Editors

Dr. Manoj Kumar Solanki Ph.D. (Microbiology) is working as a visiting scientist at the Volcani Center, Agricultural Research Organization, Israel. He was awarded a visiting scientist fellowship from the Guangxi Academy of Agriculture Sciences, China, and a senior research fellowship from the Indian Council of Agriculture Research (ICAR), India. He also worked as a Research Associate at the International Crops Research Institute for the Semi-Arid Tropics (ICRISAT), India in a Department of Biotechnology funded project. He has engaged in various research activities on plant-microbe interactions, soil microbiology, plant disease management, and enzymology, and has published several book chapters and research papers in prominent international journals.

Dr. Prem Lal Kashyap Ph.D. (Plant Pathology) is working at the ICAR-Indian Institute of Wheat and Barley Research (IIWBR), India. His major research areas include wheat pathology, biocontrol, diagnostics, microbial nanotechnology and host–parasite interactions. He has published over 60 research papers in high impact factor journals along with several book chapters, technical bulletins and books. In recognition of his significant achievements in the field of crop protection, he has received several prestigious awards including the NAAS Young Scientist Award, Dr. Basant Ram Young Scientist Award, Prof. Abrar Mustafa Memorial Award, M.K. Patel Memorial Young Scientist Award, and a NAAS Associateship at the national level.

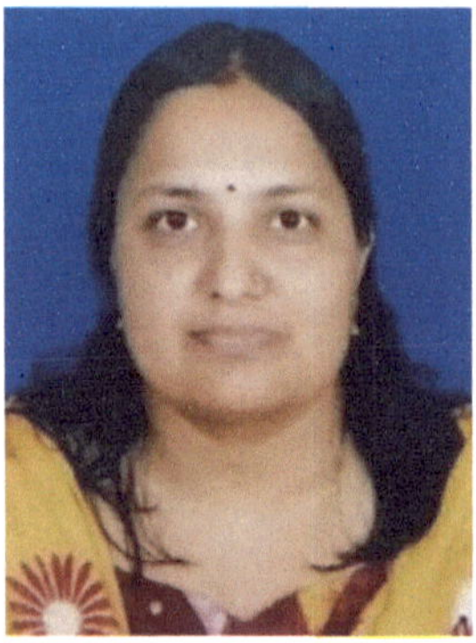

Dr. Baby Kumari MSc, Ph.D. (Biotechnology) has worked as a researcher (part of her Ph.D. work) at the Bhabha Atomic Research Centre, India. She has engaged in various research activities on plant-microbe interactions, soil microbiology, plant disease management, and nanotechnology, and has published many research papers in prominent international journals. Her research interests include plant-microbiome interactions, especially in medicinal plants, as well as nanoparticle biosynthesis and its application in plant disease management, biosensors, and bioremediation.

Phytobiomes: Role in Nutrient Stewardship and Soil Health

1

Madhumonti Saha, Abhijit Sarkar, Trisha Roy, Siddhartha Shankar Biswas, and Asit Mandal

Abstract

The enormous potential of the phytobiome for better plant growth and productivity is an essential tool for sustainable agronomic practices for eco-friendly cultivation processes. However, microbes in relation to plants, environment, omics, and their interactions are still to be clearly explored. Current predictive formulae like niche origin, ecological traits, evolution, genetics or heterosis, and resource trades are not mutually exclusive. This chapter mainly emphasizes on phytobiome related to soil fertility, nutrient cycling, plant growth, and soil health. Plant-associated phytobiome such as rhizospheric and phyllospheric played a significant role in the enhancement of plant growth and yield. These organisms form multifarious networks that are established and regulated through nutrient cycling, competition, antagonism, and chemical communication mediated by a diverse array of signaling molecules. The integration of knowledge of signaling mechanisms with that of phytobiome members and their networks will lead to a new understanding of the fate and significance of these signals at the ecosystem level. Such an understanding could lead to new biological, chemical, and breeding strategies to improve crop health and productivity. Soil organic matter (SOM) is a heterogeneous mixture of materials that range in the stage of decomposition from fresh plant residues to highly decomposed material known as humus. SOM is made of organic compounds that are highly enriched in carbon. Though half of the global population depends on fertilizer N, atmospheric N fixation by rhizospheric microbes is essential for plant productivity in low N soils. Many plants form symbiotic associations between their roots and specialized fungi in the soil

M. Saha · A. Sarkar (✉) · A. Mandal
ICAR-Indian Institute of Soil Science, Bhopal, Madhya Pradesh, India

T. Roy
ICAR-Indian Institute of Soil and Water Conservation, Dehradun, Uttarakhand, India

S. S. Biswas
ICAR-National Research Centre for Orchids, Pakyong, Sikkim, India

© Springer Nature Singapore Pte Ltd. 2020

M. K. Solanki et al. (eds.), *Phytobiomes: Current Insights and Future Vistas*,
https://doi.org/10.1007/978-981-15-3151-4_1

known as mycorrhizae; the roots provide the fungi energy in the form of carbon, while the fungi provide the plant with often-limiting nutrients such as phosphorus. Besides, some soil microbiomes play a crucial role in the solubilizing stable form of potassium and micronutrients by releasing some metabolites which act as chelating agents. We explain recent information and cracks in these areas using phytobiomes.

Keywords
Phytobiome · Soil fertility · Nutrient cycling · Soil organic matter · Soil health

1.1 Introduction

Developing countries like India need to produce a double quantity of foods to feed the population of the nation as well as the global crowds, whereas population growth touches 1.1% or 83 million annually. Feeding is not only mean full stomachs, but it mainly focused on nutritional security. The overall impact of rising demand for food production will depend on improved productivity via high-yielding varieties and sustainable use of bioenergy (FAO 2017), whereas these higher yields of agricultural crops are vigorously influenced by regular nutrient accessibility and higher soil quality for promoting plant growth. Hence, there is an excellent chance of input-output imbalance, meticulous use of chemicals, organic matter depletion, removal of nutrients, and massive exhaustion of soil system through the anthropogenic activity for sustainable food production (Godfray et al. 2010). A sustainable food production system, which delivers both adequate nutrition and appropriate energy to people in the society, also improves the natural environment intergenerationally. To maximize the sustainability of production, we require a good strategic plan, which would help to acquire knowledge regarding components that may establish and sustain a healthy, productive agroecosystem. Integration and incorporation of this information into the present crop production system may further increase the total coverage of existing farmlands (mostly rehabilitated from degraded and marginal lands). Hence, to achieve the food and nutritional security with limited increased lands, we should first concentrate the interactions among the components of phytobiomes.

The word "phytobiome" comprises *phyto* (plants) and *biome* (ecological area), which are very specific for plants and their surroundings. It includes plant interaction biology with sounding environments and living and nonliving objects (Leach et al. 2017). Living entities include microorganisms, macroorganisms, and other plants and animals, and nonliving entities include soil, air, and climate. Simply, we can explain phytobiome as a conjoint venture of the rhizosphere, phyllosphere, biosphere, spermosphere, hyphasphere/mycorrhizosphere, and detritusphere (Nelson 2004; Frey-Klett et al. 2007; Vorholt 2012; Hacquard et al. 2015; Leach et al. 2017). Thus it is easy to say that phytobiome includes all entities that have a direct and indirect influence on plants and vice versa (Beans 2017). Phytobiome extends beyond the plant microbiome and unites facets including crop improvement and production, weather data modeling and forecasting, pest pollinators and pathogens, and data generation along with dissemination (Fig. 1.1). Moreover, phytobiome considers animals, pests, physical environment, management, and other pillars of sustainable agroecosystem that boost our scientific understandings and analytical aspects about plant microbiome also leveraging the computational power of "omics" (Young and Kinkel 2017). Spanning from fundamental to applied research aspects, researcher associated with phytobiome cover the areas of biotechnology, genetics-breeding, seed technology, agronomy, soil fertility and fertilizers, microbiology, plant pathology, nematology, entomology, meteorology, crop physiology, economics, statistics, extension, and more branches of agriculture (Fig. 1.1, Young and

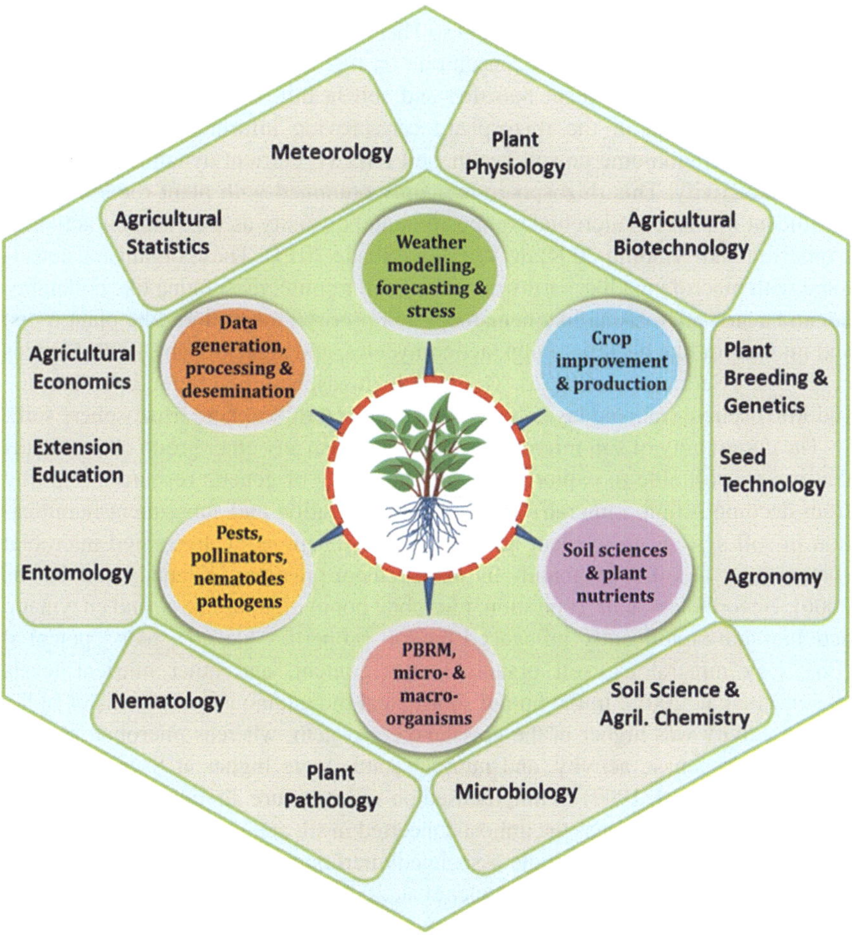

Fig. 1.1 Phytobiome components and their interrelationship with different branches of agriculture. (Conceptualized and modified from Phytobiomes: A Roadmap for Research and Translation 2016; Young and Kinkel 2017)

Kinkel 2017). However, composition, characteristics, and functions of healthy phytobiome, principally soil-plant-microbe interaction, are yet to be understood entirely.

1.2 Plant-Soil-Microbe Interactions in a Typical Phytobiome

In a typical phytobiome, phyllosphere gets more affected from the discontinuity of weather abnormalities (diurnal cycle), leaf surface, plant metabolism, environmental conditions (rain, wind, direct sunlight, and UV radiation), and nutrient availability (oligotrophic) than the rhizosphere (Leveau and Lindow 2001; Miller et al. 2001; Remus-Emsermann et al. 2012). Phyllosphere is a partial environment as compared to rhizosphere (Vorholt 2012). Though phyllosphere is also an essential aspect in the phytobiome study, nutrient cycling and soil health are controlled by the rhizosphere. Accordingly, we would like to concentrate on rhizosphere and rhizospheric engineering associated with plant nutrition and soil health. Discussion on soil-plant-microbe interaction in the rhizosphere can provide information regarding the influence of phytobiome on soil health, soil fertility, nutrient dynamics, as well as crop productivity. This rhizosphere is mainly crammed with plant roots having a significant impact on microbial composition and diversity as well as their activities (Andersen and Winding 2004; de Vries and Shade 2013). These organisms constituted with macrofauna like earthworms, ant, and termite; mesofauna like collembolan and acari; microfauna like nematodes and protists; macroflora like plant roots; and microflora like bacteria, fungi, actinomycetes, and algae (Coyne 1999; Mendes et al. 2013; Sarkar et al. 2017a). The further rhizosphere is differentiated into the endorhizosphere (mucoid layer coated root) and ectorhizosphere (rhizosphere soil).

The community of soil microbes is so diverse than any other group of organisms that we are still able to explore very little diversity of genetic resources. Starting from decomposition, mineralization, or immobilization and subsequent humification in soil system up to plant growth and the functions of diversified microbial communities are affected mainly by management practices (Kennedy and Stubbs 2006). Nevertheless, soil respiration, microbial population, biomass, and enzymatic activities are significantly influenced by soil properties like pH, redox potential (Eh), clay mineralogy, soil organic matter content, and other nutrient levels (Sessitsch et al. 2001). In the humid subtropical mountainous ecosystems of India, urease activity was higher in the grassland ecosystem, whereas microbial population, dehydrogenase activity, and nutrient status were higher at Sacred Orchard (Arunachalam et al. 1999). While focused on soil structure, literature demonstrated that microbial biomass was the utmost concerted in silt and clay-sized soil particles, especially in silt and clay particle-associated micropores (5–30 mm) (Hassink et al. 1993a; van Gestel et al. 1996; Kandeler et al. 2000). Additionally, invertase, urease, and alkaline phosphatase activities were reported to be highest in the silty and clayey type of soil. Xylanase activity, which is considered to be the indicator of fungal activity (Kandeler et al. 2000), was found to be higher in sand particles (Stemmer et al. 1998a, b; Kandeler et al. 1999; Kirchmann and Gerzabek 1999).

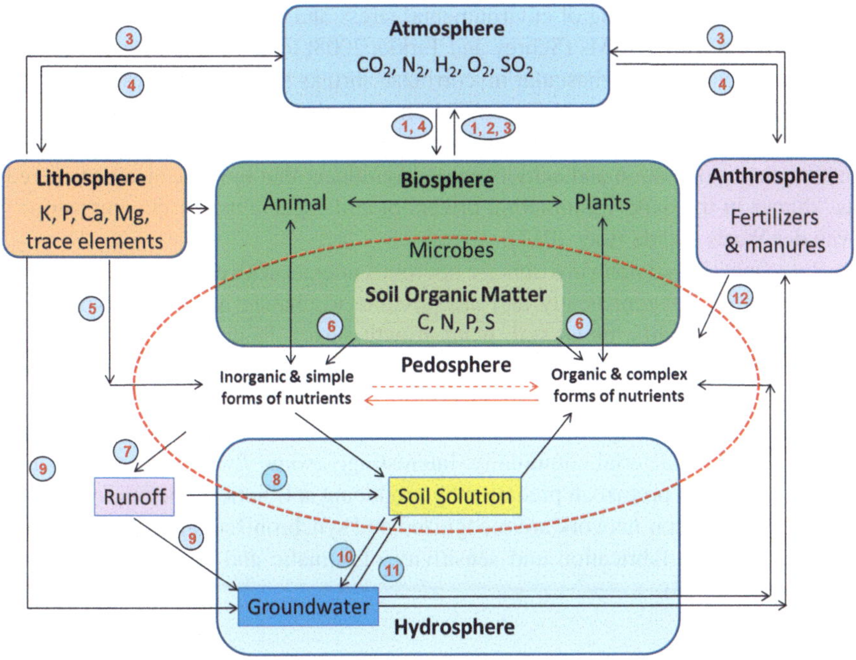

Fig. 1.2 Nutrient cycling and interconnectivity of the atmosphere, lithosphere, biosphere, pedosphere, hydrosphere, and anthrosphere in a typical phytobiome
[1]photosynthesis; [2]respiration; [3]gaseous exchange; [4]deposition; [5]weathering; [6]decomposition; [7]erosion; [8]infiltration; [9]percolation; [1]leaching; [11]capillary rise; [12]manual application; red dashed arrow, immobilization; red solid arrow, mineralization

Though the population of fungi is much lower in grassland soil and arable fields, but its crucial role in the initial fragmentation of soil organic matter could not be ignored (Brussaard et al. 1990; Hassink et al. 1993b). Nutrient cycling and interconnectivity of the atmosphere, lithosphere, biosphere, pedosphere, hydrosphere, and anthrosphere in a typical phytobiome are represented in Fig. 1.2.

1.3 Beneficial Rhizospheric Microbes and Phytobiome

Rhizospheric microbes have an immense potential in sustainable, cutting-edge crop production. Furthermore, beneficial rhizospheric microbes (BRMs) are obligatory in biogeochemical nutrient cycles that drive soil fertility and crop productivity for the span of decades (Sarkar et al. 2017a). In general, these BRMs enhance the soil and plant productivity by means of (a) secreting hormones and other regulatory growth chemicals, (b) acquisition and bioavailable felicitation of essential nutrient elements, (c) inhibition of pathogens through biocontrol agent production, and (d) abiotic stress tolerance (Yang et al. 2009).

A better understanding of environmental stress and its bio-modulation has also been possible with PBRMs (Schrey and Tarkka 2008; Ryan et al. 2009). Symbiotic association of rice and arbuscular mycorrhizae shrinks the drought stress effects by producing an antioxidative response and photosynthetic efficiency (Ruiz-Sanchez et al. 2010). Even though PBRMs and other rhizospheric microbes are insignificant in size, their population and activity are so significant that they are also considered as "dinner in the dark: illuminating drivers of soil organic matter decomposition" (van der Waals and de Boer 2017).

Rhizospheric and non-rhizospheric microbes interact with each other either synergistically or antagonistically and are possible to change microbial community structure through differential soil, plant growth stages, feeding behavior on other taxa, acidity, reducing prey abundance, etc. (de Vries and Shade 2013; Leach et al. 2017). Preferential feeding like predation of gram-negative bacteria by protozoa (Andersen and Winding 2004) and predation over harmful bacteria by nematode may change the bacterial community. Interestingly, some *Pseudomonas* sp. could be able to prevent protozoan predation by producing antibiotics. The microbial collaborations and their network are recognized and synchronized but sometimes passive, through the fabrication and sensitivity of somatic and biochemical signals (Leach et al. 2017).

1.4 Phytobiome and Nutrient Stewardship in Soil

Though plants exclusively utilize inorganic forms of nutrients, soil organic matter (SOM) plays the "key to soil fertility" and nutrient dynamics in every soil systems, because SOM is the source for almost every essential plant nutrients. The content of SOM is considered as the indicator soil quality, microbial activity, and buffering capacity of the soil. Correspondingly, SOM could also be regarded as a critical element of phytobiome. Along with SOM dynamics, soil structure and nutrient cycling in the soil are greatly influenced by microbial processes, which are frequently affected by management (Kennedy and Stubbs 2006).

1.4.1 C-Cycle and Phytobiome

Natural and agricultural-based soils are the significant source and sink for the dynamic flow of atmospheric CO_2 as a result of the soil microbe-derived respiratory flux of CO_2. Hence, concentrations of atmospheric CO_2 are more sensitive to minute changes in the soil carbon cycle. Higher plants utilized atmospheric CO_2 for photosynthesis, whereas microbes play the key role to maintain the source-sink relationship of SOC by altering the mechanisms involved in plant symbionts, detritivores, and phytopathogens. Moreover, preserve soil C by their death cells and plant roots (Amato and Ladd 1992; Lal 2004; King 2011). It has been reported that land use and management system influence soil ecosystems with the proliferation of microbial diversity which favor sequestration of C in soil (Singh et al. 2010). So, here we will discuss microorganisms responsible in the global carbon cycle as well as climate change. This detail is required to maintain practical management of the soil-plant continuum to nourish the microbes for stimulating soil carbon storage in agricultural soil.

1.4.1.1 Endurance of Soil Organic Matter and Its Microbial Decomposition

The input of organic matter into the soil system is followed by two ways: (a) aboveground plant litter, crop residue, and organic manure including soluble organic carbon which penetrate into the soil with irrigation water and rainfall and (b) belowground roots and its rhizodeposition. Rhizodepositions often consist of simple organic compounds like carbohydrates, amino acids, organic acids, alcohols, etc. Perhaps with time rhizodeposition becomes humidified, which contains complex compounds like cellulose, hemicellulose, lignin, etc. (Wallenstein and Weintraub 2008). Mycorrhizal fungi play a vigorous role in the denaturation of the humified polymer. This mycorrhizal symbiosis seems to be found in approx. 85% of all plant communities usually in herbaceous crop (Smith and Read 2008). An experimental study shows that fungal partners are satisfied with the shifting of 20% (Nakano-Hylander and Olsson 2007) or even 30% (Drigo et al. 2010) of total assimilated carbon by plants having intense effects on rhizodeposition. A fraction of the plant carbon that is moved to the mycelia is rapidly respired back to the atmosphere, which results in a crack in the soil carbon cycle.

A large portion of biomass is decomposed by heterotrophic microbial respiration causing loss of soil C. In addition, a small amount of the original C is reserved in the soil through the formation of stable organic carbon (Reynaldo et al. 2012) that results in increased SOC stocks through C sequestration over time. The main microbial modulators for soil C storage are fungi and bacteria. The fungal/bacterial ratio is closely related with C sequestration potential in soils with more massive fungal abundance being correlated with higher C stock. Greater storage of C in fungal-dominated soils can be endorsed to improved C use efficiency, more extended protection of C in living biomass, and recalcitrant portion resulting in longer resident time of C. Nevertheless, these observations are only relative, and it is still questioned whether fungal communities support soil C storage or whether soil with higher organic C favors soil fungi. Moreover, it can also be debated that fungi can negatively affect C storage due to their higher efficacy to decay recalcitrant litter (Cheng et al. 2012).

1.4.1.2 Microbial Functions in Soil Carbon Cycle Subsequent Climate Change

Responses of climate change through carbon cycle are tangled because of microbial aids, their individual effects, and collaborations with other factors (Bardgett et al. 2008; Singh et al. 2010). A simple example of response to global warming is microbial activity, organic carbon decomposition, and CO_2 release, which may be hastened in response to an increase in temperature (Davidson and Janssens 2006). This is further established by field observations that confirmed the positive correlation between higher respiration rates and elevated temperature regimes. Besides, a sign of secondary positive feedback to raise CO_2 is a result of the photosynthetic production of carbon fertilization, whereas increased atmospheric CO_2 accelerates photosynthesis (Bond-Lamberty and Thomson 2010) and the release of root exudates; consequently more labile carbon will be available for microbial decomposition and respiration. Readily available exudates after increased root deposition may "prime" the turnover of a passive pool of SOM that otherwise would not be subject to decomposition (Koyama et al. 2018). This difficulty is intensified by the diversity of soil ecosystems across the globe, which varies in their function due to their differences in forming factors including climate, organisms, relief, parent material, and time. Main interests on microbial activity in carbon cycle have been raised in peaty and frozen soils, where climatic conditions are not predicted for the addition or conservation of organic material, which may not be in favor of future climate, eventually in the release of sufficient quantities of CO_2 to the atmosphere (Srivastava et al. 2017). Further research in this area is very crucial if we can calculate the impacts and feedbacks between climate change and microbial function.

1.4.1.3 Microbial Growth Dynamics and Energy Balance

Chemotrophic soil microbes fix soil C and N by synthesizing new biomass (Sokatch 1969; Dawson 1974). The energy balance and consequent thermodynamics equations are listed subsequently (adapted from Chen et al. 2003):

$$\varepsilon X \Delta Gr = -\Delta Gs \tag{1.1}$$

where ΔGr is the energy released during respiration, ΔGs is the energy carrier (e.g., ATP) required to synthesized biomass, ε is the coefficient of energy transfer from ATP, and X is the balanced ratio of ΔGr (ATP used) and ΔGs (microorganisms involved).

Stoichiometric yield coefficient (Y, biomass formed the unit amount of C or inorganic ingredients consumed (Dawson 1974; Chen et al. 2003)) is calculated as follows:

$$Y = \frac{\alpha}{\beta(1+X)} \tag{1.2}$$

where α is the biomass (mg) formed from unit kcal energy consumption and β is the mass of organic and inorganic substances (mg) used to produce unit kcal energy:

$$\gamma \max = Yk \tag{1.3}$$

$$k = \frac{Y}{\gamma \max} \tag{1.4}$$

where γmax is the maximum specific microbial growth rate (h^{-1} mg^{-1} biomass) and k is the maximum specific substance utilization rate (mg mg^{-1} biomass h^{-1}) when ignoring microbial decay or maintenance. The rate of respiration, i.e., energy-yielding reactions, is constant and varying between 0.5 and 2.0 mg^{-1} biomass h^{-1} in many heterotrophs, autotrophs, aerobes, and anaerobes.

From this declaration, the maximum specific substance utilization rate (k) and the maximum specific microbial growth rate (γmax) are expressed as Eqs. 1.5 and 1.6, respectively:

$$k = \frac{\beta(0.5 \sim 2.0)(1+X)}{X} \tag{1.5}$$

$$\gamma \max = \frac{(0.5 \sim 2.0)\alpha}{X} \tag{1.6}$$

From Eq. 1.6, it is clear that both energy sources and microorganisms determine the maximum specific microbial growth rate (γmax). Thus, microbial growth in soil is dependent upon the types and diversity of microbes, and SOM content is described by Monod's equation (Sokatch 1969; Dawson 1974; Chen et al. 2003):

$$\gamma = \gamma \max \frac{S}{(Ks+S)} \tag{1.7}$$

where S represents limiting nutrient (mg g^{-1}), γ represents the specific growth rate (h^{-1} mg^{-1} biomass), and Ks is the half-velocity constant (Michaelis-Menten constant). Ks reproduces the affinity of the microorganisms for the growth-limiting nutrient and governs how quick the specific growth rate can reach the maximum rate and for efficient microbial work (Ks should be kept as small as possible). The Contois equation designates heterotrophic microbial growth; Eqs. 1.8 and 1.9 represent the dynamics of SOM depletion and biomass formation:

$$\frac{dA}{dt} = \frac{\gamma \max SA}{Ks+S} - kdA \tag{1.8}$$

$$\frac{dS}{dt} = -\frac{1}{Y}\left[\frac{\gamma \max SA}{Ks+S} - kdA\right] \tag{1.9}$$

where A is the biomass (mg g^{-1}) and k_d is the specific microbial decay rate (h^{-1} mg^{-1} biomass). Complete understanding of the terrestrial microbial association and particular processes that govern the degree and fate of C dynamics will increase the prospect of successful management of the terrestrial ecosystem for increasing the stable C reservoir.

1.4.2 N Cycle and Phytobiome

Nitrogen (N) is an essential nutrient required by plants for their growth and metabolism. The rate of N inputs in the agricultural system has doubled during the last century, potentially affecting both terrestrial and aquatic ecosystems. Even after excess application, it is often misplaced through volatilization, denitrification, or leaching and causes to limit its availability in most of the cultivated soils.

1.4.2.1 Nitrogen Fixation and Transformations

Although N abundantly exists in the atmosphere, the diatomic (N_2) molecule is a comparative inert. Thus, N fixation by reducing the inert N_2 molecule into NH_3 is a complex process with the expense of an enormous amount of energy (Postgate 1982). Soil N stock and its variation depend on N transformation based on microbial dynamics and environmental conditions. The N transformation is mainly controlled by soil and residue C/N ratio (Chen et al. 2003). Other than C/N ratio, soil texture, pH, fertilizer quantity, crop rotation, soil and air temperature, soil moisture, and rainfall are some of the controlling factors of N transformation and stock variations (Nave et al. 2009; Schipper and Sparling 2011). The diazotrophic microbes, which belong to the prokaryotic groups, fix atmospheric N_2 with their normal metabolic process (Galloway et al. 2008). Some of these microbes are free-living in the soil like *Cyanobacteria*, *Proteobacteria*, *Archaea*, and *Firmicutes* (Reed et al. 2011). Other organisms, including *Azotobacter* and *Azoarcus* genera, are also present at equivalent densities in the rhizosphere and non-rhizosphere soil. Bacterial genera like *Herbaspirillum* and *Azospirillum* colonize only in the higher plants' rhizosphere (Mrkovacki and Milic 2001). On the other hand, some symbionts like *Rhizobia* are capable of infecting the root and form root nodules. Interestingly, this symbiosis has often limited to legumes. Thus, N fixation related to *Rhizobium* sp. has become a more exciting issue for researchers to unveil these complex environmental phenomena. A higher impact on primary productivity and economic feasibility is also one of the pillars for a sustainable agricultural ecosystem.

Bacterial N fixation is a complex microbial process, where enzyme complex called *nitrogenase* (*dinitrogenase reductase* and *dinitrogenase* metal cofactor) takes part in the electron transport chain: the former enzyme serves as an electron donor, and the latter (substrate reduction component) accepts the energy of the electron to convert N_2 to NH_3. To generate one mole of NH_3, 16 moles of adenosine triphosphate (ATP) is required, which are obtained from the oxidation of organic molecules. The N cycle and its processes are dynamic and simultaneously proceed to maintain equilibrium to nature. A diverse pool of nitrogenous compounds including organics (proteins, urea, amines, etc.) and inorganics (NH_4^+ and NO_3^-) in soil along with gaseous forms like NO and NH_3 in troposphere determines N dynamics and bioavailability to the plants. Other than this, nitrification that is controlled by nitrifying bacteria includes nitrifiers (oxidizing ammonia to nitrite) and nitratifiers (oxidizing nitrite to nitrate). There are five common species of nitrifiers, *Nitrosomonas europaea*, *Nitrospira briensis*, *Nitrosococcus nitrous*, *Nitrosococcus oceani*, and *Nitrosolobus multiformis*, and three nitrifiers, *Nitrobacter winogradskyi*, *Nitrospina gracilis*, and *Nitrococcus mobilis*. Denitrifying microorganisms include *Thiobacillus denitrificans* and *Micrococcus denitrificans* and some species of *Serratia*, *Pseudomonas*, and *Achromobacter* (Delgado and Follett 2002).

1.4.2.2 N in Soil and Its Mineralization

Being the most extensively limiting plant nutrient in agriculture, fertilizer N is the main depending material for half of the global population for their daily hand-to-mouth activities. Total N content varies from 0.02% to 0.44% across the globe, the tropical country like India lacks SOM, causing consequently low soil N in surface soil layers. In India the total N content of surface soil ranges from 0.02% to 0.13% (Motsara 2002). Whether in the forest cover or the agricultural fields, crop-covered soils result in 10–20 times higher soil N content than the open fallow lands. Excessive cultivation leads to disintegration and decomposition of SOM, which leads to a decrease in N content in soils. Studies have shown that up to 3.5% of organic N in soil is mineralized annually. Even this rate of mineralization can take care of the normal growth of plants, except under sandy soils with inferior organic N; also crop demand of N is more than the mineralization rate (Chen et al. 2003). In a sandy loam soil, a 50% combination of sewage sludge and fertilizer could be one of the possible and best alternatives to curb fertilizer N and enhance N use efficiency (Biswas et al. 2017). As of advances in technology, the N stable isotope (^{15}N) technique has been widely used as a measuring tool in ecological studies of N cycling. This because of ^{15}N natural abundance values (δ15N) of soil samples is the net result of many biogeochemical processes that can cause ^{15}N refinement (Frank et al. 2000; Ghosh et al. 2018).

1.4.3 Phosphorus Nutrition and Phytobiome

Phosphorus (P) is one of the major nutrients essential to sustain all forms of life and is indispensable for the functioning of virtually every living cell on this planet. It is essential for energy metabolism and the principal component of adenosine diphosphate (ADP) and adenosine triphosphate (ATP). It is an important constituent of the genetic components deoxyribonucleic acid (DNA) and ribonucleic acid (RNA), and plays an important role in almost all metabolic processes in plants. An adequate amount of P is essential for root development, growth, and maturity of all the crop plants. It is second only to nitrogen (N) in terms of its limiting nature to plant growth, in almost all arable soils across the globe.

1.4.3.1 Phytobiomes and P Nutrition in Plants

The complex chemistry of P and its widespread deficiency in the world soils have driven the plant community to develop and adopt several strategies to improve the acquisition of P from marginally P-deficient soils. Changes in the root configuration, morphology, and distribution (de Souza et al. 2016), production of different organic compounds by the roots which helps in dissolution of inorganic P compounds in soil, and alliance of the plant roots with beneficial microorganisms in the rhizosphere (Zhou et al. 1992; Hasan 1996; Sharma et al. 2007) are some of the unique strategies adopted by the plants to strive over P-deficient situations. It is estimated that the amount of P fixed in world soils would be able to sustain crop production for the next 100 years if properly mined using available techniques (Goldstein et al. 1993).

1.4.3.2 Microbiome and P Nutrition in Plants

The phosphate-solubilizing microorganisms (PSMs) are essential for P nutrition of plants, and in the soil, the population of phosphate-solubilizing bacteria is higher by 2–50-fold compared to fungi. However, the fungal isolates in both solid and liquid culture media exhibited higher P-solubilization capacity compared to bacteria (Gaur et al. 1973; Banik and Dey 1982; Kucey 1983; Harrison 1987; Kucey et al. 1989). Of the various P compounds present in the soil, the Ca-P is more readily solubilized by the PSMs, while very few are reported to solubilize the Fe and Al-P which are the major compounds in acid soils like alfisols (Fig. 1.3). The PSMs are very effective in insolubilization of rock phosphates (RP) which are generally Ca-apatites and increase the P availability to crops when applied along with rock phosphates in soil. Currently, different species of bacteria such as *Azotobacter chroococcum*, *Bacillus subtilis*, *Bacillus cereus*, *Bacillus megaterium*, *Arthrobacter ilicis*, *Escherichia coli*, *Pantoea agglomerans*, *Pseudomonas putida*, *Pseudomonas aeruginosa*, *Enterobacter aerogenes*, *Microbacterium laevaniformans*, and *Micrococcus luteus* have been identified as effective P solubilizers (Kumar et al. 2014).

The microbial association thus plays a significant role in P nutrition particularly in the low P soils, and often the level of soil P determines the nature of microbes that get associated with plant roots (Gomes et al. 2018). Different maize genotypes with variable P acquisition capacity had a different association with microorganisms, and the P levels in the soil primarily guided this. The slow-growing microorganisms were more abundant in soil with low P levels, while the fast-growing

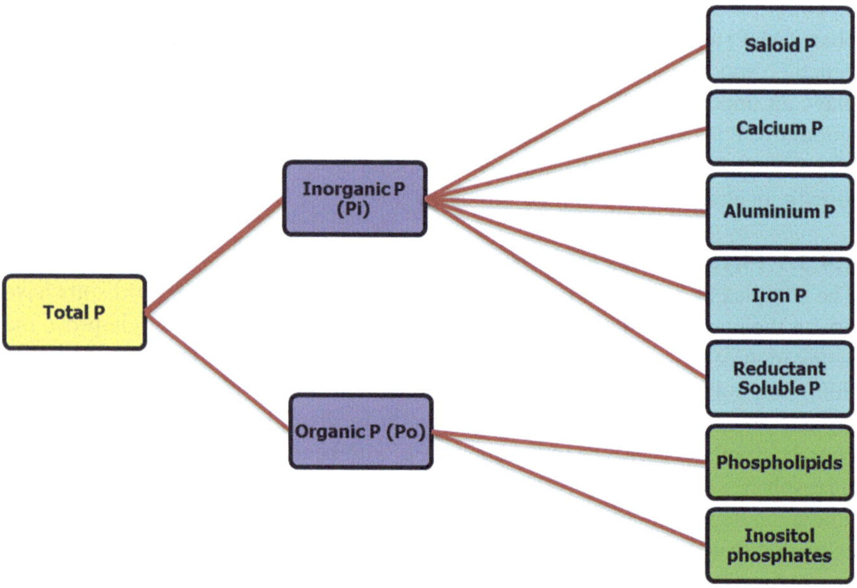

Fig. 1.3 Different fractions of soil P based on sources and chemical properties

microorganisms were predominant in the soils with high P levels (Gomes et al. 2018). In general, several bacterial species like the *Pseudomonas*, *Enterobacter*, *Azotobacter*, *Burkholderia*, *Rhizobium*, etc. are associated with P-solubilization and promote plant growth (Oliveira et al. 2009; Glick 2015; de Souza et al. 2016; Alori et al. 2017). However, the population of phosphate-solubilizing organisms was not affected by the level of P in soils (Browne et al. 2009; Gomes et al. 2018; Robbins et al. 2018).

Besides the P-solubilizing bacteria, the arbuscular mycorrhizal fungi (AMF) play an essential role in the P nutrition of plants. The hypha of the AMF acts as an extension to the plant roots and helps in exploring more soil volume, increasing the P availability to plants, and acting as an unhindered pathway for translocation of P (Schweiger and Jakobsen 1999). However, the soil P level is again critical, which regulates the abundance of AMF association with plant roots. Low soil P level increases the AMF association with plant roots, while P fertilization harms the AMF establishment. The AMF colonization in maize reduced from 29.2% to 13.7% when the soil P concentration was raised from 15 ppm to 70 ppm (Gosling et al. 2013). The plant species is an essential factor determining the association of AMF, and it is higher for crops like maize, cotton, pigeon pea, sunflower, mung bean, and sorghum while lower for oats, wheat, barley, etc. (Seymour 2009).

Sometimes even the non-P-solubilizing microbes play an essential role in P nutrition. These organisms take up sparingly soluble P compounds from the soil with the help of high-affinity P transporters, which becomes available to plants upon the death of the microbes (Gyaneshwar et al. 2002). The extremely insoluble P compounds are also cycled and become available for plant nutrition. *Escherichia coli*, a non-P solubilizer helps in the P nutrition of plants (Wanner 1996). Under P-limiting situations, the P transporters in *Rhizobium* are activated, resulting in higher accumulation of P and enhanced alkaline phosphatase activity (Al-Niemi et al. 1997).

1.4.3.3 Rhizosphere Engineering and P Nutrition

Root-induced changes in the rhizosphere also help in P acquisition and improve the P availability to plants (Hinsinger 2001). The plant root secretes a large number of substances that alter either the soil pH or chelating substances which are capable of releasing P from sparingly soluble P compounds. The secretion of organic acid molecules like citrate, oxalate, etc. also helps in P release through ligand exchange and increases P availability to plants (Shen et al. 2011). The plant roots also secrete enzymes like phosphatase or phytase, which can mobilize the organic P from the soil and make it phyto-available (Neumann and Romheld 2002; Zhang et al. 2010). Some of the new hypotheses are postulated based on the PSM-inspired chemistry by deriving polymer-coated P fertilizers to enhance PUE (Sarkar et al. 2017b; Sarkar et al. 2018). Also, nano-formulation of natural resources like phosphate rock increases the specific surface area that helped in better microbial activity and further improved PUE with or without organic acids (Roy et al. 2015, 2018a, b).

1.4.4 K-Bioavailability and Phytobiome

For achieving sustainable crop productivity, nutrients are the necessary inputs, and potassium is one of them. The flaws between depletion and the use of K fertilizers in agriculture are flaring day by day. So, it is relative to know the potassium dynamics in soil corresponding with potassium needed for the crops which offer balanced nutrition and retain its status in soils. Addressing this nutrient inequity and deficiencies, increased manufacture of potassium fertilizer is necessary for India and other developing countries. In these situations, it is required to prosper natural mineral sources of K to replace costly fertilizers. Although these minerals are not readily soluble, the combination of potassium minerals with a potassium-solubilizing microorganism (KSM) would be a better and available knowledge to increase bioavailability of K from K-bearing minerals, which could help in maintaining the ecological balance and sustaining agricultural production and environmental quality. Now, the question is how these soil organisms solubilize K. The processes involved in K solubilization are listed and described in the following (Fig. 1.4).

1.4.4.1 K-Solubilizing Microbial Species

Soil presents a variety of microbiomes including bacteria and fungi which boost the solubility of K minerals through the production of organic synthetics including chelating agents, exudates, extracellular enzymes, metabolic by-products, and both simple and compound organic acids (Meena et al. 2014; Saha et al. 2016a, b). These enzymes and organic metabolites triggered the biotic weathering of K-bearing rocks

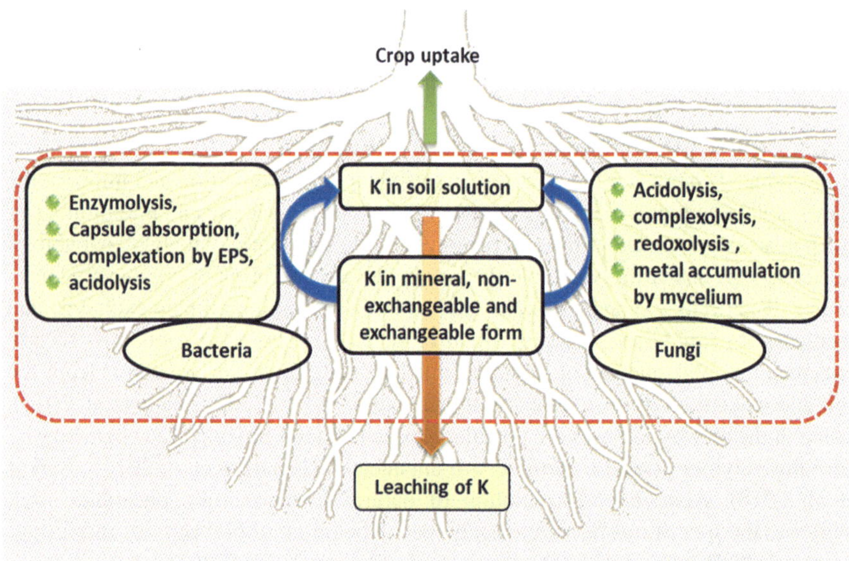

Fig. 1.4 Mechanisms of K solubilization in a typical phytobiome. (Conceptualized and modified from Meena et al. 2014; Saha et al. 2016a, b)

and minerals. Especially rhizobacteria produce 2-keto-L-gluconic acid which binds Ca and acts as a strong weathering factor for underlying rocks. Otherwise, soil microorganisms respire and produce CO_2 that reacts with water and forms carbonic acid. This carbonic acid contributes to chemical weathering by dissolving $CaCO_3$ and K minerals through dissolution reactions.

Negatively charged soil particles adsorb maximum cationic nutrients like K and retard bioavailability. However, acidification of soil by the soil microbiomes assists in ion exchange processes. In this way, K is released into the soil solution and becomes readily available to the plants. Similar charge species are also involved in ion exchange when their presence is higher in the soil solution. For example, when Ca^{2+} is present in excess in the soil solution, it can exchange two K^+ ions from the soil adsorption sites and contributes to desorption and K solubilization. These ions also transfer K, which is trapped in the interlayer spaces of the minerals to some extent. Certain bacteria produce mucilage like extracellular polysaccharides (ESP) which form a wrapper around the bacterial cell, attack the clay minerals chelating with silicon, and consequently release K from that structure.

Fungi are also considered as a biological weathering referee of rocks, minerals, and building blocks. Ectomycorrhizal hearting networks and the arbuscular structure of non-ectomycorrhizal trees, embedded in biofilms, conduct nutrients to the host. Here, biofilms assist in stimulating the weathering of minerals and thereby increasing uptake of nutrients to the plant. Some rock-decaying fungi can ooze out organic ions having low molecular weight, which forms a microscopic tunnel in the vicinity of exudates at hyphal tips within the minerals which enhances the weathering rates in that soil.

This is supported by numerous studies, whereas a wide variety of bacteria like *B. mucilaginous*, *B. edaphicus*, *B. circulans*, *Burkholderia*, *Acidithiobacillus ferrooxidans*, *Paenibacillus* spp., and *Pseudomonas* spp. have been demonstrated to solubilize K from K-containing minerals in the soils (Liu et al. 2012; Meena et al. 2014). Zhang and Kong (2014) isolated and identified that strains of *Pantoea agglomerans*, *Agrobacterium tumefaciens*, *Microbacterium foliorum*, *Myroides odoratimimus*, *Burkholderia cepacia*, *Enterobacter aerogenes*, *E. cloacae*, and *E. asburiae* remain effective in K solubilization in both solid and liquid media. Several other studies also validated the significant role of plant growth-promoting rhizobacteria (PGPR) in K solubilization and its mobilization in the plant root systems (Kumar and Singh 2001; Kukreja et al. 2004; El-Fattah et al. 2013).

1.4.5 S Cycle and Phytobiome

Sulfur (S) containing amino acids (cysteine, cystine, and methionine) is present in almost all the proteins, which makes it indispensable to more or less all living organisms. These amino acids may be present as free or in combined states. S cycle includes the transformation of S from organic to inorganic or inorganic to organic form and oxidized state to reduced state or reduced to oxidized state. These processes are mediated by various microorganisms, especially bacteria (Schoenau and Malhi 2008).

1.4.5.1 Mineralization and Immobilization of Sulfur

Sulfur mineralization is the conversion of plant-unavailable organically bound forms of sulfur (organic sulfates and carbon-bonded sulfur) to the plant-available inorganic forms of S (sulfate, SO_4^{2-}) in soil. Whereas immobilization is exactly the opposite process, where inorganic S is assimilated by living organisms (microbes) and converted to the organic forms, that plant cannot absorb. In unfertilized soil S mineralization is the dominant source of S, in S pool for plants throughout the year (Bettany et al. 1979). The main factor which decides whether an organic S source will be mineralized or immobilized is the C/S ratio. When the C/S ratio is less than 200:1, net mineralization occurs, and organic S contributes to plant-available pool of S; in case the C/S ratio of the organic matter exceeds 400:1, net S immobilization occurs, and plant-available S is converted to organic S (Schoenau and Davis 2006). Organic S consists of (1) organic sulfate (S and C are not directly bonded like thioglucosides, sulfate esters, and sulfamates; here S is bonded to C via oxygen or nitrogen) and (2) carbon-bonded sulfur (here S is directly bonded to C like sulfonic acids, sulfur-containing amino acids in proteins). Organic sulfates cover 30–70% of the total soil organic S. S-containing amino acids and organic sulfates are the labile forms of organic S; they mineralize very easily by extracellular enzyme arylsulfatases (Tabatabai and Bremner 1970). Through organic mineralization, S is converted to sulfate (SO_4^{-2}) and thiosulfate ($S_2O_3^{-2}$) which are absorbable by plants. Humus soil has a C/S ratio nearly 100:1. Thus decomposition of it results in net mineralization (Roberts et al. 1989).

1.4.5.2 Microbial Oxidation and Reduction of S

Soils where reduced forms of S (such as sulfides, elemental sulfur) are present, in those places microbial oxidation of reduced inorganic S is an important process. Through oxidation, reduced (–2, 0) forms of S (S^{2-}, S^0) are converted to higher oxidation states (+6) like sulfate (SO_4^{2-}). Both autotrophic and heterotrophic microorganisms perform microbial oxidation, including species of *Thiobacillus* (autotroph); heterotrophic bacteria like *Bacillus*, *Arthrobacter*, and *Pseudomonas*; and some fungi (Lawrence and Germida 1988). In some cases, heterotrophic sulfur oxidizers may dominate, especially in the rhizosphere (Grayston and Germida 1990). S oxidation is an acidifying process. It can be depicted by the following equation where elemental S, a reduced form of S, is oxidized to sulfuric acid:

$$2S^0 + 3O_2 + 2H_2O \rightarrow 2H_2SO_4$$

As a result of flooding, reduced aeration and high oxygen consumption produce reducing conditions, where redox potential decreases and drives the microbial conversion of sulfates to sulfide after other electron acceptors including oxygen and nitrate are depleted. This is termed as dissimilatory sulfate reduction, which is performed by the species of *Desulfovibrio* and *Desulfotomaculum* bacteria. Hydrogen sulfide gas is retained in the soil by reaction with iron to form an iron sulfide mineral, depicted in the following equation:

$$SO_4 \rightarrow H_2S(g) + Fe \rightarrow FeS(\text{mineral})$$

These sulfides may be oxidized back to sulfates if the soil becomes aerobic again (Schoenau and Malhi 2008).

1.4.6 Trace Metals and Phytobiome

The huge quantity of effluents generated from households and industries is continuously contaminating the land, river, pond, as well as groundwater aquifers (Saha et al. 2017a). Subsequent contamination of the human food chain from a significant amount of metals and pollutants is accumulated in soil following crop accumulation. As metals like iron (F)e, manganese (Mn), zinc (Zn), copper (Cu), and nickel (Ni) are essential for plant growth (Hapke 1991; Lenka et al. 2016; Saha et al. 2017b), apart their required quantity is significantly lesser than the N, P, and K. Thus, these metals along with other beneficial and harmful cationic metals are often defined as trace metals, whereas the term "heavy metal" is a cumulative term that includes both metals and metalloids having an atomic number of more than 20 (Ca) or atomic density more than 5 g cc^{-1} which also cause toxicity at very trace concentration. Moreover, trace metals like chromium (Cr), cadmium (Cd), cobalt (Co), selenium (Se), mercury (Hg), lead (Pb), and arsenic (As) are responsible for different types of malfunction or even death due to toxicity in plants as well as animals and adversely affect the soil system. However, the levels of toxicity are determined by the combination of different factors like soil pH, conductivity, and other ionic activities in soil solution and particular metal species (type) and concentration present in ecosystems (Tchounwou et al. 2012; Saha et al. 2017c). Metal toxicity induced poor plant growth, retarded carbohydrate and protein metabolism, obstructed growth hormone formations and enzyme activities, and simultaneously insisted reduced biological diversity in the rhizosphere and non-rhizospheric soils (Singh et al. 2011).

1.4.6.1 Conducive Soil Condition for Trace Metal Toxicity

Certain conditions induce trace metal toxicity for crops and humans. These trace element toxicities occur in the following (Bucher and Schenk 2000; Broadley et al. 2007; Aref 2011):

(a) Superfluous application of untreated sewage sludge, municipal waste, tannery and distillery effluents, coal combustion ashes, etc.
(b) Light-textured soil with low pH and having impeded subsurface soil layer with very low hydraulic conductivity
(c) Mineral soil having deficient soil organic matter
(d) Intensive raw minerals (e.g., waste mica, rock phosphate, pyrite, etc.) and fertilizer application
(e) Irrigation with polluted groundwater
(f) Type and concentration of trace metals in the environment

Heavy-trace metals are distributed over the solid phase, liquid phase, and also the gaseous phase, and their partitioning influenced and depends on complexation with ligands, ion exchange, adsorption-dissolution-precipitation reaction, and oxidation-reduction reaction (Vlek et al. 1974; Singh and Saha 1997; Rattan et al. 2008). These interacting properties are the controlling factors of trace metal solubility, mobility, and bioavailability in soil and water systems. Other than mechanical and chemical processes, microbial inoculation has the potentiality to reduce the metal toxicity by either decomposition or immobilizing the metals from the soil (Abou-Shanab et al. 2003; Seshadri et al. 2015).

1.4.6.2 Trace Metals' Crop Response and Phytoremediation

Negative interaction of cationic trace metals induces other trace metal deficiencies in plant tissues. Khan and Khan (2010) reported that Fe and Zn uptake was reduced from the excess application of Ni and developed chlorosis symptoms on leaves. Other reports also indicated that Cd toxicity induces poor Cr uptake (Dotaniya et al. 2017); and Zn toxicity induces Fe and Mn deficiency in plant tissues (Sivasankar et al. 2012). Correspondingly, specific metal toxicity also affects the dynamics of nutrient mineralization and nutrient release kinetics and results in poor plant growth; plants appeared brushy (Singh et al. 2016). Trace metals' (heavy metal) toxicity also retard soil enzymatic activity by replacing the native essential metals from binding sites (Bruins et al. 2000). For example, excess concentrations of Cd^{2+}, Ag^{2+}, and Hg^{2+} in the soil-solution exchange phase are likely to bind with SH groups of some sensitive enzymes, which retard the activity of the enzyme (Nies 1999). Thus, increasing the use of wastewater, municipal wastes, industrial effluents, and coal combustion ashes triggers the toxic trace metal concentration in soil, which further accelerates the metal accumulation in ecological habitats (Rajkumar et al. 2012). Apart from this, higher Cr concentration was reported in plant roots compared to shoot, because metal excluder plants can store an excessive amount of Cr in root cell vacuole (Oliveira 2012; Nematshahi et al. 2012). Contrastingly, untreated sewage sludge application enhances crop yield, soil nutrient content, organic C content, and soil biological activities (Roy et al. 2019). Integrated approach (treated organic waste and inorganic fertilizers) is the best-recommended method for sustainable soil management and crop production. Neither the excess of essential nutrients nor other trace elements are beneficial for the plant-soil-microbe system, because specific element toxicity damages the DNA structure of soil enzymes and disturbs the normal functionality in the environment (Bruins et al. 2000).

1.4.6.3 Siderophore and Phytosiderophore Engineering and Trace Metal Chemistry

Due to certain unwanted and challenging environmental conditions, the release rate of plant root exudates enhances the rhizosphere. These exudates help both ways by accelerating essential nutrient elements and inhibiting toxic elements (Marschner 1995). During Fe and Zn deficiency, graminaceous plants like barley, wheat, etc. exude phytosiderophores. While these exudates are produced from soil microbes, other higher plants are termed as siderophores. Siderophore and phytosiderophore are hexadentate non-proteinous amino acids, which act as a ligand and coordinate with Fe^{3+}, Zn^{2+}, or Mn^{2+} by the carboxylic and amino groups (Romheld 1991). In the rhizosphere, metal chelation is not specific for Fe^{3+}. Subsequently, phytosiderophores can transport a wide range of metals like Zn, Cu, Mn, Cd, and Ni (Awad and Romheld 2000). The diurnal variation of phytosiderophore release is also specific with plants and parts of plant roots. Generally, maximum phytosiderophore release and microbial activity are observed near the apical root zones due to localized microbial degradation and distribution of phytosiderophore (Marschner et al. 1986; Romheld 1991). The rate of phytosiderophore (PS) release differs among interspecies, plant to plant, and depends on daytime, sunshine hours, light intensity, and micronutrient status (Cakmak et al. 1994). However, this correlation with phytosiderophore quantity is not always consistent. Siderophores and phytosiderophores associated with metal chemistry and metal bioavailability in the soil are listed in Table 1.1. In contrast with Zn-PS and Mn-PS, the preferential uptake of Fe-PS is more prominent, which is mainly regulated by soil Fe nutritional status. The affinity of mugineic acids (MA) for heavy metal cations decreases in the order of $Cu^{2+} > Fe^{3+} > > Zn^{2+} > > Mn^{2+}$ (Dotaniya et al. 2014). Fe-PS and Zn-PS complex uptake rate corresponds to ~100 and ten times higher than that of free Fe and Zn. Thus, PS production and subsequent uptake are regulated by the transporters of the YS1/YSL protein family, which is induced by Fe deficiency (Meda et al. 2007). Due to increased nutrient availability, this process is also considered as a lifesaving mechanism for plants.

1.5 Phytobiome: An Early Indicator of Soil Health

Typically, soil health is defined as "a state of dynamic equilibrium between flora and fauna and their surrounding soil environment in which all the metabolic activities of the former proceed optimally without any hindrance, stress or impedance from the letter" (Goswami and Rattan 1992). Simultaneously the soil quality, which is analogous to soil health, is often used in the scientific literature, as scientist prefers soil quality. Soil quality is defined as "the capacity of soil to function within the ecosystem and land-use boundaries, to sustain biological productivity, maintain environmental quality, and sustain plant, animal, and human health" (Doran and Parkin 1994). On the other hand, soil health is generally dealt with by the producers (e.g., farmers). There are few conceptual differences between soil quality and soil health, where the former considers soil usefulness for a particular purpose for a long time scale and the latter considers the state of soil for a particular time (Larson and Pierce 1991).

Healthy soils are very crucial for the reliability of global ecosystems to stay integral or to recuperate from biotic and abiotic stresses as well as anthropogenic exploitation with agriculture (Ellert et al. 1997). In soil, phytobiome has the capacity to provide an integrated standard tool for soil health and quality, which cannot be obtained from physical or chemical analyses of soil. Microbiomes quickly acclimatize to environmental conditions as they have close relations with their surroundings due to their higher surface-to-volume ratio and respond rapidly to changes. Hence, the potentiality of these biomes can be adapted as an excellent indicator of soil health assessment. In some cases, alterations in microbial populations and activity

Table 1.1 Siderophores and phytosiderophores associated with metal chemistry and metal bioavailability in soil

Plants and microorganisms		Siderophore and phytosiderophore production
Plants	*Graminaceous* family	Phytosiderophores: mugineic acid (MA), deoxymugineic acid (DMA), epoxy-mugineic acid (EMA), hydroxy-mugineic acid (HMA)
Microorganisms		Siderophores:
Bacteria	*Bacillus*	Schizokinen, bacillibactin, ferrioxamines
	Nocardia, Arthrobacter	Ferrioxamines
	Escherichia coli	Enterobactin
	Staphylococcus	Staphyloferrin
	Erwinia chrysanthemi	Chrysobactin, achromobactin
	Mycobacterium tuberculosis	Mycobactin
	Pseudomonas sp.	Pyoverdines
	Bordetella sp.	Alcaligin
	Azotobacter, agrobacterium	Catecholate
Arbuscular mycorrhizae	*Cenococcum geophilum, Wilcoxinarehmii*	Ferricrocin
	Glomus etunicatum, G. mosseae, unidentified Glomus sp.	Glomuferrin
Fungi	*Rhizopus*	Rhizopherin
	Ustilagosphaerogena	Desferriferrichrome
	Fusarium roseum	Fusarinines, malonichrome, triacetylfusarinines
	Neurospora crassa	Ferricrocin
	Aspergillus fumigants, Penicillium bilaiae	Pistillarin
	Aspergillus ochraceus	Ferrichrome

Adopted and modified from Matzanke et al. (1988), Hider and Kong (2010), Dotaniya et al. (2014), Winkelmann (2017), Sarkar et al. (2017a, b), Hussein and Joo (2017)

before detecting changes in soil rather than physical and chemical properties thus provide an early signal of soil improvement or an early indication of soil degradation (Pankhurst et al. 1995). The turnover rate of microbial biomass is much faster (1–5 years) than the total soil organic matter turnover (Carter et al. 1999). Phytobiomic indicators of soil health involve various sets of microbial dimensions due to the multifunctional characteristics of microbial groups in the terrestrial ecosystem. Some indicators of soil health are as follows:

1. *Biodiversity*: genetic diversity, functional diversity, structural diversity
2. *Carbon cycling*: soil respiration, organic matter decomposition, soil enzymes, methane oxidation
3. *Nitrogen cycling*: N-mineralization, nitrification, denitrification, N-fixation
4. *Soil biomass*: microbial biomass, protozoan biomass

5. *Microbial activity*: bacterial DNA synthesis, bacterial protein synthesis, RNA measurements, bacteriophages
6. *Bioavailability*: biosensor bacteria, plasmid-containing bacteria, antibiotic-resistant bacteria, catabolic genes

In addition to the influence on nutrient cycling, microbiomes also affect the physical properties of soil by producing extracellular polysaccharides and other cellular trashes that help in maintaining soil structure. As these biochemicals function as chelating agents, they stabilize soil aggregates and affect water retention capacity, crusting, erodibility, infiltration rate, as well as susceptibility to compaction.

1.6 Future Outlook

Attempts are to be made to enhance the food, feed, and fiber production worldwide by exploring each and individual component of phytobiome and in between interactions. Due to increased food demand and high-intensity agriculture, integration and optimization of phytobiome-based pieces of knowledge, resources, and site- and condition-specific solutions become one of the most treasured materials for sustainable agriculture and soil health. Further, an advanced association of scientific society to explore this concept is a much important future aspect of phytobiome studies.

1.7 Conclusions

This chapter is an effort to enlighten the idea of phytobiome and its role in nutrient regulation, maintaining soil quality. The modern principles of different utilizations of this phytobiome have been characterized to excerpt a high prospect of their functionality and applicability for sustainable agriculture. So, these phytobiomes are evolved as plant growth-promoting microbiomes, which are important for keeping good soil health, better soil conditions, and persistent agricultural productivity. These biomes do not become alive independently but make interconnected coordination with the environment. All are engaged in the nutrient cycle of the terrestrial ecosystem by weathering and solubilizing complex and stable nutrient sources. The contest of this section is to develop new approaches for addressing the nutrient use efficiency as well as directed plant parts (within metabolomes of rhizospheric groups) to assist a significant change in the level of perception. The ultimate aid from this section will be the information about modification of the soil-plant system to support microbial chemistry or their effects on soil physico-chemical properties that are most important in promoting nutrient acquisition in plants across the diversified global agricultural condition. These provide an effective substitute for advanced crop production.

References

Al-Niemi TS, Kahn ML, McDermott TR (1997) P metabolism in the rhizobium tropici-bean symbiosis. Plant Physiol 113:1233–1242

Alori ET, Glick BR, Babalola OO (2017) Microbial phosphorus solubilization and its potential for use in sustainable agriculture. Front Microbiol 8:971

Amato M, Ladd JN (1992) Decomposition of 14C-labelled glucose and legume material in soils: properties influencing the accumulation of organic residue C and microbial biomass C. Soil Biol Biochem 24:455–466

Andersen KS, Winding A (2004) Non-target effects of bacterial biological control agents on soil Protozoa. Biol Fertil Soils 40:230–236

Aref F (2011) Concentration and uptake of zinc and boron in corn leaf as affected by zinc sulfate and boric acid fertilizers in deficient soil. Life Sci J 8(1):26–32

Arunachalam K, Arunachalam A, Melkania NP (1999) Influence of soil properties on microbial populations, activity and biomass in humid subtropical mountainous ecosystems of India. Biol Fertil Soils 30(3):217–223

Awad F, Romheld V (2000) Mobilization of heavy metals from contaminated calcareous soils by plant born, microbial and synthetic chelators and their uptake by wheat plants. J Plant Nutr 23:1847–1855

Banik S, Dey BK (1982) Phosphate solubilizing potentiality of micro-organisms capable of utilizing aluminum phosphate as sole phosphate source. Zentrabl Backteriol Prasitenkd Infektionskr Hyg II 138:17–23

Bardgett RD, Freeman C, Ostle NJ (2008) Microbial contributions to climate change through carbon cycle feedbacks. ISME J 2:805–814

Beans C (2017) Core concept: probing the phytobiome to advance agriculture. Proc Natl Acad Sci 114(34):8900–8902

Bettany JR, Stewart JWB, Saggar S (1979) The nature and forms of sulfur in organic matter fractions of soils selected along an environmental gradient. Soil Sci Soc Am J 43:481–485

Biswas SS, Singhal SK, Biswas DR, Singh RD, Roy T, Sarkar A, Ghosh A, Das D (2017) Synchronization of nitrogen supply with demand by wheat using sewage sludge as organic amendment in an Inceptisol. J Indian Soc Soil Sci 65:264–273

Bond-Lamberty B, Thomson A (2010) Temperature-associated increases in the global soil respiration record. Nature 464:579–582

Broadley MR, White PJ, Hammond JP, Zelko I, Lux A (2007) Zinc in plants. New Phytol 173(4):677–702

Browne P, Rice O, Miller SH, Burke J, Dowling DN, Morrissey JP, O'Gara F (2009) Superior inorganic phosphate solubilization is linked to phylogeny within the Pseudomonas fluorescens complex. Appl Soil Ecol 43(1):131–138

Bruins MR, Kapil S, Oehme FW (2000) Microbial resistance to metals in the environment. Ecotoxicol Environ Saf 45:198–207

Brussaard L, Bouwman LA, Geurs M, Hassink J, Zwart KB (1990) Biomass composition and temporal dynamics of soil organisms of a silt loam soil under conventional and integrated management. Neth J Agric Sci 38:283–302

Bucher AS, Schenk MK (2000) Toxicity level for phytoavailable zinc in compost-peat substrates. Scitia Hort 83(3–4):339–352

Cakmak S, Gulut KY, Marschner H, Graham RD (1994) Effect of zinc and iron deficiency on phytosiderophores release in wheat genotypes differing in zinc efficiency. J Plant Nutr 7:1–17

Carter MR, Gregorich EG, Angers DA, Beare MH, Sparling GP, Wardle DA, Voroney RP (1999) Interpretation of microbial biomass measurements for soil quality assessment in humid temperate regions. Can J Soil Sci 79:507–520

Chen G, Zhu H, Zhang Y (2003) Soil microbial activities and carbon and nitrogen fixation. Res Microbiol 154:393–398

Cheng L, Booker FL, Tu C, Burkey KO, Zhou L, Shew HD, Rufty TW, Hu S (2012) Arbuscular mycorrhizal fungi increase organic carbon decomposition under elevated CO2. Science 337:1084–1087
Coyne MS (1999) Soil microbiology: an exploratory approach. Delmar Publishers, pp 139–158; 279–289
Davidson EA, Janssens IA (2006) Temperature sensitivity of soil carbon decomposition and feedbacks to climate change. Nature 440:165–173
Dawson PSS (ed) (1974) Microbial growth. Dowden, Hutchinson and Ross, Stroudsburg/Pennsylvania
de Souza RS, Okura VK, Armanhi JS, Jorrın B, Lozano N, da Silva MJ, Gonzalez-Guerrero M, de Araujo LM, Verza NC, Bagheri HC et al (2016) Unlocking the bacterial and fungal communities assemblages of sugarcane microbiome. Sci Rep 6:28774
de Vries FT, Shade A (2013) Controls on soil microbial community stability under climate change. Front Microbiol 4:265
Delgado JA, Follett RF (2002) Carbon and nutrient cycles. J Soil Water Conserv 57:455–464
Doran JW, Parkin TB (1994) Defining and assessing soil quality. In: Doran JW, Coleman DC, Bezdicek DF, Stewart BA (eds) Defining soil quality for a sustainable environment. SSSA, Madison, pp 3–21
Dotaniya ML, Prasad D, Meena HM, Jajoria DK, Narolia GP, Pingoliya KK, Meena OP, Kumar K, Meena BP, Ram A, Das H, Sreenivasa Chari M, Pal S (2014) Influence of phytosiderophore on iron and zinc uptake and rhizospheric microbial activity. Afr J Microbiol Res 7(51):5781–5788
Dotaniya ML, Rajendiran S, Coumar MV, Meena VD, Saha JK, Kundu S, Kumar A, Patra AK (2017) Interactive effect of cadmium and zinc on chromium uptake in spinach grown on vertisol of Central India. Int J Environ Sci Technol. https://doi.org/10.1007/s13762-017-1396-x
Drigo B, Pijl AS, Duyts H, Kielak AM, Gamper HA, Houtekamer MJ et al (2010) Shifting carbon flow from roots into associated microbial communities in response to elevated atmospheric CO2. Proc Natl Acad Sci 107:10938–10942
El-Fattah DAA, Wedad EE, Mona SZ, Mosaad KH (2013) Effect of carrier materials, sterilization method, and storage temperature on survival and biological activities of Azotobacter chroococcum inoculant. Ann Agric Sci 58:111–118
Ellert BH, Clapperton MJ, Anderson DW (1997) An ecosystem perspective of soil quality. In: Gregorich EG, Carter MR (eds) Soil quality for crop production and ecosystem health. Elsevier, Amsterdam, pp 115–141
FAO (2017) The future of food and agriculture – trends and challenges, Rome
Frank DA, Groffman PM, Evans RD, Tracy BF (2000) Ungulate stimulation of nitrogen cycling and retention in Yellowstone Park grasslands. Oecologia 123:116–121
Frey-Klett P, Garbaye J, Tarkka M (2007) The mycorrhiza helper bacteria revisited. New Phytol 176:22–36
Galloway JN, Townsend AR, Erisman JW, Bekunda M, Cai Z, Freney JR, Martinelli LA, Seitzinger SP, Sutton MA (2008) Transformation of the nitrogen cycle: recent trends, questions, and potential solutions. Science 320(5878):889–892
Gaur AC, Medan M, Ostwal KP (1973) Solubilization of phosphate ores by native microflora of rock phosphates. Indian J Expt Biol 11:427–429
Ghosh A, Bhattacharyya R, Dwivedi BS, Biswas DR, Meena MC, Sarkar A, Agarwal BK, Mahapatra P, Shahi DK, Agnihotri R, Sawlani R (2018) Depth dynamics of soil N contents and natural abundances of 15N after 43 years of long-term fertilization and liming in sub-tropical Alfisol. Arch Agron Soil Sci 64(9):1290–1301
Glick BR (2015) Beneficial plant-bacterial interactions. Springer, Heidelberg
Godfray HCJ, Beddington JR, Crute IR, Haddad L, Lawrence D, Muir JF, Pretty J, Robinson S, Thomas SM, Toulmin C (2010) Food security: the challenge of feeding 9 billion people. Science 327:812–818
Goldstein AH, Rogers RD, Mead G (1993) Mining by the microbe. Bio Technol 11:1250–1254

Gomes EA, Lana UG, Quensen JF, de Sousa SM, Oliveira CA, Guo J et al (2018) Root-associated microbiome of maize genotypes with contrasting phosphorus use efficiency. Phytobiomes 2(3):129–137

Gosling P, Mead A, Proctor M, Hammond JP, Bending GD (2013) Contrasting arbuscular mycorrhizal communities colonizing different host plants show a similar response to a soil phosphorus concentration gradient. New Phytol 198(2):546–556

Goswami NN, Rattan RK (1992) Soil health-key to sustained agricultural productivity. Fert News 37(12):53–60

Grayston SJ, Germida JJ (1990) Influence of crop rhizosphere on populations and activity of heterotrophic sulfur-oxidizing microorganisms. Soil Biol Biochem 22:457–463

Gyaneshwar P, Kumar GN, Parekh LJ, Poole PS (2002) Role of soil microorganisms in improving P nutrition of plants. Plant Soil 245(1):83–93

Hacquard S, Garrido-Oter R, Gonzalez A, Spaepen S, Ackermann G, Lebeis S, McHardy AC, Dangl JL, Knight R, Ley R, Schulze-Lefert P (2015) Microbiota and host nutrition across plant and animal kingdoms. Cell Host Microbe 17:603–616

Hapke HJ (1991) Metal accumulation in the food chain and load of feed and food. In: Merian E (ed) Metals and their compounds in the environment. VCH Verlges, Weinheim

Harrison AF (1987) Soil organic phosphorus-a review of world literature. CAB International, Wallingford, p 257

Hasan R (1996) Phosphorus status of soils in India. Better Crops Int 10:4–5

Hassink J, Bouman LA, Zwart KA, Bloem J, Brussaard L (1993a) Relationships between soil texture, physical protection of organic matter, soil biota, and C and N mineralization in grassland soils. Geoderma 57:105–128

Hassink J, Bouman LA, Zwart KA, Bloem J, Brussaard L (1993b) Relationships between habitable pore space, soil biota and mineralization rates in grassland soils. Soil Biol Biochem 25:47–55

Hider RC, Kong X (2010) Chemistry and biology of siderophores. Nat Prod Rep 27(5):637–657

Hinsinger P (2001) Bioavailability of soil inorganic P in the rhizosphere as affected by root-induced chemical changes: a review. Plant Soil 237:173–195

Hussein AK, Joo JH (2017) Stimulation, purification, and chemical characterization of siderophores produced by the rhizospheric bacterial strain Pseudomonas putida. Rhizosphere 4:16–21

Kandeler E, Stemmer M, Klimanek EM (1999) Response of soil microbial biomass, urease and xylanase within particle-size fractions to long-term soil management. Soil Biol Biochem 31:261–273

Kandeler E, Tscherko D, Bruce KD, Stemmer M, Hobbs PJ, Bardgett RD, Amelung W (2000) Structure and function of the soil microbial community in microhabitats of a heavy metal polluted soil. Soil Biol Biochem 32:390–400

Kennedy AC, Stubbs TL (2006) Soil microbial communities as indicators of soil health. Ann Arid Zone 45(3-4):287–308

Khan MR, Khan MM (2010) Effect of varying concentration of nickel and cobalt on the plant growth and yield of chickpea. Aust J Basic Appl Sci 4(6):1036–1046

King GM (2011) Enhancing soil carbon storage for carbon remediation: potential contributions and constraints by microbes. Trends Microbiol 19:75–84

Kirchmann H, Gerzabek MH (1999) Relationship between soil organic matter and micropores in a long-term experiment at Ultuna, Sweden. J Plant Nutr Soil Sci 162:493–498

Koyama A, Harlow B, Kuske CR, Belnap J, Evans RD (2018) Plant and microbial biomarkers suggest mechanisms of soil organic carbon accumulation in a Mojave Desert ecosystem under elevated CO2. Soil Biol Biochem 120:48–57

Kucey RMN (1983) Phosphate solubilizing bacteria and fungi in various cultivated and virgin Albreta soils. Can J Soil Sci 63:671–678

Kucey RMN, Jenzen HH, Leggett ME (1989) Microbially mediated increases in plant available phosphorus. Adv Agron 42:199–228

Kukreja K, Suneja S, Goyal S, Narula N (2004) Phytohormone production by Azotobacter: a review. Agric Rev 25:70–75

Kumar V, Singh KP (2001) Enriching vermicompost by nitrogen fixing and phosphate solubilizing bacteria. Bioresour Technol 76:173–175
Kumar A, Choudhary CS, Paswan D, Kumar B et al (2014) Sustainable way for enhancing phosphorus efficiency in agricultural soils through phosphate solubilizing microbes. Asian J Soil Sci 9:300–310
Lal R (2004) Soil carbon sequestration impacts on global climate change and food security. Science 304:1623–1627
Larson WE, Pierce FJ (1991) Conservation and enhancement of soil quality evaluation for sustainable land management in the developing world. In: IBSRAM proceedings, no 12 vol 2, Technical Papers, Bangkok, Thailand, pp 175–203
Lawrence JR, Germida JJ (1988) Most probable number procedure to enumerate sulfur oxidizing, thiosulfate producing heterotrophs in soil. Soil Biol Biochem 20:577–578
Leach JE, Triplett LR, Argueso CT, Trivedi P (2017) Communication in the phytobiome. Cell 169:587–596
Lenka S, Rajendiran S, Coumar MV, Dotaniya ML, Saha JK (2016) Impacts of fertilizers use on environmental quality. In: National seminar on environmental concern for fertilizer usein future at Bidhan Chandra KrishiViswavidyalaya, Kalyani on February 26, 2016
Leveau JHJ, Lindow SE (2001) Appetite of an epiphyte: quantitative monitoring of bacterial sugar consumption in the phyllosphere. Proc Natl Acad Sci U S A 98:3446–3453
Liu D, Lian B, Dong H (2012) Isolation of Paenibacillus sp. and assessment of its potential for enhancing mineral weathering. Geomicrobiol J 29:413–421
Marschner H (1995) Mineral nutrition of higher plants. Academic, New York
Marschner H, Romheld V, Horst WJ, Martin P (1986) Root-induced changes in the rhizosphere-importance for the mineral-nutrition of plants. Z Pflanzenern Bodenk 149:441–456
Matzanke BF, Bill E, Trautwein AX, Winkelmann G (1988) Role of siderophores in iron storage in spores of Neurospora crassa and Aspergillus ochraceus. J Bacteriol 169(12):5873–5876
Meda AR, Scheuermann EB, Prechsl UE, Erenoglu B, Schaaf G, Hayen H, Weber G, von Wiren N (2007) Iron acquisition by phytosiderophores contributes to cadmium tolerance. Plant Physiol 143:1761–1773
Meena VS, Maurya BR, Verma JP (2014) Does rhizospheric microorganism enhance K+ availability in agricultural soils? Can J Microbiol 52:66–72
Mendes R, Garbeva P, Raaijmakers JM (2013) The rhizosphere microbiome: significance of plant beneficial, plant pathogenic, and human pathogenic microorganisms. FEMS Microbiol Rev 37:634–663
Miller WG, Brandl MT, Quinones B, Lindow SE (2001) Biological sensor for sucrose availability: relative sensitivities of various reporter genes. Appl Environ Microbiol 67:1308–1317
Motsara MR (2002) Available nitrogen, phosphorus and potassium status of Indian soils as depicted by soil fertility maps. Fertil News 47:15–21
Mrkovacki N, Milic V (2001) Use of Azotobacter chroococcum as potentially useful in agricultural application. Ann Microbiol 51:145–158
Nakano-Hylander A, Olsson PA (2007) Carbon allocation in mycelia of arbuscular mycorrhizal fungi during colonisation of plant seedlings. Soil Biol Biochem 39:1450–1458
Nave LE, Vogel CS, Gough CM, Curtis PS (2009) Contribution of atmospheric nitrogen deposition to net primary productivity in a northern hardwood forest. Can J For Res 39:1108–1118
Nelson EB (2004) Microbial dynamics and interactions in the spermosphere. Annu Rev Phytopathol 42:271–309
Nematshahi N, Lahouti M, Ganjeali A (2012) Accumulation of chromium and its effect on growth of (Allium cepa cv. Hybrid). Eur J Exp Biol 2(4):969–974
Neumann G, Romheld V (2002) Root-induced changes in the availability of nutrients in the rhizosphere. In: Waisel Y, Eshel A, Kafkafi U (eds) Plant roots, the hidden half, 3rd edn. Marcel Dekker, Inc., New York, pp 617–649
Nies DH (1999) Microbial heavy-metal resistance. Appl Microbiol Biotechnol 51:730–750
Oliveira H (2012) Chromium as an environmental pollutant: insights on induced plant toxicity. J Bot 2012:375843. 8 p

Oliveira CA, Alves VMC, Marriel IE, Gomes EA, Scotti MR, Carneiro NP, Sa NMH (2009) Phosphate solubilizing microorganisms isolated from rhizosphere of maize cultivated in an oxisol of the Brazilian Cerrado Biome. Soil Biol Biochem 41(9):1782–1787

Pankhurst CE, Hawke BG, McDonald HJ, Kirkby CA, Buckerfield JC, Michelsen P, O'Brien KA, Gupta VVSR, Doube BM (1995) Evaluation of soil biological properties as potential bioindicators of soil health. Aust J Exp Agric 35:1015–1028

Phytobiomes: A Roadmap for Research and Translation (2016) American phytopathological society, St. Paul, MN. www.phytobiomes.org/roadmap

Postgate JR (1982) Biological nitrogen fixation: fundamentals. Philos Trans R Soc Lond B 296:375–385

Rajkumar M, Sandhya S, Prasad MNV, Freitas H (2012) Perspectives of plant-associated microbes in heavy metal phytoremediation. Biotechnol Adv 30:1562–1574

Rattan RK, Datta SP, Katyal JC (2008) Micronutrient management: research achievements and future challenges. Indian J Fertil 4(12):103–118

Reed SC, Cleveland CC, Townsend AR (2011) Functional ecology of free-living nitrogen fixation: a contemporary perspective. Annu Rev Ecol Evol Syst 42:489–512

Remus-Emsermann MN, Tecon R, Kowalchuk GA, Leveau JH (2012) Variation in local carrying capacity and the individual fate of bacterial colonizers in the phyllosphere. ISME J 6:756–765

Reynaldo V, Banwart S, Black H, Ingram J, Joosten H, Milne E, Noellemeyer E (2012) The benefits of soil carbon. In UNEP Year Book, pp 19–33

Robbins C, Thiergart T, Hacquard S, Garrido-Oter R, Gans W, Peiter E, Spaepen S (2018) Root-associated bacterial and fungal community profiles of Arabidopsis thaliana are robust across contrasting soil P levels. Phytobiomes 2(1):24–34

Roberts TL, Bettany JR, Stewart JWB (1989) A hierarchical approach to the study of organic C, N, P and S in western Canadian soils. Can J Soil Sci 69:739–749

Romheld V (1991) The role of phytosiderophores in acquisition of iron and other micronutrients in gramineous species-an ecological approach. Plant Soil 130:127–134

Roy T, Biswas DR, Datta SC, Dwivedi BS, Lata BKK, Sarkar A, Agarwal BK, Shahi DK (2015) Solubilization of Purulia rock phosphate through organic acid loaded nanoclay polymer composite and phosphate solubilizing bacteria and its effectiveness as P-fertilizer to wheat. J Indian Soc Soil Sci 63(3):327–338

Roy T, Biswas DR, Datta SC, Sarkar A (2018a) Phosphorus release from rock phosphate as influenced by organic acid loaded nanoclay polymer composites in an Alfisol. Proc Natl Acad Sci India Sect B Biol Sci 88(1):121–132

Roy T, Biswas DR, Datta SC, Sarkar A, Biswas SS (2018b) Citric acid loaded nano clay polymer composite for solubilization of Indian rock phosphates: a step towards sustainable and phosphorus secure future. Arch Agron Soil Sci 64(11):1564–1581

Roy T, Biswas DR, Ghosh A, Patra AK, Singh RD, Sarkar A, Biswas SS (2019) Dynamics of culturable microbial fraction in an Inceptisol under short-term amendment with municipal sludge from different sources. Appl Soil Ecol. https://doi.org/10.1016/j.apsoil.2018.12.024

Ruiz-Sanchez M, Aroca R, Munoz Y, Polon R, Ruiz-Lozano JM (2010) The arbuscular mycorrhizal symbiosis enhances the photosynthetic efficiency and the antioxidative response of rice plants subjected to drought stress. J Plant Physiol 167:862–869

Ryan RP, Monchy S, Cardinale M, Taghavi S, Crossman L, Avison MB, Berg G, van der Lelie D, Dow JM (2009) The versatility and adaptation of bacteria from the genus Stenotrophomonas. Nat Rev Microbiol 7(7):514–525

Saha M, Maurya BR, Meena VS, Bahadur I, Kumar A (2016a) Identification and characterization of potassium solubilizing bacteria (KSB) from indo-Gangetic plains of India. Biocatal Agric Biotech 7:202–209

Saha M, Maurya BR, Bahadur I, Kumar A, Meena VS (2016b) Can potassium-solubilising bacteria mitigate the potassium problems in India? In: Meena V, Maurya B, Verma J, Meena R (eds) Potassium solubilizing microorganisms for sustainable agriculture. Springer, New Delhi

Saha JK, Rajendiran S, Coumar MV, Dotaniya ML, Kundu S, Patra AK (2017a) Soil pollution-an emerging threat to Indian agriculture. Springer, Singapore

Saha JK, Rajendiran S, Coumar MV, Dotaniya ML, Kundu S, Patra AK (2017b) Soil and its role in the ecosystem. In: Saha JK, Rajendiran S, Coumar MV, Dotaniya ML, Kundu S, Patra AK (eds) Soil pollution – an emerging threat to agriculture. Springer, Singapore, pp 11–36

Saha JK, Rajendiran S, Coumar MV, Dotaniya ML, Kundu S, Patra AK (2017c) Industrial activities in India and their impact on agroecosystem. In: Saha JK, Rajendiran S, Coumar MV, Dotaniya ML, Kundu S, Patra AK (eds) Soil pollution – an emerging threat to agriculture. Springer, Singapore, pp 229–249

Sarkar A, Saha M, Meena VS (2017a) Plant beneficial rhizospheric microbes (PBRM): prospects for increasing productivity and sustaining the resilience of soil fertility. In: Meena VS et al (eds) Agriculturally important microbes for sustainable agriculture. Springer, Singapore, p 1. https://doi.org/10.1007/978-981-10-5589-8_1

Sarkar A, Biswas DR, Datta SC, Manjaiah KM, Roy T (2017b) Release of phosphorus from laboratory made coated phosphatic fertilizers in soil under different temperature and moisture regimes. Proc Natl Acad Sci India Sect B Biol Sci 87(4):1299–1308

Sarkar A, Biswas DR, Datta SC, Roy T, Moharana PC, Biswas SS, Ghosh A (2018) Polymer coated novel controlled release rock phosphate formulations for improving phosphorus use efficiency by wheat in an Inceptisol. Soil Tillage Res 180:48–62

Schipper LA, Sparling GP (2011) Accumulation of soil organic C and change in C:N ratio after establishment of pastures on reverted scrubland in New Zealand. Biogeochem 104(1–3):49–58

Schoenau JJ, Davis JG (2006) Optimizing soil and plant responses to land-applied manure nutrients in the Great Plains of North America. Can J Soil Sci 86:587–595

Schoenau JJ, Malhi SS (2008) Sulfur forms and cycling processes in soil and their relationship to sulfur fertility. In: Sulfur: a missing link between soils, crops, and nutrition, (sulfuramissingl), pp 1–10

Schrey SD, Tarkka MT (2008) Friends and foes: streptomycetes as modulators of plant disease and symbiosis. Antonie Van Leeuwenhoek 94(1):11–19

Schweiger PF, Jakobsen I (1999) The role of mycorrhizas in plant P nutrition: fungal uptake kinetics and genotype variation. In: Plant nutrition-molecular biology and genetics. Springer, Dordrecht, pp 277–289

Seshadri B, Bolan NS, Naidu R (2015) Rhizosphere-induced heavy metal (loid) transformation in relation to bioavailability and remediation. J Soil Sci Plant Nutr 15:524–548

Sessitsch A, Weilharter A, Gerzabek MH, Kirchmann H, Kandeler E (2001) Microbial population structures in soil particle size fractions of a long-term fertilizer field experiment. Appl Environ Microbiol 67(9):4215–4224

Seymour N (2009) Mycorrhizae and their influence on P nutrition. Aust Grain 19(2):23

Sharma K, Dak G, Agrawal A, Bhatnagar M, Sharma R (2007) Effect of phosphate solubilizing bacteria on the germination of Cicer arietinum seeds and seedling growth. J Herb Med Toxicol 1:61–63

Shen J, Yuan L, Zhang J, Li H, Bai Z, Chen X, Zhang F (2011) Phosphorus dynamics: from soil to plant. Plant Physiol 156(3):997–1005. https://doi.org/10.1104/pp.111.175232

Singh MV, Saha JK (1997) 27th annual report of the all India Coordinated research project of micro- and secondary- nutrients and pollutant elements in soils and plants. Indian Institute of Soil Science, Bhopal, p 108

Singh BK, Bardgett RD, Smith P, Recay DS (2010) Microorganisms and climate change: terrestrial feedbacks and mitigation options. Nat Rev Microbiol 8:779–790

Singh R, Gautam N, Mishra A, Gupta R (2011) Heavy metals and living systems: an overview. Indian J Pharmacol 43(3):246–253

Singh M, Dotaniya ML, Mishra A, Dotaniya CK, Regar KL (2016) Role of biofertilizers in conservation agriculture. In: Bisht JK, Meena VS, Mishra PK, Pattanayak A (eds) Conservation agriculture: an approach to combat climate change in Indian Himalaya. Springer, Singapore, pp 113–134

Sivasankar R, Kalaikandhan R, Vijayarengan P (2012) Phytoremediating capability and nutrient status of four plant species under zinc stress. Int J Res Plant Sci 2(1):8–15

Smith SE, Read DJ (2008) The Mycorrhizal Symbiosis. Academic Press, San Diego

Sokatch JR (1969) Bacterial physiology and metabolism. Academic, London

Srivastava P, Singh R, Tripathi S, Singh P, Singh S, Singh H, Raghubanshi AS, Mishra PK (2017) Soil carbon dynamics under changing climate-a research transition from absolute to relative roles of inorganic nitrogen pools associated microbial processes: a review. Pedosphere 27(5):792–806

Stemmer M, Gerzabek MH, Kandeler E (1998a) Invertase and xylanase activity of bulk soil and particle-size fractions during maize straw decomposition. Soil Biol Biochem 31:9–18

Stemmer M, Gerzabek MH, Kandeler E (1998b) Organic matter and enzyme activity in particle size fractions of soils obtained after low-energy sonication. Soil Biol Biochem 30:9–17

Tabatabai MA, Bremner JM (1970) Arylsulfatase activity of soils. Soil Sci Soc Am Proc 34:225–229

Tchounwou PB, Yedjou CG, Patlolla AK, Sutton DJ (2012) Heavy metals toxicity and the environment. EXS 101:133–164

Turner BL, Paphazy MJ, Haygarth PM, McKelvie ID (2002) Inositol phosphates in the environment. Philos Trans R Soc Lond B Biol Sci 357:449–469

van der Wal A, de Boer W (2017) Dinner in the dark: illuminating drivers of soil organic matter decomposition. Soil Biol Biochem 105:45–48

van Gestel M, Merckx R, Vlassek K (1996) Spatial distribution of microbial biomass in microaggregates of a silty-loam soil and the relation with the resistance of microorganisms to soil drying. Soil Biol Biochem 28:503–510

Vlek PLG, Blom TJM, Beek J, Lindsay WL (1974) Determination of the solubility product of various iron hydroxides and Jarosite by chelation method. Soil Sci Soc Am Proc 38:429–432

Vorholt JA (2012) Microbial life in the phyllosphere. Nat Rev Microbiol 10:828–840

Wallenstein MD, Weintraub MN (2008) Emerging tools for measuring and modeling the in situ activity of soil extracellular enzymes. Soil Biol Biochem 40:2098–2106

Wanner BL (1996) Phosphorus assimilation and control of phosphate regulon. In: Neidhart FC, Curtiss R, Ingraham JL, Schaechter M, Umbarger HE (eds) Escherichia coli and Salmonella: cellular and molecular biology. ASM Press, Washington, pp 1357–1381

Winkelmann G (2017) A search for glomuferrin: a potential siderophore of arbuscular mycorrhizal fungi of the genus Glomus. Biometals 30(4):559–564

Yang J, Kloepper JW, Ryu CM (2009) Rhizosphere bacteria help plants tolerate abiotic stress. Trends Plant Sci 14(1):1–4

Young AY, Kinkel L (2017) Editorial: welcome to phytobiomes. Phytobiomes J 1:3–4

Zhang C, Kong F (2014) Isolation and identification of potassium-solubilizing bacteria from tobacco rhizospheric soil and their effect on tobacco plants. Appl Soil Ecol 82:18–25

Zhang F, Shen J, Zhang J, Zuo Y, Li L, Chen X (2010) Rhizosphere processes and management for improving nutrient use efficiency and crop productivity: implications for China. Adv Agron 107:1–32

Zhou K, Binkley D, Doxtader KG (1992) A new method for estimating gross phosphorus mineralization and immobilization rates in soils. Plant Soil 147:243–250

Role of a Quorum Sensing Signal Acyl-Homoserine Lactone in a Phytobiome

2

Pushparani D. Philem, Avinash Vellore Sunder, and Sila Moirangthem

Abstract

A phytobiome is influenced by its many members, which includes plants, soils, microbes, animals (insects), and the constantly fluctuating environment. Plant-microbe interactions represent one of the most impactful relationships inside a phytobiome and may have a beneficial, harmful, or neutral effect on one another. This chapter provides a comprehensive analysis of the complex network of signal exchange between microbes and plant in a phytobiome, via the quorum-sensing circuit with a special focus on N-acyl homoserine lactones (AHLs) signaling. Incorporating the current understanding of this plant-microbe dynamic by tracing their signals is one of the major tools to customize a sustainable phytobiome. There are still many gaps to cover such as understanding a system-level communication of the phytobiome and the molecular nitty-gritty of signal transport within plants and molecular pathways coordinating plant physiological changes. Future advances would depend on the collaborative effort of interdisciplinary scientist groups backed by advance "omic" techniques to link all the biotic and abiotic components and understand the synchronized dynamic of a phytobiome.

Keywords

Phytobiome · Quorum sensing · N-Acyl homoserine lactones · Signaling · Microorganism

P. D. Philem (✉)
Biochemical Sciences Division, CSIR-National Chemical Laboratory, Pune, Maharastra, India

A. V. Sunder
Department of Chemical Engineering, IIT Bombay, Mumbai, Maharastra, India

S. Moirangthem
Tissue Culture Laboratory, Research Silviculture and Training Division, Imphal, Manipur, India

© Springer Nature Singapore Pte Ltd. 2020

M. K. Solanki et al. (eds.), *Phytobiomes: Current Insights and Future Vistas*, https://doi.org/10.1007/978-981-15-3151-4_2

2.1 Introduction

Given the ever-increasing food demand and energy crisis in the world today, increasing biomass yields for greater food supply and low-cost, low-maintenance energy crop production without the use of harmful synthetic agrochemicals is the need of the hour. In recent years, there has been a paradigm shift in the understanding of the nature of plant-associated community, from the perspective that a plant is strongly interconnected with different biotic factors such as plants, animals, and microorganisms and abiotic factors like soil, light, and water and that sustainable improvement in agro-productivity and agro-quality would require a system-level approach, taking into account all the optimum biotic and abiotic set of influences on a plant (Leach et al. 2017). Phytobiome is defined as the totality of a plant, its physical environment, and the entire population of microorganism in, on, and around the plant. The composition of biotic community associated with the plant depends on the host genotype, local niche, or the plant compartment (spermosphere, endosphere, rhizosphere, phyllosphere) and environment such as soil type (Hacquard et al. 2015; Leach et al. 2017). For example, phyllosphere, the aerial plant part, experiences a highly fluctuating environment and nutrient availability and hosts a more distinct microbiome as compared to a more environmentally stable rhizosphere (Remus-Emsermann et al. 2012).

Chemical communication plays a major role in the growth, development, and evolution of plants (Witzany 2006). For example, plants release volatile compounds to invite insects to help them spread their pollen for reproduction (Baldwin 2010). A good understanding of the integrity and the basic administration in a phytobiome would help to design strategies toward sustainable agriculture and environmental preservation. This would require a sound understanding of the communication among the various members of a phytobiome. However, since signals can be quite often co-opted, modified, or even destroyed by another member of the community, there is the need for a system-level analysis of communication within the phytobiome (Leach et al. 2017).

Microbes represent the largest biotic component of a phytobiome, which also exist in the closest proximity, that is, on, around, and inside of a plant. Plants and microbes have co-evolved over time, interacting both in symbiosis and pathogenesis, and so has the communication signaling network between the two kingdoms (De-la-Peña and Loyola-Vargas 2014). The microbial world communicates via signaling molecules based on the population size-driven phenomenon of quorum sensing (QS) (Fuqua and Winans 1994). Bacterial QS machinery consists of an autoinducer, its cognate receptor protein, and a synthase gene for signal synthesis. QS-controlled bacterial phenotypes include bioluminescence, biofilm formation, motility, conjugation, sporulation, production of antibiotics, and the expression of virulence factors (Williams 2007). Disruption of the QS system or quorum quenching (QQ) takes place by (1) signal degradation by enzymes and physical agents, (2) inhibition of signal synthesis, and (3) receptor blocking (Uroz et al. 2009). Given this myriad of roles of QS in both pathogenic and symbiotic interaction, QS and QQ find multiple applications in agriculture and medicine (Grandclément et al. 2015).

Common bacterial QS signal molecules, such as *N*-acyl homoserine lactones (AHLs), diffusible signal factors (DSF), and signaling peptides, are among the best-studied communication signals in the phytobiome (Leach et al. 2017). AHLs produced by Gram-negative bacteria are the most reported autoinducer signal molecules. AHLs can be defined by the length of their acyl chain, with short-chain (C_4-C_6), medium-chain (C_6-C_{10}), and long-chain (C_{10}-C_{16}) AHLs. AHLs are produced by both pathogenic bacteria and plant growth-promoting rhizobacteria (PGPR). Table 2.1

Table 2.1 Different QS signaling molecules and their effect on plants along with the probable mechanism of action

QS molecule	Producers	Probable mechanism of action	Effect on plant	References
N-Acyl homoserine lactones (AHLs)	Gram-negative bacteria	LuxR receptor/G--protein signaling Induction of salicylic acid and oxylipin-dependent pathways	Seed germination and plant development (C_6HSL) Root elongation ($3OC_6$HSL, $3OC_8$HSL) Primary root growth, lateral root formation, and root hair development	Moshynets et al. (2019) Jin et al. (2012) Ortíz-Castro et al. (2008)
			Increased defense and systemic resistance against fungal pathogens "Priming" for induced resistance and rapid and robust response to abiotic stresses (C_{12}HSL, OC_{14}HSL)	Schuhegger et al. (2006) Schenk et al. (2014) and Schenk and Schikora (2015)
Aryl HSLs	*Rhodopseudomonas palustris*, *Bradyrhizobium*	RpaR, BraR	Not known	Schaefer et al. (2008) and Ahlgren et al. (2011)
Diffusible signal factor (DSF)	*Stenotrophomonas maltophilia*, *Xanthomonas*, *Burkholderia cenocepacia*, *Pseudomonas aeruginosa*	RpcF/RpfG two-component system	Promotes seed germination and plant growth, enhanced biocontrol, elicit innate immunity in plants	Alavi et al. (2013) and Venturi and Keel (2016)
Cyclodipeptides/ diketopiperazines	*Pseudomonas aeruginosa*	Auxin signal mimics	Promote plant growth, lateral root development	Ortiz-Castro et al. (2011)
Autoinducing peptides (AIPs)	Gram-positive bacteria	Not known		Monnet et al. (2016)

provides a list of QS signal molecules reported for their effect on plants and possible mechanism of action. Many reports have shown that plants can detect and respond to AHLs, and the responses differ based on the acyl chain length and substitution of the AHL molecule and the plant species-AHL combination (Mathesius et al. 2003; Schuhegger et al. 2006; Schikora et al. 2011; Schenk et al. 2014). This chapter summarizes the current knowledge of (a) how AHL signaling in a phytobiome impacts plants, b) how plants manipulate AHL signals to their advantage, c) the basic dynamic of the plant-microbe signal exchange, and finally d) how this knowledge can be incorporated in the bigger scheme of creating a healthier phytobiome.

2.2 Plant Senses and Responds to AHL

The first report on the impact of bacterial AHLs on plant physiology was a differential proteome analysis of *M. truncatula* roots, studying their response to the exposure of $3OC_{12}$-HSL and $3OC_{16:1}$-HSL, in nanomolar (nM) concentrations (Mathesius et al. 2003). This work revealed complex functional responses to AHL application in the plant, with significant changes in the expression of over 150 proteins. The proteins detected were associated with defense, stress responses, energetic and metabolic activities, transcriptional regulation, protein processing, cytoskeletal activities, and plant hormone responses. In another proteomic study done on the responses of *A. thaliana* seedlings to $3OC_8$-HSL treatment (Miao et al. 2012), 53 proteins showed alteration in expression pattern, and 34 of these proteins were recognized to be associated with energy and carbohydrate metabolism, protein synthesis, plant defense, signal transduction, cytoskeleton remodeling, etc. The chloroplast was observed to be the most sensitive intracellular organelle to $3OC_8$-HSL treatment. Such findings strongly suggest that plant displayed extensive functional responses to bacterial AHLs that might be important for inter-kingdom interactions. In a recent proteomic study on the effect of exogenous AHLs on *Arabidopsis thaliana* seedlings under salinity stress, AHL application was found to impart salt tolerance and growth, and it might involve 97 proteins associated with defense/stress/detoxification, photosynthesis, protein metabolism, signal transduction, transcription, cell wall biogenesis, energy metabolisms, etc. (Ding et al. 2016).

2.3 AHL Impact on Plant Root Growth and Architecture

Mathesius and coworkers (2003) reported the activation of auxin-inducible GH3 promoter upon AHL application, from their proteomic analysis of *M. truncatula* roots post AHL treatment. In a later study, von Rad et al. (2008) also reported C_6-HSL-induced changes in the expression of several plant hormone-related genes, resulting in a higher expression of auxin and lower cytokinin concentrations. Moreover, they observed a significant increase in primary root growth in the presence of short-chain AHLs (C_4-C_6HSL) at 1 nM and a concentration above 10 micromolar (μM), respectively, in *A. thaliana*, whereas the long-chain C_{10}-HSL remained neutral. The authors attributed this different plant response to the different acyl chain length of AHLs to the difference in their hydrophobicity. Long-chain AHLs with higher hydrophobicity are not transported from root to the shoot, and their accumulation in the root makes them toxic, thereby inhibiting the growth of root tissue. Since auxin induces root growth, the change in auxin/cytokinin ratio upon AHL treatment might be a potential mechanism behind short-chain AHL-induced root growth (von Rad et al. 2008).

Modification of overall root architecture of *A. thaliana* on exposure to μM concentration of long-chain AHLs, especially C_{10}-HSL, has been observed as a result of inhibition of primary root growth and promotion of lateral root and root hair growth (Ortíz-Castro et al. 2008). This change in root morphology was associated with alterations in cell division and differentiation in the primary root meristem. Although auxin-treated *Arabidopsis* seedlings show similar changes in root morphology, C_{10}-HSL action was independent of the auxin-regulated process. Enhanced lateral root and root hair growth could increase the absorptive capacity of the plant and also provide a larger surface area for bacterial colonization. These findings suggest the advantage of long-chain AHL-producing rhizobacteria to form a symbiotic association with the plant host.

Short-chain C_6-HSL has been shown (Schenk et al. 2012) to increase the shoot biomass and cause primary root elongation in *Arabidopsis*, with eventual decrease and loss of activity with an increase in acyl chain length (Fig. 2.1). Liu et al. (2012) demonstrated promotion of primary root growth in *Arabidopsis* by AHLs with modification at the C3 position of their branched chain, $3OC_6$-HSL and $3OC_8$-HSL treatment, in a concentration-dependent manner similar to C_3 unsubstituted C_6-HSL. On the other hand, long-chain AHLs, C_{12}-HSL, and C_{14}-HSL, promoted lateral root

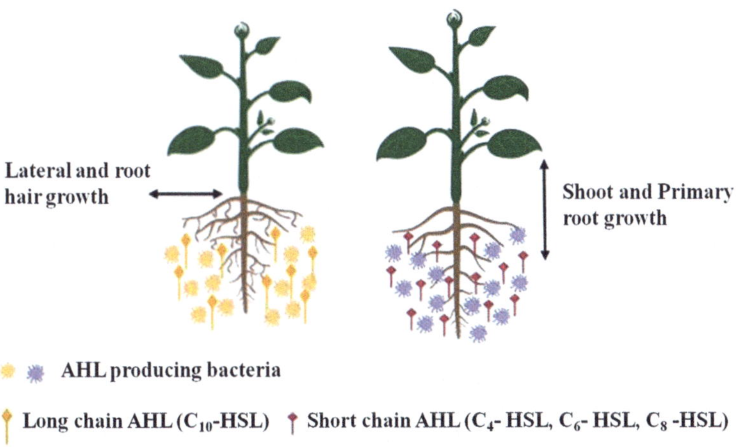

Fig. 2.1 AHL influences plant root growth and morphology in *Arabidopsis*. Changes varied based on AHL acyl chain length. Short-chain AHLs (C_6-HSL, C_8-HSL) causes the growth of primary root, whereas long-chain AHL (C_{10}-HSL) causes the growth of lateral root and root hair (von Rad et al. 2008; Bai et al. 2012; Liu et al. 2012)

growth and inhibited root elongation, suggesting separate signaling pathways for the long- and short-chain AHLs (Liu et al. 2012).

A different model plant, mung bean, was studied by Bai et al. (2012) for AHL treatment response. They found enhanced auxin-mediated adventitious root formation post AHL treatment. AHLs with C3 modification, particularly $3OC_{10}$-HSL, showed higher activity than their unmodified analogs such as C_8-HSL, C_{10}-HSL, and C_{12}-HSL. This preference for specific AHL structures may be the result of spatial specificity, required by some plant proteins (Bai et al. 2012). In contrast to these reports, there was no significant effect on the growth of yam and barley plants when exposed to C_6-HSL, C_8-HSL, and C_{10}-HSL (Götz-Rösch et al. 2015). These findings strongly suggest that different plants respond differently to AHLs, which may be a result of variations in AHL signaling cascades (Götz-Rösch et al. 2015). Figure created using BioRender.

2.4 AHL as Plant Strengthener

Plants possess a wide range of defenses that get activated in response to various stress, pathogens, and parasites (Gimenez et al. 2018). Induced resistance in plants is an important defense mechanism where defenses are preconditioned by prior infection or treatment resulting in resistance (or tolerance) against future infection (Vallad and Goodman 2004). The prior attack acts as a warning signal, preparing the plant for a stronger defense. The sensibilization mechanism, before an attack, called the priming phase, can be induced by a diverse range of low-molecular-weight metabolites and natural compounds, including AHLs (Mauch-Mani et al. 2017).

Tomato plants get primed via SA- and ethylene-dependent defense genes, upon exposure to AHL-producing rhizobacteria, *Serratia liquefaciens* MG1, and *Pseudomonas putida* IsoF, against the fungal pathogen *Alternaria alternata* (Schuhegger et al. 2006). A similar AHL-mediated antifungal activity that involved on SA- and ethylene-related defense reactions was reported in cucumber (*Cucumis sativus*) by *Serratia plymuthica* HRO-C48 against the damping-off disease caused by *Pythium aphanidermatum*, as well as in bean and tomato plants against the gray mold fungus *Botrytis cinerea* infection (Pang et al. 2009).

Among long-chain AHLs, 3-oxo-C_{14}-HSL conferred systemic resistance in monocotyledonous (barley), and dicotyledonous plant (*Arabidopsis*) plants toward biotrophic and hemibiotrophic fungus, but not against necrotrophic fungal pathogens (Schikora et al. 2011). This resistance is associated with an increased expression of mitogen-activated protein kinases, AtMPK3, and AtMPK6, which consequently increases the expression of defense-related transcription factors, WRKY22 and WRKR29, and the Pathogenesis-Related1 gene (PR1). However, 3-oxo-C_{14}-HSL did not show any impact on root and shoot growth as reported for long-chain AHLs (von Rad et al. 2008). A similar effect of 3-oxo-C_{14}-HSL imparting systemic resistance was reported by Zarkani et al. (2013) in *A. thaliana* when inoculated with *Sinorhizobium meliloti* strains producing oxo-C_{14}-HSL. They observed that neither 3-oxo-C_{14}-HSL-negative *S. meliloti* nor

3-oxo-C_8-HSL-positive *Rhizobium etli* produced a similar effect. These findings revealed that the resistance induced by AHLs does not necessarily require a host-symbiont relationship to start with, as *A. thaliana* is not a symbiotic or nodule-forming plant.

Schenk et al. (2014) reported the priming effect of AHL on exposure to oxo-C_{14}-HSL leading to increased deposition of lignin, phenolic compounds, and callose on the cell wall, reinforcing the cell wall to increase resistance to pathogens. A second possible mechanism of resistance by oxo-C_{14}-HSL could be the enhanced stomatal closure observed due to AHL-induced accumulation of cis-12-oxo-phytodienoic acid (OPDA) and salicylic acid (SA) in plant leaves. Stomatal closure is an important part of innate immunity. Another interesting mechanism of resistance by AHL-induced priming is the elevated level of hypersensitive response (HR) after infection. Barley primed with *S. meliloti* expR+ AHL showed accumulation of reactive oxygen species (ROS) and increased expression of Peroxidase7 as an HR, following infection with *Blumeria graminis* (Hernández Reyes et al. 2014). Given these various mechanisms of resistance induced by AHL priming, AHL-induced resistance seems to differ from the systemic acquired and the induced systemic resistances, providing new insight into inter-kingdom communication (Schenk et al. 2014). Moshynets et al. (2019) used C_6-HSL as a seed primer for winter wheat (*Triticum aestivum* L.), which was found to improve germination levels significantly, biomass at filtering stage, crop structure, and productivity at maturity of the crop. Thus, AHLs could find application in improving growth and productivity of cereal crops.

2.5 Effect of AHL on Symbiosis and Nitrogen Cycle

Many strains of nodulating rhizobia such as *Rhizobium leguminosarum* bv. viciae, *R. leguminosarum*, *S. meliloti* Rm41, and *A. tumefaciens* are associated with quorum sensing gene regulation systems (Gonzalez and Marketon 2003). The process of legume-host symbiosis is a complex interplay of communication between various signal molecules that includes flavonoids from root exudates that attract the rhizobia, exopolysaccharides for bacterial attachment, Nod factors for initiation of root cell division to form the infection threads, and AHLs. Once bacteria reach the host, a further process of symbiosis depends on reaching a threshold cell density, coordinated by the QS regulatory system (Gonzalez and Marketon 2003; Sanchez-Contreras et al. 2007; Downie 2010).

A few QS regulatory systems directly linked to symbiosis are discussed here. Association of root nodule formation with an unidentified AHL was initially reported in *Rhizobium leguminosarum* by Gray et al. (1996). Rosemeyer et al. (1998) identified *luxI*- and *luxR*-homologous genes, *raiI*, and *raiR* in *Rhizobium etli* CNPAF512 and found *raiI* to have a restrictive effect on the number of nodules in the host plant *Phaseolus vulgaris* without affecting the nitrogen-fixing capacity per nodule. On the other hand, the second quorum sensing locus, *cinI*, and *cinR* in *Rhizobium etli* CNPAF512 was essential for nodule bacteroid differentiation and involved in symbiotic nitrogen fixation (Daniels et al. 2002). However, the same genes showed varying implications in different *Rhizobium* strains. Mutation of *cinR and cinI* in *R. leguminosarum* had little or no effect on the growth of the bacterium and in the symbiosis with peas (Edwards et al. 2009).

QS-deficient mutants of *Mesorhizobium huakuii* showed an inability to form root nodules (Gao et al. 2006). *MrtI-MrtR* pair was identified as *LuxI-LuxR* homologs in *M. tianshanense* and played an indispensable role in root hair adherence and nodulation on the host plant *Glycyrrhiza uralensis*; QS-deficient mutants showed the absence of nodulation (Zheng et al. 2006; Cao et al. 2009). Increase in several nodules was observed in *M. truncatula* after treatment with 1 μM 3-oxo-C_{14}-HSL with no effect on lateral root number (Veliz-Vallejos et al. 2014). Study on ExpR/Sin QS system in *Sinorhizobium meliloti* by Gurich and González (2009) revealed that QS stays active during all the growth phases till symbiosis is established, after which the bacteroids' metabolic functions focus on nitrogen fixation.

Besides nitrogen fixation, nitrogen mineralization is also responsible for providing plants the utilizable form of nitrogen, where microbes enzymatically break down the organic nitrogen sources such as chitin, to the simpler inorganic form of nitrogen such as ammonia. DeAngelis et al. (2008) studied the link between QS and extracellular enzyme activity of 533 bacterial isolates, dominated by α-proteobacterial isolates such as *Agrobacterium rhizogenes*, *Inquilinus ginsengisoli*, and *Burkholderia* sp. in the rhizosphere of *Avena* and observed compromised chitinase or protease activity in all but one isolates upon disruption of QS. This strongly suggests that QS plays an important role in the regulation of exoenzyme production by bacteria in soil, and thus targeting QS for disease control should also take into account the resultant risk of disturbing soil nitrogen cycle.

2.6 AHL-Induced Plant Diseases

Phytopathogenic bacteria employ QS regulatory system to control virulence and may carry one or more QS regulons. The most well-characterized bacterial signal molecules involved in plant pathogenesis include AHLs, DSF, and 3-hydroxy palmitic acid methyl ester or methyl 3-hydroxypalmitate (Sibanda et al. 2016).

The LuxR homologs *aviR, avsR, and avsI* in *Agrobacterium vitis* are required for causing necrosis in grapes and hypersensitive-like response on tobacco (Zheng et al. 2003; Hao et al. 2005; Hao and Burr 2006). Ferluga et al. (2007) reported a LuxR family-type regulator, OryR, in *Xanthomonas oryzae* pv. *oryzae* (Xoo), essential for virulence in rice, and it required macerated rice for activity. The plant pathogen, *Pantoea ananatis* SK-1 depends on the quorum sensing system for the biosynthesis of extracellular polymeric substances (EPS), biofilm formation and causes center rot disease in onion (Morohoshi et al. 2007).

Liu et al. (2008) demonstrated a greater role for QS in pathogenesis as a "master regulator," controlling other regulatory systems that further coordinate to control virulence genes. The study monitored the progression of soft rot in potato caused by *P. atrosepticum* using virulence factors and plant cell wall-degrading enzymes (PCWDEs) under the control of the QS system. Differential transcription of up to

26% of the *P. atrosepticum* genome was observed in a QS mutant. They also identified many novel components of the QS regulon namely type I, II secretion systems associated with secretion of PCWDE; type III secretion system and coronafacoyl-amide conjugates for manipulation of plant defenses; T6SS (Type VI secretion system) and VirS, a novel potential regulator essential for complete virulence.

It is interesting how some plant pathogens exploit changes in plant chemicals to mount an optimal attack. This is a common phenomenon among plant pathogens carrying the "orphan or the solo" LuxR homologs, which do not possess a cognate LuxI homolog (Subramoni and Venturi 2009). LuxR solos occur in both AHL-producing and AHL-nonproducing bacteria, allowing them to respond to signals from other bacteria or eukaryotes. These LuxR solos may act as important interspecies and inter-kingdom signals (Ferluga and Venturi 2009; Subramoni and Venturi 2009). For example, OryR protein of *Xanthomonas oryzae* pv. *oryzae*, solubilized in the presence of rice extract, however became insoluble in the presence different AHLs. This indicates the presence of a molecule in rice extract that might interact with and stabilize OryR (Ferluga et al. 2007) and that this molecule could be closely related to AHLs, as OryR possesses an AHL-binding motif.

Similarly, XccR, the LuxI homolog in *Xanthomonas campestris* pv. *campestris*, causes infection in cabbage. XccR associates with an unknown plant factor and binds to a lux-box present in the promoter of the proline aminopeptidase (*pip*) virulence gene (Zhang et al. 2012). It is probable that OryR, XccR solos, and their plant partner participate in inter-kingdom signaling, and identification of the plant compounds would set novel insights into inter-kingdom signaling (Subramoni and Venturi 2009).

2.7 AHL Uptake in Plant and Possible Signaling Pathways for Plant Response to AHL

There are a few reports addressing this basic question of whether AHLs coordinate with other signals upon perception by plant or are taken up by plants by some physical mechanism (Schikora et al. 2016). Götz and coworkers (2007) tested the uptake of radiolabeled C_6-HSL, C_8-HSL, and C_{10}-HSL in barley (*Hordeum vulgare* L.) and yam (*Pachyrhizus erosus* L.). Ultra-performance liquid chromatography (UPLC) and Fourier transform ion cyclotron resonance mass spectrometry (FTICR-MS) analysis revealed short AHLs, only C_6-HSL, and C_8-HSL, transported from roots into barley shoots and C_6-HSL in case of yam shoots. *A. thaliana* also transported C_6-HSL from root to leaves in a time-dependent manner, while C_{10}-HSL remained accumulated in root region (von Rad et al. 2008). Similarly, it was observed that the long-chain AHLs, oxo-C_{14}-HSL, were not taken up within the *Arabidopsis* plant.

Moreover, it showed no effect on root growth, whereas short-chain AHL C_6-HSL got systemically translocated from root to shoot and affected growth-promoting effect (Schikora et al. 2011). A possible reason behind the difference in uptake of AHLs might be the reduced motility with an increase in acyl chain length or higher hydrophobicity of longer AHLs (von Rad et al. 2008; Schikora et al. 2011).

However, Sieper et al. (2014) reported systemic transport of C_8-HSL and C_{10}-HSL into the barley shoots with a molecular analysis of the process backed by autoradiography analysis, sensor strain assays, and monoclonal antibodies. They suggested a probable active or semi-active process of translocation of AHLs with the involvement of ATP-binding cassette (ABC) transporters based on the inhibition of AHL uptake in the presence of orthovanadate, an inhibitor of ABC transporters. All these reports point toward different signaling pathways and receptors for different AHLs. Götz-Rösch et al. (2015) observed a distinct pattern of selected detoxification enzymes activity in barley and yam plants, upon C_6-, C_8-, and C_{10}-HSL treatment. The activities of glutathione S-transferase and ROS scavenging enzymes were mostly tissue specific. C_6-HSL was readily transported into all barley plant parts without degradation with the most prominent influence on the leaf-located enzymes.

On the other hand, the yam plant showed no such change in detoxification enzyme activity, which might be a result of the absence of AHL-transport machinery toward the shoot in yam bean. In both the plants, C_{10}-HSL was almost broken down completely before it entered the shoot in its initial form. Interestingly, the metabolization was faster in yam plant than in barley, which may be related to the fact that symbiotic yam beans are naturally exposed to AHL (Götz-Rösch et al. 2015).

Other studies have aimed at deciphering the molecular mechanisms behind the plant-AHL interactions, such as AHL perception mechanism in plant and signal transduction pathways involved (Jin et al. 2012; Song et al. 2011; Liu et al. 2012; Zhao et al. 2016). G-protein signaling is considered one of the most conserved signaling pathways in plants and is associated with the transduction of many extracellular signals. It is involved in many physiological functions in plants like stomatal opening (Urano and Jones 2013), phytochrome-dependent cell death (Zhang et al. 2012), stress signaling (Tuteja and Sopory 2008), response to plant hormones, and agronomical traits like nitrogen fixation, seed size, and number (Pandey et al. 2019). In typical G-protein signaling, the G-protein complex comprised of Gα, Gβ, and Gγ subunits remains inactive with GDP bound to the Gα subunit. In animals, the G-proteins get activated via a membrane protein called G-protein-coupled receptor (GPCR) upon binding with a ligand, leading to the exchange of GTP with GDP and the dissociation of GTP-Gα from the Gβγ subunit. The free Gα and Gβγ subunits further initiate different intracellular signaling pathways (Tuteja 2009). However, in plants, the G-protein signaling is reported to be independent of GPCR, and the plant G-proteins release GDP spontaneously and thus are self-activating (Urano and Jones 2013). Although some authors think this hypothesis is supported by genetic and physiological findings, it is still not clear whether the hypothesis applies to all G-proteins in all plant species (Pandey et al. 2019). Song et al. (2011) hypothesized that AHLs in plants is received by GPCRs, which activate the G-protein α subunit and Ca^{2+} channels to allow Ca^{2+} influx. C_4-HSL treatment on *A. thaliana* roots was shown to cause a transient rise in cytosolic Ca^{2+} as a result of influx from the extracellular matrix. The authors suggest Ca^{2+} signaling might help plant cells to sense QS signals. However, considering the possibility of the absence of GPCR, AHL might also interact directly with intracellular receptors without the help of transmembrane proteins (Jin et al. 2012; Song et al. 2011).

Many G-protein-coupled receptors (GPCRs) have been reported to mediate AHL-induced root elongation in *A. thaliana*. Expression of GPCRs, Cand2, Cand7

(Jin et al. 2012), GCR1, and GCR2 (Bian et al. 2011) increased in *A. thaliana* root in response to 1 μM $3OC_6$-HSL and 10 μM $3OC_8$-HSL treatment. GPA1 was also added to this list of GPCRs affecting $3OC_6$-HSL and $3OC_8$-HSL action on root growth in *A. thaliana* in a concentration-dependent manner (Liu et al. 2012). Involvement of H_2O_2 and NO-dependent cGMP signaling pathways behind this AHL-induced auxin-dependent lateral root formation has also been proposed (Bai et al. 2012). A recent report on $3OC_6$-HSL enhancing root elongation involves the transcriptional factor AtMYB44 in *A. thaliana* (Zhao et al. 2016).

2.8 Plant Interference with QS Signals

Not only do plants respond to AHL and QS systems in various ways, from reaping symbiotic benefits to enhancing disease resistance, they also evolve strategies to interfere with the microbial signaling systems for their advantage, by producing signal mimics, certain plant metabolites, and signal-degrading enzymes (Fig. 2.2).

2.8.1 Plant Interferes with QS via AHL Mimics

AHL mimic compounds are reported to be produced by many plants to create confusion in microbial communication (Bauer and Mathesius 2004). Halogenated furanones from *Delisea pulchra*, a marine red alga, were reported to have strong

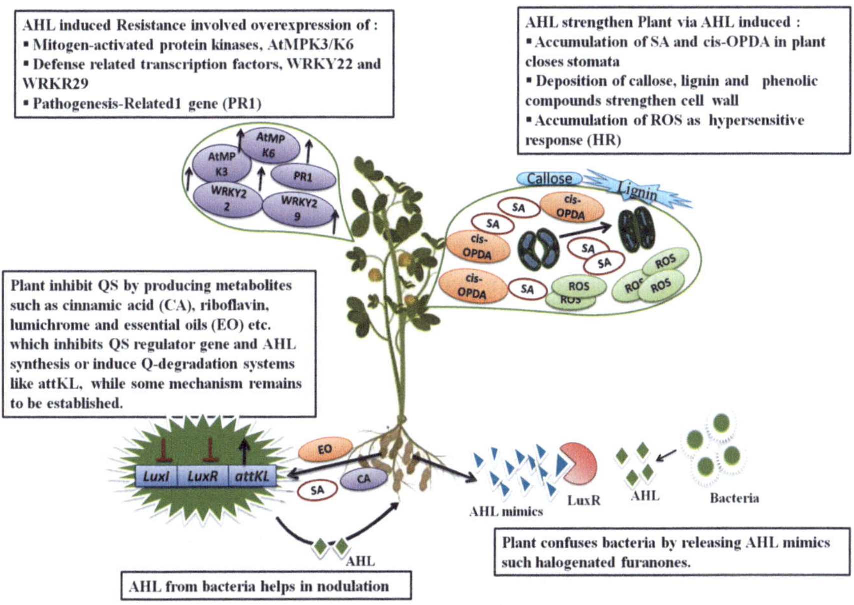

Fig. 2.2 Schematic presentation of how plants resist and manipulate QS signaling to their advantage. Figure created using BioRender

inhibitory effects against AHL-controlled phenotypes in *Serratia liquefaciens*, *Escherichia coli*, and *Proteus mirabilis* models (Givskov et al. 1996; Gram et al. 1996). Later, the structural similarity of furanones to AHLs was confirmed alongside their possible mode of action of displacing AHLs competitively from the AHL LuxR receptor (Manefield et al. 1999). Further clarity into the mechanism of furanones came from the finding that halogenated furanones destabilized LuxR protein, accelerating its degradation (Manefield et al. 2002).

The leaves and stem of rice plants produce AHL mimics, which could activate three different bacterial AHL biosensors and were very sensitive to the AiiA lactonase. Given the highly specific nature of lactonases, these rice compounds may be another AHL or a compound of close structure (Degrassi et al. 2007). Teplitski et al. (2000) reported the presence of AHL mimics in pea seedling exudates using various AHL reporter strains, *C. violaceum* CV026, *P. aureofaciens* 30-84I, and *S. liquefaciens* MG44. Activity analysis of different HPLC fractions revealed the presence of both inhibitory and stimulatory substances in pea plant. AHL mimic activity was also shown by seedlings of other plant species, rice, soybean, tomato, crown vetch, and *Medicago truncatula*, suggesting AHL mimic synthesis is a common response among higher plants (Helman and Chernin 2015). Gao et al. (2003) reported around 15 to 20 separable unidentified QS mimic compounds in *M. truncatula* young seedling, capable of specifically stimulating or inhibiting responses in QS reporter bacteria, *Vibrio harveyi* BB170, *Pseudomonas putida* pAS-C8, *C. violaceum* CV026, and *E. coli* strains JM109, p(SB401), p(SB536), and p(SB1075). However, the precise chemical structure of these QS mimics in pea, rice, and *M. truncatula* is not yet known, except for the possibility of a structure close to that of AHLs. It is also suggested that such mimic compounds might be degradation products of bacterial AHLs as the result of the action of plant AHL-degrading enzymes such as lactonases (Mathesius and Watt 2010). Pérez-Montaño et al. (2013) reported AHL mimic molecules produced by *Oryza sativa* (rice) and *P. vulgaris* (bean) plants that specifically interfere with the QS-regulated biofilm formation of two plant-associated bacteria, *Sinorhizobium fredii* SMH12 and *Pantoea ananatis* AMG501. These mimic compounds are considered non-AHL-type mimics as they lack a lactone ring typical of AHLs. Seed and root exudates of *Arachis hypogaea* L. (peanut) contain unidentified long-chain AHL-like mimics and short inhibitory chain AHL-like compounds, which are tools for manipulating the bacterial behavior in the rhizosphere (Nievas et al. 2017).

2.8.2 Plant Metabolites and Enzymatic Interference of QS

Many plant secondary metabolites interfere directly and indirectly with bacterial QS. Keshavan et al. (2005) coined the term quorum sensing-interfering (QSI) compounds for the second group of plant signals that do not resemble AHLs. They observed inhibition of violacein production in *C. violaceum* CV026 by Alfalfa seed and seedling exudates without affecting cell growth, indicating interference of QS. Structural characterization of a molecule purified from the exudates revealed it as L-canavanine, an arginine analog. The Alfalfa seed exudates also inhibited the synthesis of exopolysaccharide EPS II, QS-regulated phenotype, required for establishing symbiosis in *S. meliloti*, the natural symbiont of Alfalfa. A similar effect was demonstrated by synthetic L-canavanine, which did not affect protein synthesis or synthesis of AHL in *S. meliloti.* A possible mechanism of action may be the incorporation of L-canavanine in place of L-arginine in the QS regulator protein, thus impairing protein folding. Alternatively, some bacteria such as *Streptococcus faecalis* can degrade L-canavanine to homoserine (Kalyankar et al. 1958), which if converted to AHL mimic scan also contribute to the QS inhibition property of L-canavanine.

Salicylic acid (SA) produced by plants downregulates the virulence genes and activates the quorum degradation system, *attKLM* operon in *Agrobacterium*, which includes the *att*M gene encoding a lactonase (Yuan et al. 2007). Thus, SA might act as a powerful plant defense signal acting against *Agrobacterium* and other QS bacteria in the phytobiome. A similar interference with QS mechanism was observed when plant phenolic acids, cinnamic acid (CA), and SA inhibited expression of QS regulator gene; QS regulated virulence genes and reduced AHL level in *Pectobacterium aroidearum* and *P. carotovorum* (Joshi et al. 2016). Cinnamaldehyde is a naturally occurring compound in the bark of cinnamon trees and widely used in food industries. Niu et al. (2006) reported inhibition of 3-hydroxy-C_4-HSL- and 3OC_6-HSL-mediated QS by very low concentrations of synthetic cinnamaldehyde. p-Coumaric acid, a natural phenolic compound produced by plants, showed both stimulatory and inhibitory effects on QS in a concentration- and strain-dependent manner, whereas garlic extract inhibited QS receptors LuxR, AhyR, and TraR, which become toxic at higher concentrations (Bodini et al. 2009). It is interesting that coumaric and cinnamic acids have also been implicated in the synthesis of aryl HSL signals in certain nodulating rhizobacteria (Schaefer et al. 2008; Ahlgren et al. 2011).

The vitamin riboflavin and its derivative, lumichrome, are synthesized by bacteria such as *B. subtilis*, *E. coli*, *Photobacterium phosphoreum*, and *Actinobacillus pleuropneumoniae* (Vitreschak et al. 2002), a majority of plants (Roje 2007), and algae like *Chlamydomonas* (Palacios et al. 2014). Riboflavin and lumichrome purified from the green alga *Chlamydomonas reinhardtii* can bind to and activate the AHL receptor LasR and thus interfere with bacterial QS (Rajamani et al. 2008). Change in response to riboflavin and lumichrome after mutation of LasR residues required for AHL binding indicated a similar binding site for all the three signals. These molecules may have a larger and complex role, given the fact that they are produced by plants, bacteria, and algae (Mathesius and Watt 2010).

Medicinal plants also contribute a vast repertoire of secondary metabolites with anti-QS effects. Curcumin, the phenolic bioactive compound of *Curcuma longa* (turmeric), is a popular example. It can attenuate *Pseudomonas aeruginosa* (PAO1) virulence by interfering with AHL production, biofilm formation, and inhibiting elastase and protease activity (Rudrappa and Bais 2008). It is associated with reduction in biofilm formation by uropathogens, *E. coli*, *Proteus mirabilis*, *P. aeruginosa*, and *Serratia marcescens* that cause persistent urinary tract infection via QS-dependent virulence mechanisms (Packiavathy et al. 2014). Some eminent examples of medicinal plants with anti QS activity include *Terminalia chebula*, *Ocimum sanctum*, *Amphipterygium adstringens*, and *Centella asiatica* (Bouyahya et al. 2017). The primary bioactive compounds present in these medicinal plants are polyphenols, terpenoids, flavonoids, quinines, tannins, anthocyanins, polyamines, cytokinins, and polysaccharides (Bouyahya et al. 2017).

Many plant volatiles such as essential oils (EO) and their components compounds also act as a signal and affect bacterial QS (Ahmad et al. 2015). *Eucalyptus globulus* and *E. radiate* essential oils contain eucalyptol and limonene as the main bioactive component, respectively, which inhibited violacein pigment of the reporter strain *C. violaceum* CV026 without changing cell growth parameters, indicating anti-quorum sensing activity (Luís et al. 2016). A similar effect was seen in EOs from different species of *Piper* (Olivero et al. 2011). Khan et al. (2009) reported the anti-QS property of an unknown component of clove oil, whereas eugenol, the main constituent of clove oil, could not inhibit QS activity, indicating the anti-QS property of other minor components of the oil such as α-caryophyllene and β-caryophyllene. Citrus reticulate EOs also showed inhibition of biofilm formation and AHL production (Luciardi et al. 2016). *Thymus vulgare* EO inhibited the expression of the AHL-related flagella gene *flg*A required for initial bacterial attachment for biofilm formation by the foodborne pathogen *P. fluorescens* KM121 strain (Myszka et al. 2016).

The significance of the AHL-based signaling systems is reflected in the abundance of quorum quenching activities in plants and animals, which have the role of destroying the signaling activity of bacterial pathogens (Schikora et al. 2016). Plants can interfere with bacterial QS by producing enzymes as well. Delalande et al. (2005) reported rapid disappearance of AHLs from the root region of germinating *Lotus* plants, and a possible enzymatic mechanism is suspected. AHLs are reported to alter *Arabidopsis* postembryonic root development (Ortíz-Castro et al. 2008). Fatty acid amide hydrolase expressed by *Arabidopsis* could degrade AHLs, as evident from the increased resistance to AHLs in overexpressing mutants, and AHLs induced susceptibility to developmental changes in fatty acid amide hydrolase mutant (Ortíz-Castro et al. 2008).

2.9 Role of Endophyte in a Phytobiome

Endophytes are microbes that symbiotically live whole or at least a part of their life cycle within a living plant without showing any signs of infection. However, commonly known endophytes are mostly considered commensals and do not affect host plant functioning, while less common are either mutualistic with beneficial effects or antagonistic as latent pathogens (Hardoim et al. 2015). The beneficial effects of an endophyte include the promotion of plant growth, health, and contribution to host defense response against a pathogen and abiotic stresses (Khare et al. 2018). One of the most notable properties of endophytes is their ability to either synthesize or induce a host plant to synthesize metabolites that promote plant growth and help them adapt better to the environment (Varma et al. 2017). Thus, they promise eco-friendly sources of useful bioactive compounds such as antibiotics and anticancer drugs (Hardoim et al. 2015). Moreover, endophytic make up of edible plants becomes a concern as it is known that some plant-friendly endophytes could be potential human pathogens (van Overbeek et al. 2014).

Although it is accepted that the interaction of endophyte and its host plant may be multidimensional (Khare et al. 2018), many fundamental questions remain to be answered. How do plants recruit endophytes (Kandel et al. 2017)? How does the endophyte survive the first line of defense of the host plant? How do they influence host gene expression and how different members of the endophyte community interact and influence one another?

There are a few reports on QS activity among endophytes and about its influence on the host plant. Hudson et al. (2010) isolated and identified several culturable endophytic strains from stem tissue of sugarcane and from xylem fluids of grapevine. Five of six sugarcanes and fourteen of fifteen grapes with identified endophytes showed a significant response in at least one AHL-dependent biosensor. AHL lactonase activity has been reported in endophytic population isolated from potato tuber peel (Ha et al. 2018) and *Pterocarpus santalinus* (Rajesh and Rai 2014). Two AHL-based quorum sensing systems, SplIR and SpsIR, have been reported from endophytic *Serratia* sp. strain G3 (Liu et al. 2011). While similar to the free-living strain of *Serratia* HRO-C48, antifungal activity, indole-3-acetic acid and exoenzyme production in the G3 strain are regulated by QS, however, QS control of swimming motility and biofilm formation was a strain-specific phenotype, reflecting a lifestyle driven by the evolution of the QS systems in free-living and endophytic strains. Jiang et al. (2014) reported production of $3OC_8$-HSL by *Pantoea agglomerans* YS19, an endophytic diazotrophic bacterium isolated from rice, and revealed the importance of $3OC_8$-HSL in promoting bacterial growth and also in the formation of a multicellular aggregate structure called symplasmata, which is a characteristic of the bacterium.

Pseudomonas sp. strain GM79, an endophyte isolated from cottonwood (*Populus deltoides*), possesses an orphan OryR homolog PipR, which was shown to require an unknown signal derived from the cottonwood leaf macerates instead of an AHL (Schaefer et al. 2016). The signal was later identified as a chemical derived from ethanolamine activity (Coutinho et al. 2018). Similar roles of plant signals for activation of different solo/orphan Lux R homologs have been reported in many other free-living bacteria (Ferluga et al. 2007; Ferluga and Venturi 2009; Subramoni and Venturi 2009). Thus, the current understanding of endophytes signifies their immense potential and advantage in improving the quality of a phytobiome. At the same time, the need for a deeper understanding of the physiology of endophytes is also obvious, given the limited number of reports and a nonsystemic approach of study used so far (Hardoim et al. 2015).

2.10 Conclusion and Forthcoming Prospects

Managing the phytobiome may offer an important biotechnological area for improving agricultural yield and quality with a minimum participation of harmful agrochemicals. Phytobiome studies may provide more precise insights into the mechanisms and consequences of disease (and resistance) and identify microbial indicators of disease (and resistance) progress. In the field, it can be assumed that communication signals are exposed to many interfering and mimic signals from the neighbors, and hence a system-level understanding of phytobiome is the appropriate approach for successful application of this understanding.

One of the pivotal points of control to bring desired traits in a phytobiome is the manipulation of the interspecies and intraspecies communication between plants and microbes, either by introducing selected beneficial microbes as microbial biofertilizers and biopesticides or by genetic engineering of plants and microbes. A known plasmid pME6863 carrying a lactonase coding gene *aiiA* from *Bacillus* sp. A24 to *Pseudomonas fluorescens* P3 exhibited the ability to degrade AHLs and significantly reduced soft potato rot caused by *Erwinia carotovora* and crown gall of tomato caused by *Agrobacterium tumefaciens* to a similar level as *Bacillus* sp. A24. Tobacco and potato plants transformed with AHL lactonase from *Bacillus* sp. showed enhanced resistance to *E. carotovora* infection.

The use of advanced "omic" technologies, such as metabolomics and proteomics, allows a more reliable analysis of phytobiome interactions in unprecedented detail and provides insights into the resistance mechanisms that consider both simultaneous attacks of various pathogens and the interplay with beneficial microbes. In this context, the metagenomic study of unculturable microorganisms would also provide an indispensable solution to this whole complex equation in a phytobiome. Moreover, a multidisciplinary analysis should be incorporated by collaborative discussions among different disciplines such as biologists, ecologists, plant pathologists and agricultural entomologists. It should be soon possible to provide a systemic picture of how and to what extent plants can shape their own detrimental or beneficial community.

References

Ahlgren NA, Harwood CS, Schaefer AL, Giraud E, Greenberg EP (2011) Aryl homoserine lactone quorum sensing in stem-modulating photosynthetic bradyrhizobia. Proc Natl Acad Sci U S A 103:7183–7188

Ahmad A, Viljoen AM, Chenia HY (2015) The impact of plant volatiles on bacterial quorum sensing. Lett Appl Microbiol 60(1):8–19

Alavi P, Muller H, Cardinale M, Zachow C, Sanchez MB, Martinez JL, Berg G (2013) The DSF quorum sensing system controls the positive influence of *Stenotrophomonas maltophilia* on plants. PLoS One 8:e67103

Bai X, Todd DC, Desikan R, Yang Y, Hu X (2012) N -3-Oxo-Decanoyl- L -Homoserine-lactone activates auxin-induced adventitious root formation via hydrogen peroxide- and nitric oxide-dependent cyclic GMP signaling in mung bean. Plant Physiol 158(2):725–736

Baldwin IT (2010) Plant volatiles. Curr Biol 20(9):R392–R397

Bauer WD, Mathesius U (2004) Plant responses to bacterial quorum sensing signals. Curr Opin Plant Biol 7(4):429–433

Bian X, Klemm RW, Liu TY, Zhang M, Sun S, Sui X, Liu X, Rapoport TA, Hu J (2011) Structures of the atlastin GTPase provide insight into homotypic fusion of endoplasmic reticulum membranes. Proc Natl Acad Sci 108(10):3976–3981

Bodini SF, Manfredini S, Epp M, Valentini S, Santori F (2009) Quorum sensing inhibition activity of garlic extract and p-coumaric acid. Lett Appl Microbiol 49(5):551–555

Bouyahya A, Dakka N, Et-touys A, Abrini J, Bakri Y (2017) Medicinal plant products targeting quorum sensing for combating bacterial infections. Asian Pac J Trop Med 10(8):729–743

Cao H, Yang M, Zheng H, Zhang J, Zhong Z, Zhu J (2009) Complex quorum-sensing regulatory systems regulate bacterial growth and symbiotic nodulation in *Mesorhizobium tianshanense*. Arch Microbiol 191(3):283–289

Coutinho BG, Mevers E, Schaefera AL, Pelletier DA, Harwood CS, Clardy J, Greenberg EP (2018) A plant-responsive bacterial-signaling system senses an ethanolamine derivative. Proc Natl Acad Sci U S A 115(39):9785–9790

Daniels R, De Vos DE, Desair J, Raedschelders G, Luyten E, Rosemeyer V, Verreth C, Schoeters E, Vanderleyden J, Michiels J (2002) The cin quorum-sensing locus of *Rhizobium etli* CNPAF512 affects growth and symbiotic nitrogen fixation. J Biol Chem 277(1):462–468

DeAngelis KM, Lindow SE, Firestone MK (2008) Bacterial quorum sensing and nitrogen cycling in rhizosphere soil. FEMS Microbiol Ecol 66(2):197–207

Degrassi G, Devescovi G, Solis R, Steindler L, Venturi V (2007) Oryza sativa rice plants contain molecules that activate different quorum-sensing N-acyl homoserine lactone biosensors and are sensitive to the specific AiiA lactonase. FEMS Microbiol Lett 269(2):213–220

Delalande L, Faure D, Raffoux A, Uroz S, D'Angelo-Picard C, Elasri M, Carlier A, Berruyer R, Petit A, Williams P, Dessaux Y (2005) N-hexanoyl-L-homoserine lactone, a mediator of bacterial quorum-sensing regulation, exhibits plant-dependent stability and may be inactivated by germinating *Lotus corniculatus* seedlings. FEMS Microbiol Ecol 52(1):13–20

De-la-Peña C, Loyola-Vargas VM (2014) Biotic interactions in the rhizosphere: a diverse cooperative enterprise for plant productivity. Plant Physiol 166(2):701–719

Ding L, Cao J, Duan Y, Li J, Yang Y, Yang G, Zhou Y (2016) Retracted: proteomic and physiological responses of *Arabidopsis thaliana* exposed to salinity stress and N-acyl-homoserine lactone. Physiol Plant 158(4):414–434

Downie JA (2010) The roles of extracellular proteins, polysaccharides and signals in the interactions of rhizobia with legume roots. FEMS Microbiol Rev 34(2):150–170

Edwards A, Frederix M, Wisniewski-Dyé F, Jones J, Zorreguieta A, Downie JA (2009) The cin and rai quorum-sensing regulatory systems in *Rhizobium leguminosarum* are coordinated by ExpR and CinS, a small regulatory protein coexpressed with CinI. J Bacteriol 191(9):3059–3067

Ferluga S, Venturi V (2009) OryR is a LuxR-family protein involved in Interkingdom signaling between pathogenic *Xanthomonas oryzae* pv. oryzae and Rice. J Bacteriol 191(3):890–897

Ferluga S, Bigirimana J, Hofte M, Venturi V (2007) A LuxR homologue of *Xanthomonas oryzae* pv. oryzae is required for optimal rice virulence. Mol Plant Pathol 8(4):529–538

Fuqua WC, Winans SC (1994) A LuxR-LuxI type regulatory system activates *Agrobacterium* Ti plasmid conjugal transfer in the presence of a plant tumor metabolite. J Bacteriol 176:2796–2806

Gao M, Teplitski M, Robinson JB, Bauer WD (2003) Production of substances by *Medicago truncatula* that affect bacterial quorum sensing. Mol Plant-Microbe Interact 16(9):827–834

Gao Y, Zhong Z, Sun K, Wang H, Zhu J (2006) The quorum-sensing system in a plant bacterium *Mesorhizobium huakuii* affects growth rate and symbiotic nodulation. Plant Soil 286(1–2):53–60

Gimenez E, Salinas M, Manzano-Agugliaro F (2018) Worldwide research on plant defense against biotic stresses as improvement for sustainable agriculture. Sustainability 10(2):391

Givskov M, de Nys R, Manefield M, Gram L, Maximilien RI, Eberl LE, Molin S, Steinberg PD, Kjelleberg S (1996) Eukaryotic interference with homoserine lactone-mediated prokaryotic signalling. J Bacteriol 178(22):6618–6622

Gonzalez JE, Marketon MM (2003) Quorum sensing in nitrogen-fixing rhizobia. Microbiol Mol Biol Rev 67(4):574–592

Götz C, Fekete A, Gebefuegi I, Forczek ST, Fuksová K, Li X, Englmann M, Gryndler M, Hartmann A, Matucha M, Schmitt-Kopplin P (2007) Uptake, degradation and chiral discrimination of N-acyl-D/L-homoserine lactones by barley (*Hordeum vulgare*) and yam bean (*Pachyrhizus erosus*) plants. Anal Bioanal Chem 389(5):1447–1457

Götz-Rösch C, Sieper T, Fekete A, Schmitt-Kopplin P, Hartmann A, Schröder P (2015) Influence of bacterial N-acyl-homoserine lactones on growth parameters, pigments, antioxidative capacities and the xenobiotic phase II detoxification enzymes in barley and yam bean. Front Plant Sci 6:205

Gram L, de Nys RO, Maximilien RI, Givskov M, Steinberg P, Kjelleberg S (1996) Inhibitory effects of secondary metabolites from the red alga *Delisea pulchra* on swarming motility of *Proteus mirabilis*. Appl Environ Microbiol 62(11):4284–4287

Grandclément C, Tannières M, Moréra S, Dessaux Y, Faure D (2015) Quorum quenching: role in nature and applied developments. FEMS Microbiol Rev 40(1):86–116

Gray KM, Pearson JP, Downie JA, Boboye BE, Greenberg EP (1996) Cell-to-cell signaling in the symbiotic nitrogen-fixing bacterium *Rhizobium leguminosarum*: autoinduction of a stationary phase and rhizosphere-expressed genes. J Bacteriol 178(2):372–376

Gurich N, González JE (2009) Role of quorum sensing in *Sinorhizobium meliloti*-alfalfa symbiosis. J Bacteriol 191(13):4372–4382

Ha NT, Minh TQ, Hoi PX (2018) Biological control of potato tuber soft rot using N -acyl-L-homoserine lactone-degrading endophytic bacteria. Curr Sci 115:1921–1927

Hacquard S, Garrido-Oter R, González A, Spaepen S, Ackermann G, Lebeis S, McHardy AC, Dangl JL, Knight R, Ley R, Schulze-Lefert P (2015) Microbiota and host nutrition across plant and animal kingdoms. Cell Host Microbe 17(5):603–616

Hao G, Burr TJ (2006) Regulation of long-chain N-acyl-homoserine lactones in *Agrobacterium vitis*. J Bacteriol 188(6):2173–2183

Hao G, Zhang H, Zheng D, Burr TJ (2005) LuxR Homolog avhR in *Agrobacterium vitis* affects the development of a grape-specific necrosis and a tobacco hypersensitive response. J Bacteriol 187(1):185–192

Hardoim PR, van Overbeek LS, Berg G, Pirttilä AM, Compant S, Campisano A, Döring M, Sessitsch A (2015) The hidden world within plants: ecological and evolutionary considerations for defining functioning of microbial endophytes. Microbiol Mol Biol Rev 79(3):293–320

Helman Y, Chernin L (2015) Silencing the mob: disrupting quorum sensing as a means to fight plant disease. Mol Plant Pathol 16(3):316–329

Hernández Reyes C, Schenk ST, Neumann C, Kogel KH, Schikora A (2014) N-acyl-homoserine lactones-producing bacteria protect plants against plant and human pathogens. Microb Biotechnol 7(6):580–588

Hudson AO, Ahmad NH, Buren R Van, Savka MA (2010) Sugarcane and grapevine endophytic bacteria: isolation, detection of quorum sensing signals and identification by 16S v3 rDNA sequence analysis. 801–806

Jiang J, Wu S, Wang J, Feng Y (2014) AHL-type quorum sensing and its regulation on symplasmata formation in *Pantoea agglomerans* YS19. J Basic Microbiol 55:1–10

Jin G, Liu F, Ma H, Hao S, Zhao Q, Bian Z, Jia Z, Song S (2012) Two G-protein-coupled-receptor candidates, Cand2 and Cand7, are involved in Arabidopsis root growth mediated by the bacterial quorum-sensing signals N-acyl-homoserine lactones. Biochem Biophys Res Commun 417(3):991–995

Joshi JR, Burdman S, Lipsky A, Yariv S, Yedidia I (2016) Plant phenolic acids affect the virulence of *Pectobacterium aroidearum* and *P. carotovorum* ssp. brasiliense via quorum sensing regulation. Mol Plant Pathol 17(4):487–500

Kalyankar GD, Ikawa M, Snell EE (1958) The enzymatic cleavage of canavanine to homoserine and hydroxyguanidine. J Biol Chem 233(1):175–171

Kandel SL, Joubert PM, Doty SL (2017) Bacterial endophyte colonization and distribution within plants. Microorganisms 5:9–11

Keshavan ND, Chowdhary PK, Haines DC, González JE (2005) L-Canavanine made by *Medicago sativa* interferes with quorum sensing in *Sinorhizobium meliloti*. J Bacteriol 187(24):8427–8436

Khan MS, Zahin M, Hasan S, Husain FM, Ahmad I (2009) Inhibition of quorum sensing regulated bacterial functions by plant essential oils with special reference to clove oil. Lett Appl Microbiol 49(3):354–360

Khare E, Mishra J, Arora NK (2018) Multifaceted interactions between endophytes and plant: developments and prospects. Front Microbiol 9:1–12

Leach JE, Triplett LR, Argueso CT, Trivedi P (2017) Review communication in the Phytobiome. Cell 169(4):587–596

Liu H, Coulthrust SJ, Pritchard L, Hedley PE, Ravensdale M, Humphris S, Burr T, Takkle G, Brurburg MB, Birch PR, Salmon GPC, Toth IK (2008) Quorum sensing coordinates brute force and stealth modes of infection in the plant pathogen *Pectobacterium atrosepticum*. PLoS Pathog 4(6). https://doi.org/10.1371/journal.ppat.1000093

Liu X, Jia J, Popat R, Ortori CA, Li J, Diggle SP, Gao K, Cámara M (2011) Characterisation of two quorum sensing systems in the endophytic *Serratia plymuthica* strain G3: differential control of motility and biofilm formation according to life-style. BMC Microbiol 11:26

Liu F, Bian Z, Jia Z, Zhao Q, Song S (2012) The GCR1 and GPA1 participate in promotion of Arabidopsis primary root elongation induced by N-acyl-homoserine lactones, the bacterial quorum-sensing signals. Mol Plant-Microbe Interact 25(5):677–683

Luciardi MC, Blázquez MA, Cartagena E, Bardón A, Arena ME (2016) Mandarin essential oils inhibit quorum sensing and virulence factors of *Pseudomonas aeruginosa*. LWT-Food Sci Technol 68:373–380

Luís Â, Duarte A, Gominho J, Domingues F, Duarte AP (2016) Chemical composition, antioxidant, antibacterial and anti-quorum sensing activities of *Eucalyptus globulus* and *Eucalyptus radiata* essential oils. Ind Crop Prod 79:274–282

Manefield M, de Nys R, Naresh K, Roger R, Givskov M, Peter S, Kjelleberg S (1999) Evidence that halogenated furanones from *Delisea pulchra* inhibit acylated homoserine lactone (AHL)-mediated gene expression by displacing the AHL signal from its receptor protein. Microbiology 145(2):283–291

Manefield M, Rasmussen TB, Henzter M, Andersen JB, Steinberg P, Kjelleberg S, Givskov M (2002) Halogenated furanones inhibit quorum sensing through accelerated LuxR turnover. Microbiology 148(4):1119–1127

Mathesius U, Watt M (2010) Rhizosphere signals for plant–microbe interactions: implications for field-grown plants. In: Progress in botany, vol 72. Springer, Berlin/Heidelberg, pp 125–161

Mathesius U, Mulders S, Gao M, Teplitski M, Caetano-Anollés G, Rolfe BG, Bauer WD (2003) Extensive and specific responses of a eukaryote to bacterial quorum-sensing signals. Proc Natl Acad Sci U S A 100(3):1444–1449

Mauch-Mani B, Baccelli I, Luna E, Flors V (2017) Defense priming: an adaptive part of induced resistance. Annu Rev Plant Biol 68:485–512

Miao C, Liu F, Zhao Q, Jia Z, Song S (2012) A proteomic analysis of arabidopsis thaliana seedling responses to 3-oxo- octanoyl-homoserine lactone, a bacterial quorum-sensing signal. Biochem Biophys Res Commun 427:293–298

Monnet V, Juillard V, Gardan R (2016) Peptide conversations in Gram-positive bacteria. Crit Rev Microbiol 42:339–351

Morohoshi T, Nakamura Y, Yamazaki G, Ishida A, Kato N, Ikeda T (2007) The plant pathogen *Pantoea ananatis* produces N -acylhomoserine lactone and causes center rot disease of onion by quorum sensing. J Bacteriol 189(22):8333–8338

Moshynets OV, Babenko LM, Rogalsky SP, Iungin OS, Foster J, Kosakivska IV, Potters G, Spiers AJ (2019) Priming winter wheat seeds with the bacterial quorum sensing signal N-hexanoyl-L-homoserine lactone (C6-HSL) shows potential to improve plant growth and seed yield. PloS One 14(2):e0209460

Myszka K, Schmidt MT, Majcher M, Juzwa W, Olkowicz M, Czaczyk K (2016) Inhibition of quorum sensing-related biofilm of *Pseudomonas fluorescens* KM121 by *Thymus vulgare* essential oil and its major bioactive compounds. Int Biodeterior Biodegrad 114:252–259

Nievas F, Vilchez L, Giordano W, Bogino P (2017) Arachis hypogaea L. produces mimic and inhibitory quorum sensing like molecules. Antonie Van Leeuwenhoek 110(7):891–902

Niu C, Afre S, Gilbert ES (2006) Subinhibitory concentrations of cinnamaldehyde interfere with quorum sensing. Lett Appl Microbiol 43(5):489–494

Olivero J, Pajaro N, Stashenko E (2011) Anti-quorum sensing activity of essential oils isolated from different species of the genus *Piper*. Vitae 18(78):77–82

Ortíz-Castro RA, Martínez-Trujillo MI, López-Bucio JO (2008) N-acyl-L-homoserine lactones: a class of bacterial quorum-sensing signals alter post-embryonic root development in *Arabidopsis thaliana*. Plant Cell Environ 31(10):1497–1509

Ortiz-Castro RA, Diaz-Perez C, Martinez-Trujillo MI, del Rio RE, Campos-Garcia J, Lopez-Bucio J (2011) Transkingdom signaling based on bacterial cyclodipeptides with auxin activity in plants. Proc Natl Acad Sci U S A 108:7253–7258

Packiavathy IA, Priya S, Pandian SK, Ravi AV (2014) Inhibition of biofilm development of uropathogens by curcumin–an anti-quorum sensing agent from *Curcuma longa*. Food Chem 148:453–460

Palacios OA, Bashan Y, de Bashan LE (2014) Proven and potential involvement of vitamins in interactions of plants with plant growth-promoting bacteria—an overview. Biol Fertil Soils 50(3):415–432

Pandey PK, Yu J, Omranian N, Alseekh S, Vaid N, Fernie AR, Nikoloski Z, Laitinen RAE (2019) Plasticity in metabolism underpins local responses to nitrogen in populations. Plant Direct 3(11):e00186

Pang Y, Liu X, Ma Y, Chernin L, Berg G, Gao K (2009) Induction of systemic resistance, root colonisation and biocontrol activities of the rhizospheric strain of *Serratia plymuthica* are dependent on N-acyl homoserine lactones. Eur J Plant Pathol 124(2):261–268

Pérez-Montaño F, Jiménez-Guerrero I, Sánchez-Matamoros RC, López-Baena FJ, Ollero FJ, Rodríguez-Carvajal MA, Bellogín RA, Espuny MR (2013) Rice and bean AHL-mimic quorum-sensing signals specifically interfere with the capacity to form biofilms by plant-associated bacteria. Res Microbiol 164(7):749–760

Rajamani S, Bauer WD, Robinson JB, Farrow JM III, Pesci EC, Teplitski M, Gao M, Sayre RT, Phillips DA (2008) The vitamin riboflavin and its derivative lumichrome activate the LasR bacterial quorum-sensing receptor. Mol Plant-Microbe Interact 21(9):1184–1192

Rajesh PS, Rai VR (2014) Biochemical and biophysical research communications molecular identification of aiiA homologous gene from endophytic *Enterobacter* species and in silico analysis of putative tertiary structure of AHL-lactonase. Biochem Biophys Res Commun 443(1):290–295

Remus-Emsermann MNP, Tecon R, Kowalchuk GA, Leveau JHJ (2012) Variation in local carrying capacity and the individual fate of bacterial colonizers in the phyllosphere. ISME J 6(4):756–765

Roje S (2007) Vitamin B biosynthesis in plants. Phytochemistry 68(14):1904–1921

Rosemeyer V, Michiels J, Verreth C, Vanderleyden J (1998) LuxI-and luxR-homologous genes of rhizobium etli CNPAF512 contribute to synthesis of autoinducer molecules and nodulation of Phaseolus vulgaris. J Bacteriol 180(4):815–821

Rudrappa T, Bais HP (2008) Curcumin, a known phenolic from *Curcuma longa*, attenuates the virulence of *Pseudomonas aeruginosa* PAO1 in whole plant and animal pathogenicity models. J Agric Food Chem 56(6):1955–1962

Sanchez-Contreras M, Bauer WD, Gao M, Robinson JB, Downie JA (2007) Quorum-sensing regulation in rhizobia and its role in symbiotic interactions with legumes. Philos Trans R Soc Lond B Biol Sci 362(1483):1149–1163

Schaefer AL, Greenberg EP, Oliver CM, Oda Y, Huang JJ, Bittan-Banin G, Peres CM, Schmidt S, Juhaszova K, Sufrin JR, Harwood CS (2008) A new class of homoserine lactone quorum sensing signals. Nature 454:595–600

Schaefer AL, Oda Y, Coutinho BG, Pelletier DA, Weiburg J, Venturi V, Greenberg EEP, Harwood CS (2016) A LuxR Homolog in a cottonwood tree endophyte that activates gene expression in response to a plant signal or specific peptides. mBio 7(4):1–8

Schenk ST, Schikora A (2015) AHL-priming functions via oxylipin and salicylic acid. Front Plant Sci 5:784

Schenk TS, Stein E, Kogel HK, Schikora A (2012) Arabidopsis growth and defense are modulated by bacterial quorum sensing molecules. Plant Signal Behav 7(2):178–181

Schenk ST, Hernández-Reyes C, Samans B, Stein E, Neumann C, Schikora M, Reichelt M, Mithöfer A, Becker A, Kogel KH, Schikora A (2014) N-acyl-homoserine lactone primes plants for cell wall reinforcement and induces resistance to bacterial pathogens via the salicylic acid/oxylipin pathway. Plant Cell 26:2708–2723

Schikora A, Schenk ST, Stein E, Molitor A, Zuccaro A, Kogel KH (2011) N-acyl-homoserine lactone confers resistance towards biotrophic and hemibiotrophic pathogens via altered activation of AtMPK6. Plant Physiol 157:1407–1418
Schikora A, Schenk ST, Hartmann A (2016) Beneficial effects of bacteria-plant communication based on quorum sensing molecules of the N -acyl homoserine lactone group. Plant Mol Biol 90(6):605–612
Schuhegger R, Ihring A, Gantner S, Bahnweg G, Knappe C, Vogg G, Hutzler P, Schmid M, Van Breusegem F, Eberl LE, Hartmann A (2006) Induction of systemic resistance in tomato by N-acyl-L-homoserine lactone-producing rhizosphere bacteria. Plant Cell Environ 29(5):909–918
Sibanda S, Theron J, Shyntum DY, Moleleki LN, Coutinho TA (2016) Characterization of two LuxI/R homologs in Pantoea ananatis LMG 2665T. Can J Microbiol 62(11):893–903
Sieper T, Forczek S, Matucha M, Krämer P, Hartmann A, Schröder P (2014) N-acyl-homoserine lactone uptake and systemic transport in barley rest upon active parts of the plant. New Phytol 201(2):545–555
Song S, Jia Z, Xu J, Zhang Z, Bian Z (2011) N-butyryl-homoserine lactone, a bacterial quorum-sensing signaling molecule, induces intracellular calcium elevation in Arabidopsis root cells. Biochem Biophys Res Commun 414(2):355–360
Subramoni S, Venturi V (2009) LuxR-family 'solos': bachelor sensors/regulators of signalling molecules. Microbiology 155(5):1377–1385
Teplitski M, Robinson JB, Bauer WD (2000) Plants secrete substances that mimic bacterial N-acyl homoserine lactone signal activities and affect population density-dependent behaviors in associated bacteria. Mol Plant-Microbe Interact 13(6):637–648
Tuteja N (2009) Signaling through G protein coupled receptors. Plant Signal Behav 4(10):942–947
Tuteja N, Sopory S (2008) Plant signaling in stress: G-protein coupled receptors, heterotrimeric G-proteins and signal coupling via phospholipases. Plant Signal Behav 3:79–86
Urano D, Jones AM (2013) "Round up the usual suspects": a comment on nonexistent plant G protein-coupled receptors. Plant Physiol 161(3):1097–1102
Uroz S, Dessaux Y, Oger P (2009) Quorum sensing and quorum quenching: the yin and yang of bacterial communication. Chembiochem 10(2):205–216
Vallad EG, Goodman MR (2004) Systemic acquired resistance and induced systemic resistance in conventional agriculture. Crop Sci 44:1920–1934
van Overbeek LS, van Doorn J, Wichers JH, van Amerongen A, van Roermund HJW, Willemsen PTJ (2014) The arable ecosystem as battleground for emergence of new human pathogens. Front Microbiol 5:1–17
Varma PK, Uppala S, Pavuluri K, Chandra KJ, Chapala MM, Kumar KV (2017) Endophytes: role and functions in crop health. In: Plant-microbe interactions in agro-ecological perspectives. Springer, Singapore, pp 291–310
Veliz-Vallejos DF, van Noorden GE, Yuan M, Mathesius U (2014) A *Sinorhizobium meliloti*-specific N-acyl homoserine lactone quorum-sensing signal increases nodule numbers in *Medicago truncatula* independent of autoregulation. Front Plant Sci 5:551
Venturi V, Keel C (2016) Signaling in the rhizosphere. Trends Plant Sci 21:187–198
Vitreschak AG, Rodionov DA, Mironov AA, Gelfand MS (2002) Regulation of riboflavin biosynthesis and transport genes in bacteria by transcriptional and translational attenuation. Nucleic Acids Res 30(14):3141–3151
von Rad U, Klein I, Dobrev PI, Kottova J, Zazimalova E, Fekete A, Hartmann A, Schmitt-Kopplin P, Durner J (2008) Response of *Arabidopsis thaliana* to N-hexanoyl-DL-homoserine-lactone, a bacterial quorum sensing molecule produced in the rhizosphere. Planta 229(1):73–85
Williams P (2007) Quorum sensing, communication and cross-kingdom signaling in the bacterial world. Microbiology 153(12):3923–3938
Witzany G (2006) Plant communication from biosemiotic perspective: differences in abiotic and biotic signal perception determine the content arrangement of response behavior. Context determines the meaning of meta-, inter-and intraorganismic plant signaling. Plant Signal Behav 1(4):169–178

Yuan ZC, Edlind MP, Liu P, Saenkham P, Banta LM, Wise AA, Ronzone E, Binns AN, Kerr K, Nester EW (2007) The plant signal salicylic acid shuts down expression of the vir regulon and activates quormone-quenching genes in agrobacterium. Proc Natl Acad Sci U S A 104(28):11790–11795

Zarkani AA, Stein E, Röhrich CR, Schikora M, Evguenieva-Hackenberg E, Degenkolb T, Vilcinskas A, Klug G, Kogel KH, Schikora A (2013) Homoserine lactones influence the reaction of plants to rhizobia. Int J Mol Sci 14(8):17122–17146

Zhang J, Kan J, Zhang J, Guo P, Chen X, Fang R, Jia Y (2012) Synergistic activation of the pathogenicity-related proline iminopeptidase gene in *Xanthomonas campestris* pv. Campestris by HrpX and a LuxR homolog. Appl Environ Microbiol 78(19):7069–7074

Zhao Q, Li M, Jia Z, Liu F, Ma H, Huang Y, Song S (2016) AtMYB44 positively regulates the enhanced elongation of primary roots induced by N-3-oxo-hexanoyl-homoserine lactone in *Arabidopsis thaliana*. Mol Plant-Microbe Interact 29(10):774–785

Zheng D, Zhang H, Carle S, Hao G, Holden MR, Burr TJ (2003) A luxR homolog, aviR, in *Agrobacterium vitis* is associated with induction of necrosis on grape and a hypersensitive response on tobacco. Mol Plant-Microbe Interact 16(7):650–658

Zheng H, Zhong Z, Lai X, Chen WX, Li S, Zhu J (2006) A LuxR/LuxI-type quorum-sensing system in a plant bacterium, *Mesorhizobium tianshanense*, controls symbiotic nodulation. J Bacteriol 188(5):1943–1949

Plant Microbiomes: Understanding the Aboveground Benefits

3

Mohini Prabha Singh, Pratiksha Singh, Rajesh Kumar Singh, Manoj Kumar Solanki, and Sumandeep Kaur Bazzer

Abstract

Soil and plant root are known as the microbial reservoir, and these microbes are found broadly in the plant rhizosphere and tissues. Phytobiome generally exists as epiphytic, endophytic, and rhizospheric that undertakes a critical role in plant development. These microbiomes may shape networks, to stabilize the function among different kinds of plant-associated factors to propagate or transmit in a different part of the plant. Microbial networks linked with plant health give crucial beneficial insights to look upon. The present section covers the features of such microbial networks that build the phytobiome. The chapter highlights their ability to better uptake nutrients or plant growth regulators in a stressed environment and further extends an evolution of studies depicting the supporting components that shape the phylogenetic and plant-related networks. The chapter advocates the possibility to understand the techniques by which plants select and connect with their microbiomes and affect plant improvement and well-being, thereby laying the foundation of novel microbiome-driven systems to the advancement of sustainable agriculture. The microbiome is unpredictably

M. P. Singh (✉)
Punjab Agricultural University, Ludhiana Punjab, India

P. Singh · R. K. Singh
Guangxi Key Laboratory of Sugarcane Biotechnology and Genetic Improvement (Guangxi), Ministry of Agriculture, Sugarcane Research Institute, Chinese Academy of Agricultural Sciences, Guangxi Academy of Agricultural Sciences, Nanning China

M. K. Solanki
Department of Food Quality & Safety, Institute for Post-harvest and Food Sciences, The Volcani Center, Agricultural Research Organization, Rishon LeZion Israel

S. K. Bazzer
Department of Crops, Soil, and Environmental Sciences, University of Arkansas, Fayetteville AR, USA

© Springer Nature Singapore Pte Ltd. 2020

M. K. Solanki et al. (eds.), *Phytobiomes: Current Insights and Future Vistas*,
https://doi.org/10.1007/978-981-15-3151-4_3

engaged with plant well-being providing extra qualities to the plant. To understand the guideline of plant characteristic articulation, henceforth plant execution, and how this impacts the biological systemic network, it is required to get well versed with phytobiome and its usefulness. In the present section, the significance of the phytobiome to plant genomics is tended to describe the phytobiome in assembly to the environment of the outline with attention on natural surroundings happening subterranean at the plant-soil between face, where the center is around the job of exudates as currency in this framework.

Keywords
Microbial communities · Rhizosphere · Endophytes · Phyllosphere

3.1 Introduction

Nature allows the coexistence of healthy and asymptomatic plants with diverse microbes such as archaea, bacteria, fungi, and protists where a complex microbial consortium is formed to impact plant growth and productivity (Vorholt 2012; Kumari et al. 2019; Solanki et al. 2019). Phytobiome has either neutral or helpful roles in the plants' fitness (Mendes et al. 2013). The useful impacts on plants include disease suppression (Ritpitakphong et al. 2016; Solanki et al. 2012), plant immunization (Van der Ent et al. 2009), induction of systemic resistance (Zamioudis et al. 2015), increased nutrient acquisition (Van der Heijden et al. 2016), increased tolerance to abiotic stresses (Rolli et al. 2015; Wang et al. 2018), adaptation to environmental variations (Haney et al. 2015), or enhancement of the mycorrhizal colonization (Garbaye 1994). Microorganisms can also target agricultural productivity by providing nutrient availability/acquisition (Kavamura et al. 2013). Lack of precise methodologies has led to limited access to nonculturable microbial groups, and thus, most of the work relies on single microbial groups associated with plants (Andreote et al. 2009). Mycorrhizal association with a plant (Chagnon et al. 2013) and microbial diazotrophs (Raymond et al. 2004) are the few examples that need to be explored in-depth. Nevertheless, an inclusive map of this system laid stress on the interactions happening between diverse groups of microbes, permitting the term "microbiome." Joshua Lederberg coined the term "microbiome" for the first time and described it to be the "ecological community of commensal microorganisms, symbionts or pathogens that occupy a space in our body" (Lederberg and McCray 2001). New terminology for "microbiome" was suggested by Boon et al. (2014) that relates to host-associated genes in a defined surrounding, thereby bypassing the abundance of the microbial community of low significance. Plant-associated microbial groups work in multidimensional ways as host plant delivers unique metabolic adeptness to attract beneficial microbial niches that can have a positive (mutualistic), neutral (communalistic), or deleterious (pathogenic) effect on plant health (Thrall et al. 2007). Microbes are the main component of plant functional traits such as soil formation, organic matter decomposition, nutrient mobilization, and improvement in plant productivity (deBello et al. 2010). Rhizospheric prokaryotes are known as plant helpers due to their beneficial activities such as nitrogen (N_2) fixation (Martinez-Romero 2006), solubilization of insoluble minerals, and stimulation of phytohormones (Hardoim et al. 2008). Genome duplication (polyploidization) is defined as macroevolutionary events of host that can change microbiome structure. The phytobiome exerts influences on plant trait expression through upstream and downstream regulation of nutritional uptake, thus supervising plant's performance. To unlock the subtleties inside the ecosystem, and the regulation of plant trait expression, impacts of the microbiome are needed to be observed. Bernedsen et al. (2012) reported that plant microbiome interface aligned as "microbe-soil-microbe-plant-microbe interface" rather than the "soil-microbe-plant interface." Plant genome is itself a complex system, and microbial interaction is coined as the plant's "second genome" because it extends the plants' genetic compendium extensively.

This chapter also contains a detailed description of beneficial phytobiome interactions. Three microbial groups (bacteria, fungi, and protists) that abundantly originate on plant tissues are deliberated, and diverse mechanisms used to cooperate and compete in planta are defined. Nevertheless, the activity of microbiomes is a new systematic approach that is required to understand the multidimensional actions of microbial communities (Bashiardes et al. 2018). To some degree, microbiome applications would include an emphasis on enlightening basic components that can improve crop production such as management of plant nutrients, soil health, and environmental safety (Syed Ab Rahman et al. 2018).

3.2 Soil Microbiome Characterization

Phytobiomes are discrete that comprise unfavorable pathogens, potential endophytes, and helpful symbionts (Rosenblueth and Martínez-Romero 2006; Wang et al. 2017; Malviya et al. 2019). Be that as it may, traditionally, the microbial assorted variety was assessed by segregating and refined on various supplement media and development conditions. Microbial metabolism fulfills the nutritional and regulatory prerequisites of plants (Lugtenberg and Kamilova 2009). These healthful necessities, for the most part, incorporate nitrogen, phosphorous, and iron. Moreover, these elements also control plant growth by stimulating the production of plant growth regulators. Screening of the most suitable bacteria would require culture-based methods (Taulé et al. 2012). On a routine basis, the procedures normally contain an agar plate assay or a broth medium to multiply the microbes. These assays also help in locating genetic components of microbes. However, these protocols failed to explore the microbial diversity of nonculturable microbiota.

For investigating the entire microbiome, the very first effort is initiated with sequencing of a conserved gene region such as the 16S rRNA gene that is widely applied for microbial identification (Mullis et al. 1987). To contemplate and comprehend the microbiome in a brief span, thorough upgrades have been accomplished by this technique, thereby yielding metagenomics. These techniques incorporate beginning with the entire metagenome examining, trailed by refinement, partition, and sequencing and lastly information investigation and elucidation. Particularly, the sequencing innovation is experiencing fast improvement, as it gives wide and top to bottom perspectives on metagenomics, and now it is extensively named as high-throughput sequencing (HTS) or cutting-edge sequencing technology. HTS methods incorporate the utilization of the AB SOLiD System (Life Technologies), the HiSeq 2000 (Illumina), and the 454 Genome Sequencer (Roche Diagnostics) (Yergeau et al. 2014). Besides, other propelled methods, for example, DNA/RNA-SIP and DNA arrays (PhyloChip and practical quality exhibits), likewise have prospective highlights in the examination of microbiomes, mostly their useful parts (Uhlik et al. 2013). At present, there is a change from metagenomics to metatranscriptomics, as the latter helps in understanding the numerous microbial functions and structure (Turner et al. 2013).

In the metatranscriptomics approach, complementary DNA analysis aligned with quantitative reverse transcription-PCR and RNA-SIP explored the microbial functionality associated with the soil and rhizosphere (Uhlik et al. 2013). RNA-SIP significantly is used to crack the complexity in interactions especially between root-derived carbon and microbiome so as to provide sequence as first and second utilizers of carbon within the microbiome. This method is dissimilar to DNA-SIP because it provides higher amounts of labeling and does not rely on cell multiplication. Challenges coming with these cutting-edge innovations include choosing either mRNA or rRNA alone and accomplishing more extensive inclusion of environmental RNA pool that gives naturally vital information through the sequencing. Peiffer et al. (2013) demonstrated noteworthy community contrasts among 27 maize innate lines (a genetic variant of a single species) with a normal enhanced population in the maize rhizosphere. Metaproteomics, on the other hand, has a different approach as it focuses on the dynamic function of the phytobiome and extracts samples of metaproteome and performs peptide fingerprinting by mass spectrometry (Kolmeder and de Vos 2014; Lakshmanan et al. 2014). Using metagenomic and metaproteomic (existing and future) information is an essential process-driven methodology and should be supplemented by different strategies to decide the diversity and functional relatedness of the rhizospheric microbiome (Keiblinger et al. 2012).

Molecular methods (molecular fingerprinting) and plate count anomaly (culture-dependent methods) demonstrate the entire bacteria community structure (Amann et al. 1995). Therefore, both approaches are utilized, for thoughtful knowledge of separate classification and communication with host plants. Dini-Andreote and van Elsas (2013) have, however, stressed the present need for a change in outlook from HTS (or comprehensive endeavors) to investigations of basic studies.

3.3 Structural and Compositional Factors in Plant-Associated Microbial Network

3.3.1 Plant-Associated Bacterial and Archaeal Microbiomes

Plant-associated bacterial population detected on plants does not look arbitrary; relatively numerous components participate in controlling the structure of microbiomes such as soil type (Lundberg et al. 2012), plant compartment (Leff et al. 2015), host genotype/species (Tkacz et al. 2015), plant invulnerable framework (Horton et al. 2014), plant attribute variety/developmental stage (Donn et al. 2015), and residence time/season (Shi et al. 2015). Hacquard et al. (2015) described that *Proteobacteria*, *Actinobacteria*, and *Bacteroidetes* are the major bacterial phyla that exist in both substrata such as above- and belowground plant tissues, and they influence the plant metabolism. Broad cover among root- and leaf-related network individuals has been portrayed at OTU (operational taxonomic unit) level determination in different plants such as *Arabidopsis thaliana*, wild mustard, grapevine, and agave (Wagner et al. 2016), and microbiota reconstitution experiments with germ-free *A. thaliana* approved that root- and leaf-related bacterial networks have reciprocal relocation. Regardless of the conspicuous elementary uniformities saw between *A. thaliana* leaf- and root-related bacterial networks, it is observed that related microbiota individuals are particular and adjusted to their separate related plant organs (Bai et al. 2015). Among all phytobiomes, the nonpathogenic segregation of archaea has been depicted. The plant endophytic archaeal taxa of the phyla *Thaumarchaeota*, *Crenarchaeota*, and *Euryarchaeota* have plant-associated functional significance (Müller et al. 2015).

3.3.2 The Fungal Microbiota of Plants

Ascomycota and *Basidiomycota* are two noteworthy phyla that colonize both above- and belowground plant tissues (Hardoim et al. 2015). In roots, even though arbuscular (*Glomeromycota* phylum) and ectomycorrhizal growths have been for the most part contemplated, ongoing network profiling information demonstrates that other endophytic organisms too make up for root microbiota (Toju et al. 2013). The structure of fungal communities on plants relies upon different kinds of soil, plant parts, plant genotypes, or seasons (Coince et al. 2014) and is subjected to stochastic variations (Wang et al. 2013) and reacts distinctively to ecological elements (Thomson et al. 2015). Thus, mostly dispersal restriction and atmosphere clarify the worldwide biogeographic conveyance of growths and have been recommended to compel contagious dispersal, supporting high endemism in parasitic populaces (Talbot et al. 2014). Steady with that, the synchronous examination of both contagious and bacterial networks related to plants recommended a more prominent significance of biogeography for organizing parasitic networks contrasted with bacterial networks (Hacquard. 2016). Regardless of using molecular markers such as 16S rRNA and ITS, their loci need to be elucidated (Peay et al. 2016).

Recently, Yunshi et al. (2018) quantified the prokaryotic and fungal groups within the phyllosphere and rhizosphere of six spruce (*Picea* spp.) tree species through illumine amplicon sequencing. In brief, this microbial quantification experiment is performed in a common garden, and linkages among phenotypic characters of their plant hosts and bacterial/archaeal and fungal community are analyzed. Correlation results among plant microbiome and different phenotypic characters of host plants (such as leaf morphology, water content, water storage ability, dry biomass, nitrogen, etc.) which suggests that plant genotype played a significant role to shape its microbiota by improving plant phenotypes.

3.3.3 Plant-Associated Protists: The Outcasted Fraction of the Plant Microbiota

Protists are a vital constituent of the soil microbiome, and method progresses now extended to our thoughts of the real taxonomic and efficient diversity of soil protists. The Stramenopiles-Alveolata-Rhizaria (SAR) group is known as a large group of plant-associated protists (Ruggiero et al. 2015) and especially those having a place with the *Oomycota* (Stramenopiles) and *Cercozoa* (Rhizaria) lineage. Inside *Oomycota*, a couple of individuals having a place with the genera *Peronospora*, *Phytophthora*, *Pythium* (and other wool buildup genera), or *Albugo* frequently exist in the plant roots or leaves (Agler et al. 2016). Root colonization by oomycetes (i.e., *Pythium oligandrum*) provides positive benefits to the host (Van Buyten and Hofte 2013). Even though plant tissue-associated oomycete network profiling stays scanty, an exceptionally low decent variety is demonstrated with individuals from the *Pythiaceae* family being the most spoken about to be present on plant tissues (Sapp et al. 2018). Inside *Cercozoa*, one of the prevailing protistan bunches in biological systems, network profiling information uncovered a surprisingly high diversity in plant roots and leaves (Ploch et al. 2016), also giving a piece of strong evidence that indicates the plant stress tolerance and metabolic behavioral changes governed by special community structure. Thus, *Oomycota* and *Cercozoa* individuals are significantly important for holobiont wellness. Recent reports concluded that plant microbe linkages are outlined well under the evolutionary measure and it helps to unlock the complex interactions of plant and microbes in more depth, and plant-microbe or plant bacterial interaction is new as compared to the bacterial and other kingdom interactions (Lücking et al. 2009; Hassani et al. 2018).

3.4 Microbial Currency: Exudates

Plants and microbes release certain chemicals called exudates, which help them to communicate with each other and to accelerate the disease tolerance against biotic and abiotic factors, stabilize the plant and microbial growth during nutrient scarcity, and remediate the toxic elements. Microbes utilized exudates as a food source, particularly carbon and other acids. This section discussed the two-way interaction of plant and microbial exudate that is influenced by plant and microbial metabolism. Huang et al. (2014) reported the significance of plant root exudates to regulate the microbial structure in the plant rhizosphere that is influenced by plant variety, growth stages, disease-suppressive soils, root exudate composition, and plant hormone signaling. Plant-microbe interaction is a complex system that is mediated by numerous compounds, and these compounds are released under specific conditions. These compounds play an indispensable role to shape the microbial community and unified the microbes and their functions up to species level. For example, legumes and rhizobia symbiosis is signaled by flavonoids, plant mycorrhizal association is stimulated by strigolactones, malic acid regulates the quorum sensing (QS) of plant microbial helpers and major chemoattractants of microbes such as sugars and amino acids attract the beneficial microbial niches toward the plant roots to protect the plant against the multiple stresses. However, various protein molecules are released from the root in the rhizosphere that are less explored to understand their mechanism in plant fitness. Besides, root exudates played intermediate role in several other interactions such as plant attract the nematodes, and these nematodes are the vectors of rhizobia that enhanced the nodulation of root to fix the nitrogen, plant nodulation efficiency enhanced by the interaction of rhizobia with PGPR and arbuscular mycorrhiza (Huang et al. 2014). A multitude of rhizospheric interactions is mediated by root exudates, which are depicted in Fig. 3.1.

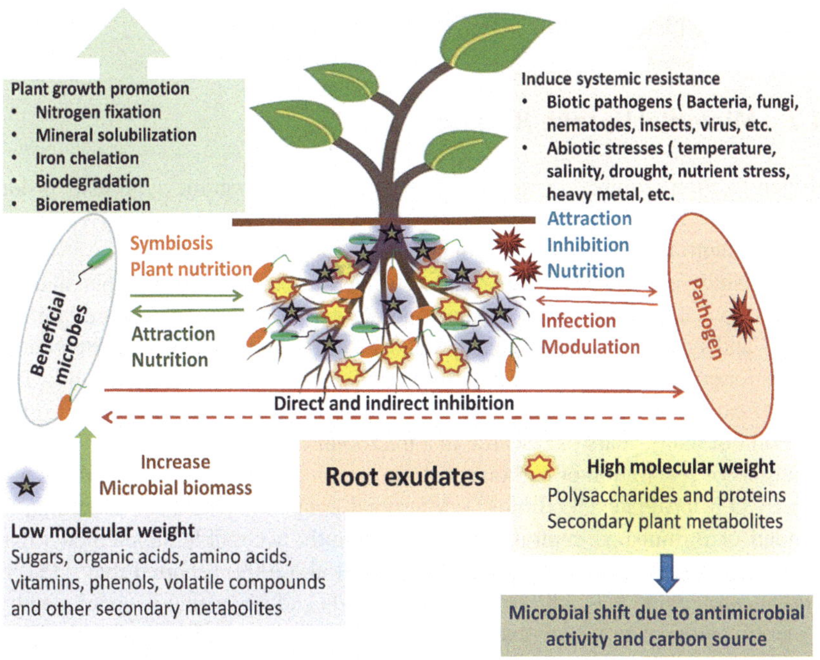

Fig. 3.1 The complex plant microbe system of rhizosphere that is mediated by plant root exudates. Root exudates improve plant health status directly and indirectly as well

3.4.1 Plant Uptake and Release

Plant exudate components and assembly are specific to plant species and incorporate high-molecular-weight particles (e.g., sugar molecules, proteins, and fatty acid) and low-molecular-weight signaling molecules (e.g., natural compounds, metabolites, and amino acids) (Badri and Vivanco 2009). Jaeger et al. (1999) reported that plant root exudate contains sugar molecules and amino acids that help bacteria and other microbes to attract toward the plant root. Exudates assist numerous jobs such as stimulate the antagonism, allelopathic particles, and pathogen/herbivore safeguards. A large number of these exudates likewise fill in as a vitality hotspot for the microbiome; prokaryotes can use plant exudates as nutrient sources. For instance, grass *Sorghum halepense* excretes the exudate sorgoleone from root hairs having allelopathic properties (Kagan et al. 2003) which can be used as microbial nutrients (Gimsing et al. 2009). The different elements of plant exudate repeat the significance it fills in as numerous monetary forms of the phytobiome.

Roots participate in taking up the nutrients and signaling molecules from the rhizosphere while at the same time saving these supplements and concoction signaling molecules into this equivalent space required for evoking defense reactions. Terpenoids, flavonoids, and isoflavonoids contain a large number of the plant's antimicrobial barriers. Isoprenoids being the most diverse primary metabolite is required to control cellular processes such as photosynthesis (as phytopigments) and seed growth stimulation (as gibberellic and abscisic acids), and allelopathic molecules also protect the plants from the pathogens (Hardoim et al. 2008).

3.4.2 Microbial Uptake/Release

Nitrogen fixation requires a constant need for rhizobium-legume symbiosis inside the biosphere. The ability of this methodology to enhance agricultural yield has produced attention in knowledge to manipulate this process for better use. A number of the study approaches were procured inside the examiner of rhizobia, and the precise knowledge collected from these numerous tools is used to focus on the genome- and systems-level procedures (diCenzo et al. 2019). Exudates are essential methods for correspondence within the environment for the microbial network. The uptake of exudates such as sugars, organic, and amino acids has been a noteworthy focal point of many years of research in a microbial environment utilizing different estimations of respiration or carbon substrate use measures, for example, those utilizing ECO MicroPlates™ (Biolog®). Microbial community behaves according to expanded or diminished centralizations of promptly accessible supplements that require insignificant vitality to absorb. Moreno et al. (2009) reported that a PGPB *Pseudomonas putida* KT2440 utilized amino acids and sugar that is concluded by the identification of proteins that regulate the amino acid and sugar uptake. The microbiome of rhizosphere has a perplexing task in nutrient cycling and includes a horde of nutrient transformations in soils. Microorganisms act as a catalyst for chemical changes in the soil during biogeochemical cycling. These changes plant supplement (N and P) uptake, soluble metal (Ca^{2+}, Mg^{2+}, K^+, and Na^+) uptake, and micronutrient uptake (Zn^{2+}, $Fe^{2/3+}$, Cu^+, and Mn^{2+}) (Stevenson and Cole. 1999).

Nitrogen-fixing prokaryotes played a vital role to fix nitrogen gas (N_2) into ammonia (NH3), and this method helps the plants. Howard and Rees (1996) reported that physiological and genetic drivers N_2-fixing prokaryotes have a highly conserved protein complex that is nitrogenase, and it is used to assess the abundance of N_2-fixers in diverse ecological zones (Zehr et al. 2003). Specific bacterial metabolites worked as important plant hormones, for example, indole-3-acidic corrosive (IAA) engaged with managing plant hormone flagging (e.g., 1-aminocyclopropane-1-carboxylate (ACC) deaminase). These hormone flagging particles can advance plant development.

When wheat is inoculated with rhizosphere bacteria expressing ACC deaminase activity, expanded root improvement, and consequently expanded nutrient uptake, has been recorded (Shaharoona et al. 2008; Honma and Shimomura. 1978). ACC deaminase-producing bacteria manage the ethylene production in the plant, in this manner limiting effects of different ecological anxieties, which typically trigger expanded ethylene generation (Hardoim et al. 2008). Microbial exudates involve a significant part of antimicrobial and antifungal compounds. Several prokaryotic and eukaryotic microorganisms are able to secrete or discharge antimicrobial substances, but among all very few are cultivable (Piel 2011); due to this to understand the complex functions of these substances, metatranscriptomics and metabolomics tools need to be applied. It is found that 35% of *Escherichia coli* strains produce the antimicrobial compound known as calcium. The discoveries bolster the theory that antimicrobial cooperations inside microbial networks serve to look after diversity; this thought was created utilizing recreation models (Czaran et al. 2002).

Notwithstanding oozing antimicrobials that encourage plant resistance, microorganisms additionally release low-molecular-weight compounds to the plants, and plant sensors identify the microbes as a pathogen or beneficial and then trigger the reaction (Boller and He 2009). In this way, microbial metabolites can act straightforwardly on different microorganisms inside the microbiome in suppressive components or can act specifically on the plant to revitalize secure reactions, regularly activating plant exudation. As modern biotechnological tools improve our knowledge to measure these microbial metabolites in the soil matrix, multiple functions of the same microbiological substances are discovered.

3.5 Ecological Considerations for Utilizing Plant's Benefits in the Farmer's Field

An effective microbial inoculant needs to attack the pathogens and survive in varying abiotic conditions, also to set up good cooperation with the host that incorporates molecular passivity with the plant resistant framework. All through the developing season, microbial network experiences progression in both over the ground and subterranean (Edwards et al. 2015; (Copeland et al. 2015) portions of the plant. Along these lines, regardless of whether PGP inoculants colonize the plant at first, their constancy after some time isn't ensured. Estimating the determination of bacterial inoculants in the soil presents technical difficulties, as the inoculant should be distinguished from a complex network. Strategies such as culture-based count utilizing re-separation of antibiotic-resistant inoculants or culture-autonomous estimation of relative bounty of the inoculant's 16S rRNA quality in the soil, by means of DGGE (Schreiter et al. 2014), amplicon sequencing (Haney et al. 2015), or metagenomic sequencing (Krober et al. 2014), are used to determine the persistence of microbes.

Ecological components impacting root exudate organization and amount include raised dimensions of CO_2, dry season, and nutrient deprivation (especially nitrogen and phosphorus). Increased carbon allocated in roots is observed in CO_2-fertilization experiments, resulting in shifts in exudate composition and concentration that differ with plant species (Cheng and Gershenson 2007). These species-explicit effects can result in increased yield, in no net profitability increment, or can be unfavorable to plant development and generation. For example, positive biomass reactions in rye and clover to CO_2 preparation were observed, while maize demonstrated no net biomass advantage (Phillips et al. 2006). Be that as it may, maize showed expanded exudation of a few amino acids under CO_2 treatment. These discoveries are not amazing thinking about that the C_4 photosynthetic pathway encourages development under elevated amounts of CO_2; in any case, the effects of expanded arrival of amino acids into the rhizosphere by the C_4 grass (maize) may assume a job in a large number of criticisms between different plants and organisms (Klironomos 2002).

3.5.1 Impact on Plant Functions

Plants participate in nutrient exchange and exudate correspondence depending on the molecule and energy required for the plant (alone or through help from the microbiome) to acquire or release exudate currency. An active transport system using ATP-restricting tape transporter participates in root exudation creation and fixation (Badri et al. 2009). Low-molecular-weight particles such as amino acids can be discharged through membrane diffusion or through protein channels (Badri and Vivanco 2009).

Plants utilize those microbes which can communicate with increased levels of N-acyl-L-homoserine lactones. AHL-degrading enzymes in the presence of a pathogen subsequently suppress gene expression of pathogens (Reading and Sperandio 2006). Plants in the same manner also help in AHL degradation inside the microbiome (Teplitski et al. 2000). Fluorescent pseudomonads which are fundamental to the rhizosphere of the different clusters of the plant are used to deliver the antimicrobials 2,4-diacetyl phloroglucinol (2,4-DAPG) and phenazine (Phz) derivatives (Mavrodi et al. 2011). These antimicrobials are of wide range and act against a number of plant pathogens that are contagious leading to their suppression (Raaijmakers et al. 2009). 2,4-DAPG and Phz derivatives are evident in the rhizosphere and are associated with the suppression of disease in wheat called as take-all in wheat. These plant hormones administrate the plant performance and its marking abilities to adjust to exudate profiles and enhance the plant's immunity (Doornbos et al. 2012).

Most of the microorganisms are capable of producing and controlling the majority of plant hormones (Friesen et al. 2011), thereby modifying plant physiological pathways. In plant rhizospheres, 80% of bacterial taxa accounts for IAA production (Loper and Schroth 1986). In plants, root development is accelerated with a low concentration of IAA (Glick 1999), while its high concentration represses plant's development, hence making plants prone to pathogen's attack (Sarwar and Kremer 1995). An example of this is seen in *Sorghum halepense*; microbes of the invasive grass secrete high concentrations of IAA in contrast to other prokaryotes that secrete lower levels of IAA (Rout et al. 2013). Environmental stress and plant phenology drive the changes in plants that need to increase or decrease the hormones. PGPB are competent cells having multiple genes required for plant-microbial association (Hardoim et al. 2008). Yield expansion, organic methodologies, intercropping, and other cultural practices are utilized for possible farming production. New strategies are formulated to modulate the plant microbiome in an ideal course (Fig. 3.1). Distinctive microbiota is induced by diverse agro-management in viticulture (natural, biodynamic, or biodynamic with green compost) (Longa et al. 2017). It is observed that in an integrated management system, soil has diminished bacterial species richness as compared to organic management, even though microbial composition was similar to organically and biodynamically managed soils (Hendgen et al. 2018).

3.5.2 Impact on Bacterial Functions

Microbiome present in the rhizosphere possesses varied phenotypic expressions due to root exudates. Inhabitant microflora of plants perform different functions such as chemotaxis, stress tolerance (Amador et al. 2010), polychlorinated biphenyl degradation (Toussaint et al. 2012), modulation of genes involved in competence and sporulation (Mader et al. 2002), and biofilm formation on plant roots (Rudrappa et al. 2008). Plant exudates such as terpenoids, flavonoids, and isoflavonoids protect and control the internal structure of plants and outward surface of roots from microbial inhabitant (Hardoim et al. 2008). Microbiome symbionts can be epiphytes or endophytes, which survive for a short growth period and may encompass not only the pathogens but plant growth-promoting microbes using a mechanism of hormone signaling. Organic farming impacts the community composition on soil and roots of winter wheat (Hartman et al. 2018). The structure of bacterial communities is taken care of by tillage. Root bacteria respond to management types, whereas fungal communities respond to both. Different agricultural practices are parameters in affecting the microbial structure with differences in soil, roots, bacteria, and fungi and hence bringing around 10% of the variation in microbial communities.

3.6 Formation of Biofilm

Microbial communities act as a unit and secrete polymeric substances to produce a network known as biofilms (Stoodley et al. 2002). Microbes, when present in a biofilm as consortia, are highly protected from their competitor, antimicrobial agents, enzyme degradation, and acquisition of new genes through horizontal gene transfer (Van Acker et al. 2014; Nadell et al. 2009; Zhang et al. 2014). *Enterobacter* spp., a root-occupying bacterial endophyte, when forming a biofilm, inhibits the entry of root-colonizing pathogen *Fusarium graminearum* (Mousa et al. 2016). Bacteria commonly produce biofilms on fungi, but it is rarely seen on the hyphae of ascomycete fungi. An example of this is seen in *Pseudomonas fluorescens* BBc6 which formulates biofilm on the hyphal region of the ectomycorrhiza *Laccaría bicolor* specifically at its root tip, thereby establishing ectomycorrhizal beneficial symbiosis and promoting bacterial biofilm on fungal host surfaces (Guennoc et al. 2017).

3.7 Molecular Communications

The mechanism of quorum sensing is used by microbes to sense their counterparts. Gram-negative microorganism secretes signaling molecule N-acyl-l-homoserine lactone (AHL) to screen out their populace densities (Eberl 1999). Regulation and secretion of signaling molecules are evident in *Saccharomyces cerevisiae* and *Candida albicans* (human fungal pathogens) that secrete farnesol to control filamentation (Oh et al. 2001), constrain biofilm formation, and activate oxidative stress responses or drug efflux (Sharma and Prasad 2011). Quorum sensing mechanisms are not defined thoroughly for plant-associated fungi. Signaling compounds such as volatile organic compounds (VOCs), oxalic acid, trehalose, glucose, or thiamine accelerate fungal bacterial associations (Schmidt et al. 2016).

3.8 Ecology of the Microbiome

The plant microbiome is localized in three different regions, namely, rhizosphere, endosphere, and phyllosphere (Hirsch and Mauchline 2012). Rhizosphere presents a microbial community in requirement with plant metabolism; endosphere presents those microorganisms which interact with host closely and inhabit inner part of plant tissues asymptomatically (Hardoim et al. 2008); phyllosphere, on the contrary, is composed of those microbes which inhabit plant surfaces (Lambais et al. 2006). Irrespective of different plant habitats, specific microbes are present in all, also known as "keystone" species which interact with other microbes within the networks and affect the microbial structure (Bakker et al. (2014).

Rhizosphere and endosphere account for root microbiome that was key for the advancement of land plants and underlay crucial ecosystem processes. It is reported that nearly 30 angiosperm species affect root bacterial diversity and composition (Fitzpatrick et al. 2018). A competitive interaction gets affected when there is a similarity in root microbiomes between hosts among plant species. Climatic parameters such as drought affect the root microbiome composition, by elevating the *Actinobacteria* population. In the endosphere, *Streptomyces* are associated with host drought tolerance influencing drought response crosswise over host plant species bringing host-specific changes.

3.8.1 Rhizosphere and Rhizoplane

Plant health is influenced by the rhizosphere using next-generation and third-generation technologies (Hiltner 1904). Rhizospheric soil shows a significant difference in contrast to bulk soil due to abiotic and biotic stresses impacted by the atmosphere. Properties such as higher water holding capacity, expanded nutrient availability, and diverse microbial biomass mark its importance than bulk soil (Schade and Hobbie 2005). Spatiotemporal movements are observed in the rhizosphere microbiome (Kaplan et al. 2013); however, it is still to affirm how much abiotic stresses impact the microbiomes. Protection from a wide range of pathogens both aboveground and belowground is provided by microbiomes. For example, induction of systemic resistance (ISR) is initiated where jasmonic acid-inducible genes are secreted in leaves (Pineda et al. 2010).

3.8.2 Epiphytes and Endophytes

The epiphyte and endophyte microbial communities in root involve the acknowledgment and selection of those microbiomes that establish a homeostatic association with the plant. Technologies such as metabolomics and metatranscriptomics are used to observe microbial members that colonize in adherent (epiphytic) or internal (endophytic) parts of plants. Microbes colonize the outside root surfaces. For instance, secondary metabolite root exudates were released due to an ISR response in maize that appoints PGPB *P. putida*, based on chemotaxis inclinations (Neal et al. 2012). Plant chemical exudate is secreted, and valuable PGPB is selected in a plant-mediated reaction as observed in tomato, where natural acids are the major chemotactic operator (De Weert et al. 2002), while in rice, amino acids serve the purpose (Bacilio-Jimenez et al. 2003). Root microbes that laid distinctive qualities such as that code for the sort IV pilus and twitching motility (Bohm et al. 2007), isoflavonoid efflux siphon (Palumbo et al. 1998), and DNA improvements influence colony aggregation (Dekkers et al. 1998).

Endophytes are inhabitant to both wild and domesticated crops including intrusive species (Compant et al. 2008; Rout and Chrzanowski 2009). Biotechnology and agriculture ensure to utilize plant developing qualities of microbiomes as seen in phosphate-solubilizing *Bacillus* strains, where apart from secreting protein ACC deaminase, these also show plant development advancements (Baig et al. 2012). Thus, a microbiome serves a double attribute working together with mycorrhizal parasites to improve plant development advancement (Zaidi and Khan 2005).

Plant phenology correlates with endophyte microbiome composition shifts (van Overbeek and van Elsas 2008) which further depends upon colonization and similarity (Hardoim et al. 2008). This collaboration inclines more toward mutualism than parasitism. Most of known plant endophytes and epiphytes are horizontally transmitted (Friesen et al. 2011). This empowers host-to-host exchange of endosymbionts without the association of plant sexual reproduction. Endophytes also show vertical transmission depending upon host wellness and present more host benefits than horizontal transmission (Clay and Schardl 2002; Sachs et al. 2004). But the environmental hypothesis proposes that the presence of an accessible host allows for horizontally transmitted life forms. It is observed that horizontally transmitted endophytes were positively related to plant thickness reliance, while vertically transmitted endophytes did not demonstrate this pattern (Rudgers et al. 2009).

3.8.3 Phyllosphere Region

Phyllosphere is regarded as a third segment of the plant microbiome that colonizes the outside region of the external area of plant tissues specifically when describing the leaf surface (Vorholt 2012). The microbiomes in the phyllosphere perform nitrogen fixation, securing plants against attacking pathogens and biosynthesizing phytohormones (Kishore et al. 2005). These can be beneficial in carbon sequestration (Bulgarelli et al. 2013), and they can also participate in sustainable agricultural practices. Fungi (filamentous and yeasts), bacteria, and algae make phyllosphere network, and at lower frequencies, protozoa and nematodes are seen (Lindow and Brandl 2003). The bacterial population is the most abundant group of microorganisms present in the phyllosphere at numbers ranging from 10^5 to 10^7 cells for each cm^2 (Andrews and Harris 2000). These microbes can thrive in harsh environmental conditions such as limited availability of nutrients and variable conditions of humidity, UV radiation, pH, and temperature (Andrews and Harris 2000). The phyllosphere community is created with the help of various hotspots as air, soil, and water (Bulgarelli et al. 2013). Agricultural plants also show specificity to phyllosphere microbiomes as seen in beans, cucumber, grasses, lettuce, and maize (Rastogi et al. 2012). Plant genotype plays a significant effect on the composition of phyllosphere microbiomes (Bokulich et al. 2014). The microbial population shows intraspecific variations in its composition which are due to nutritional heterogeneity observed in regions on the leaf surface where heterogeneous carbon sources such as glucose, fructose, and sucrose are utilized near the stomata and surface appendages (Vorholt 2012). At times, this heterogeneity is observed when microbial cells aggregate to form a biofilm and hence defending themselves from unfavorable conditions (Lindow and Brandl 2003).

Proteobacteria, *Actinobacteria*, *Bacteroidetes*, and *Firmicutes* are major phyla that account for the microbial community in the phyllosphere region (Vorholt 2012). Hence, this core is accepted to be made out of individuals exhibiting a co-developmental history with plant species, with the host physiology being complementary to the features found inside the microbial cells. Microbes such as protists largely act as predators on the bacterial community (Flues et al. 2017). Environmental conditions such as low nutrients, high UV, changing temperature, and humidity help in selecting consistent biological traits and low functional diversity at the community level for phyllosphere microbiomes as observed in next-generation sequencing (Lambais et al. 2017). The phyllosphere is dominated by oxygen-consuming organoheterotrophs, and metabolic diversity exists with regard to utilizable carbon compounds.

3.9 Competitive Interactions Among Plant Microbiota Members

Plant microbiomes show competitive behavior with closely or distantly related microbiota and affect microbial structure, its stability, and homeostasis.

3.9.1 Resource Competition

Microorganisms utilize limited resources and therefore compete indirectly with other microbes. For example, using advanced techniques, microorganisms sequestrate iron using the emission of siderophores, thereby affecting the growth of the opponent microbes present in their niche (Little et al. 2008). When advantageous *Pseudomonas* spp. secrete iron-chelating molecules, it suppresses the disease caused by fungal pathogens indicating that nutrient sequestration is a trait of biocontrol agents to outpower pathogens (Mercado-Blanco and Bakker 2007). In tomato plants, resource competition is said to be an essential factor connecting the bacterial network and pathogen attack on plants (Wei et al. 2015). In resource competition, individual microbes in a group share resources, but the ones who use the resources in an uncooperative can evade paying the price of cooperation while reaping the benefits of utilizing the resource, thereby increasing their fitness (Riehl and Frederickson 2016) and bringing the situation of distress to the commons. Nitrogen-deficient soil could harbor plants with rhizosphere having microbes that can capture nutrients for their usage. For instance, actively growing roots could signal for microorganisms that are capable of producing extracellular enzymes releasing nitrogen bound in soil organic matter (Lemanceau et al. 2017). In prokaryotes, mineralization processes are density-dependent and need a quorum of producers to sufficiently enter key nutrients in the soil. Such producers are taxonomically diverse microbiota that can biosynthetically produce the specific enzymes which are secreted into the soil. In this scenario, selection in the rhizosphere could favor microhabitats to promote coordinated group behaviors that enhance plant access to nitrogen or phosphorus upon cell turnover, while the microorganisms benefit from having an abundant supply of carbon and other nutrients from plant roots (Fig. 3.1). Phosphorous is also made available to plants using microbial taxa which could mobilize phosphorus in the soil via the production of extracellular compounds (Alori et al. 2017). Iron is an important plant nutrient which can be obtained through the production of siderophores (Radzki et al. 2013).

3.9.2 Contact-Dependent Competition

Plant microbiome participates in direct antagonistic cooperations interceded by the bacterial type VI secretion framework, a molecular weapon used by certain microscopic organisms (generally *Proteobacteria*) to convey effectors/toxins into both eukaryotic and prokaryotic cells (Records 2011). A few examples of contact-dependent competition are discussed as in the case of the plant pathogen *Agrobacterium tumefaciens* that utilizes a puncturing type VI secretion system to convey DNase effectors upon contact with a bacterial competitor in vitro and on the leaves of *Nicotiana benthamiana*. Moreover, the bacterial kind III secretion system can also be used in *Burkholderia rhizoxinica*, which uses this mechanism to control the productivity of its beneficial interaction with the contagious host, *Rhizopus* microspores. Physical parameters bring a change in plant-associated microbes. Soil condition (organic matter, nitrogen, and moisture content) identification helps in changing the macrophage activation potential of *Echinacea purpurea* and determining these changes in activity that relates to the shifts (Haron et al. 2019). Increasing soil organic matter in the root extracts of *E. purpurea* may increase the macrophage activation. Bacterial communities also differed significantly between root materials having varying levels of organic matter. The activity of *E. purpurea* roots is changed by the soil's organic matter level. Use of bacterial preparation (e.g., probiotics) is reported to impact human health; similarly, *Echinacea* too shows therapeutic effects and is impacted by development conditions that change its related bacterial community (Haron et al. 2019).

3.9.3 Antimicrobial Compound Secretion

Various plant-related microorganisms appeared to emit chemical compounds that stifle the development of microbial rivals (Raaijmakers and Mazzola 2012). Filamentous eukaryotes are outstanding in delivering a large number of antifungal activity of secondary metabolites that have a low molecular weight spotted against phylogenetically distinct organisms (e.g., acetylgliotoxin and hyalodendrin) (Coleman et al. 2011). The secondary metabolites so obtained become activated in co-culture and remained inactive in pure culture. Netzker et al. (2015) indicate their specific role in competitive interactions. Antagonistic collaborations among microscopic organisms have been reported to be imperative in the organizing of soil-, coral-, or plant-related bacterial networks (Maida et al. 2016).

Strikingly, the investigation of adversarial collaborations among bacterial segregates from the rhizosphere, the roots, and the phyllosphere of the healing plant *Echinacea purpurea* proposes that plant-related microorganisms compete against one another through the discharge of antimicrobials (Maida et al. 2016). The microbiome related to plants has a robust influence on their strength and yield. The bacterial pathogen, *Candidatus Liberibacter asiaticus* (Las), causes Huanglongbing (HLB) disease and lives inside the phloem of citrus plants, including the root system. It has been proposed that Las negatively affects citrus microbiome. At the same time, the natural microbial flora of citrus also impacts the association between Las and citrus (Riera et al. 2017), i.e., two bacteria closely related to Las *Agrobacterium tumefaciens* and *Sinorhizobium meliloti* were found. Among them, *Burkholderia metallica* strain A53 and *Burkholderia territorii* strain A63 are within the β-proteobacteria class, whereas *Pseudomonas granadensis* strain 100 and *Pseudomonas geniculata* strain 95 are within the γ-proteobacteria class. It was observed that four bacterial strains *Burkholderia territorii* A63, *Burkholderia metallica* A53, *Pseudomonas geniculate* 95, and *Bacillus pumilus* 104 showed antagonistic action against the pathogen *Phytophthora nicotianae* (citrus root) on the basis of dual culture assays. Some of the antimicrobial-producing strains, *Burkholderia metallica* A53 and *Burkholderia territorii* stress A63, from a mandarin rhizosphere, belong to the *Burkholderia cepacia* complex (BCC) and its agricultural applications are restricted because of its high risk to human health (Depoorter et al. 2016). It remains to be determined whether *Burkholderia metallica* strain A53 and *Burkholderia territorii* strain A63 can cause human diseases. Both *Burkholderia metallica* A53 and *Burkholderia territorii* strain A63 can modulate citrus immune system beneath greenhouse situations while applied as a soil drench. Additionally, the *Burkholderiaceae* family changed into determined to be key taxa within the citrus microbiome of healthy trees in comparison to that of HLB-symptomatic trees in the discipline (Zhang et al. 2017).

3.9.4 Predation

Bacterial mycophagy comprises of microscopic organisms' capacity to effectively develop at the expense of living contagious hyphae (De Boer et al. 2004). Mycophagous microbes colonize saprotrophic rhizosphere parasites and feed as auxiliary consumers on root-determined carbon (Rudnick et al. 2015). Some oomycetal species of family *Trichoderma* or *Pythium* can parasite or irritate other growths or oomycetes and can be utilized as biocontrol operators (Benitez et al. 2004). Root-related bacteria can prey on other microscopic organisms as described for *Bdellovibrio* spp. Protist predation on microscopic organisms is also well studied, and recent microbiota reconstitution tests in microcosm demonstrate a reasonable impact of *Cercomonads* (*Rhizaria*: *Cercozoa*) on the structure and the capacity of the leaf microbiota (Flues et al. 2017). Their results show that Alphaproteobacteria and Betaproteobacteria are less impervious to grazing and that predation rebuilds the bacterial system in leaves and impacts bacterial metabolic center capacities. Microbial assortment related to aphid inhabitants was characterized at species and intraspecies scales using a methodological structure (Guyomar et al. 2018). Utilizing this approach, on metagenomics read sets, high genomic diversity in different symbiont taxa can be uncovered in both between and within their hosts. The complete functional diversity related with host and microbiota was the first time it can be accessed using metatranscriptomics datasets which also helps in isolating the transcriptome of each member of the holobiont (Meng et al. 2018).

3.9.5 Genetic Management of Valuable Plant-Microbe Interactions

Host hereditary qualities add to plant microbiome assembly. Plants identify microorganisms through pattern recognition pattern that binds to the microbe-associated molecular pattern (MAMPs), setting off a basal barrier adequate to stop the development of most pathogenic organisms (Böhm et al. 2014). Plants can probably separate pathogens from nonpathogens and react by opposing microbial development, overlooking it, or effectively supporting it on or inside plant tissues. The transcriptional reaction of *Arabidopsis* leaves varies when vaccinated with various nonpathogenic individuals from its regular microbiota (Böhm et al. 2014). While *Methylobacterium extorquens* actuates no transcriptional reaction, *Sphingomonas melonis* initiates the defense-related genes that somewhat cover with those activated by the pathogen *Pseudomonas syringae* DC3000. This characterizes a mechanism of plant defense priming (Martinez-Medina et al. 2016) driven by the plant microbiome. The reaction example to nonpathogenic microorganisms can vary both crosswise over plant species (Ofek-Lalzar et al. 2014) and crosswise over promotions inside a single species (Haney et al. 2015). While some *Arabidopsis* accessions are colonized by and build up a valuable association with *Pseudomonas fluorescens*, different promotions effectively restrain the development of similar strains in their foundations. Given the basic capacity of defense phytohormones in the invulnerable plant framework, it is not astonishing that the plant microbiome organization is impacted by defense phytohormone flagging. Tests by a set of mutants with transformed protection phytohormone synthesis and notion stated that salicylic acid and salicylic acid-mediated events have an effect on the root microbiome composition at multiple taxonomic levels (Lebeis et al. 2015). The plant microbiome structure changes upon infection (Agler et al. 2016). Antifungal characteristics are enhanced in barley following infection with *Fusarium graminearum*, conceivably using changes in exudate arrangement (Dudenhöffer et al. 2016). An investigation of tomato plants tested with the pathogen *Ralstonia solanacearum* uncovered that the root exudation profile changed upon pathogen infection, expanding the discharge of phenolic mixes. Plant protection systems additionally sway different drivers of plant – organism cooperations, similar to plant sustenance (Hacquard et al. 2015). Present-day molecular methodologies are likewise being connected to understanding nitrogen-fixing symbioses in non-nodulating plants. Utilization of double host-organism transcriptomics depicted that the limit of a nitrogen-fixing *Burkholderia* strain to frame microaerobic biofilms on sugarcane roots is shown to have diminished motility and immunogenicity, trailed by metabolic adjustment to the sugar-rich plant condition. The plant does not enact an invulnerable reaction but, however, changes its root morphology and supplies the bacterium with photosynthates (Paungfoo-Lonhienne et al. 2016), a reaction pattern that is undifferentiated from the procedure of infection by BNF in legumes (Cao et al. 2017). These examples endorse that the coordination of defense and nutrition is crucial to driving microbiome characteristics.

3.10 Implication of the Soil Microbiome on Sustainable Agriculture and Food Security

Conveying sustenance security, the way toward expanding nourishment creation, and improving nourishment quality to support populace development without trading off ecological well-being have been known as a worldwide green revolution (Gupta 2012). Sustainable agriculture improvement is expected to relieve these issues. A definitive objective of economical agribusiness, as per the US National Research Council, is to create cultivating frameworks that are gainful, beneficial, vitality saving, and environmentally solid, preserving natural materials, and that guarantee nourishment well-being and quality. This can be achieved by substituting risky agrochemicals (chemical fertilizers and pesticides) with environmentally friendly beneficial microorganisms, which could improve the sustenance of yields and animals and furthermore present protection from biotic (pathogens and pests) and abiotic (pollution and climatic change) stresses. The potential microbial segregates are detailed utilizing different natural and inorganic bearers through either solid or liquid fermentation technologies (outlined in Table 3.1).

Table 3.1 Marketable products of plant growth-promoting rhizobacteria in plant health and disease management (Lakshmanan et al. 2014)

Bioagent	Trade name/formulation
Agrobacterium radiobacter strain K1026	Nogall
A. radiobacter strain K84	Galltrol, Diegall
Azospirillum brasilense/Azotobacter chroococcum	Gmax Nitromax
A. brasilense	Azo-Green
B. subtilis MB1600	BaciGold, HiStick N/T, Subtilex
B. subtilis strain FZB24	Rhizo-Plus, Serenade, Rhapsody, Taegro, Tae-Technical
Bacillus chlororaphis 63-28	AtEze
Bacillus cereus BPO1	Pix plus
Bacillus pumilus GB 34	Concentrate; YieldShield
B. pumilus QST2808	Sonata ASO, Ballard
B. subtilis GB03	Companion, System 3, Kodiak, Kodiak HB, Epic
Bacillus amyloliquefaciens GB99	Quantum 4000
Bacillus licheniformis SB3086	EcoGuard, Green Releaf
Burkholderia cepacia	Blue Circle, Deny, Intercept
P. fluorescens A506	BlightBan A506, Conquer, Victus
Pseudomonas syringae ESC-100	Bio-Save 10, 11, 100, 110,1000, and 10 LP
Pseudomonas chlororaphis	Cedomon
Pseudomonas cepacia	Intercept
Streptomyces griseovirdis K61	Mycostop
B. subtilis + *B. amyloliquefaciens*	Bio Yield
Pseudomonas spp. + *Azospirillum* spp.	BioJet

Further improvement of microbial confines, and the plan procedure is required through broad research to present them in sustainable agricultural practices. Applications of microbial consortia are described in Table 3.2. Aside from the application of individual organisms, distinguishing sound and practically diverse microbiomes and their application for improving harvest yield poses another big challenge to meet.

3.11 Conclusions

Several illustrations depict the significance of understanding the multitude of plant microbiome relations that pay to plant versatility in a specified environment. Acknowledgment of the plant microbiome as a coordinated part of the plant genome develops the environmental idea of "feedback." Disproportional accumulation of microbiome parasites (communicated as pathogenic impacts) prompts negative feedback, whereas the disproportional combination of microbiome mutualists stimulates positive feedback. An enhanced information about these interactions and how changes in biodiversity affect ecosystem functions (plant yield, biogeochemical pools, and fluxes) may be a vital feature for explaining plant microbiome boom and gene expression forms. Fast microbial generation time and the prevalence of horizontal gene transfer give probable systems to the improvement of localized genetic differences, or ecotypes, to emerge because of the impacts of local plant species and networks. As the plant-microbiome interaction unfolds, a new emerging methodology incorporates the study of microbial biology, microbiomes, and transcriptomes into plant genetics. The vast diversity documented in the rhizosphere microbiome is linked with the useful genes responsible for important nutrient changes, similar to those involved in N_2-fixation. The age of expansive confine accumulations and the investigation of engineered microbial networks in the mix with plant genetic properties will enable us to connect this hole and to direct reductionist, theory-driven tests in progressively complex environmental settings up to handle field tests. These developments can convert our expertise of plant-microbe interactions in nature and agriculture and could make contributions extensively to the next green revolution. The key player(s) regarding microbiome structure have not been recognized. As needs are, there is a major break in the identification of the molecular segments associated with the collaboration among the host plant and the microbial populace. Also, these ongoing microbiome examinations attempted only to distinguish its structure and multifaceted nature instead of to decide how these microbial gatherings are adjusting the plant phenome, which is basic to investigate its usage. Likewise, there would be a cross talk using signal transduction among aboveground and belowground plant tissues that can be modified by an outer biotic or abiotic stress impacting the rhizospheric microbiome.

Table 3.2 Application of bacterial consortia (Compant et al. 2019)

Plant and growth conditions	Consortia/origin of bacteria	Stress	Consortia effect	References
Arabidopsis thaliana, growth chamber, non-sterile soil	*Xanthomonas* sp. WCS2014-23, *Stenotrophomonas* sp. WCS2014-113, *Microbacterium* sp. WCS2014-259/field soil with endemic *Arabidopsis* plants	*Hyaloperonospora arabidopsidis*	Less fungal spores and higher plant fresh weight	Berendsen et al. (2018)
Solanum lycopersicum cv. Moneymaker, growth chamber	*Bacillus megaterium* SOGA_2, *Curtobacterium ceanosedimentum* SOGA_3, *Curtobacterium* sp. SOGA_6, *Massilia aurea* SOGA_7, *Pseudomonas coleopterorum* SOGA_5, 11, and 12, *Pseudomonas psychrotolerans* SOGA_13, *Pseudomonas rhizosphaerae* SOGA_14 and 19, *Frigoribacterium faeni* SOGA_17, *Xanthomonas campestris* SOGA_20/phyllosphere of field-grown tomato plants	*Pseudomonas syringae* pv. tomato	Fewer pathogen DNA copies on leaf disks	Berg and Koskella (2018)
Solanum tuberosum cv. Lady Clair, cv. *Victoria*, cv. Bintje, leaf disks in petri dishes	Double or triple combinations of *Pseudomonas* spp. R32, R47, R76, R84, S04, S19, S34, S35, S49/rhizosphere and phyllosphere of field grown potatoes	*Phytophthora infestans*	Reduced fungal sporangiophore development	de Vrieze et al. (2018)
Lycopersicon esculentum cv. Jiangshu, greenhouse pots with soil	*Pseudomonas* spp. CHA0, PF5, Q2–87, Q8R1-96, 1M1-96, MVP1-4, F113, Phl1C2/pea, wheat, cotton, tomato, sugar beet, tobacco	*Ralstonia solanacearum*	Reduced disease severity and pathogen abundance	Hu et al. (2016)
Blue maize CAP15–1 TLAX/greenhouse pots with vermiculite	*Hyaloperonospora*, *Pseudomonas putida* KT2440, *Sphingomonas* sp. OF178, *Azospirillum brasilense* Sp7, *Acinetobacter* sp. EMM02/unknown	Desiccation	Increase of shoot and root dry weight, plant height and plant diameter	Molina-Romero et al. (2017)
Capsicum annuum, *Vitis vinifera* cv. Barbera, growth chamber, greenhouse	*Acinetobacter* sp. S2 and *Bacillus* sp. S4, *Sphingobacterium* sp. S6, *Enterobacter* sp. S7 and *Delftia* sp. S8/*Vitis vinifera* rhizosphere and endosphere	Drought	Increased fresh root, aerial biomass, and photosynthesis	Rolli et al. (2015)
Nicotiana attenuate, field	*Arthrobacter nitroguajacolicus* E46, *Bacillus mojavensis* K1, *Pseudomonas frederiksbergensis* A176, *Arthrobacter nitroguajacolicus* E46, *Bacillus cereus* CN2, *Bacillus megaterium* B55, *Bacillus mojavensis* K1, *Pseudomonas azotoformans* A70, *Pseudomonas frederiksbergensis* A176, *Bacillus megaterium* B55, *Pseudomonas azotoformans* A70/tobacco plants	Natural wilt disease	Less dead plants	Santhanam et al. (2015)

References

Agler MT, Ruhe J, Kroll S, Morhenn C, Kim ST, Weigel D, Kemen EM (2016) Microbial hub taxa link host and abiotic factors to phytobiome variation. PLoS Biol 14(1):e1002352

Alori ET, Glick BR, Babalola OO (2017) Microbial phosphorus solubilization and its potential for use in sustainable agriculture. Front Microbiol 8:971

Amador CI, Canosa I, Govantes F, Santero G (2010) Lack of CbrB in Pseudomonas putida affects not only amino acids metabolism but also different stress responses and bio-film development. Environ Microbiol 12:1748–1761

Amann RI, Ludwig W, Schleifer KH (1995) Phylogenetic identification and in situ detection of individual microbial cells without cultivation. Microbiol Rev 59(1):143–169

Andreote FD, Azevedo JL, Araújo WL (2009) Assessing the diversity of bacterial communities associated with plants. Braz J Microbiol 40:417–432

Andrews JH, Harris RF (2000) The ecology and biogeography of microorganisms on plant surfaces. Annu Rev Phytopathol 38:145–180

Bacilio-Jimenez M, Aguilar-Flores S, Ventura-Zapata E, Perez-Campos E, Bouquelet S, Zenteno E (2003) Chemical characterization of root exudates from rice (*Oryza sativa*) and their effects on the chemotactic response of endophytic bacteria. Plant Soil 249:271–277

Badri DV, Vivanco JM (2009) Regulation and function of root exudates. Plant Cell Environ 32:666–681

Badri DV, Quintana N, El Kassis EG, Kim K, Choi YH, Sugiyama A, Verpoorte R, Martinoia E, Manter DK, Vivanco JM (2009) An ABC transporter mutation alters root exudation of phytochemicals that provoke an overhaul of natural soil microbiota. Plant Physiol 151:2006–2017

Bai Y, Muller DB, Srinivas G, Garrido-Oter R, Potthoff E, Rott M, Dombrowski N, Münch PC, Spaepen S, Remus-Emsermann M, Hüttel B, McHardy AC, Vorholt JA, Schulze-Lefert P (2015) Functional overlap of the Arabidopsis leaf and root microbiota. Nature 528:364–369

Baig KS, Arshad M, Shaharoona B, Khalid A, Ahmed I (2012) Comparative effectiveness of *Bacillus* spp. possessing either dual or single growth-promoting traits for improving phosphorus uptake, growth and yield of wheat (*Triticum aestivum* L.). Ann Microbiol 62:1109–1119

Bakker MG, Schlatter DC, Otto-Hanson L, Kinkel LL (2014) Diffuse symbioses: roles of plant-plant, plant-microbe and microbe-microbe interactions in structuring the soil microbiome. Mol Ecol 23:1571–1583

Bashiardes S, Godneva A, Elinav E, Segal E (2018) Towards utilization of the human genome and microbiome for personalized nutrition. Curr Opin Biotechnol 51:57–63

Benitez T, Rincon AM, Limon MC, Codon AC (2004) Biocontrol mechanisms of *Trichoderma* strains. Int Microbiol 7:249–260

Berendsen RL, Vismans G, Yu K, Song Y, de Jonge R, Burgman WP, Burmølle M, Herschend J, Bakker AHM, Pieterse CMJ (2018) Disease-induced assemblage of a plant-beneficial bacterial consortium. ISME J 12:1496–1507

Berg M, Koskella B (2018) Nutrient- and dose-dependent microbiome-mediated protection against a plant pathogen. Curr Biol 28:2487–2492

Bernedsen RL, Pieterse CMJ, Bakker AHM (2012) The rhizosphere micro-biome and plant health. Trends Plant Sci 17:478–486

Bohm M, Hurek T, Reinhold-Hurek B (2007) Twitching motility is essential for endophytic rice colonization by the N2-fixing endophyte *Azoarcus* sp. strain BH72. Mol Plant Microbe Interact 20:526–533

Böhm H, Albert I, Fan L, Reinhard A, Nürnberger T (2014) Immune receptor complexes at the plant cell surface. Curr Opin Plant Biol 20:47–54

Bokulich NA, Thorngate JH, Richardson PM, Mills DA (2014) Microbial biogeography of wine grapes is conditioned by cultivar, vintage, and climate. Proc Natl Acad Sci 111:E139–E148

Boller T, He SY (2009) Innate immunity in plants: an arms race between pattern recognition receptors in plants and effectors in microbial pathogens. Science 324:742–744

Boon E, Meehan CJ, Whidden C, Wong DHJ, Langille MGI, Beiko RG (2014) Interactions in the microbiome: communities of organisms and communities of genes. FEMS Microbiol Rev 38:90–118

Bulgarelli D, Spaepen SS, Themaat EVL, Shulze-Lefert P (2013) Structure and functions of the bacterial microbiota of plants. Annu Rev Plant Biol 64:807–838

Cao Y, Halane MK, Gassmann W, Stacey G (2017) The role of plant innate immunity in the legume-rhizobium symbiosis. Annu Rev Plant Biol 68:535–561

Chagnon PL, Bradley RL, Maherali H, Klironomos JN (2013) A trait-based framework to understand life history of mycorrhizal fungi. Trends Plant Sci 18:484–491

Cheng W, Gershenson A (2007) Carbon fluxes in the rhizosphere. In: Cardon ZG, Whitbeck JL (eds) The rhizosphere – an ecological perspective. Academic, San Diego, pp 31–56

Clay K, Schardl C (2002) Evolutionary origins and ecological consequences of endophyte symbiosis with grasses. Am Nat 160:99–127

Coince A, Cordier T, Lengelle J, Defossez E, Vacher C, Robin C, Buée M, Marçais B (2014) Leaf and root-associated fungal assemblages do not follow similar elevational diversity patterns. PLoS One 9:e100668

Coleman JJ, Ghosh S, Okoli I, Mylonakis E (2011) Antifungal activity of microbial secondary metabolites. PLoS One 6:e25321

Compant S, Kaplan H, Sessitsch A, Nowak J, Barka EA, Clement C (2008) Endo-phytic colonization of *Vitis vinifera* L. by *Burkholderia phytofirmans* strain PsJN: from the rhizosphere to inflorescence tissues. FEMS Microbiol Ecol 63:84–93

Compant S, Samad A, Faist H, Sessitsch A (2019) A review on the phytobiome: ecology, functions, and emerging trends in microbial application. J Adv Res: 1–9

Copeland JK, Yuan L, Layeghifard M, Wang PW, Guttman DS (2015) Seasonal community succession of the phyllosphere microbiome. Mol Plant-Microbe Interact 28:274–285

Czaran TL, Hoekstra RF, Pagie L (2002) Chemical warfare between microbes promotes biodiversity. Proc Natl Acad Sci 99:786–790

deBello F, Lavorel S, Diaz S, Harrison PA (2010) Towards an assessment of multiple ecosystem processes and services via functional traits. Biodivers Conserv 19:2873–2893

De Boer W, Leveau JH, Kowalchuk GA, Gunnewiek PJK, Abeln EC, Figge MJ, Sjollema K, Janse JD, van Veen JA (2004) *Collimonas fungivorans* gen. nov., sp. nov., a chitinolytic soil bacterium with the ability to grow on living fungal hyphae. Int J Syst Evol Microbiol 54:857–864

de Vrieze M, Germanier F, Vuille N, Weisskopf L (2018) Combining different potato associated *Pseudomonas* strains for improved biocontrol of *Phytophthora infestans*. Front Microbiol 9:2573

De Weert S, Vermeiren H, Mulders IHM, Kuiper I, Hendrickx N, Bloemberg GV, Vanderleyden J, Mot RD, Lugtenberg BJJ (2002) Flagella-driven chemotaxis towards exudate components is an important trait for tomato root colonization by *Pseudomonas fluorescens*. Mol Plant Microbe Interact 15:1173–1180

Dekkers LC, Phoelich CC, van der Fits L, Lugtenberg JJ (1998) A site-specific recombinase is required for competitive root colonization by *Pseudomonas fluorescens* WCS365. Proc Natl Acad Sci 95:7051–7056

Depoorter E, Bull MJ, Peeters C, Coenye T, Vandamme P, Mahenthiralingam E (2016) Burkholderia: an update on taxonomy and biotechnological potential as antibiotic producers. Appl Microbiol Biotechnol 100:5215–5229

diCenzo GC, Zamani M, Checcucci A, Fondi M, Griffitts JS, Finan TM, Mengoni A (2019) Multidisciplinary approaches for studying rhizobium–legume symbioses. Can J Microbiol 65:1–33

Dini-Andreote F, van Elsas JD (2013) Back to the basics: the need for ecophysiological insights to enhance our understanding of microbial behaviour in the rhizosphere. Plant Soil 373:1–15

Donn S, Kirkegaard JA, Perera G, Richardson AE, Watt M (2015) Evolution of bacterial communities in the wheat crop rhizosphere. Environ Microbiol 17:610–621

Doornbos R, van Loon L, Bakker P (2012) Impact of root exudates and plant defense signaling on bacterial communities in the rhizosphere. A review. Agron Sustain Dev 32:227–234

Dudenhöffer JH, Scheu S, Jousset A (2016) Systemic enrichment of antifungal traits in the rhizosphere microbiome after pathogen attack. J Ecol 104:1566–1575
Eberl L (1999) N-acyl homoserinelactone-mediated gene regulation in gram-negative bacteria. Syst Appl Microbiol 22:493–506
Edwards J, Johnson C, Santos-medellın C, Lurie E, Kumar N (2015) Structure, variation, and assembly of the root-associated microbiomes of rice. Proc Natl Acad Sci 112:E911–E920
Fitzpatrick CR, Copelandc J, Wang PW, Guttman DS, Peter M, Kotanen PM, Johnson MTJ (2018) Assembly and ecological function of the root microbiome across angiosperm plant species. Proc Natl Acad Sci 115:E1157–E1165
Flues S, Bass D, Bonkowski M (2017) Grazing of leaf associated *Cercomonads* (Protists: Rhizaria: Cercozoa) structures bacterial community composition and function. Environ Microbiol 19:3297–3309
Friesen ML, Porter SS, Stark SC, von Wettberg EJ, Sachs JL, Martinez-Romero E (2011) Microbially mediated plant functional traits. Annu Rev Ecol Evol Syst 42:23–46
Garbaye J (1994) Helper bacteria – a new dimension to the mycorrhizal symbiosis. New Phytol 128:197–210
Gimsing AL, Blum J, Dyan FE, Locke MA, Sejer LH, Jacobsen CS (2009) Mineralization of the allelochemical sorgoleone in soil. Chemosphere 76:1041–1047
Glick BR (1999) The enhancement of plant growth by free-living bacteria. Can J Microbiol 41:109–117
Guennoc CM, Rose C, Labbe J, Deveau A (2017) Bacterial biofilm formation on soil fungi: a widespread ability under controls. Biol Res 130740
Gupta VV (2012) Beneficial microorganisms for sustainable agriculture. Microbiol Aust 3:113–115
Guyomar C, Legeai F, Jousselin E, Mougel C, Lemaitre C, Simon J-C (2018) Multi-scale characterization of symbiont diversity in the pea aphid complex through metagenomic approaches. Microbiome 6(1):181
Hacquard S (2016) Disentangling the factors shaping microbiota composition across the plant holobiont. New Phytol 209:454–457
Hacquard S, Garrido-Oter R, González A, Spaepen S, Ackermann G, Lebeis S, McHardy AC, Dangl JL, Knight R, Ley R, Schulze-Lefert P (2015) Microbiota and host nutrition across plant and animal kingdoms. Cell Host Microbe 17:603–616
Haney CH, Samuel BS, Bush J, Ausubel FM (2015) Associations with rhizosphere bacteria can confer an adaptive advantage to plants. Nat Plant 1:15051
Hardoim PR, van Overbeek LS, van Elsas JD (2008) Properties of bacterial endophytes and their proposed role in plant growth. Trends Microbiol 16:463–471
Hardoim PR, van Overbeek LS, Berg G, Pirttila AM, Compant S, Campisano A, Döring M, Sessitsch A (2015) The hidden world within plants: ecological and evolutionary considerations for defining functioning of microbial endophytes. Microbiol Mol Biol Rev 79:293–320
Haron MH, Tyler HL, Chandra S, Moraes RM, Jackson CR, Pugh ND, Pasco DS (2019) Plant microbiome-dependent immune enhancing action of echinacea purpurea is enhanced by soil organic matter content. Sci Rep 9(1):136
Hartman K, Van der Heijden MGA, Wittwer RA, Banerjee S, Walser JC, Schlaeppi K (2018) Cropping practices manipulate abundance patterns of root and soil microbiome members paving the way to smart farming. Microbiome 6:14
Hassani MA, Durán P, Hacquard S (2018) Microbial interactions within the plant holobiont. Microbiome 6:58
Hendgen M, Hoppe B, Döring J, Friedel M, Kauer R, Frisch M, Dahl A, Kellner H (2018) Effects of different management regimes on microbial biodiversity in vineyard soils. Science Report 8:9393
Hiltner L (1904) Uber neuere erfahrungen und probleme auf dem gebiet der bodenbakteriologie und unter besonderer berucksichtigung der grundungung und brache. Arbeiten der Deutschen Landwirtschaftlichen Gesellschaft 98:59–78
Hirsch PR, Mauchline TH (2012) Who's who in the plant root microbiome? Nat Biotechnol 30:961–962

Honma M, Shimomura T (1978) Metabolism of 1-aminocyclopropane-1-carboxylic acid. Agric Biol Chem 42:1825–1831
Horton MW, Bodenhausen N, Beilsmith K, Meng DZ, Muegge BD, Subramanian S, Vetter MM, Vilhjálmsson BJ, Nordborg M, Gordon JI, Bergelson J (2014) Genome-wide association study of *Arabidopsis thaliana* leaf microbial community. Nat Commun 5:5320
Howard JB, Rees DC (1996) Structural basis of biological nitrogen fixation. Chem Rev 96:2965–2982
Hu J, Wei Z, Friman VP, Gu SH, Wang XF, Eisenhauer N, Yang TJ, Ma J, Shen QR, Xu YC, Jousset A (2016) Probiotic diversity enhances rhizosphere microbiome function and plant disease suppression. mBio 7(6):1790–1816
Huang F, Chaparro JM, Reardon KF et al (2014) Rhizosphere interactions: root exudates, microbes, and microbial communities 1 Xing. Botany 92:267–275. https://doi.org/10.1139/cjb-2013-0225
Jaeger CHIII, Lindow SE, Miller W, Clark E, Firestone MK (1999) Mapping of sugar and amino acid availability in soil around roots with bacterial sensors of sucrose and tryptophan. Appl Environ Microbiol 65:2685–2690
Kagan IA, Rimando AM, Dyan FE (2003) Elucidation of the biosynthetic pathway of the allelochemical sorgoleone using retro biosynthetic NMR analysis. J Biol Chem 278:28607–28611
Kaplan D, Maymon M, Agapakis CM, Lee A, Wang A, Prigge BA, Volkogon M, Hirsch AM (2013) A survey of the microbial community in the rhizosphere of two dominant shrubs of the Negev Desert highlands, *Zygophyllum dumosum* Boiss. and *Atriplex halimus*, using cultivation-dependent and -independent methods. Am J Bot 100:1713–1725
Kavamura VN, Santos SN, Silva JL, Parma MM, Avila LA, Visconti A, Zucchi TD, Taketani RG, Andreote FD, Melo IS (2013) Screening of Brazilian cacti rhizobacteria for plant growth promotion under drought. Microbiol Res 168:183–191
Keiblinger KM, Wilhartitz IC, Schneider T, Roschitzki B, Schmid E, Eberl L, Riedel K, Zechmeister-Boltenstern S (2012) Soil metaproteomics: comparative evaluation of protein extraction protocols. Soil Biol Biochem 54:14–24
Kishore GK, Pande S, Podile AR (2005) Biological control of late leaf spot of peanut (*Arachis hypogaea*) with chitinolytic bacteria. Phytopathology 95:1157–1165
Klironomos JN (2002) Feedback with soil biota contributes to plant rarity and invasiveness in communities. Nature 417:67–70
Kolmeder CA, de Vos WM (2014) Metaproteomics of our microbiome: developing insight in function and activity in man and model systems. J Proteome 97:3–16
Krober M, Wibberg D, Grosch R, Eikmeyer F, Verwaaijen B, Chowdhury SP, Hartmann A, Pühler A, Schlüter A (2014) Effect of the strain Bacillus *amyloliquefaciens* FZB42 on the microbial community in the rhizosphere of lettuce under field conditions analyzed by whole metagenome sequencing. Front Microbiol 5:252
Kumari B, Mallick MA, Solanki MK et al (2019) Plant growth-promoting rhizobacteria (PGPR): modern prospects for sustainable agriculture. In: Ansari RA, Mahmood I (eds) Plant health under biotic stress, vol II. Springer Netherlands, Dordrecht
Lakshmanan V, Selvaraj G, Bais HP (2014) Functional soil microbiome: belowground solutions to an aboveground problem. Plant Physiol 166:689–700
Lambais MR, Crowley DE, Cury JC, Bull RC, Rodrigues RR (2006) Bacterial diversity in tree canopies of the Atlantic forest. Science 312:1917
Lambais MR, Barrera SE, Santos EC, Crowley DE, Jumpponen A (2017) Phyllosphere metaproteomes of trees from the Brazilian Atlantic forest show high levels of functional redundancy. Microb Ecol 73:123–134
Lebeis SL, Paredes SH, Lundberg DS, Breakfield N, Gehring J, McDonald M, Malfatti S, del Rio TG, Jones CD, Tringe SG, Dangl JL (2015) PHYTOBIOME: salicylic acid modulates colonization of the root microbiome by specific bacterial taxa. Science 349(6250):860–864
Lederberg J, McCray AT (2001) 'Ome sweet omics' - a genealogical treasury of words. Scientist 15:8

Leff JW, Del Tredici P, Friedman WE, Fierer N (2015) Spatial structuring of bacterial communities within individual *Ginkgo biloba* trees. Environ Microbiol 17:2352–2361

Lemanceau P, Barret M, Mazurier S, Mondy S, Pivato B, Fort T, Vacher C (2017) Plant communication with associated microbiota in the spermosphere, rhizosphere and phyllosphere. In: Beards G (ed) How plants communicate with their biotic environment. Academic, London, pp 101–133

Lindow SE, Brandl MT (2003) Microbiology of the phyllosphere. Appl Environ Microbiol 69:1875–1883

Little AEF, Robinson CJ, Peterson SB, Raffa KF, Handelsman J (2008) Rules of engagement: interspecies interactions that regulate microbial communities. Annu Rev Microbiol 62:375–401

Longa CMO, Nicola L, Antonielli L, Mescalchin E, Zanzotti R, Turco E, Pertot I (2017) Soil microbiota respond to green manure in organic vineyards. J Appl Microbiol 123:1547–1560

Loper JE, Schroth MN (1986) Influence of bacterial source of indole-3-acetic acid on root elongation of sugar beet. Phytopathology 76:386–389

Lücking R, Huhndorf S, Pfister DH, Plata ER, Lumbsch HT (2009) Fungi evolved right on track. Mycologia 101:810–822

Lugtenberg B, Kamilova F (2009) Plant-growth-promoting rhizobacteria. Annu Rev Microbiol 63:541–556

Lundberg DS, Lebeis SL, Paredes SH, Yourstone S, Gehring J, Malfatti S, Malfatti S, Tremblay J, Engelbrektson A, Kunin V, Del Rio TG, Edgar RC, Eickhorst T, Ley RE, Hugenholtz P, Tringe SG, Dangl JL (2012) Defining the core Arabidopsis thaliana root microbiome. Nature 488:86–90

Mader U, Antelmann H, Buder T, Dahl MK, Hecker M, Homuth G (2002) *Bacillus subtilis* functional genomics: genome-wide analysis of the DegS-DegU regulon by transcriptomics and proteomics. Mol Genet Genomics 268:455–467

Maida I, Chiellini C, Mengoni A, Bosi E, Firenzuoli F, Fondi M, Fani R (2016) Antagonistic interactions between endophytic cultivable bacterial communities isolated from the medicinal plant *Echinacea purpurea*. Environ Microbiol 18:2357–2365

Malviya MK, Solanki MK, Li C-N et al (2019) Beneficial linkages of endophytic Burkholderia anthina MYSP113 towards sugarcane growth promotion. Sugar Tech:1–12. https://doi.org/10.1007/s12355-019-00703-2

Martinez-Medina A, Flors V, Heil M, Mauch-Mani B, Pieterse CM, Pozo MJ, Ton J, van Dam NM, Conrath U (2016) Recognizing plant defense priming. Trends Plant Sci 21:818–822

Martinez-Romero E (2006) Dinitrogen-fixing prokaryotes. In: Dworkin M, Falkow S, Rosenberg E, Schleifer KH, Stackebrandt E (eds) The prokaryotes. Springer, New York, pp 793–817

Mavrodi DV, Mavrodi OV, Parejko JA, Weller DM, Thomashow LS (2011) The role of 2,4-diacetylphloroglucinol- and phenazine-1-carboxylic acid-producing *Pseudomonas* spp. in natural protection of wheat from soilborne pathogens. In: Maheshwari DK (ed) Bacteria in agrobiology: plant nutrient management. Springer, Berlin, pp 60–63

Mendes R, Garbeva P, Raaijmakers JM (2013) The rhizosphere microbiome: significance of plant beneficial, plant pathogenic, and human pathogenic microorganisms. FEMS Microbiol Rev 37(5):634–663

Meng A, Marchet C, Corre E, Peterlongo P, Alberti A, Da Silva C, Wincker P, Pelletier E, Probert I, Decelle J, Le Crom S, Not F, Bittner L (2018) A *de novo* approach to disentangle partner identity and function in holobiont systems. Microbiome 6(1):105

Mercado-Blanco J, Bakker PAHM (2007) Interactions between plants and beneficial Pseudomonas spp.: exploiting bacterial traits for crop protection. Antonie Van Leeuwenhoek 92:367–389

Molina-Romero D, Baez A, Quintero-Hernández V, Castañeda-Lucio M, Fuentes-Ramírez LE, Bustillos-Cristales MDR, Rodríguez-Andrade O, Morales-García YE, Munive A, Muñoz-Rojas J (2017) Compatible bacterial mixture, tolerant to desiccation, improves maize plant growth. PLoS One 12(11):e0187913

Moreno R, Martines-Gomariz M, Yuste L, Gil C, Rojo F (2009) The *Pseudomonas putida* Crc global regulator controls the hierarchical assimilation of amino acids in a complete medium: evidence from proteomic and genomic analyses. Proteomics 9:2910–2928

Mousa WK, Shearer C, Limay-Rios V, Ettinger CL, Eisen JA, Raizada MN (2016) Root-hair endophyte stacking in finger millet creates a physicochemical barrier to trap the fungal pathogen *Fusarium graminearum*. Nat Microbiol 1:16167
Müller H, Berg C, Landa BB, Auerbach A, Moissl-Eichinger C, Berg G (2015) Plant genotype-specific archaeal and bacterial endophytes but similar *Bacillus* antagonists colonize Mediterranean olive trees. Front Microbiol 6:138
Mullis KB, Erlich HA, Arnheim N, Horn GT, Saiki RK, Scharf SJ (1987) Process for amplifying, detecting, and/or-cloning nucleic acid sequences. US Patent Application No. US4683195A
Nadell CD, Xavier JB, Foster KR (2009) The sociobiology of biofilms. FEMS Microbiol Rev 33:206–224
Neal AL, Ahmad S, Gordon-Weeks R, Ton J (2012) Benzoxazinoids in root exudates of maize attract *Pseudomonas putida* to the rhizosphere. PLoS One 7:e35498
Netzker T, Fischer J, Weber J, Mattern DJ, Konig CC, Valiante V, Schroeckh V, Brakhage AA (2015) Microbial communication leading to the activation of silent fungal secondary metabolite gene clusters. Front Microbiol 6:299
Ofek-Lalzar M, Sela N, Goldman-Voronov M, Green SJ, Hadar Y, Minz D (2014) Niche and host-associated functional signatures of the root surface microbiome. Nat Commun 5:4950
Oh KB, Miyazawa H, Naito T, Matsuoka H (2001) Purification and characterization of an autoregulatory substance capable of regulating the morphological transition in *Candida albicans*. Proc Natl Acad Sci 98:4664–4668
Palumbo JD, Kado CI, Phillips DA (1998) An isoflavonoid-inducible efflux pump in Agrobacterium tumefaciens is involved in competitive colonization of roots. J Bacteriol 180:3107–3113
Paungfoo-Lonhienne C, Lonhienne TGA, Yeoh YK, Donose BC, Webb RI, Parsons J, Liao W, Sagulenko E, Lakshmanan P, Hugenholtz P, Schmidt S, Ragan MA (2016) Crosstalk between sugarcane and a plant-growth promoting *Burkholderia* species. Sci Rep 6:37389
Peay KG, Kennedy PG, Talbot JM (2016) Dimensions of biodiversity in the earth mycobiome. Nat Rev Microbiol 14:434
Peiffer JA, Spor A, Koren O, Jin Z, Tringe SG, Dangl JL, Buckler ES, Ley RE (2013) Diversity and heritability of the maize rhizosphere microbiome under field conditions. Proc Natl Acad Sci 110(16):6548–6653
Phillips DA, Fow TC, Six J (2006) Root exudation (net efflux of amino acids) may increase rhizodeposition under elevated CO2. Glob Chang Biol 12:561–567
Piel J (2011) Approaches to capturing and designing biologically active small molecules produced by uncultured microbes. Annu Rev Microbiol 65:431–453
Pineda A, Zheng S, van Loon JJA, Pieterse CMJ, Dicke M (2010) Helping plants to deal with insects: the role of beneficial soil-borne microbes. Trends Plant Sci 15:507–514
Ploch S, Rose LE, Bass D, Bonkowski M (2016) High diversity revealed in leaf-associated Protists (Rhizaria: *Cercozoa*) of *Brassicaceae*. J Eukaryot Microbiol 63:635–641
Raaijmakers JM, Mazzola M (2012) Diversity and natural functions of antibiotics produced by beneficial and plant pathogenic bacteria. Annu Rev Phytopathol 50:403–424
Raaijmakers JM, Paulitz TC, Steinberg C, Alabouvette C, Moenne-Loccoz Y (2009) The rhizosphere: A playground and battlefield for soilborne pathogens and beneficial microorganisms. Plant Soil 321:341–361
Radzki W, Gutierrez Mañero FJ, Algar E, Lucas García JA, García-Villaraco A, Ramos Solano B (2013) Bacterial siderophores efficiently provide iron to iron-starved tomato plants in hydroponics culture. Antonie Van Leeuwenhoek 104:321–330
Rastogi G, Sbodio A, Tech JJ, Suslow TV, Coaker GL, Leveau JHJ (2012) Leaf microbiota in an agroecosystem: spatiotemporal variation in bacterial community composition on field-grown lettuce. ISME J 6:1812–1822
Raymond J, Siefert JL, Staples CR, Blankenship RE (2004) The natural history of nitrogen fixation. Mol Biol Evol 21:541–554
Reading NC, Sperandio V (2006) Quorum sensing: the many languages of bacteria. FEMS Microbial Lett 254:1–11

Records AR (2011) The type VI secretion system: a multipurpose delivery system with a phage-like machinery. Mol Plant-Microbe Interact 24:751–757

Riehl C, Frederickson ME (2016) Cheating and punishment in cooperative animal societies. Philos Trans R Soc Lond Ser B Biol Sci 371:20150090

Riera N, Handique U, Zhang Y, Dewdney MM, Wang N (2017) Characterization of antimicrobial-producing beneficial bacteria isolated from Huanglongbing escape citrus trees. Front Microbiol 8:2415

Ritpitakphong U, Falquet L, Vimoltust A, Berger A, Metraux JP, L'Haridon F (2016) The microbiome of the leaf surface of Arabidopsis protects against a fungal pathogen. New Phytol 210:1033–1043

Rolli E, Marasco R, Vigani G, Ettoumi B, Mapelli F, Deangelis ML, Gandolfi C, Casati E, Previtali F, Gerbino R, Pierotti CF, Borin S, Sorlini C, Zocchi G, Daffonchio D (2015) Improved plant resistance to drought is promoted by the root-associated microbiome as a water stress-dependent trait. Environ Microbiol 17:316–331

Rosenblueth M, Martínez-Romero E (2006) Bacterial endophytes and their interactions with hosts. Mol Plant Microbe Interact 19:827–837

Rout ME, Chrzanowski TH (2009) The invasive *Sorghum halepense* harbors endophytic N2-fixing bacteria and alters soil biogeochemistry. Plant Soil 315:163–172

Rout ME, Chrzanowski TH, DeLuca TH, Westlie TK, Callaway RM, Holben WE (2013) Bacterial endophytes enhance invasive plant competition. Am J Bot 100:1726–1737

Rudgers JA, Afkhami MA, Rua MA, Davitt SH, Hammer S, Huguet VM (2009) A fungus among us: broad pattern of endophyte distribution in the grasses. Ecology 90:1531–1539

Rudnick MB, van Veen JA, de Boer W (2015) Baiting of rhizosphere bacteria with hyphae of common soil fungi reveals a diverse group of potentially mycophagous secondary consumers. Soil Biol Biochem 88:73–82

Rudrappa T, Czymmek KJ, Pare PW, Bais HP (2008) Root-secreted malic acid recruits beneficial soil bacteria. Plant Physiol 148:1547–1556

Ruggiero MA, Gordon DP, Orrell TM, Bailly N, Bourgoin T, Brusca RC, Cavalier-Smith T, Guiry MD, Kirk PM (2015) A higher-level classification of all living organisms. PLoS One 10:e0119248

Sachs JL, Mueller UG, Wilcox TP, Bull JJ (2004) The evolution of cooperation. Q Rev Biol 79:135–160

Santhanam R, Van LT, Weinhold A, Goldberg J, Oh Y, Baldwin IT (2015) Native root-associated bacteria rescue a plant from a sudden-wilt disease that emerged during continuous cropping. Proc Natl Acad Sci 112:5013–5020

Sapp M, Ploch S, Fiore-Donno AM, Bonkowski M, Rose LE (2018) Protists are an integral part of the *Arabidopsis thaliana* microbiome. Environ Microbiol 20:30–43

Sarwar M, Kremer RJ (1995) Enhanced suppression of plant growth through production of L-tryptophan-derived compounds by deleterious rhizobacteria. Plant Soil 172:261–269

Schade J, Hobbie SE (2005) Spatial and temporal variation in the islands of fertility in the Sonoran Desert. Biogeochemistry 73:541–553

Schmidt R, Etalo DW, de Jager V, Gerards S, Zweers H, de Boer W, Garbeva P (2016) Microbial small talk: volatiles in fungal-bacterial interactions. Front Microbiol 6:1495

Schreiter S, Sandmann M, Smalla K, Grosch R (2014) Soil type dependent rhizosphere competence and biocontrol of two bacterial inoculant strains and their effects on the rhizosphere microbial community of field-grown lettuce. PLoS One 9:e103726

Shaharoona B, Naveed M, Arshad M, Zahir ZA (2008) Fertilizer-dependent efficiency of pseudomonads for improving growth, yield and nutrient use efficiency of wheat (*Triticum aestivum* L.). Appl Microbiol Biotechnol 79:147–155

Sharma M, Prasad R (2011) The quorum-sensing molecule farnesol is a modulator of drug efflux mediated by ABC multidrug transporters and synergizes with drugs in Candida albicans. Antimicrob Agents Chemother 55:4834–4843

Shi SJ, Nuccio E, Herman DJ, Rijkers R, Estera K, Li JB, da Rocha UN, He Z, Pett-Ridge J, Brodie EL, Zhou J, Firestone M (2015) Successional trajectories of rhizosphere bacterial communities over consecutive seasons. mBio 6:e00746-15

Solanki MK, Kumar S, Pandey AK et al (2012) Diversity and antagonistic potential of Bacillus spp. associated to the rhizosphere of tomato for the management of Rhizoctonia solani. Biocontrol Sci Tech 22:203–217. https://doi.org/10.1080/09583157.2011.649713

Solanki MK, Wang F-Y, Wang Z et al (2019) Rhizospheric and endospheric diazotrophs mediated soil fertility intensification in sugarcane-legume intercropping systems. J Soils Sediments 19:1911–1927. https://doi.org/10.1007/s11368-018-2156-3

Stevenson FJ, Cole MA (1999) Cycles of soil: carbon, nitrogen phosphorus, sulphur and micronutrients, 2nd edn. Wiley, New York

Stoodley P, Sauer K, Davies DG, Costerton JW (2002) Biofilms as complex differentiated communities. Annu Rev Microbiol 56:187–209

Syed Ab Rahman SF, Singh E, Pieterse CMJ, Schenk PM (2018) Emerging microbial biocontrol strategies for plant pathogens. Plant Sci 267:102–111

Talbot JM, Bruns TD, Taylor JW, Smith DP, Branco S, Glassman SI, Erlandson S, Vilgalys R, Liao HL, Smith ME, Peay KG (2014) Endemism and functional convergence across the North American soil mycobiome. Proc Natl Acad Sci 111:6341–6346

Taulé C, Mareque C, Barlocco C, Hackembruch F, Reis VM, Sicardi M, Battistoni F (2012) The contribution of nitrogen fixation to sugarcane (*Saccharum officinarum* L.), and the identification and characterization of part of the associated diazotrophic bacterial community. Plant Soil 356:35–49

Teplitski M, Robinson JB, Bauer WD (2000) Plants secrete substances that mimic bacterial N-acyl homoserine lactone signal activities and affect population density-dependent behaviors in associated bacteria. Mol Plant-Microbe Interact 13:637–648

Thomson BC, Tisserant E, Plassart P, Uroz S, Griffiths RI, Hannula SE, Buee M, Mougel C, Ranjard L, Van Veen JA, Martin F, Bailey MJ, Lemanceau P (2015) Soil conditions and land use intensification effects on soil microbial communities across a range of European field sites. Soil Biol Biochem 88:403–413

Thrall PH, Hochberg ME, Burdon JJ, Bever JD (2007) Coevolution of symbiotic mutualists and parasites in a community context. Trends Ecol Evol 22:120–126

Tkacz A, Cheema J, Chandra G, Grant A, Poole PS (2015) Stability and succession of the rhizosphere microbiota depends upon plant type and soil composition. ISME J 9:2349–2359

Toju H, Yamamoto S, Sato H, Tanabe AS (2013) Sharing of diverse mycorrhizal and root-endophytic fungi among plant species in an oak-dominated cool-temperate forest. PLoS One 8:e78248

Toussaint JP, Pham TTM, Barriault D, Sylvestre M (2012) Plant exudates promote PCB degradation by a rhodococcal rhizobacteria. Appl Microbiol Biotechnol 95:1589–1603

Turner TR, Ramakrishnan K, Walshaw J, Heavens D, Alston M, Swarbreck D, Osbourn A, Grant A, Poole PS (2013) Comparative metatranscriptomics reveals kingdom level changes in the rhizosphere microbiome of plants. ISME J 7:2248–2258

Uhlik O, Leewis MC, Strejcek M, Musilova L, Mackova M, Leigh MB, Macek T (2013) Stable isotope probing in the metagenomics era: a bridge towards improved bioremediation. Biotechnol Adv 31:154–165

Van Acker H, Van Dijck P, Coenye T (2014) Molecular mechanisms of antimicrobial tolerance and resistance in bacterial and fungal biofilms. Trends Microbiol 22:326–333

Van Buyten E, Hofte M (2013) Pythium species from rice roots differ in virulence, host colonization and nutritional profile. BMC Plant Biol 13:203

Van der Ent S, Van Hulten M, Pozo MJ, Czechowski T, Udvardi MK, Pieterse CM, Ton J (2009) Priming of plant innate immunity by rhizobacteria and beta-aminobutyric acid: differences and similarities in regulation. New Phytol 183:419–431

Van der Heijden MG, de Bruin S, Luckerhoff L, van Logtestijn RS, Schlaeppi K (2016) A widespread plant-fungal-bacterial symbiosis promotes plant biodiversity, plant nutrition and seedling recruitment. ISME J 10:389–399

Van Overbeek L, van Elsas JD (2008) Effects of plant genotype and growth stage on the structure of bacterial communities associated with potato (*Solanum tuberosum* L.). FEMS Microbiol Ecol 64:283–296

Vorholt JA (2012) Microbial life in the phyllosphere. Nat Rev 10:828–840
Wagner MR, Lundberg DS, del Rio TG, Tringe SG, Dangl JL, Mitchell-Olds T (2016) Host genotype and age shape the leaf and root microbiomes of a wild perennial plant. Nat Commun 7:12151
Wang JJ, Shen J, Wu YC, Tu C, Soininen J, Stegen JC, He J, Liu X, Zhang L, Zhang E (2013) Phylogenetic beta diversity in bacterial assemblages across ecosystems: deterministic versus stochastic processes. ISME J 7:1310–1321
Wang Z, Solanki MK, Pang F et al (2017) Identification and efficiency of a nitrogen-fixing endophytic actinobacterial strain from sugarcane. Sugar Tech 19:492–500. https://doi.org/10.1007/s12355-016-0498-y
Wang Z, Solanki MK, Yu Z-X et al (2018) Draft genome analysis offers insights into the mechanism by which Streptomyces chartreusis WZS021 increases drought tolerance in sugarcane. Front Microbiol 9:3262. https://doi.org/10.3389/fmicb.2018.03262
Wei Z, Yang T, Friman VP, Xu Y, Shen Q, Jousset A (2015) Trophic network architecture of root-associated bacterial communities determines pathogen invasion and plant health. Nat Commun 6:8413
Yergeau E, Sanschagrin S, Maynard C, St-Arnaud M, Green CW (2014) Microbial expression profiles in the rhizosphere of willows depend on soil contamination. ISME J 8(2):344–358
Yunshi L, Xiukun W, Tuo C, Wanfu W, Guangxiu L, Wei Z, Shiweng L, Minghao W, Changming Z, Huaizhe Z, Gaosen Z (2018) Plant phenotypic traits eventually shape its microbiota: a common garden test. Front Microbiol 9:2479
Zaidi A, Khan MS (2005) Interactive effect of rhizospheric microorganisms on growth, yield and nutrient uptake of wheat. J Plant Nutr 28:2079–2092
Zamioudis C, Korteland J, Van Pelt JA, van Hamersveld M, Dombrowski N, Bai Y, Hanson J, Van Verk MC, Ling HQ, Schulze-Lefert P, Pieterse CM (2015) Rhizobacterial volatiles and photosynthesis-related signals coordinate MYB72 expression in Arabidopsis roots during onset of induced systemic resistance and iron-deficiency responses. Plant J 84:309–322
Zehr JP, Jenkins BD, Short SM, Steward GF (2003) Nitrogenase gene diversity and microbial community structure: A cross-system comparison. Environ Microbiol 5:539–554
Zhang M, Silva MCP, De Mares MC, van Elsas J (2014) The mycosphere constitutes an arena for horizontal gene transfer with strong evolutionary implications for bacterial-fungal interactions. FEMS Microbiol Ecol 89:516–526
Zhang Y, Xu J, Riera N, Jin T, Li J, Wang N (2017) Huanglongbing impairs the rhizosphere-to-rhizoplane enrichment process of the citrus root-associated microbiome. Microbiome 5:97

Plant Mycobiome: Current Research and Applications

4

Ajit Kumar Dubedi Anal, Shalini Rai, Manvendra Singh, and Manoj Kumar Solanki

Abstract

Plant mycobiome studies are one of the leading burning aspects of the twenty-first century for ecological management and sustainable agricultural. A plant-associated fungal community plays an important role in maintaining ecological fitness by cycling organic matter and channeling nutrients across the trophic levels. Several reports highlighted the need for plant mycobiome studies for better disease management, ecological practices, and the use of eco-friendly methods for crop production. In this context, plant mycobiome revealed the effect of the fungal community on the composition of other microbial communities associated with the plant, plant growth, and plant responses against the pathogens. Fungal biodiversity, functionality, and associative interaction with other microbiome organisms and plants are revealed by high-throughput sequencing methods that broaden our view on understanding the fungal importance to plants. The present chapter discussed the modern tools and techniques utilized to study fungal diversity and community structure by the use of different kinds of OMICS approaches such as ITS rDNA gene or specific functional gene sequencing, transcriptomics, proteomics, and metabolomics. Here, the chapter focused on the current research, development of new techniques and approaches that can provide an integrative insight of the role of fungal communities in the plant

A. K. D. Anal
Amity Institute of Microbial Technology, Amity University, Noida, Uttar Pradesh, India

S. Rai (✉)
Department of Biotechnology, Society of Higher Education and Practical Application (SHEPA), Varanasi, Uttar Pradesh, India

M. Singh
ICAR-National Research Centre on Litchi, Muzaffarpur, Bihar, India

M. K. Solanki
Department of Food Quality & Safety, Institute for Post-harvest and Food Sciences, The Volcani Center, Agricultural Research Organization, Rishon LeZion, Israel

© Springer Nature Singapore Pte Ltd. 2020

M. K. Solanki et al. (eds.), *Phytobiomes: Current Insights and Future Vistas*,
https://doi.org/10.1007/978-981-15-3151-4_4

microbiome. Plant mycobiome and their diversity are important to predict plant growth and survival against the pathogen and multitrophic interactions between the organisms to identify their functional cores in regard to plant health and forecast which fungal community is likely to affect plant fitness and produce useful secondary metabolites.

Keywords
Mycobiome · Sequencing technique · Plant pathogen · Biofertilizer

4.1 Introduction

The plant mycobiome represents the plant-associated fungal community that have multiple functional roles, in response to plant, ecological management, and environmental indications (Finzi et al. 2011; Pagano et al. 2017). The plant-associated fungal community is most commonly associated with both terrestrial and aquatic ecosystems by continuous degradation of organic matter and cycling nutrients across the trophic levels of the food web (Delgado-Baquerizo et al. 2013; Toju et al. 2018). These communities are cosmopolitan in different ecological habitats, such as natural soils, decaying wood, and plant material, living plants (endophytes), water-related environments, etc. (Rai et al. 2019; Hardoim et al. 2015). Most of these fungal species are saprophytes, and they are capable to decompose complex polymers such as cellulose, chitin, and lignin by different hydrolytic enzymes (Benitez et al. 2004; Anil and Lakshmi 2010; Błaszczyk et al. 2014). Several factors, like plant host, plant density, environmental conditions, nutrient availability, and interactions with other external microbiomes (e.g., soil fungi and bacteria), are contributed to maintaining the plant mycobiome composition (Bahram et al. 2015; Nilsson et al. 2018).

The plant mycobiome is usually comprised of five main functional groups of fungal communities, viz., saprotrophic, pathogenic, epiphytic, endophytic, and mycorrhizal fungi (ectomycorrhizal fungi, arbuscular mycorrhizal fungi, ericoid mycorrhizal fungi, and orchid mycorrhizal fungi) (Porras-Alfaro and Bayman 2011; Beckers et al. 2016). The plant-associated fungal community is participating in mineral conduction, carbon distribution, water acquisition, tolerance to abiotic and biotic stresses, and interplant competition (Bucher et al. 2009; Druzhinina et al. 2011; Cai et al. 2013). They applied different mechanisms, such as parasitism, stimulation of plant growth, competition for nutrients, cross-protection, microbial compound production, and induction of systemic resistance against abiotic and biotic stresses in plant mycobiome to maintain ecological fitness and fungal diversity (Kubicek et al. 2011; Porras-Alfaro and Bayman 2011; Hardoim et al. 2015; Cai et al. 2015).

Several reports highlighted the increasing interest of plant mycobiome studies to explore hidden mechanisms and reshape the rhizospheric microbiome for biofertilization, which stimulates plant growth, protection from stress, parasitism, antibiosis, induced plant systemic resistance, and rhizoremediation (Ikeda et al. 2010; Hu et al. 2018). The mycobiome research will draw new insights into plant fungi interaction that can help to understand the complex networking and ecological function of fungi. In this context, molecular identification revealed the cultivable and uncultivable fungal communities of mycobiome that gives information about mycobiome diversity, structure, compositions, and signaling mechanism between the host plant and associated microbiome (Bayman 2006; Chacon et al. 2007; Tellenbach et al. 2010). However, due to limited knowledge of plant-associated mycobiome composition, diversity, and functionality, these subjects need to explore up to depth. The modern molecular approaches such as high-throughput sequencing methods and omics techniques (epigenomics, genomics, transcriptomics, proteomics, and metabolomics) are utilized to identify the function of fungal communities with plant and soil as well as with other microbes in different habitats (Bálint et al. 2016; Gohl et al. 2016; Kchouk et al. 2017; Castle et al. 2018).

This chapter focuses on plant mycobiome compartments (rhizosphere, phyllosphere, and endosphere) and interactions between plants and microorganisms with major emphasis on fungal fractions of the microbiome within each compartment. The chapter discussed the significant application of plant mycobiomes to improve plant health and how artificially engineered mycobiome can help the crop to improve productivity. This study provides a summary of recent and cutting-edge techniques that can help to classify the fungal communities, their biodiversities, and plant-associated functions.

4.2 Plant-Associated Microbial Communities

The soil-inhabiting fungal communities play a vital role in plant health management. They suppress plant diseases through physiological restrictions of pathogen establishment in plant tissues (Mendes et al. 2011; Mukherjee et al. 2012; Classen et al. 2015) and help to convey resistance to the system against "invaders" (Pérez-Jaramillo et al. 2018). Several additional factors have been recognized to mycobiome in close association with plants such as ability to provide nutrient mobilization, like phosphorus solubilization and nitrogen fixation, their support in nutrient uptake from the soil, suppression of abiotic and biotic stresses, host health, and ecological fitness (Molla et al. 2012; Oros and Naár 2017). The plant-associated microbial communities usually comprise five main functional groups of fungal communities, saprotrophic, pathogenic, epiphytic, endophytic, and mycorrhizal, which performs different functions in ecological balancing (Mendes et al. 2013). The major studied compartments of plant-associated microbial communities are established microbial cells and perform distinct functions, called rhizosphere, phyllosphere, and endosphere (Herrera et al. 2010; Porras-Alfaro and Bayman 2011).

4.2.1 Rhizosphere Fungal Microbiome Communities

Lorenz Hiltner coined the term rhizosphere in 1904 (Curl and Truelove 1986), which refers to the bulk of soil surrounding the plant roots. Rhizosphere provides opportunities to organisms to competitively colonize plant roots by secreting root exudates, which are active carbon compounds, and uptake of mobile nutrients and water (Hartmann et al. 2008; Singh and Sharma 2012). The biodiversity and community structure of fungal communities are maintained by the coevolutionary process of rhizosphere and plant roots which perform a significant role in soil biological processes, soil carbon sequestration, and nutrient channelization through decomposed matter cycling in natural systems (Hinsinger et al. 2009; Lambers et al. 2009). The plant rhizosphere system generally enhanced the biomass and soil microorganism's activities by releasing root exudates, which respond with chemotaxis and grow faster (Hartmann et al. 2009). The rhizosphere microbial community composition and diversity are affected by plant growth, as root exudates change during the plant's life cycle and seasonal environment responses (Baetz and Martinoia 2014).

The rhizospheric fungal communities are closely interconnected to plant health, fitness, and growth, decaying plant residues, providing nutrients by cycling minerals, and facilitating their roles in antagonizing pathogens (Ehrmann and Ritz 2014). The beneficial fungal microbes include mycorrhizal fungi (Zhang et al. 2018; Turrini et al. 2018), *Trichoderma* strains (Kotasthane et al. 2015; Li et al. 2015), *Penicillium* sp. (Babu et al. 2015), and other endophytic fungi like *Fusarium* sp., *Colletotrichum* sp., *Cladosporium* sp., and *Dendrobium moniliforme* (Chadha et al. 2015; Shah et al. 2019) which are an indispensable component of agroecosystems. In response to microbial activities, plants fundamentally control rhizospheric fungi through the production of carbon and its derivative compounds and bioactive metabolites (Ellouze et al. 2014). Several examples are elucidating the role of fungal communities, as biocontrol, activity against pathogenic microorganisms (Kowalchuk and Veen 2004, Chapelle et al. 2016). *Trichoderma* is a hyper-diverse genus of rhizospheric fungal community that gained importance due to their antagonistic capability to plant pathogen by employing different mechanisms of parasitism, stimulation of plant growth, competition for nutrients, and induction of systemic resistance against abiotic and biotic stresses (Rai et al. 2016). These fungal communities positively influence plant productivity and protect the plants from oxidative stress by synthesizing antioxidant enzymes (peroxidase, catalase, superoxide) and nonenzymatic antioxidants (glutathione, ascorbate, and α-tocopherol). In another context, certain rhizospheric fungal communities can also negatively influence plant productivity by causing disease, for example, *Fusarium* species, *Verticillium* spp., and *Macrophomina* spp., which affects many crops (Tetali et al. 2015). Most studies have focused on rhizospheric bacterial communities. However, very less information is known about the interaction of fungal communities in the plant rhizosphere. Recently developed molecular techniques, metagenomics, transcriptomics, proteomics, and next-generation sequencing studies showed colonization of plant roots by fungal communities that establish information about the changes in expression of plant genes in response to stress, biomolecule synthesis, photorespiration, photosynthesis, and carbohydrate metabolism.

4.2.2 The Phyllosphere Fungal Microbiome Communities

The phyllosphere microbiome is associated with a diverse group of microorganisms, such as viruses, bacteria, algae, yeasts, filamentous fungi, protozoa, and nematodes (Lindow and Brandl 2003; Porras-Alfaro and Bayman 2011). Bacteria are the most dominant and abundant microorganisms in the phyllosphere community (Lindow and Brandl 2003; Vorholt 2012), while fungi are comparatively less abundant. The structure of phyllospheric communities is governed by immigration, survival, growth of the microbial colony, and leaf physicochemical properties (Gomes et al. 2018; Yao et al. 2019). The phyllosphere niche has a high significance in sustainable agriculture and an environmental process that provides confirmation for interactions of phyllospheric microbial communities that affect the health of the natural plant and the quality and productivity of agricultural crops (Boddy et al. 2008). Filamentous fungi largely occurring as dormant spores in phyllospheric fungal communities rather than active mycelia and population size range between 10^2 and 10^8 colony-forming unit/gram of leaf (Andrews and Harris 2000; de Jager et al. 2001). The most abundant fungal communities found on leaves are considered *Cladosporium*, *Alternaria*, *Penicillium*, *Acremonium*, *Mucor*, and *Aspergillus* (Inácio et al. 2002; Porras-Alfaro and Bayman 2011). Filamentous fungi appear to occur ubiquitously as endophytes, and their diversity is particularly observed in long-lived tropical leaves. Fungal endophytes are the second most abundant phyllospheric fungal communities that protect against pathogens and increase abiotic tolerance (Arnold et al. 2000, 2003; Schweitzer et al. 2006). The diversity of cultured yeasts appears as the genera *Cryptococcus*, *Sporobolomyces*, and *Rhodotorula*, either singly or with multiple species in the phyllosphere (Inácio et al. 2002; Glushakova and Chernov 2004). To explore the phyllospheric fungal species, culture-dependent approaches are applied over 340 genetically distinct taxa of two tropical forests, but still, culture-independent approaches have not yet been reported to characterize fungal diversity.

4.2.3 Endosphere Fungal Microbiome Communities

Although less than 100,000 fungal species have been described as an indispensable role of endophytes encompasses a large, hidden component of fungal biodiversity (Arnold 2007; Rodriguez et al. 2009), the dominating classes of fungal endophyte communities included *Sordariomycetes*, *Dothideomycetes*, *Eurotiomycetes*, *Leotiomycetes*, and *Pezizomycetes* (Jumpponen and Jones 2009; Atugala and Deshappriya 2015) which are associated with grasses and many woody plant tissues and roots (Alberton et al. 2010; Bhagobaty and Joshi 2012). Endophytic fungal communities are important components of plant microbiomes that live within plant tissues without causing disease symptoms. Fungal endophytes reside symbiotically inside the plant tissue or interact with other microbial groups that colonize in the plant tissues, e.g., mycorrhizal fungi, pathogens, epiphytes, and saprotrophs (Porras-Alfaro and Bayman 2011). Some fungal endophytes play an important role in plant growth, resistance to abiotic and biotic stresses, and disease in the plant by producing useful, antagonistic, and signaling molecules (Gao et al. 2010; Khan et al. 2011; Gautam et al. 2013). Endophytes have a significant effect on the plant fitness, growth, and development by modulating different pathways of plant-like phytohormone (Khan et al. 2012, 2015), antimicrobial compound (Prabukumar et al. 2015), and secondary metabolites (brefelcin A, mevinolin, 2-(3,4-dihydroxyphenyl) ethanol, cytochalasins, polyketides, terpenoids, flavonoids, and steroids) producing pathway (Guo et al. 2008; Balbi and Devoto 2008; Aramsirirujiwet et al. 2016).

Because most of the important root endophytes are not culturable, that's why molecular techniques have been important for identification (Roe et al. 2010;) and confirmed that many of these endophytes coexist with other functional groups in the microbiome. Rapid advances in DNA and RNA sequencing technologies and comparative genomics, transcriptomics, and proteomics approach now facilitate us to study fungal communities in an integrative way, including exploring the taxonomic and phylogenetic profiles of fungal communities in the rhizosphere. In this sequence, their functional and ecological attributes put forward an open discussion about the role of fungal endophytes in the plant microbiome composition, structure, and their diversity, important for plant growth and survival and interactions with another plant-associated microbiome.

4.3 Manipulation of the Plant Microbiome Toward Improved Plant Health

The rhizosphere, phyllosphere, and endosphere microorganisms both directly and indirectly influence the composition, structure, diversity, and productivity of natural plant communities (Mendes et al. 2018). Microbial species richness provides information about the plant diversity and productivity (Sharma et al. 2015). On purpose manipulation of the plant microbiome may be an alternative way to advance the agriculture sustainability (Bai et al. 2015). This would be done by manipulating rhizosphere, phyllosphere, and endosphere microorganisms with useful traits.

The recent advancement in rhizosphere engineering, manipulation of the rhizospheric microbiome, can aid in the management of soil health and ecological stability (Broberg et al. 2018). The diversity of microorganisms in the soil is increased through crop rotation that promotes high flexibility to plant pathogens (Hwang et al. 2009). The application of certain microorganisms or the introduction of inoculants is one of the emerging strategies for plant microbiome manipulation, and through this approach, a beneficial community effectively replaces the plant pathogenic microbes. The co-inoculation of several beneficial strains, including endophytes, reduces the time of niche (Compant et al. 2010).

Manipulation of the phyllosphere microbiome can provide a new strategy for enhancing plant growth and health. Falk et al. (1995) suggested that the severity of powdery mildew infections caused by *Uncinula necator* on grapevines was reduced by releasing the conidia of the mycoparasite fungus *Ampelomyces quisqualis*. The literature describes several success stories of the use of *Trichoderma* spp. in mycobiome engineering for biotic stress management (Błaszczyk et al. 2014) and to combat with abiotic stresses by multiple beneficial effects on plant growth and stress tolerance (Singh and Sharma 2012; Chepsergon et al. 2014). Perazzolli et al. (2014) showed that the naturally occurring microbiomes of grapevine leaves could reduce signs of powdery mildew on controlled conditions. A number of reports indicated that growth-promoting and stress-managing fungal communities are applied in rhizosphere, as in the form of liquid and solid formulations to produce active and viable conidia of fungal strains, which are comparatively more tolerant to adverse conditions and provide crop growth in terms of root expansion and nutrient uptake (Waghunde et al. 2016).

4.4 Modern Molecular Tools to Study Plant Mycobiome

The traditional techniques used to identify plant mycobiome relied upon cultural and morphological based approaches. These methods are often time-consuming, laborious, and species-dependent, require extensive knowledge of classical taxonomy, and have the inability to accurately quantify the pathogen (Goud and Termorshuizen 2003). For this reason, the availability of fast, sensitive, and accurate methods is required for detection and identification purpose. These limitations have led to the development of different molecular approaches with improved accuracy and reliability. A variety of molecular methods have been used to detect, identify, and quantify plant pathogenic fungi. Here, we discuss the modern molecular tools and technique to study plant mycobiome, their applicability, and their implementation (Table 4.1).

Still, most of the studies on plant mycobiome are based on a culturable technique that involved first isolation of organisms into pure culture and then identification of them, but majority of uncultivable organism is overlooked due to their not or slow-growing properties in cultural media. To overcome this constraint, several techniques were developed, viz., direct amplification of DNA from the surface-sterilized plant material and identification of fungal species, multilocus barcode approaches, and next-generation sequencing studies that investigated fungal communities and overall fungal diversity in plant mycobiome. To fingerprint, the most dominant fungal species at different taxonomic levels usually applied denaturing gradient gel electrophoresis (DGGE) or single-strand conformation polymorphism (SSCP). Quantifying fungal biomass in plant tissues can be achieved by real-time PCR techniques that greatly depend on the type of fungus as well as cell size, number, and type and require extensive calibration for each species using pure cultures (Tellenbach et al. 2010). Next-generation sequencing technologies and systems biology allow simultaneous extension and examination of all members of microbial

Table 4.1 Different molecular tools used in plant mycobiome study

Method	Plant mycobiome organisms	Purpose of the study	References
Conventional PCR	*Sclerotium rolfsii*, *Colletotrichum capsici*	Diversity analysis by using ITS region	Torres-Calzada et al. (2011) and Jeeva et al. (2010)
Cooperational-PCR (Co-PCR)	Grapevine fungi	Sensitive and specific detection of microorganism	Martos et al. (2011)
PCR-DGGE	*Phytophthora* species	Detection of multiple species	Rytkönen et al. (2011)
Real-time PCR	*Colletotrichum acutatum*, *Fusarium oxysporum* f. sp. *ciceris*, *Discula destructiva*, *F. poae*, *Pythium vexans*	Quantification of a minimal amount of pathogenic microorganisms	Samuelian et al. (2011), Jiménez-Fernández et al. (2011), Zhang, N. et al. (2011), Kulik et al. (2011), and Tewoldemedhin et al. (2011)
Restriction fragment length polymorphism (RFLP)	*Pythium myriotylum*	Differentiation of pathogenic and nonpathogenic strains	Gómez-Alpizar et al. (2011)
Nested-PCR	Numerous fungi	Used for detection and/or characterization	Hong et al. (2010); Aroca and Raposo (2007), and Grote et al. (2002)
Multiplex PCR	*Podosphaera xanthii*, *Golovinomyces cichoracearum*, and *Phytophthora lateralis*	Detection and differentiation	Guglielmo et al. (2007) and Chen et al. (2008)
Reverse transcription (RT)-PCR	*Mycosphaerella graminicola*, *Oidiumneo lycopersici*	Detection of viable populations	Guo et al. (2005) and Matsuda et al. (2005)
In situ PCR	*Blumeria graminis*	Identification and diversity analysis	Bindslev et al. (2002)
PCR-ELISA	*Didymella bryoniae*, *Phoma* species, *Phytophthora*, and *Pythium*	Detection and differentiation at species level	Somai et al. (2002) and Bailey et al. (2002)
Magnetic capture hybridization (MCH) PCR	*Nectria galligena*	Detection at species level	Langrell and Barbara (2001)
Isothermal amplification methods	*Fusarium graminearum*, *Phytophthora ramorum*, and *P. kernoviae*	Rapid detection and diversity analysis	Abd-Elsalam et al. (2011) and Tomlinson et al. (2007, 2010)

(continued)

Table 4.1 (continued)

Method	Plant mycobiome organisms	Purpose of the study	References
Random amplified polymorphic DNA (RAPD)	*Fusarium* spp. and *Elsinoë* spp.	Genetic diversity analysis	Arici and Koc (2010), Lievens et al. (2007); Hyun et al. (2009), and Hiremani and Dubey (2019)
Amplified fragment length polymorphism(AFLP)	*Cladosporium fulvum*, *Pyrenopeziza brassicae*, *Aspergillus carbonarius*, *A. ochraceus*, *Colletotrichum gossypii*, *C. Gossypii* var. *Cephalosporioides*	Differentiate fungal isolates at several taxonomic levels	Schmidt et al. (2004) and Silvar et al. (2005)
Microsatellites	*Ascochyta rabiei*, *Ceratocystis fimbriata*, *Macrophomina phaseolina*, *Puccinia graminis*, *P. triticina*, *Sclerotinia subarctica*, *S. sclerotiorum*, and *Magnaporthe grisea*	Genetic diversity of plant pathogenic fungi within species and genetic map construction	Jana et al. (2005), Bayraktar et al. (2007), Szabo (2007), Szabo and Kolmer (2007), Winton et al. (2007), and Zheng et al. (2008)
DNA arrays	*Fusarium* species and *Penicillium* species	Differentiation of toxin-producing and nonproducing species. *Cox* I high-density oligonucleotide microarray identification	Nicolaisen et al. (2005) and Chen et al. (2009)
Micro RNA (miRNAs)	*Fusarium* head blight (FHB) on wheat	Genomic study of wheat	Kharbikar et al. (2019)
Induced mutagenesis	*Phytophthora nicotianae*	Identification of *Phytophthora* blight-resistant mutants	Kumari et al. (2019)
DNA barcoding.	*Fusarium proliferatum*	Identification at the species level	Nayyar et al. (2018)
Transcriptome analysis	*Phytophthora capsici*, *P. infestans*	To assess the change in mRNA levels for selected genes and pathogenic variability	Kandel (2014) and Muthuswamy et al. (2018)

communities and their interactions by using "omics approaches" (metagenomics, transcriptomics, proteomics, and metabolomics).

The discovery of emerging molecular technology helps to gather the information about the uncultivable fungal species, but nutritional modes and ecological associations of that fungal taxa are still challenging for the researchers (Tedersoo and Nilsson 2016). As DNA sequencing technologies progressed from sequencing single specimens to parallel Sanger sequencing in the early 2000s, it brought inspiration to researchers to reveal the unseen mycobiota and their diversity, structure, functioning, and significance. Sanger and Maxam-Gilbert sequencing technologies were considered as first-generation (the 1990s), most common sequencing technologies used by researchers because of its high efficiency and low radioactivity until the emergence of a new era of sequencing technologies opening new perspectives for genome exploration and analysis (Pareek et al. 2011). The second-generation sequencing methods were developed in the 2000s and marked the beginning of high-throughput sequencing (HTS) which were characterized by the need to prepare amplified sequencing banks before starting the sequencing of amplified DNA clones to analyzed the hidden fungal communities (Vezzi 2012; Qiang-long et al. 2014). The second-generation sequencing included pyrosequencing, Illumina sequencing, Ion Torrent semiconductor sequencing, and SOLiD (Supported Oligonucleotide Ligation and Detection) sequencing (Kchouk et al. 2017). NGS technologies continue to improve, and the number of sequencers increases these last years. However, the third-generation sequencing performs at the level of single molecules and produce higher read lengths than the earlier generations that overcome the problem of the necessity to create the amplification libraries. Also, third-generation sequencing can produce long reads of several kilobases for the resolution of the assembly problem and repetitive regions of complex genomes which is useful in metabarcoding and community analysis of mycobiome (Song et al. 2015; Rhoads and Au 2015; Nilsson et al. 2018). This chapter briefly discusses the application of different HTS, which elucidate the study of fungal community associations related to taxonomic profiling of fungal communities and their mycobiome structure, function, signaling mechanism, and ecosystem functioning.

4.5 Major Applications of Plant Mycobiome

Recently, plant-associated fungal communities have been taken a greater interest as bioinoculant to enhance plant growth in different kinds of biotic and abiotic stresses. Thus the rhizospheric engineering involves the development of new strategies to reshape the rhizospheric microbiome for biofertilization and induces root growth stimulation, antibiosis, induced plant systemic resistance, and parasitism.

These microorganisms follow several mechanisms to support plant growth which include the production of phytohormones and the mobilization of organic matter and minerals, like carbon, nitrogen, phosphorus, and iron (Tkacz and Poole 2015). The mechanisms of suppression of pathogens demonstrate several direct interactions with plant pathogens, as well as indirect interaction, which include stimulation of the plant immune system or systemic resistance (Lugtenberg and Kamilova 2009).

4.5.1 Plant Growth Promotion

Plant-associated fungal community structure and metabolism are altered during development and environmental changes, as the nutrients are provided in the microbiome by the plants. So in these consequences, secretion of the nutrients exuded by different plants, the specific fungal pathways responding to them, and the mechanisms by which plant-associated fungal community activates or repress the root colonization process in the response of exudate (Contreras-Cornejo et al. 2016; Zeilinger et al. 2016). Fungal community interactions in plant microbiota likely play determinative roles in colonization of the host plant, plant growth promotion, nutrient uptake, microbiome succession over plant development, and the ability of mutualists to suppress pathogen growth (López-Bucio et al. 2015; Guler et al. 2016; You et al. 2016). Several researchers reported *Trichoderma* colonized plants to secrete different bioactive compounds like auxins, ethylene, gibberellins, plant enzymes, antioxidants, etc. for signaling molecule between mycobiome and compounds like phytoalexins and phenols that provide tolerance to abiotic stresses and enhance the branching capacity of the root system (Brotman et al. 2013; López-Bucio et al. 2015).

Application of plant growth-promoting fungi *Trichoderma longibrachiatum* T6 enhances the tolerance of wheat to salt stress through the improvement of the antioxidative defense system and gene expression (Zhang et al. 2016). The plant microbiota also influences the composition of plant secondary metabolites and the resulting development of different metabolites. Studies on the influence of the microbiome on the taste of strawberries (Zabetakis et al. 1999; Verginer et al. 2010) and the production of metabolites in medicinal plants (Köberl et al. 2013; Schmidt et al. 2014) have been reported. In a study on *Arabidopsis thaliana*, the rhizosphere microbiome could be linked to insect feeding behavior, which was most probably a result of microbiome-driven changes in the leaf metabolome (Badri et al. 2013). Peñuelas et al. (2014) showed that the removal of the floral microbiome of *Sambucus nigra* resulted in a reduced floral terpene emission, which plays an important role in pollination and consequently in fruit and seed production.

4.5.2 Disease Management

The plant microbiome is effectively involved in pathogen suppression, particularly the plant mycobiome that acts as a protective armor against phytopathogens (Weller et al. 2002). The various mechanisms are involved in direct interactions with phytopathogens as well as indirect interactions via mechanisms of parasitism, competition for nutrients, and induction of systemic resistance against abiotic and biotic stresses (Lugtenberg and Kamilova 2009). The recent research has shown that the plant mycobiome is involved not only in managing with biotic stress but also in protection against abiotic stress (Bragina et al. 2013). Several reports indicated that *Trichoderma* spp. have beneficial effects on plant growth and biotic stress management in different horticulture crops like radish, cucumber, pepper, bottle gourd, periwinkle, bitter gourd, chrysanthemum, lettuce, and tomato (Donoso et al. 2008; Bae et al. 2009; Brotman et al. 2013; Contreras-Cornejo et al. 2016; Zeilinger et al. 2016). Kotasthane et al. (2015) reported the antagonistic potential of *Trichoderma* spp. against *Sclerotium rolfsii* and *Rhizoctonia solani* and their plant growth promotion response toward the growth of cucumber, bottle gourd, and bitter gourd. The ongoing research is an effort to elucidate the molecular basis of plant mycobiome to gather broad perspectives regarding their functioning and applicability for growth promotion and defense activation of the plant in disease management (Table 4.2).

4.6 Conclusion and Future Prospects

This review was undertaken to explore the current information on plant mycobiome management and focused on developing technologies to overcome the constraints related to mycobiome engineering. Recent advancement in molecular technologies and ongoing research have revealed the greater diversity and complexity of plant mycobiome rather than previously imagined, while an amalgamation of biochemical, molecular, and genetic approaches has led to new visions into the exact mechanism of signal induction and transduction processes of secondary metabolites from fungal communities in plants and other organisms. The ongoing research, technological advancement in molecular biology, and high-throughput omics included genomics, transcriptomics, proteomics, and metabolomics that provide information to elucidate the role of the mycobiome gene that aids in improving plant performance. Therefore, the better understanding of metagenomics, genetic sequence information of microbiome, obtained from next-generation sequencing data has substantial potential for the discovery of diversity and functional perspective of the sequenced genes of the microbiome-related organisms. Looking forward toward the potential of NGS data to demonstrate the functional diversity of mycobiome will require an elucidation of basic biology of fungi through traditional culturing approaches and sequencing of individual fungal genomes to demonstrate the evolutionary studies in natural habitat. Here, this chapter reviews omics-based approaches that are driving forward the current understanding of plant-associated fungal gene functions and describes how these technologies have helped unravel key bacterial genes and pathways that mediate pathogenic, beneficial, and commensal host interactions.

Table 4.2 Beneficial and pathogenic mycobiome association with different crops

Antagonist	Pathogen	Counter mechanism	Host plant	References
Trichoderma hamatum and *T. harzianum*	*Fusarium oxysporum* f. sp. *cepae*	Induction of antifungal compounds	Onion	Adèle et al. (2019)
Glomus sp. and *Trichoderma* sp.	*Cryphonectria parasitica*	Induction of secondary metabolites	Chestnut	Murolo et al. (2019)
T. viride	*Macrophomina phaseolina*	Antagonist mechanism by mycelial destruction	Mung bean	Shahid and Khan (2019)
Trichoderma spp., *T. brevicompactum*, *T. gamsii*, and *T. harzianum*	*F. oxysporum*, *Pythium* spp., and *Rhizoctonia solani* (Kuhn.)	Production of the enzymatic machinery	Tomato	Biam et al. (2019) and Bader et al. (2019)
Piriformospora indica	*Fusarium* spp.	Induction of antimicrobial product, enzymatic mechanisms, and plant growth promotion	Sugarcane, potato, peas, maize, soybean	Aslam et al. (2019)
Trichoderma harzianum strain *Th22*	*F. graminearum*	Activation of defense-related genes	Maize	Saravanakumar et al. (2018)
T. harzianum,	*M. phaseolina, Fusarium oxysporum* f. sp. *cumini*	Induction of secondary metabolites	Guar, moth bean, mung bean, and sesame	Mawar et al. (2019)
T. viride and *Ampelomyces quisqualis*	*Oidium euonymi-japonici*	Antagonistic mechanism by mycelial destruction and mycolytic enzyme production	*Euonymus japonicus*	Ahanger et al. (2018)
T. viride isolate NRCL T-01	*F. solani*	Mycolytic enzyme production	Litchi	Kumar et al. (2018)
T. harzianum	*M. phaseolina* (Tassi) Goid	Induction of secondary metabolites	Mungbean	Thombre and Kohire (2018)

(continued)

Table 4.2 (continued)

Antagonist	Pathogen	Counter mechanism	Host plant	References
Massarina igniaria, *Periconia macrospinosa*, *Noosia bankssiae*, *Flavomyces fulophazii*	–	Plant growth promotion and phytohormone production	*Oryza sativa*	Vergara et al. (2018)
Acremonium, *Arthrinium*, *Botryotinia*, *Chaetomium*, *Dictyosporium*, *Humicola*, *Ilyonectria*, *Mucor*, *Myrothecium*, *Penicillium*, *Periconia*, *Thielavia*	*Alternaria panax*, *Fusarium oxysporum*, *Fusarium solani*, *Phoma herbarum*, and *Mycocentrospora acerina*	Induction of antimicrobial product, enzymatic mechanisms, and plant growth promotion	*Panax notoginseng*	Zheng et al. (2017) and Nandhini et al. (2018)
T. harzianum	*F. graminearum*	Antagonistic activity by mycelial destruction and nutrient competition	Maize	Saravanakumar et al. (2018)
Penicillium chrysogenum and *P. crustosum*	*Staphylococcus aureus*, *Salmonella typhimurium*, and *Candida albicans*	Induction of antimicrobial product, enzymatic mechanisms, and plant growth promotion	*Teucrium polium*	Hassan (2017)
T. harzianum	*F. oxysporum* f. sp. *gladioli* and *Meloidogyne incognita*	Interaction and mycolytic enzyme production	Gladiolus	Khan et al. (2017)
Curvularia geniculata	–	Plant growth promotion and phytohormone production	*Parthenium hysterophorus*	Priyadharsini and Muthukumar (2017)
T. asperellum ZJSX5003	*F. graminearum*	Antagonistic and biocontrol potential	Maize	Li et al. (2016)
T. harzianum, *T. viride*, *A. flavus*, *A. falcatum*, and *A. niger*	*Colletotrichum falcatum*	Mycoparasitism and mycolytic enzyme production	Sugarcane	Suresh and Nelson (2016)

(continued)

Table 4.2 (continued)

Antagonist	Pathogen	Counter mechanism	Host plant	References
T. viride	*Pestalotia theae* and *F. solani*.	Hyphal swelling and distortion	Tea	Naglot et al. (2015)
A. niger and *T. harzianum*	*Fusarium* spp., *R. solani* and *M. phaseolina*	Mycoparasitism and mycolytic enzyme production	Cowpea and mungbean	Ikram and Dawar (2014)
Trichoderma isolates	*F. fujikuroi*	Antibiosis and biocontrol potential	Rice	Bhramaramba and Nagamani (2013)
Trichoderma isolates (TvO, TvG, ThC, ThR, and ThM)	*R. solani*, *Sclerotium rolfsii*, *M. phaseolina*, *F. oxysporum*, *A. niger*	Induction of secondary metabolites	Okra, maize cauliflower, and groundnut	Pan et al. (2013) and Gajera et al. (2011)
Trichoderma viride	*Macrophomina* spp.	Antagonistic and hyphal swelling and distortion	Jute	Srivastava et al. (2010)
Trichoderma spp.	*Lasiodiplodia theobromae*	Induction of secondary metabolites	*Elaeocarpus munronii*	Priya and Nagaveni (2009)
Glomus mosseae, *T. harzianum*, and *P. simplicissimum*	*Rhizoctonia solani*	Induction of antimicrobial product, enzymatic mechanisms, and plant growth promotion	*Cucumis sativus*	Chandaniea et al. (2009)

References

Abd-Elsalam K, Bahkali A, Moslem M, Amin OE, Niessen L (2011) An optimized protocol for DNA extraction from wheat seeds and Loop-Mediated Isothermal Amplification (LAMP) to detect *Fusarium graminearum* contamination of wheat grain. Int J Mol Sci 12(6):3459–3472. ISSN 1422-0067

Adèle L, Bunbury-Blanchette A, Walker K (2019) *Trichoderma* species show biocontrol potential in dual culture and greenhouse bioassays against *Fusarium* basal rot of onion. Biol Control 130:127–135

Ahanger RA, Qazi NA, Bhat HA et al (2018) Indian Phytopathol 71:377

Alberton O, Kuyper TW, Summerbell RC (2010) Dark septate root endophytic fungi increase growth of Scots pine seedlings under elevated CO_2 through enhanced nitrogen use efficiency. Plant Soil 328:459–470

Andrews JH, Harris RF (2000) The ecology and biogeography of microorganisms on plant surfaces. Annu Rev Phytopathol 38:145–180

Anil K, Lakshmi T (2010) Phosphate solubilisation potential and phosphate activity of rhizospheric *Trichoderma* spp. Braz J Microbiol 41:787–795

Aramsirirujiwet Y, Gumlangmak C, Kitpreechavanich V (2016) Studies on antagonistic effect against plant pathogenic fungi from endophytic fungi isolated from *houttuynia cordata* thunb. and screening for siderophore and Indole-3-acetic acid production. KKU Res J 21(1):55–66

Arici SE, Koc NK (2010) RAPD–PCR analysis of genetic variation among isolates of *Fusarium graminearum* and *Fusarium culmorum* from wheat in Adana Turkey. Pak J Biol Sci 13(3):138–142. ISSN 1028-8880

Arnold AE (2007) Understanding the diversity of foliar endophytic fungi: progress, challenges and frontiers. Fungal Biol Rev 21:51–56

Arnold EA, Mejia LC, Kyllo D, Rojas E, Maynard Z, Robbins N, Herre EA (2003) Fungal endophytes limit pathogen damage in a tropical tree. Proc Natl Acad Sci U S A 100:15649–15654

Arnold AE, Maynard Z, Gilbert GS, Coley PD, Kursar TA (2000) Are tropical fungal endophytes hyperdiverse? Ecol Lett 3:267–274

Aroca A, Raposo R (2007) PCR-based strategy to detect and identify species of *Phaeoacremonium* causing grapevine diseases. Appl Environ Microbiol 73(9):2911–2918

Aslam MM, Karanja J, Bello SK (2019) *Piriformospora indica* colonization reprograms plants to improved P-uptake, enhanced crop performance, and biotic/abiotic stress tolerance. Physiol Mol Plant Pathol 106:232–237

Atugala DM, Deshappriya N (2015) Effect of endophytic fungi on plant growth and blast disease incidence of two traditional rice varieties. J Natl Sci Found Sri Lanka 43(2):173–187

Babu AG, Kim SW, Yadav DR, Hyum U, Adhikari M, Lee YS (2015) *Penicillium menonorum*: a novel fungus to promote growth and nutrient management in cucumber plants. Microbiology 43(1):49–56

Bader AN, Salerno GL, Covacevich F, Consolo VF (2019) Native *Trichoderma harzianum* strains from Argentina produce indole-3 acetic acid and phosphorus solubilization, promote growth and control wilt disease on tomato (*Solanum lycopersicum* L.). J King Saud Univ Sci. https://doi.org/10.1016/j.jksus.2019.04.002

Badri DV, Zolla G, Bakker MG et al (2013) Potential impact of soil microbiomes on the leaf metabolome and on herbivore feeding behavior. New Phytol 198:264–273

Baetz U, Martinoia E (2014) Root exudates: the hidden part of plant defense. Trends Plant Sci 19:90–98

Bae H, Sicher RC, Kim MS, Kim SH et al (2009) The beneficial endophyte *Trichoderma hamatum* isolate DIS 219b promotes growth and delays the onset of the drought response in Theobroma cacao. J Exp Bot 60:3279–3295

Bahram M, Peay KG, Tedersoo L (2015) Local-scale biogeography and spatiotemporal variability in communities of mycorrhizal fungi. New Phytol 205:1454–1463

Bai Y, Muller DB, Srinivas G, Garrido-Oter R, Potthoff E, Rott M, Dombrowski N, Munch PC, Spaepen S, Remus-Emsermann M et al (2015) Functional overlap of the *Arabidopsis* leaf and root microbiota. Nature 528:364–369

Bailey AM, Mitchell DJ, Manjunath KL, Nolasco G, Niblett CL (2002) Identification to the species level of the plant pathogens *Phytophthora* and *Pythium* by using unique sequences of the ITS1 region of ribosomal DNA as capture probes for PCR Elisa. FEMS Microbiol Lett 207(2):153–158. ISSN 0378-1097

Balbi V, Devoto A (2008) Jasmonate signalling network in *Arabidopsis thaliana*: crucial regulatory nodes and new physiological scenarios. New Phytol 177(2):301–318

Bálint M et al (2016) Millions of reads, thousands of taxa: microbial community structure and associations analyzed via marker genes. FEMS Microbiol Rev 40:686–700

Bayman P (2006) Diversity, scale and variation of endophytic fungi in leaves of tropical plants. In: Bailey MJ, Lilley AK, Timms-Wilson TM (eds) Microbial ecology of aerial plant surfaces. CABI, Oxfordshire, pp 37–50

Bayraktar H, Dolar FS, Tor M (2007) Determination of genetic diversity within *Ascochyta rabiei* (pass.) labr., the cause of ascochyta blight of chickpea in Turkey. J Plant Pathol 89(3):341–347. ISSN 1125-4653

Beckers B, Op De Beeck M, Weyens N, Van Acker R, Van Montagu M, Boerjan W et al (2016) Lignin engineering in field-grown poplar trees affects the endosphere bacterial microbiome. Proc Natl Acad Sci 113:2312–2317

Benitez T, Rincon AM, Limon MC, Codon AC (2004) Biocontrol mechanisms of *Trichoderma* strains. Int Microbiol 7:249–260

Bhagobaty RK, Joshi SR (2012) Enzymatic activity of fungi endophytic on five medicinal plant species of the Pristine sacred forests of Meghalaya, India. Biotechnol Bioprocess Eng 17:33–40

Bhramaramba S, Nagamani A (2013) Antagonistic *Trichoderma* isolates to control bakanae pathogen of rice. Agric Sci Digest 33:104–108

Biam M, Majumder D, Papang H (2019) *In vitro* efficacy of native *Trichoderma* isolates against *Pythium* spp. and *Rhizoctonia solani* (Kuhn.) causing damping off disease in tomato (*Solanum lycopersicum* Miller). Int J Curr Microbiol App Sci 8(02):566–579

Bindslev L, Oliver RP, Johansen B (2002) In situ PCR for detection and identification of fungal species. Mycol Res 106(3):277–279. ISSN 0953-7562

Błaszczyk L, Siwulski M, Sobieralski K, Lisiecka J, Jędryczka M (2014) *Trichoderma* spp.—application and prospects for use in organic farming and industry. J Plant Prot Res 54:309–317

Boddy L, Frankland J, Van West P (2008) Ecology of saprotrophic basidiomycetes (British Mycological Society symposia series), vol 28. Academic, London

Bragina A, Cardinale M, Berg C, Berg G (2013) Vertical transmission explains the specific *Burkholderia* pattern in Sphagnum mosses at multi-geographic scale. Front Microbiol 4:394

Broberg M, Doonan J, Mundt F, Denman S, McDonald JE (2018) Integrated multi-omic analysis of host-microbiota interactions in acute oak decline. Microbiome 6:21

Brotman Y, Landau U, Cuadros-Inostroza A, Takayuki T, Fernie AR et al (2013) *Trichoderma*-plant root colonization: Escaping early plant defense responses and activation of the antioxidant machinery for saline stress tolerance. PLoS Pathog 9(3):e1003221. https://doi.org/10.1371/journal.ppat.1003221

Bucher M, Wegmuller S, Drissner D (2009) Chasing the structures of small molecules in arbuscular mycorrhizal signaling. Curr Opin Plant Biol 12:500–507

Cai F, Yu G, Wang P, Wei Z, Fu L, Shen Q, Chen W (2013) Harzianolide, a novel plant growth regulator and systemic resistance elicitor from *Trichoderma harzianum*. Plant Physiol Biochem 73:106–113

Cai F, Chen W, Wei Z, Pang G, Li R, Ran W, Shen Q (2015) Colonization of *Trichoderma harzianum* strain SQR-T037 on tomato roots and its relationship to growth, nutrient availability and soil microflora. Plant Soil 388:337–350

Castle SC et al (2018) DNA template dilution impacts amplicon sequencing-based estimates of soil fungal diversity. Phytobiomes J 2:100–107

Chacon MR, Rodriguez-Galan O, Beritez T, Sousa S, Rey M, Llobell A, Delgado-Jarana J (2007) Microscopic and transcriptome analyses of early colonization of tomato roots by *Trichoderma harzianum*. Int Microbiol 10:19–27

Chadha N, Mishra M, Rajpal K, Bajaj R, Choudhary DK, Varma A (2015) An ecological role of fungal endophytes to ameliorate plants under biotic stress. Arch Microbiol 197:869–881. https://doi.org/10.1007/s00203-015-1130-3

Chandaniea WA, Kubota M, Hyakumachi M (2009) Interactions between the arbuscular mycorrhizal fungus *Glomus mosseae* and plant growth-promoting fungi and their significance for enhancing plant growth and suppressing damping-off of cucumber (*Cucumis sativus* L.). Appl Soil Ecol 41:336–341

Chapelle E, Mendes R, Bakker PA, Raaijmakers JM (2016) Fungal invasion of the rhizosphere microbiome. ISME J 10(1):265–268

Chepsergon J, Mwamburi L, Kassim MK (2014) Mechanism of drought tolerance in plants using *Trichoderma* spp. IJSR 3:1592–1595

Chen RS, Chu C, Cheng CW, Chen WY, Tsay JG (2008) Differentiation of two powdery mildews of sunflower (*Helianthus annuus*) by a PCR-mediated method based on ITS sequences. Eur J Plant Pathol 121(1):1–8. ISSN 0929-1873

Chen W, Seifert KA, Lévesque CA (2009) A high density COX1 barcode oligonucleotide array for identification and detection of species of *Penicillium* subgenus *Penicillium*. Mol Ecol Resour 9:114–129. ISSN 1755-098X

Classen AT, Sundqvist MK, Henning JA, Newman GS, Moore JAM, Cregger MA et al (2015) Direct and indirect effects of climate change on soil microbial and soil microbial-plant interactions: what lies ahead? Ecosphere 6(8):130

Contreras-Cornejo HA, Macías-Rodríguez L, del-Val E, Larsen J (2016) Ecological functions of *Trichoderma* spp. and their secondary metabolites in the rhizosphere: interactions with plants. FEMS Microbiol Ecol 92:fiw036
Compant S, Clément C, Sessitsch A (2010) Plant growth-promoting bacteria in the rhizo- and endosphere of plants: their role, colonization, mechanisms involved and prospects for utilization. Soil Biol Biochem 42(5):669–678
Curl EA, Truelove B (1986) The rhizosphere, Advanced series in agriculture sciences, vol 15. Springer-Verlag, Berlin
De Jager ES, Wehner FC, Korsten L (2001) Microbial ecology of the mango phyllo- plane. Microb Ecol 42:201–207
Delgado-Baquerizo M, Maestre FT, Gallardo A, Bowker MA et al (2013) Decoupling of soil nutrient cycles as a function of aridity in global drylands. Nature 502:672–676. https://doi.org/10.1038/nature12670
Donoso EP, Bustamante RO, Carú M, Niemeyer HM (2008) Water deficit as a driver of the mutualistic relationship between the fungus *Trichoderma harzianum* and two wheat genotypes. Appl Environ Microbiol 74:1412–1417
Druzhinina IS, Seidl-Seiboth V, Herrera-Estrella A, Horwitz BA, Kenerley CM, Monte E, Mukherjee PK, Zeilinger S, Grigoriev IV, Kubicek CP (2011) *Trichoderma*: the genomics of opportunistic success. Nat Rev Microbiol 9:749–759
Ehrmann J, Ritz K (2014) Plant: soil interactions in temperate multi-cropping production systems. Plant Soil 376:1–29
Ellouze W, Esmaeili Taheri A, Bainard LD, Yang C, Bazghaleh N, Navarro-Borrell A et al (2014) Soil fungal resources in annual cropping systems and their potential for management. Bio Med Res Int 15:436
Falk SP, Gadoury DM, Pearson RC, Seem RC (1995) Partial control of grape powdery mildew by the mycoparasite Ampelomyces quisqualis. Plant Dis 79(5):483–490
Finzi AC, Austin AT, Cleland EE et al (2011) Coupled biochemical cycles: responses and feedbacks of coupled biogeochemical cycles to climate change: examples from terrestrial ecosystems. Front Ecol Environ 9:61–67
Gajera H, Rakholiya K, Vakharia D (2011) Bioefficacy of *Trichoderma* isolates against *Aspergillus niger* Van Tieghem inciting collar rot in groundnut (*Arachis hypogaea* L.). J Plant Prot Res 51:240–247
Gao FK, Dai CC, Liu XZ (2010) Mechanisms of fungal endophytes in plant protection against pathogens. Afr J Microbiol Res 4:1346–1351
Gautam AK, Kant M, Thakur Y (2013) Isolation of endophytic fungi from *Cannabis sativa*and study their antifungal potential. Arch Phytopathol Plant Protect 46:627–635
Gohl DM et al (2016) Systematic improvement of amplicon marker gene methods for increased accuracy in microbiome studies. Nat Biotechnol 34:942–949
Gómez-Alpizar L, Saalau E, Picado I, Tambong JT, Saborio F (2011) A PCR-RFLP assay for identification and detection of *Pythium myriotylum*, causal agent of the cocoyam root rot disease. Lett Appl Microbiol 52(3):185–192. ISSN 0266-8254
Goud JC, Termorshuizen AJ (2003) Quality of methods to quantify microsclerotia of *Verticillium dahliae* in soil. Eur J Plant Pathol 109(6):523–534. ISSN 0929-1873
Grote D, Olmos A, Kofoet JJ, Tuset E, Bertolini E, Cambra M (2002) Specific and Sensitive detection of *Phytophthora nicotianae by* simple and nested-PCR. Eur J Plant Pathol 108(3):197–207
Guglielmo F, Bergemann SE, Gonthier P, Nicolotti G, Garbelotto M (2007) A multiplex PCR-based method for the detection and early identification of wood rotting fungi in standing trees. J Appl Microbiol 103:1490–1507. ISSN 1364-5072
Glushakova AM, Chernov IY (2004) Seasonal dynamics in a yeast population on leaves of the common wood sorrel *Oxalisacetosella* L. Microbiology 73:184–188
Gomes T, Pereira JA, Benhadi J, Lino-Neto T, Baptista P (2018) Endophytic and epiphytic phyllosphere fungal communities are shaped by different environmental factors in a mediterranean ecosystem. Microb Ecol 76(3):668–679

Guler NS, Pehlivan N, Karaoglu SA, Guzel S, Bozdeveci A (2016) *Trichoderma atroviride* ID20G inoculation ameliorates drought stress-induced damages by improving antioxidant defence in maize seedlings. Acta Physiol Plant 38:132

Guo JR, Schnieder F, Beyer M, Verreet JA (2005) Rapid detection of *Mycosphaerella graminicola* in wheat using reverse transcription-PCR assay. J Phytopathol 153(11–12):674–679. ISSN 0931-1785

Guo B, Wang Y, Sun X, Tang K (2008) Bioactive natural products from endophytes: a review. Appl Biochem Microbiol 44:136–142

Hardoim PR, van Overbeek LS, Berg G, Pirttila AM, Compant S, Campisano A et al (2015) The hidden world within plants: ecological and evolutionary considerations for defining functioning of microbial endophytes. Microbiol Mol Biol Rev 79:293–320

Hartmann A, Rothballer M, Schmid M (2008) Lorenz Hiltner, a pioneer in rhizosphere microbial ecology and soil bacteriology research. Plant Soil 312:7–14

Hartmann A, Schmid M, Van Tuinen D, Berg G (2009) Plant-driven selection of microbes. Plant Soil 321:235–257

Hassan SED (2017) Plant growth-promoting activities for bacterial and fungal endophytes isolated from medicinal plant of *Teucrium polium* L. J Adv Res 8:687–695

Herrera J, Khidir HH, Eudy DM, Porras-Alfaro A, Natvig DO, Sinsabaugh RL (2010) Variation in root-associated fungal endophytes: some taxonomic consistency at a transcontinental scale. Mycologia 102:1012–1026

Hinsinger P, Bengough GA, Vetterlein D, Young IM (2009) Rhizosphere: biophysics, biogeochemistry and ecological relevance. Plant Soil 321:117–152. https://doi.org/10.1007/s11104-008-9885-9

Hiremani NS, Dubey SC (2019) Genetic diversity of Fusarium oxysporum f. sp. lentis populations causing wilt of lentil in India. Indian Phytopathol. https://doi.org/10.1007/s42360-019-00126-9

Hong SY, Kang MR, Cho EJ, Kim HK, Yun SH (2010) Specific PCR detection of four quarantine *Fusarium* species in Korea. Plant Pathol J 26(4):409–416. ISSN 1598-2254

Hu W, Strom N, Haarith D, Chen S, Bushley KE (2018) Mycobiome of cysts of the soybean cyst nematode under long term crop rotation. Front Microbiol 9:386

Hwang SF, Ahmed HU, Gossen BD, Kutcher HR, Brandt SA, Strelkov SE, Chang KF, Turnbull GD (2009) Effect of crop rotation on soil pathogen population and dynamics and canola seedlings establishment. Plant Pathol J 8(3):106–112

Hyun JW, Yi SH, MacKenzie SJ, Timmer LW, Kim KS, Kang SK, Kwon HM, Lim HC (2009) Pathotypes and genetic relationship of worldwide collections of *Elsinoe* spp. causing scab diseases of citrus. Phytopathology 99(6):721–728. ISSN 0031-949X

Ikeda S, Okubo T, Anda M, Nakashita H, Yasuda M et al (2010) Community- and genome-based views of plant-associated bacteria: plant-bacterial interactions in soybean and rice. Plant Cell Physiol 51:1398–1410

Ikram N, Dawar S (2014) Impact of biocontrol agents in combination with *Prosopis juliflora* (Swartz) DC. In controlling the root-infecting fungi of leguminous crops. Arch Phytopathol Plant Protect 47(8):930–937

Inácio J, Pereira P, Carvalho M, Fonseca Á, Amaral-Collaço MT, Spencer-Martins I (2002) Estimation and diversity of phylloplane mycobiota on selected plants in a Mediterranean-type ecosystem in Portugal. Microb Ecol 44:44–353

Jana T, Sharma TR, Singh NK (2005) SSR-based detection of genetic variability in the charcoal root rot pathogen *Macrophomina phaseolina*. Mycol Res 109(1):81–86. ISSN 0953-7562

Jeeva ML, Mishra AK, Vidyadharan P, Misra RS, Hegde V (2010) A species-specific polymerase chain reaction assay for rapid and sensitive detection of *Sclerotium rolfsii*. Australas Plant Pathol 39(6):517–523. ISSN 0815-3191

Jiménez-Fernández D, Montes-Borrego M, Jimenez-Diaz RM, Navas-Cortes JA, Landa BB (2011) In planta and soil quantification of *Fusarium oxysporum* f. sp. *ciceris* and evaluation of Fusarium wilt resistance in chickpea with a newly developed quantitative polymerase chain reaction assay. Phytopathology 101(2):250–262. ISSN: 0031-949X

Jumpponen A, Jones KL (2009) Massively parallel 454 sequencing indicates hyperdiverse fungal communities in temperate Quercus macrocarpa phyllosphere. New Phytol 184:438–448

Kandel KP (2014). Transcriptomic studies of the early stages of potato infection by *Phytophthora infestans*. PhD thesis, University of Dundee

Kchouk M, Gibrat JF, Elloumi M (2017) Generations of sequencing technologies: from first to next generation. Biol Med 9:3

Khan AL, Hamayun M, Kim YH, Kang SM, Lee IJ (2011) Ameliorative symbiosis of endophyte (*Penicillium funiculosum* LHL06) under salt stress elevated plant growth of *Glycine max* L. Plant Physiol Biochem 49:852–861

Khan AL, Hamayun M, Kang SM, Kim YH, Jung HY, Lee JH, Lee IJ (2012) Endophytic fungal association via gibberellins and indole acetic acid can improve plant growth under abiotic stress: an example of *Paecilomyces formosus* LHL10. BMC Microbiol 12:3

Khan AL, Hussain J, Al-Harrasi A, Al-Rawahi A, Lee IJ (2015) Endophytic fungi: resource for gibberellins and crop abiotic stress resistance. Crit Rev Biotechnol 35:62–74

Khan MR, Shahid S, Mohidin FA, Mustafa U (2017) Interaction of *Fusarium oxysporum* f. sp. *gladioli* and *Meloidogyne incognita* on gladiolus cultivars and its management through corm treatment with biopesticides and pesticides. Biol Control 115:95–104

Kharbikar LL, Shanware AS, Saharan MS et al (2019) *Fusarium graminearum* microRNA-like RNAs and their interactions with wheat genome: a much needed study. Ind Phytopathol:1–9. https://doi.org/10.1007/s42360-019-00139-4

Köberl M, Schmidt R, Ramadan EM et al (2013) The microbiome of medicinal plants: diversity and importance for plant growth, quality and health. Front Microbiol 4:400

Kotasthane A, Agrawal T, Kushwah R, Rahatkar OV (2015) *In-vitro* antagonism of *Trichoderma* spp. against *Sclerotium rolfsii* and *Rhizoctonia solani* and their response towards growth of cucumber, bottle gourd and bitter gourd. Eur J Plant Pathol 141:523–543

Kowalchuk GA, Veen JAHV (2004) The significance of microbial diversity in agricultural soil for suppressiveness of plant diseases and nutrient retention. Physiol Behav 100(5):519–524

Kubicek CP, Herrera-Estrella A, Seidl-Seiboth V, Martinez DA et al (2011) Comparative genome sequence analysis underscores mycoparasitism as the ancestral life style of *Trichoderma*. Genome Biol 12:R40

Kulik T, Jestoi M, Okorski A (2011) Development of TaqMan assays for the quantitative detection of *Fusarium avenaceum/Fusarium tricinctum* and *Fusarium poae* esyn1 genotypes from cereal grain. FEMS Microbiol Lett 314(1):49–56. ISSN 0378-1097

Kumar V, Anal AKD, Nath V (2018) Biocontrol fitness of an indigenous *Trichoderma viride*, isolate NRCL T-01 against *Fusarium solani* and *Alternaria alternata* causing diseases in Litchi (*Litchi chinensis*). Int J Curr Microbiol App Sci 7(03):2647–2662

Kumari V, Singh A, Chaudhary HK et al (2019) Identification of Phytophthora blight resistant mutants through induced mutagenesis in sesame (*Sesamum indicum* L.). Indian Phytopathol 72:71

Lambers H, Mougel C, Jaillard B, Hinsinger P (2009) Plant-microbe-soil interactions in the rhizosphere: an evolutionary perspective. Plant Soil 321:83–115

Langrell SRH, Barbara DJ (2001) Magnetic capture hybridization for improved PCR detection of *Nectria galligena* from lignified apple extracts. Pl Mol Biol Rep 19:5–11

Li R-X, Cai F, Pang G, Shen QR, Li R, Chen W (2015) Solubilisation of phosphate and micronutrients by *Trichoderma harzianum* and its relationship with the promotion of tomato plant growth. PLoS One 10:e0130081

Li Y, Sun R, Yu J, Saravanakumar K, Chen J (2016) Antagonistic and biocontrol potential of *Trichoderma asperellum* ZJSX5003 against the maize stalk rot pathogen *Fusarium graminearum*. Indian J Microbiol 56:318–327. https://doi.org/10.1007/s12088-016-0581-9

Lievens B, Claes L, Vakalounakis DJ, Vanachter ACRC, Thomma BPHJ (2007) A robust identification and detection assay to discriminate the cucumber pathogens *Fusarium oxysporum* f. sp *cucumerinum* and f. sp *radicis-cucumerinum*. Environ Microbiol 9(9):2145–2161. ISSN 1462-2912

Lindow SE, Brandl MT (2003) Microbiology of the phyllosphere. Appl Environ Microbiol 69(4):1875–1883. https://doi.org/10.1128/AEM.69.4.1875-1883

López-Bucio J, Pelagio-Flores R, Herrera-Estrell A (2015) *Trichoderma* as biostimulant: exploiting the multilevel properties of a plant beneficial fungus. Sci Hortic 196:109–123

Lugtenberg B, Kamilova F (2009) Plant-growth-promoting rhizobacteria. Annu Rev Microbiol 63:541–556

Martos S, Torres E, El Bakali MA, Raposo R, Gramaje D, Armengol J, Luque J (2011) Co-operational PCR coupled with Dot Blot Hybridization for the detection of *Phaeomoniella chlamydospora* on infected grapevine wood. J Pathol 159(4):247–254

Matsuda Y, Sameshima T, Moriura N, Inoue K, Nonomura T, Kakutani K, Nishimura H, Kusakari S, Tamamatsu S, Toyoda H (2005) Identification of individual powdery mildew fungi infecting leaves and direct detection of gene expression of single conidium by polymerase chain reaction. Phytopathology 95(10):1137–1143. ISSN 0031-949X

Mawar R, Tomer AS, Singh D (2019) Demonstration of efficacy of bio-control agents in managing soil-borne diseases of various crops in arid region of India. Indian Phytopathol 72:699–703. https://doi.org/10.1007/s42360-018-0071-6

Mendes R, Kruijt M, Bruijn I, Dekkers E, Voort M, van der Schneider JHM, Piceno YM, Santis TZ, Andersen GL, PAHM B, Raaijmakers JM (2011) Deciphering the rhizosphere microbiome for disease-suppressive bacteria. Science 332:1097–1100

Mendes R, Garbeva P, Raaijmakers JM (2013) The rhizosphere microbiome: significance of plant beneficial, plant pathogenic, and human pathogenic microorganisms. FEMS Microbiol Rev 37:634–663

Mendes LW, Raaijmakers JM, de Hollander M, Mendes R, Tsai SM (2018) Influence of resistance breeding in common bean on rhizosphere microbiome composition and function. ISME J 12:212–224

Molla AH, Haque MM, Haque MA, Ilias GNM (2012) *Trichoderma* enriched biofertilizer enhances production and nutritional quality of tomato (*Lycopersicon esculentum* Mill.) and minimizes NPK fertilizer use. Agric Res 1(3):265–272

Mukherjee M, Mukherjee PK, Horwitz BA, Zachow C, Berg G, Zeilinger S (2012) *Trichoderma*-plant-pathogen interactions: advances in genetics of biological control. Indian J Microbiol 52:522–529

Murolo S, Concas J, Romanazzi G (2019) Use of biocontrol agents as potential tools in the management of chestnut blight. Biol Control 132:102–109

Muthuswamy A, Kakkattil Balakrishnan V, Palaniyandi U et al (2018) Pathogenic variability in *Phytophthora capsici* from black pepper (*Piper nigrum* L.) as revealed by transcriptome analysis. Indian Phytopathol 71:495

Naglot A, Goswami S, Rahman I, Shrimali DD, Yadav KK, Gupta VK, Rabha AJ, Gogoi HK, Veer V (2015) Antagonistic potential of native *Trichoderma viride* strain against potent tea fungal pathogens in North East India. Plant Pathol J 31:278–289

Nandhini M, Rajini SB, Udayashankar AC, Niranjana SR, Lund OS, Shetty HS, Prakash HS (2018) Diversity, plant growth promoting and downy mildew disease suppression potential of cultivable endophytic fungal communities associated with pearl millet. Biol Control 127:127–138

Nayyar BG, Woodward S, Mur AJ, Akram A, Arshad SMM, Naqvi S, Akhund S (2018) Identification and pathogenicity of *Fusarium* species associated with sesame (*Sesamum indicum* L.) seeds from the Punjab, Pakistan. Physiol Mol Plant Pathol 102:128–135

Nicolaisen M, Justesen AF, Thrane U, Skouboe P, Holmstrom K (2005) An oligonucleotide microarray for the identification and differentiation of trichothecene producing and nonproducing *Fusarium* species occurring on cereal grain. J Microbiol Methods 62(1):57–69. ISSN 0167-7012

Nilsson RH, Anslan S, Bahram M, Wurzbacher C, Baldrian P, Tedersoo L (2018) Mycobiome diversity: high-throughput sequencing and identification of fungi. Nat Rev Microbiol 17(2):95–109

Oros G, Naár Z (2017) Mycofungicide: *Trichoderma* based preparation for foliar applications. Am J Plant Sci 8(02):113–125

Pagano MC, Correa EJA, Duarte NF, Yelikbayev B, O'Donovan A, Gupta VK (2017) Advances in eco-efficient agriculture: the plant-soil mycobiome. Agriculture 7:14. https://doi.org/10.3390/agriculture7020014

Pan S, Mukherji R, Bhagat S (2013) Evaluation of *Trichoderma* spp. against soil borne plant pathogens. Ann Plant Protect Sci 21:197–198

Pareek CS, Smoczynski R, Tretyn A (2011) Sequencing technologies and genome sequencing. J Appl Genet 52:413–435

Penuelas J, Farre-Armengol G, Llusia J et al (2014) Removal of floral microbiota reduces floral terpene emissions. Sci Rep 4:6727

Perazzolli M, Antonielli L, Storari M, Puopolo G, Pancher M, Giovannini O et al (2014) Resilience of the natural phyllosphere microbiota of the grapevine to chemical and biological pesticides. Appl Environ Microbiol 80(12):3585–3596. https://doi.org/10.1128/AEM.00415-14

Pérez-Jaramillo JE, Carrión VJ, de Hollander M, Raaijmakers JM (2018) The wild side of plant microbiomes. Microbiome 6:143

Porras-Alfaro A, Bayman P (2011) Hidden fungi, emergent properties: endophytes and microbiomes. Annu Rev Phytopathol 49:291–315

Prabukumar S, Rajkuberan C, Ravindran K, Sivaramakrishnan S (2015) Isolation and characterization of endophytic fungi from medicinal plant *Crescentia cujete* L. and their antibacterial, antioxidant and anticancer properties. Int J Pharm Pharm Sci 7:316–321

Priya KS, Nagaveni HC (2009) Screening of *Trichoderma* spp. against *Lasiodiplodia theobromae* causing fruit rot of *Elaeocarpus munronii*. Indian J Plant Protect 37(1/2):166–169

Priyadharsini P, Muthukumar T (2017) The root endophytic fungus *Curvularia geniculate* from *Parthenium hysterophorus* roots improves plant growth through phosphate solubilization and phytohormone production. Fungal Ecol 27:69–77

Qiang-long Z, Shi L, Peng G, Fei-shi L (2014) High-throughput sequencing technology and its application. J Northeast Agric Univ 21:84–96

Rai S, Kashyap PL, Kumar S et al (2016) Identification, characterization and phylogenetic analysis of antifungal *Trichoderma* from tomato rhizosphere. Springerplus 5:1939. https://doi.org/10.1186/s40064-016-3657-4

Rai S, Slanki MK, Solanki AC, Surapathrudu K (2019) *Trichoderma* as biocontrol agent: molecular prospectus and application. In: Ansari RA, Mahmood I (eds) Plant health under biotic stress. Springer, Singapore. https://doi.org/10.1007/978-981-13-6040-4_7

Rhoads A, Au KF (2015) PacBio sequencing and its applications. Genomics Proteomics Bioinformatics 13:178–289

Rodriguez RJ, White JF, Arnold AE, Redman RS (2009) Fungal endophytes: diversity and functional roles. New Phytol 182:314–330

Roe AD, Rice AV, Bromilow SE, Cooke JEK, Sperling FAH (2010) Multilocus species identification and fungal DNA barcoding: insights from blue stain fungal symbionts of the mountain pine beetle. Mol Ecol Resour 10:946–959

Rytkönen AP, Lilja A, Hantula J (2011) PCR-DGGE method for in planta detection and identification of *Phytophthora* species. F Pathol 42(1):22–27

Samuelian SK, Greer LA, Savocchia S, Steel CC (2011) Detection and monitoring of *Greeneria uvicola* and *Colletotrichum acutatum* development on grapevines by real time PCR. Plant Dis 95(3):298–303. ISSN 0191-2917

Saravanakumar K, Dou K, Lu Z, Wang X, Li Y, Chen J (2018) Enhanced biocontrol activity of cellulase from *Trichoderma harzianum* against *Fusarium graminearum* through activation of defense-related genes in maize. Physiol Mol Pl Pathol 103:130–136

Schmidt H, Taniwaki MH, Vogel RF, Niessen L (2004) Utilization of AFLP markers for PCR-based identification of Aspergillus carbonarius and indication of its presence in green coffee samples. J Appl Microbiol 97(5):899–909. ISSN 1364-5072

Schmidt R, Koberl M, Mostafa A et al (2014) Effects of bacterial inoculants on the indigenous microbiome and secondary metabolites of chamomile plants. Front Microbiol 5:64

Schweitzer JA, BaileyBangert RK, Hart SC, Whitham TG (2006) The role of plant genetics in determining above – and below – ground microbial communities. In: Bailey MJ, Lilley AK,

Timms-Wilson TM, Spencer-Phillips PTN (eds) Microbial ecology of the aerial plant surface. CABI International, Wallingford, pp 107–119

Shah N, Meisel JS, Pop M (2019) Embracing ambiguity in the taxonomic classification of microbiome sequencing data. Front Genet 10:1022. https://doi.org/10.3389/fgene.2019.01022

Shahid S, Khan MR (2019) Evaluation of biocontrol agents for the management of root-rot of mung bean caused by *Macrophomina phaseolina*. Indian Phytopathol 72:89

Sharma KK, Singh US, Sharma P, Kumar A, Sharma L (2015) Seed treatments for sustainable agriculture – a review. J Appl Nat Sci 7(1):521–539

Silvar C, Díaz J, Merino F (2005) Real-time polymerase chain reaction quantification of *Phytophthora capsici* in different pepper genotypes. Phytopathology 95(12):1423–1429. ISSN 0031-949X

Singh M, Sharma OP (2012) *Trichoderma*- A savior microbe in the era of climate change. IJABR 2(4):784–786

Somai BM, Keinath AP, Dean RA (2002) Development of PCR-ELISA detection and differentiation of *Didymella bryoniae* from related *Phoma* species. Plant Dis 86(7):710–716. ISSN 0191-2917

Song Z et al (2015) Effort versus reward: preparing samples for fungal community characterization in high-throughput sequencing surveys of soils. PLoS One 10:e0127234

Srivastava RK, Singh RK, Kumar N, Singh S (2010) Management of *Macrophomina* disease complex in jute (*Corchorus olitorius*) by *Trichoderma viride*. J Biol Control 24:77–79

Suresh N, Nelson R (2016) Isolation of antagonistic fungi and evaluation of antifungal activity of the separated metabolite against the red rot of sugarcane pathogen. Eu J Exp Bol 6(1):15–21

Szabo LJ (2007) Development of simple sequence repeats markers for the plant pathogenic rust fungus, *Puccinia graminis*. Mol Ecol Notes 7:92–94

Szabo LJ, Kolmer JA (2007) Development of simple sequence repeats markers for the plant pathogenic rust fungus, *Puccinia triticina*. Mol Ecol Notes 7(4):708–710. https://doi.org/10.1111/j.1471-8286.2007.01686.x

Tedersoo L, Nilsson RH (2016) Molecular mycorrhizal symbiosis (ed Martin F). Wiley, Hoboken, pp 301–322

Tellenbach C, Grunig CR, Sieber TN (2010) Suitability of quantitative real-time PCR to estimate the biomass of fungal root endophytes. Appl Environ Microbiol 76:5764–5772

Tetali S, Karpagavalli S, Pavani SL (2015) Management of dry root rot of black gram caused by *Macrophomina phaseolina* (Tassi) Goid. using bio agent. Plant Arch 15(2):647–650

Tewoldemedhin YT, Mazzola M, Botha WJ, CFJ S, McLeod A (2011) Characterization of fungi (Fusarium and Rhizoctonia) and oomycetes (Phytophthora and Pythium) associated with apple orchards in South Africa. Eur J Plant Pathol 130(2):215–229

Thombre B, Kohire O (2018) Integrated management of Macrophomina blight of mungbean (Vigna radiata L.) caused by Macrophomina phaseolina (Tassi) Goid. Indian Phytopathol 71:423–429. https://doi.org/10.1007/s42360-018-0055-6

Toju H, Peay KG, Yamamichi M, Narisawa K, Hiruma K et al (2018) Core microbiomes for sustainable agroecosystem. Nat Plants 4(9):247–257

Tkacz A, Poole P (2015) Role of root microbiota in plant productivity. J Exp Bot 66(8):2167–2175. https://doi.org/10.1093/jxb/erv157

Tomlinson JA, Barker I, Boonham N (2007) Faster, simpler, more-specific methods for improved molecular detection of *Phytophthora ramorum* in the field. Appl Environ Microbiol 73(12):4040–4047. https://doi.org/10.1128/AEM.00161-07

Tomlinson JA, Dickinson MJ, Boonham N (2010) Rapid detection of *Phytophthora ramorum* and *P. kernoviae* by two-minute DNA extraction followed by isothermal amplification and amplicon detection by generic lateral flow device. Phytopathology 100:143–149

Torres-Calzada C, Tapia-Tussell R, Quijano-Ramayo A, Martin-Mex R, Rojas-Herrera R, Higuera-Ciapara I, Perez-Brito D (2011) A species-specific polymerase chain reaction assay for rapid and sensitive detection of Colletotrichum capsici. Mol Biotechnol 49:48–55

Turrini A, Avio L, Giovannetti M, Agnolucci M (2018) Functional Complementarity of Arbuscular Mycorrhizal Fungi and Associated Microbiota: The Challenge of Translational Research. Front Plant Sci 9:1407. https://doi.org/10.3389/fpls.2018.01407

Vergara C, Araujoa KEC, Alvesa LS, de Souza SR, Santos LA et al (2018) Contribution of dark septate fungi to the nutrient uptake and growth of rice plants. Braz J Microbiol 49:67–78

Verginer M, Siegmund B, Cardinale M et al (2010) Monitoring the plant epiphyte *Methylobacterium extorquens* DSM 21961 by real-time PCR and its influence on the strawberry flavor. FEMS Microbiol Ecol 74:136–145

Vezzi F (2012). Next generation sequencing revolution challenges: search, assemble, and validate genomes. Ph.D, Universita degli Studi di Udine, Italy

Vorholt JA (2012) Microbial life in the phyllosphere. Nat Rev Microbiol 10(12):828–840. https://doi.org/10.1038/nrmicro2910

Waghunde RR, Shelake RM, Sabalpara AN (2016) *Trichoderma*: A significant fungus for agriculture and environment. Afr J Agric Res 11:1952–1965

Weller DM, Raaijmakers JM, Gardener BBM, Thomashow LS (2002) Microbial populations responsible for specific soil suppressiveness to plant pathogens. Annu Rev Phytopathol 40:309–348

Winton LM, Krohn AL, Leiner RH (2007) Microsatellite markers for *Sclerotinia subarctica* nom. prov., a new vegetable pathogen of the High North. Mol Ecol Notes 7(6):1077–1079. ISSN 1471-8278

Yao H, Sun X, He C, Maitra P, Li CX, Guo DL (2019) Phyllosphere epiphytic and endophytic fungal community and network structures differ in a tropical mangrove ecosystem. Microbiome 7:57. https://doi.org/10.1186/s40168-019-0671-0

You J, Zhang J, Wu M, Yang L, Chen W, Li G (2016) Multiple criteria-based screening of *Trichoderma* isolates for biological control of *Botrytis cinerea* on tomato. Biol Control 101:31–38

Zabetakis I, Moutevelis-Minakakis P, Gramshaw JW (1999) The role of 2-hydroxypropanal in the biosynthesis of 2,5-dimethyl-4- hydroxy-2H-furan-3-one in strawberry (*Fragaria × ananassa, cv. Elsanta*) callus cultures. Food Chem 64:311–314

Zeilinger S, Gruber S, Bansalb R, Mukherjee PK (2016) Secondary metabolism in *Trichoderma*-Chemistry meets genomics. Fungal Biol Rev 30:74–90

Zhang N, Tantardini A, Miller S, Eng A, Salvatore N (2011) TaqMan real-time PCR method for detection of Discula destructiva that causes dogwood anthracnose in Europe and North America. Eur J Plant Pathol 130(4):551–558. ISSN 0929–1873

Zhang F, Liu Z, Gulijimila M, Wang Y, Fan H, Wang Z (2016) Functional analysis of the 1-aminocyclopropane-1-carboxylate deaminase gene of the biocontrol fungus *Trichoderma asperellum* ACCC30536. Can J Plant Sci 96:265–275

Zhang S, Xu B, Zhang J, Gan Y (2018) Identification of the antifungal activity of *Trichoderma longibrachiatum* T6 and assessment of bioactive substances in controlling phytopathogens. Pestic Biochem Physiol 147:59–66

Zheng Y, Zhang G, Lin FC, Wang ZH, Jin GL, Yang L, Wang Y, Chen X, Xu ZH, Zhao XQ, Wang HK, Lu JP, Lu GD, Wu WR (2008) Development of microsatellite markers and construction of genetic map in rice blast pathogen *Magnaporthe grisea*. Fungal Genet Biol 45(10):1340–1347. ISSN 1087-1845

Zheng YK, Miao CP, Chen HH, Huang FF, Xia YM, Chen YW, Zhao LX (2017) Endophytic fungi harbored in Panax notoginseng: diversity and potential as biological control agents against host plant pathogens of root-rot disease. J Ginseng Res 41:353–360

Role of Soil Fauna: En Route to Ecosystem Services and Its Effect on Soil Health

5

Apurva Mishra and Dharmesh Singh

Abstract

Soils have an amazingly assorted network of soil fauna that contrast in their versatile procedures, and consequently, in the capacities, they satisfy in soils. Soil fauna works as an extremely proficient means, which aids microorganisms in inhabiting and broadening their venture furthermore into the soil skylines. Lives of soil fauna as soil settlers, indicators, and architects have been underlined, but recent research and worldwide natural concerns are urging the scientific community to look further into the controls of soil fauna for sustainable management of soils in these turbulent times. Molecular data alone is insufficient for many investigations about soil fauna, and hence there is a need for sincere efforts in increasing expertise in the classical taxonomy of soil fauna. This specialization, along with data on the soil fauna's biogeography, their relationship to over-the-ground problem areas, and land management schemes, will be acute for accepting how soil fauna will communicate and react to numerous worldwide alterations. This chapter discusses the soil fauna on the soil grounds as well as below grounds, tight associations of abiotic factors with the soil biodiversity, their roles in ecosystem events and functions, and, finally, the arrangement of ecosystem benefits for human prosperity. A proper and researched thought about soil faunal species, their identities, geographic extents, and an understanding of their roles in soil and land management will help in giving balanced forecasts for the working of future biological communities that are under siege due to inevitable environmental climate change.

A. Mishra
Academy of Scientific and Innovative Research (AcSIR), Ghaziabad, India

Environmental Biotechnology and Genomics Division, CSIR-National Environmental Engineering Research Institute, Nagpur, Maharashtra, India

D. Singh (✉)
Environmental Biotechnology and Genomics Division, CSIR-National Environmental Engineering Research Institute, Nagpur, Maharashtra, India

© Springer Nature Singapore Pte Ltd. 2020

M. K. Solanki et al. (eds.), *Phytobiomes: Current Insights and Future Vistas*,
https://doi.org/10.1007/978-981-15-3151-4_5

Keywords
Soil fauna · Soil biodiversity · Soil health · Ecosystem engineers · Faunal indicators

5.1 Introduction

Soil speaks as a standout among the essential supplies of biodiversity. It reflects the environmental metabolism since all the biogeochemical procedures of distinctive biological systems are consolidated inside it. It is an intricate framework comprising biotic and abiotic segments, which coordinates with the essential habitat and ensures biological activity and diversity, assisting ecosystem services (Robinson et al. 2013). Inside the composite assembly of soil, biotic and abiotic parts communicate in governing the natural debasement of materials and nutrient cycling (Morgado et al. 2018). Soil fauna or the biotic portion of the soil is an imperative store of biodiversity and assumes a basic role in several soil biological system capacities. Soil environmental conditions are degrading due to the increase in human activities, leading to a reduction in the abundance of plants and animal communities responsible for a balanced environment (Morgado et al. 2018). The consequence of this biodiversity loss is a counterfeit environment that needs steady human intervention and requires additional expenses, while regular ecosystems are optimally controlled by a group of plants and animals and a thorough supply of energy and nutrients—a form of control gradually vanishing with land-use change and industrial growth. As a result, agricultural practices that permit a mix of production targets and ecologically amicable administrations work sustainably, ensuring healthy soil biodiversity, which is fundamental to preventing soil fauna networks from declining in agricultural lands and also helping us in understanding their role toward ecosystem services.

5.2 Structure of Soil Ecosystem

5.2.1 Soil Organization

Amid biological communities, soil stands as the focal organization, incorporating a huge number of geochemical and ecological capacities (Coleman and Whitman 2005; Crawford et al. 2005; Wall et al. 2010, Delgado-Baquerizo et al. 2019). Soils' position exists at the interface media (Fig. 5.1), located at the intersection of the four main life-supporting provinces: lithosphere, hydrosphere, atmosphere, and the biosphere (living matter). This provides exceptionally dynamic and multiphasic character to the soil, where numerous estimated aggregates are connected and balanced out inside a perplexing lattice of solid, fluid, and gaseous parts communicating at different scales (Lal 2016; Parker 2010).

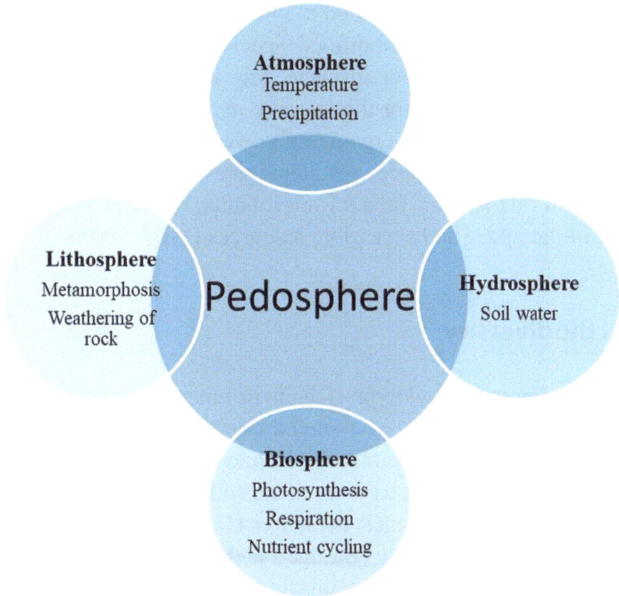

Fig. 5.1 Pedosphere presenting biotic and abiotic interactions in the soil matrix (Modified from Fitzpatrick [1984] and Coleman and Crossley [1996])

Soil particles do not make a nonstop and conservative mass; instead, the soil volume is framed by pores, chambers, channels, and splits that give an appropriate domain to soil fauna and lead to the development of plant roots. This physical organization in soil additionally directs the motion of gases and fluids inside the soil making different land- or water-capable situations, heterogeneously loaded up with soil "solution" and mostly occupied with gases, which are critical for the soil fauna (Lavelle 2012). A wide variety of dissolved and suspended matter, including inorganic and organic materials, contributes toward soil "solution" composition (Lavelle and Spain 2001). Soil components engaged on the solid stage diffuse to the fluid stage. In this manner, minerals and organic matter end up being accessible to the larger part of soil-living creatures and plants. Soil gaseous stage permits oxygen (O_2) utilization and carbon dioxide (CO_2) generation amid biological exercises. As the soil O_2 gets used up, usually there is a trade-off between the soil and atmospheric O_2 with the help of a concentration gradient, with CO_2 transition happening in reverse. Relative humidity of terrestrial ecosystems remains near saturation, which is indispensable to most soil fauna, for instance, in the case of springtails. Nematodes, on the other hand, can survive at different levels of relative humidity (Lavelle and Spain 2001)

Notwithstanding the huge predisposition toward the above-ground portion of the soil ecosystem, it is the below ground, the subterranean, where indispensable biodiversity focused on the pore spaces resides (Beare et al. 1995). Pore spaces have an expanded surface territory that makes a large number of active microenvironments where neighborhood species are exposed to low, competitive exclusion, and concurrence is set between these neighborhood species through resource partitioning (Haynes 2014; Lavelle 1997). For instance, low water potential causes low pore connectivity, hence responsible for the increase in the bacterial abundance (Carson et al. 2010). Competition acts as organizing power for biota consensus in the soil, limited to finer scales, for instance, soil aggregates for microbial networks or the soil pore space and rhizospheric area for microfauna, where prospective contenders may utilize a similar space and accessible resources (MEA 2005).

5.2.2 Soil Biodiversity

On earth, soils are the most phylotype-rich environments (Giller 1996). No place in nature is plausible enough to discover such a large number of organisms as in soil ecosystems (Hågvar 1998). Unfortunately, despite the immense efforts by scientists in the past few years to portray and comprehend soil networks, the taxonomic shortage for soil biodiversity is the biggest mysteries (Decaëns 2010) and little is known about the soil's cryptic biodiversity (Bardgett and van der Putten 2014). In spite of this fact, one perspective as of now remains undisputable: soil biodiversity is obligatory for an optimum level of soil functioning that eventually supports all soil-based ecosystems and goods (Barrios 2007). Subsequently, enriching the information about soil biodiversity is foremost to completely comprehend the fundamentals of soil health, adequately oversee soil-based environmental benefits, and foresee future patterns and situations for the recent period (Bardgett and van der Putten 2014).

Soil organisms are hyperdiverse and amazingly unpredictable and incorporate soil communities from all major scientific categorizations (Parker 2010; Wardle 2006). They are typically arranged depending on their body size, whose variety inside soil networks traverses a few orders of size (Barrios 2007; Lavelle 2012; Parker 2010). A greater part of decent soil variety is created by microbiota or microscopic organisms such as bacteria, archaea, and fungi; however, it includes an exceptional range of microfauna, mesofauna, macrofauna, and even megafauna (Bardgett 2002; Lavelle 2012; Orgiazzi et al. 2016; Wurst et al. 2012). Also, it incorporates a tremendous assortment of photosynthetic entities like lichens, bryophytes, and vascular plants, and their root systems have significant roles in soil ecosystem organization (Orgiazzi et al. 2016). Soil microbiota contributes, for the most part, to decay forms, favoring carbon (C), nitrogen (N_2), and other biogeochemical cycling, yet also assume a vital role in defeating disease and in plant growth and development (Wurst et al. 2012). These life forms make vital cooperative collaborations with plants, such as enhancing nutrient uptake (e.g., N_2 fixation, phosphorus redistribution), as well as manage plant hormones (Wurst et al. 2012). Soil organic matter (SOM) is critical for the biological processes and supply of nutrients, in every variety of soils. Two to ten percent of the soil mass is SOM, which is essential for the comprehended function, including physical, chemical, and biological functions. SOM has been instigated in plants, which can be categorized into "living" and "dead" components during different stages of decomposition and varies in age from fresh residues to those that are years old in the form of resistant organic matter. Carbon and other organic particles, such as hydrogen (H_2), oxygen (O_2), and little amounts of N_2, phosphorous, sulfur, potassium, calcium (Ca), and magnesium (Mg), constitute SOM. Nearly 5–10% of below-ground SOM contains roots, fauna, and microorganism, that is, living, and this living pool of microbial components is referred to as microbial biomass, which is considered essential for decomposition of organic matter, nutrient cycling, chemical degradation, and stabilization of soil (Ha et al. 2008). During early plant development, phosphorus is required for cell growth. Soil phosphorous can be distributed into three pools, each varying in its accessibility to plants:

1. Soil organic phosphorus bound to organic compounds
2. Inorganic compounds (phosphorus joined with Ca, Mg, iron (Fe), aluminum (Al), or clay minerals)
3. Natural and inorganic phosphorus compounds related to living cells

Soil phosphorus moves between each of these pools using mineralization (separate of natural matter), immobilization, and redistribution of phosphorus between microorganisms, plants, and natural matter. Phosphorus is immobilized (made inaccessible to plants) when it is consolidated into the living microbial biomass. Redistribution of phosphorus happens when phosphorus is discharged from microbial cells and moved into different phosphorus pools. Phosphorus mineralization and immobilization happen at the same time in soil. While mineralization of natural soil phosphorus to inorganic phosphorus expands the accessibility of phosphorus to plants and microorganisms, an extensive amount (between 15 and 80%) of soil phosphorus stays in natural structure and remains inaccessible to plants (Ha et al. 2008).

Soil microfauna incorporates organisms that are $\leq$100 μm (e.g., nematodes, protozoa, and rotifers) in size. Only one taxon, protozoa, is discovered completely inside this class (Ruggiero et al. 2015). Mainly, mites, nematodes, rotifers, tardigrades, and copepod scavengers fall inside this group (Kennedy and Gewin 1997). Soil microfauna feeds on microscopic organisms such as fungi and algae, yet they additionally represent predator and saprophytic gatherings too. Through their actions they manage: (1) nutrient cycling by enhancing the accessibility of nutrients to different species (e.g., excretion), (2) population and actions of microscopic organisms and fungi, and (3) scattering of pivotal rhizospheric microbiota (Wurst et al. 2012). Their impacts on plants are likewise—when in direct contact with roots, they benefit from the root excretions and modify the plant's defenses or hormones. The role of plant roots in soil processes cannot be separated from the varied group of organisms, which flourish around them. These comprise parasites too, herbivores and predators, nematodes or worms as microbial grazers, and free-living microbes feeding on root exudates. Secondary metabolites produced by plants such as iridoid glycosides are present in root exudates. Other volatile organic compounds (VOCs) are also released by green leaves and above-ground roots that attract pollinators. Below-ground roots also release VOCs that dissuade soil-inhabiting herbivores.

Likewise, plants are not just providers of litter for decomposers but rather they assume a functional job in pulling in valuable soil invertebrates. For instance, plants draw in entomopathogenic nematodes to kill the rhizospheric herbivores, giving innoculum to the rhizospheric bacterial population, aggravating the correspondence between destructive microorganisms, and furthermore, in adjusting rhizodeposition and root architecture. This eventually results in a better communication between soil fauna and plant roots (Briones 2018). All these synthetic signs discharged by the plants are coordinated in a way that profits their very own development and health.

Soil mesofauna ranges from 100 μm to 2 mm (gatherings include taxa like Acari, Collembola, Tardigrada, Protura, Diplura, and Enchytraeidae) in size. Microarthropods, for example, mites and springtails, are the fundamental organisms of this group. They could be classified into herbivores, bacterivores, or fungivores. At times, they can additionally gorge on higher trophic dimension life forms. They live near the air and water existing in soil and are thus exceptionally reliant on soil aeration and humidity. Soil mesofauna helps in nutrient cycling and management of pests and disease control by their preferential mode of feeding, acts as food for other soil organisms, and aids in soil fauna distribution (Wurst et al. 2012).

Soil macrofauna (>2 mm in size) is fundamentally in charge of litter comminution and redistribution, and predation on other soil-abiding organisms frequently called as "ecosystem engineers." Macroarthropods (e.g., isopods, spiders, bugs) alongside annelids and gastropods are the fundamental congregations of soil macrofauna. These creatures are in charge of modifying the structure of their natural surrounding by contributing to various soil capacities, like physical disintegration and transport of litter residues to lower layers of soil that eventually help in nutrient cycling by microfauna and microbes, water penetration (e.g., by burrowing practices), and pest and disease control (rich biodiversity and predation; Wurst et al. 2012). They have an immediate constructive outcome on plant development and yield, yet they could be the reason for some harmful impacts on crops.

Soil megafauna (soil fauna whose size surpasses 20 mm): Individuals from this group incorporate an extensive range of invertebrates (worms, snails, myriapods) and vertebrates (insectivores, terrestrial and aquatic rodents, and reptiles). Moles, voles, gophers, snakes, and burrowing owls are few examples of soil megafauna, which help in the breakdown of the complex substances in decomposing plants and animals so that living plants and other organisms can utilize them. This comprises soil organisms as promoters in biogeochemical cycles, with carbon (C) cycle being the most prominent followed by nitrogen (N_2) and sulfur (S) cycles.

Soil life forms can be characterized additionally as per their usefulness, which makes a difference to explain about their environmental roles in soil systems. Turbé et al. (2010) recommended the utilization of three widely inclusive practical gatherings: chemical engineers, biological regulators, and ecosystem engineers (Fig. 5.2). Creatures directly participating in carbon and other biogeochemical cycles, including nitrogen, phosphorus, and sulfur cycles, and helping in soil decomposition and transformation can be classified as the chemical engineers. Biological regulators are in charge of monitoring the changing aspects of soil inhabitants, thereby advancing the strength and constancy of soil biological systems (Fusaro et al. 2018). Ecosystem

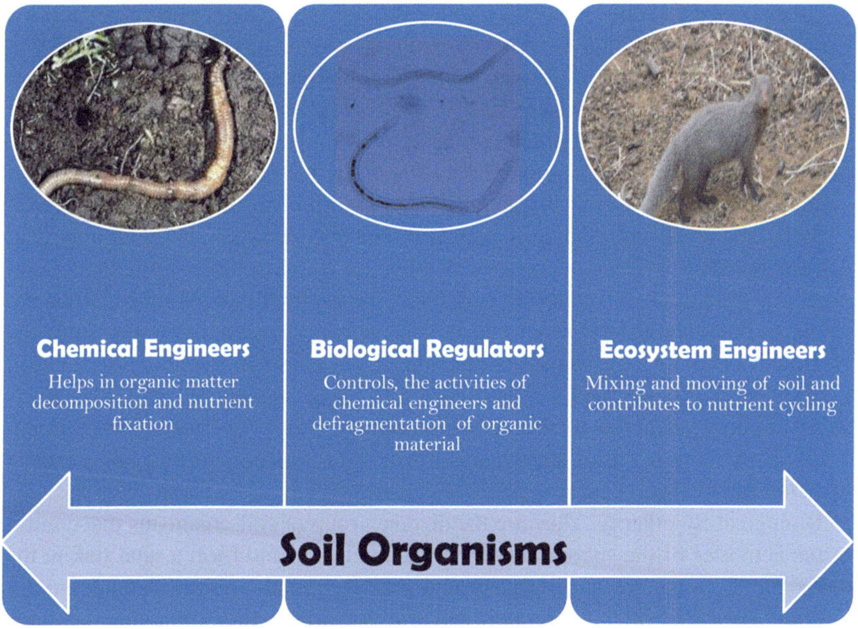

Fig. 5.2 Classification of soil beneficiaries and their appropriate role in the maintenance of agroecosystem

engineers are in charge of maintaining the soil structure by advancing the formation of stable aggregates (unit of soil structure), the pore network construction, and the advancement of complex biostructures.

In spite of a long history of studies focusing on the enumeration of soil fauna (Menta 2012), it is still extremely hard to give precise biomass estimates for soil fauna. This is somewhat due to their inconsistency in time and space (Menta 2012) and also due to contrasts in the methodologies used for their inspection (Wall et al. 2001). Also, most of the studies are from the temperate zones while other climactic/ ecological regions have been genuinely disregarded (Turbé et al. 2010).

5.3 Soil Fauna Organization in the Ecosystem

Soil fauna is viewed as an all-inclusive imperative part necessary for reusing organic matter, soil vitality, and nutrient cycling (Jeffery and Gardi 2010). In conjunction, they are the main players in supporting and managing environment services. Hence, soil biologists frequently organize soil fauna in terms of their broad functions divided into three categories: chemical engineers, biological regulators, and ecosystem engineers (Fig. 5.2).

1. Chemical Engineers: They are the key components of the soil food web. Bacteria and fungi make up the largest portion of this group and also include algae and

viruses (viruses could impact C cycling through viral shunt where, by bacterial cell lysis, they build up the concentration of labile C in the soil ecosystem that eventually serves toward increment in microbial generation and respiration in soils; refer to Williamson et al. [2017] and Trubl et al. [2018] for more details on this topic). These microscopic engineers help in the decomposition of organic material and transforming complex forms of carbon and nitrogen into CO_2 and simpler nutrients (Coleman and Crossley 1996). For their survival and growth, they depend on optimum soil moisture, suitable atmospheric conditions, and pore spaces between soil particles. Significant quantities of organic matter and animal manure in soil environments certify their prevalence. Moreover, they are key parts of a soil food web or sustenance networks as their action ensures the growth of living organisms, commencing from plants to the animals that feed upon them, responsible for organic turnover. Rough evaluations of soil biodiversity demonstrate a few thousand invertebrate species for every site and additionally ambiguous levels of microbial and protozoan variety that contributes in each of the trophic level (Turbé et al. 2010).

2. Biological Regulators: They are the diverse group of soil organisms that control the activities of the subordinate chemical engineers and form a vital link in the food web. Some act as plant pests and parasites, while others stimulate microflora. Their movement around the soil assists fragmentation of organic material, providing more surface area and hence enhancing the availability of the nutrients to microbes. In this group, protists are the smallest organisms, which live in the water layer surrounding soil particles and control bacterial populations through feeding. Nematodes, microarthropods, and protists are the most abundant biological regulators.
3. Ecosystem Engineers: Of major significance in the ecosystem, soil advancement, and support are the alleged "ecosystem engineers," as these species control, either straightforwardly or in a roundabout way, the accessibility of resources to other co-inhabitants of soil (Davies et al. 2019; Wright et al. 2002). These life forms physically change, keep up, and make new habitat for other life forms. Typically, there are two types of ecosystem engineers—firstly, the allogenic ones that modify the soil background by transforming living or nonliving things from one physical state to another, via mechanical or other means, and, secondly, the autogenic ones that change the soil background using their physical structures, that is, their existing and/or dead tissue (Byers et al. 2006; Lavelle and Spain 2001; Wright et al. 2002). A precedent of physical "ecosystem engineers" is plant roots that make substantial voids (spaces) in the soil through root rot (Byers et al. 2006). Termites and earthworms assume a noteworthy role in moving, blending, and circulating air through the soil through their tunneling.

Soil fauna is a very important entity, and the greater part is exceedingly versatile, stretching from herbivores to omnivores and even include carnivores. Contingent upon the accessible nutrient supply, a good chunk of soil fauna can modify their feeding systems from a more noteworthy to a lesser degree with instances where numerous carnivorous species amid low food accessibility can head toward the transformation and organic matter turnover (Brose and Scheu 2014). The cooperation between soil fauna and ecosystem are various, and unpredictable, due to the huge diversity and prevalent shifts.

The level of connection amid soil fauna and the soil itself can vary significantly between taxa and is subjected to soil existence cycle (Wall et al. 2001). Specifically, in conjunction with their physical forms and the ecological roles, it is conceivable to arrange soil fauna into four fundamental gatherings: incidentally dormant geophiles, briefly dynamic geophiles, periodical geophiles, and geobionts (Fig. 5.3). It is noticed that this classification does not have any taxonomical importance, yet, rather, they are helpful after contemplating the existential procedures of soil invertebrates.

Incidentally, dormant geophiles are life forms residing in the soil for a certain period of their lifecycle, for example, during hibernation or while experiencing transformation, at a point when security from climatic forces is essential (Menta 2012). Because of their relative dormancy, these life forms affect the natural capacity of the soil, even though they can act as prey for different creatures in this critical

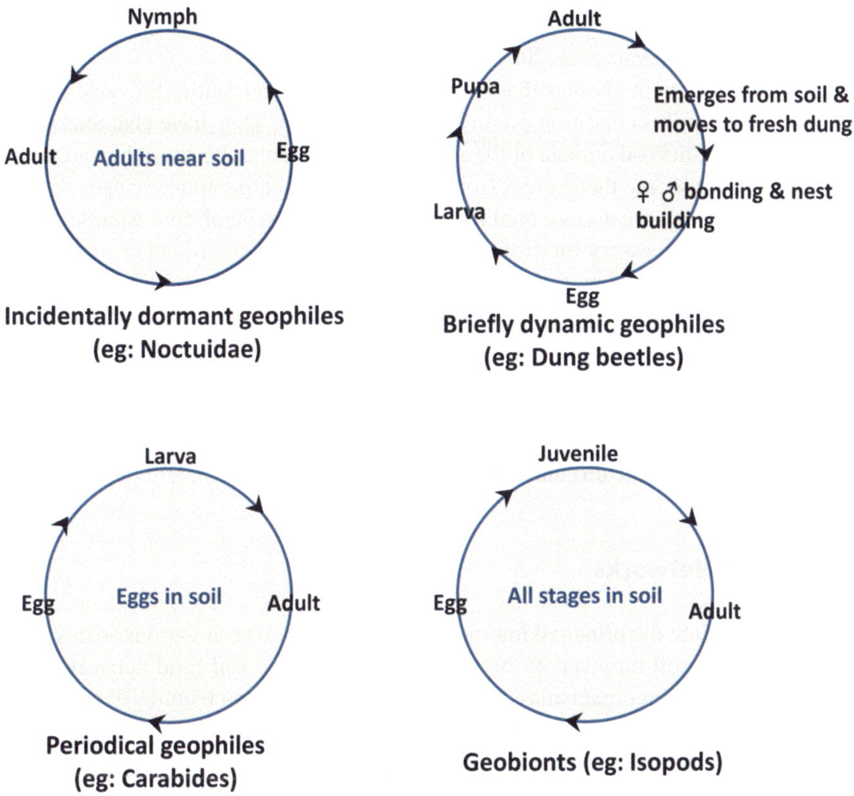

Fig. 5.3 The primary four fundamental gatherings of soil invertebrates contingent upon their life techniques and how nearly they are connected with the soil

time of their lifecycle. They spend time in the soil to let pass adverse atmospheric conditions, or for metamorphosis; for instance, caterpillars of butterfly families like Noctuidae, Geometridae, or Sphingidae bury themselves in the ground at the time of nymphosis. These incidentally dormant geophiles exert no mechanical force on soil (Menta 2012).

Briefly dynamic geophiles reside in the soil in a steady state for an extensive period of their life (i.e., for at least one or more developmental phases and rise out of the earth as an adult). A large portion of these creatures is insects, for example, Neuroptera, Diptera, Coleoptera, and Lepidoptera. Organisms with "pupal" phase in their life cycle assume a minor role in the soil amid this stage, while the "larval" phase is considerably more imperative for the soil ecology, particularly at times where populace thickness is high (Wall et al. 2001, 2008). Most larvae in the soil can go about as both detritivores and predators.

Periodical geophiles reside for a significant portion of their life in the dirt, for the most part as hatchlings; however, for the duration of their lives, they at times return to the dirt to perform different exercises, for example, chasing, egg laying, or to flee from threats. A few Coleoptera gatherings (e.g., carabides, scarabeids, cicindelids) pass their larval phase in the litter or the upper layers of mineral soil, and when they reach a mature stage, they utilize soil as a nourishment source, a shelter, and for different purposes (Wall et al. 2001, 2008).

Geobionts are living beings that exceptionally spend their entire lifecycle in soil and cannot leave this condition even for a limited period. They have characteristics that counteract survival outside of the soil environment either because they are deficient in ways to secure themselves from desiccation or temperature changes, or are lacking the sensory organs essential to survive above the ground, for example, lacking appendages, necessary for discovering nourishment above ground or for protection from predators. A few types of myriapods, isopods, Acari, mollusks, and the larger part of Collembola, Diplura, and Protura have a place with this gathering (Wall et al. 2001). These distinctive sorts of connections between soil life forms and the soil decide a separated level of weakness among different gatherings and as a result of any conceivable effect on soil condition. In case, if soil pollution happens due to some human interference, geobionts will be the most severely affected group as they cannot leave the dirt and must consume all their time on earth there.

5.3.1 Soil Networks

Soil organisms are the principal intermediates of soil working at various scales. The investigation of soil biodiversity began through surveying soil food networks as a feeding web between organisms—one of the major integrating features of soil communities. Natural ecosystems contain numerous species that are associated with their feeding connections over various trophic levels to make a complex food web (Digel et al. 2014). Consumer and producer interactions can be envisioned as a binding principle for food webs in different ecosystems with hierarchical

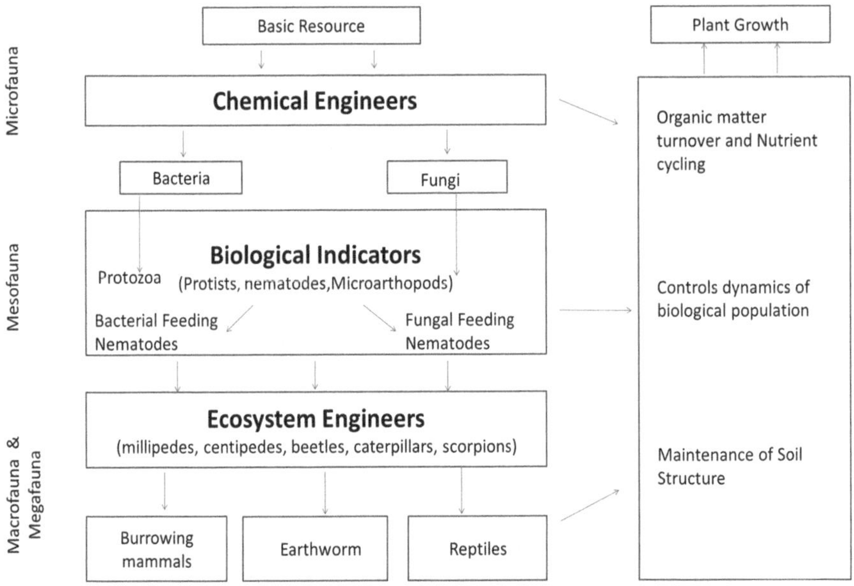

Fig. 5.4 Soil network showing linkages between the different functional groups of soil fauna, along with their role in different soil biological functions

associations acting as a backbone (Brose and Scheu 2014; Digel et al. 2014; Fontaine et al. 2011). Figure 5.4 delineates the various hierarchical associations of the determinants of soil forms (Beare et al. 1995).

The arrangement of elements influencing soil working is resolved at both spatial and temporal scales (Wu and Wang 2019). Different organisms, including higher plants and creatures, assume considerable roles in this regard.

The working of the soil framework is additionally controlled by:

- The deterioration rates of dead organic materials and the equalization between mineralization (which discharges nutrients accessible to plants and microorganisms) and humification (which frames stores of soil organic matter or SOM and colloidal organic mixes)
- The level of synchronization of nutrient discharge with plant request
- The soil physical structure, which decides the rates and examples of gas trade, soil water movement into and through the soil, and disintegration rates
- The texture of the soil (percent of sand, silt, and clay), which impacts the action of soil life forms and henceforth the soil biological working

Soil food networks are created by two fundamental frameworks: one, herbivory based ("green" food networks) and the other detritus based ("brown" food networks; Briones 2014). In herbivory-based food networks, plant roots establish the fundamental basal reserve, grazed by plant-feeding nematodes and insects, which in turn are grazed by a predaceous network ranging within a few trophic levels (Parker 2010). The brown food network is composed of recalcitrant, nonliving organic materials, supported by the detritus pool. Two vitality channels have been distinguished inside brown food networks: (1) the bacterial vitality channel, with microorganisms as essential decomposers, and (2) the fungal vitality channel, where they are the decomposers (Ramsden and Kervalishvili 2008). This bacterial–fungal vitality channel idea by and large adopts a sensibly well-separated network of detritivores/microbivores (Coleman 1996). Moreover, expanding confirmations recommend that omnivory and feeding versatility is summed up inside soil food networks (Jordán 2009), not only at higher trophic levels but also inside detritivores/microbivores organizations (Gupta and Malik 1996). This entire system, counting the distinctive vitality channels, is balanced out by a broad number of trophic and nontrophic collaborations at different spatial and temporal scales (Ramsden and Kervalishvili 2008). Interactions between networks happen at predator levels (Parker 2010), as well as among autotrophs and decomposers, either connected by uneven metabolic abilities (Giller 1996) or resource competition (Hågvar 1998; Scheu 2002). Nontrophic interactions, for example, the movement of soil engineers, are additionally thought to advance assorted variety and decrease competitive interactions, thus contributing to balance soil food networks (Thompson et al. 2012).

5.3.2 Soil Health and Faunal Indicators

Soil health suggests the limit of soil to work as an indispensable living framework, a dynamic framework. It includes the knowledge of soil as a powerful living organism that works comprehensively relying on its condition or state. Soil health relies upon the joined impacts of three noteworthy connecting parts. These are the chemical, physical, and biological attributes of the soil (Cardoso et al. 2013). Soil health is upgraded by the management and land-utilization choices that consider the various functions of soil. It is weakened by choices that look just on single capacities and momentary arrangements, for example, expanding yet not sustaining crop productivity.

Soil organisms apply a noteworthy power over many soil forms through their impacts on the deterioration of dead natural material, nutrient cycling, the change and transport of soil materials, and the development and support of soil structure (Griffiths et al. 2018). Though occasionally not easily perceived, the biological activity of soils is to a great extent gathered in the topsoil, from a couple of centimeters to a depth of 30 cm. The living segment of SOM comprises of:

- Dissolved organic matter made up of soluble root exudates
- Particulate organic matter, including each of fresh residues and living organisms
- Humus content
- Resistant organic matter

Dead organic matter is at first mostly devoured by macrofauna, comminuting and redistributing it into smaller sections. These partially degraded organic residues are then accessible to meso- and microfauna and microorganisms. Through its tunneling movement, soil macrofauna mixes organic matter profoundly into the soil and stimulates the movement of microorganisms. In this manner, soil organisms take an interest in a variety of procedures fundamental to the working ecosystems. For instance, they assume an imperative role in the cycling of SOM and nutrients in the soil, water purification, agrochemicals detoxification, and in the change of soil structure.

Among the capacities directed by soil organisms that help in maintaining the soil health are:

- Decay: separating litter, making hummus, and nutrient cycling
- Atmospheric N_2 conversion into organic forms and reconverting organic N_2 into gaseous N_2
- Enzyme synthesis, hormones, vitamins, and other organic substances for plant growth and development
- Soil structure modification, influencing porosity, water transitions, and organic matter conveyance, and advancing further root development
- Overpowering as well as feeding on soil-borne plant pathogens and plant-parasitic nematodes

More significance has been accustomed to individuals from soil fauna as pointers of health. This aggregate includes the invertebrate network that lives absolutely or amid something like a period of the existing cycle in the soil (Cardoso et al. 2013; Coyle et al. 2017; Jordán 2009). They assume roles in organizing procedures of terrestrial ecosystems, the disintegration of plant residues, and setting up connections at various dimensions with microorganisms. Hence, they effectively participate in procedures that influence the soil properties and quality, and hence are great indicators of changes in the soil (Lavelle and Spain 2001). For example, nematodes, the simplest metazoan, act as bioindicators of soil health. They demonstrate a high and varied sensitivity to pollutants, and because of their trophic diversity, nematode gatherings do reproduce not only their fate but also the state of the bacterial, fungal, and protozoan communities. Because of such characteristics, they are the potentially remarkable bioindicators for soil disturbance and health (Bongers and van de Haar 1990).

A decent variety, biomass, abundance, and thickness of soil fauna have been utilized as indicators of natural or anthropogenic effects on terrestrial ecosystems since these are entirely associated with physical, chemical, and microbiological soil traits (Eggleton et al. 2005; Wardle et al. 2004), and these characteristics of the soil fauna are considered in any prudent methodology utilized for evaluation of the soil health.

Associations among plants and soil's chemical engineers play a critical role in plant network advancement, assorting plant variety, biogeochemical cycling, and supporting general soil structure. Plant roots and microorganisms interact with one another through molecular crosstalk, which is useful, impartial, and sometimes detrimental (Kardol et al. 2006). Plants influence the creation of their dirt network, and, consequently, the soil network influences the efficiency and arrangement of the plant network. There are two noteworthy pathways of soil input. One is immediate, using root herbivores, pathogens, and symbionts, and the second pathway is more roundabout, through the impact of the dirt decomposer subsystem on the supply of supplements (Miethling et al. 2000). In many biological communities, plant development is constrained by the measure of nutrients discharged by microbes and parasites, such as ammonium (NH_4^+), which rely mainly upon the microbial driven disintegration rate. Plants being nonmotile regularly face nutrient deficiencies in their environment, but they have measures to overcome this and obtain micronutrients required for their development and growth. Plant–soil feedback (PSFs) is one such process in which they change the biotic and abiotic potentials of soil, in which they grow, which then alters the plant ability to grow in that soil in the future. Alteration in the measure of quality and quantities of organic substrates flowing into the soil as exudates and litter affects the substrate availability for the microorganism and, eventually, the macroorganisms residing in the soil. Soil invertebrates, along with micro-, meso-, and macrofauna, contribute to plant–soil feedback function, which can lead to the negative, positive, and neutral effects on plant growth. Plant invasion activities are influenced by the PSF, by their impacts on the plant growth. PSFs are mostly involved in the nutrient availability, litter decomposition transfer, soil pathogen accumulation, and interruptions in mutualistic associations in conjunction with soil micro- and mesofaunal populations (Yang et al. 2018 and Zhang et al. 2019). In such similar ways, soil fauna can modulate the soil health that is directly or indirectly related to the associated plant health and can influence the plant invasions and plant network advancement in natural systems by chance, speed, and other different consequences.

5.4 Soil Fauna in Ecosystem Functions and Services

Understanding of how the fauna in soil networks reacts to natural change and how this impacts over-the-ground forms has progressed extensively in the last few decades (Albert et al. 2016; Bragazza et al. 2013; Cheeke et al. 2012; Wurst et al. 2012). A prevalent loss of phylotypes is being observed because of the different management practices, anthropological disintegration, pollution, and widespread urbanization. These have resulting impacts on biological system capacity and ecosystem services, which end up as broadly perceived and identified with bigger concerns such as a reduction in biodiversity, the transformation of habitable lands into deserts, and increased levels of greenhouse gases in the atmosphere (Koch et al. 2013). An examination by Jeffery and Gardi (2010) showcased prospective susceptibility of micro- and mesofaunal biodiversity and ecosystem services to ecological pressures, together with land management practice. Global trials and amalgamations have kept on tending on the evaluation of the role of soil fauna in ecosystem processes and specifically have prompted expanded proof for their commitment to

C cycling. Worldwide tests from different geographical areas demonstrate that soil biota are the main controllers of deterioration rates at biome scale (Wall et al. 2008) and indicate that mainly the invertebrate populations in the soil fauna were in charge for approximately 27% normal improvement of litter decay crosswise over biome scales (GarcíaPalacios et al. 2013; Ma et al. 2019).

The network of organisms living in soil conveys out an extremely wide scope of biochemical and biophysical forms that directs the working of the soil itself and that can likewise influence the neighboring ecosystems (Maltby et al. 2017; Faber and Van Wensem 2012). Huge numbers of these capacities additionally give basic advantages to human society. The vast majority of these services are supporting services or services that are not specifically utilized by people in any case, which underlie the provisioning of every other service. These incorporate, for instance, nutrient cycling and soil arrangement. Furthermore, soil biodiversity is associated with all the fundamental regulatory services—to be specific, the regulation of atmospheric configuration and temperature, water and air quality, pest and disease occurrence in agriculture and natural environments, and human infections. Soil life forms may likewise control or diminish ecological contamination. At last, soil life forms additionally add to provisioning services that straightforwardly benefit individuals; for instance, the genetic resource of soil microorganisms can be utilized for creating novel pharmaceuticals.

Each function adds directly or indirectly up to services. For example, nutrient cycling underlies crop production, while water transfer and storage is affected by soil engineering, and soil biodiversity offers a source of species that may add to pest control, the upgrading of new medicines, or decontamination (Ritz et al. 2012). Different capacities exhibited by soil and soil biodiversity contribute, in a way, to human prosperity; for example, decomposition of soil organic matter, which adds to carbon storage and climate control.

Ecosystem services are characterized as the advantages that individuals get from nature, fundamental for natural wellbeing and human prosperity (MEA 2005). The Millennium Ecosystem Assessment (MA) and the Common International Classification of Ecosystem Services (CICES) classifies ecosystem services into: (1) *provisioning*, which incorporates the generation of products by biological systems (e.g., fibers, water, food, and energy); (2) *regulating*, which incorporates the support of a few procedures identified with climate, water and air quality, and pest and disease management; (3) *supporting*, vital for the execution of all remaining administrations, for example, soil arrangement, nutrient cycling, essential generation, and natural surroundings arrangement; and (4) *cultural*, which incorporates nonmaterial advantages like amusement, ecotourism, and social legacy.

The disintegration of the natural entities by soil fauna is pivotal for the working of a biological community in light of its significant role in giving biological community administrations for plant development and its essential efficiency (Brevik et al. 2019). A comprehensive list of various soil functions performed by soil fauna along with the ecosystem services can be found in Table 5.1.

Soil ecosystem services rely upon soil environmental organization (soil biotic and abiotic parts and the connections inside and among them) and soil environment

Table. 5.1 Relationship between soil ecosystem functions and services with relevant indicators and responsible soil fauna

Kind of service	Ecosystem service	Functions of soil biota	Indicators	Soil fauna	References
Regulating and supporting services	Water quality	Soil organization and its preservation; decomposition and mineralization of organic compounds; resource for predatory microfauna; biofilm formation	Bulk density, porosity, particle size distribution, soil and water retention capacity, water holding capacity	Microfood web; litter transformers; root/rhizosphere biota; N-fixers: mycorrhizae, free-living microbes, microbial pathogens; root herbivores: bacteria, archaea, fungi, actinomycetes; Protista: predators (bacterivores); Nematoda: plant feeders, predators (bacterivores, fungivores); Collembola: saprovores, predators; Acari: saprovores, predators; litter transformers: mesofauna—Enchtriaeids, earthworms, termites	Metcalfe et al. (2014)
	Soil erosion control	Regulation of soil hydrological processes; associative nitrogen (N)-fixing with legumes; N fixing			
	Climate regulation	Gas exchange and carbon sequestration			Hammer et al. (2016) and Metcalfe et al. (2014)
	Attenuation of pollutant	Soil detoxification; immobilization of nutrients; mutualistic and commensal association; parasite of nematodes and organic matter	Macro- and micronutrients, total organic matter, total organic carbon	Ecosystem engineers: macrofauna—earthworm, macroarthropods, burrowers, saprovores, Arthropoda, Enchytraeids; biocontrollers: predators, parasites (hyper), microbial pathogens, root herbivores	Hope et al. (2014), Baron et al. (2014) and Wu et al. (2010)
	Soil structural maintenance (pest and disease control)	Decomposition of organic matter; synthesis of organic compounds; mediate transport of water and ions from soil into the plant root; C allocation below ground and its regulation; suppression of pests and diseases	Emissions of greenhouse gas, phosphorous overflow, nitrate leaching		Jouquet et al. (2014); Evans et al. (2011); Leal et al. (2014)
	Biodiversity conservation	Sources of food and medicine; symbiotic and asymbiotic relationship with plants and their roots; plant growth control (positive + negative)	Soil toxicity testing, soil food webs, soil respiration		Leal et al. (2014), Pringle et al. (2010), Shi and Shofler (2014), and Maleque et al. (2009)

Provisioning services	Food, water, fiber, and energy supply	Consumption of organic matter; create channels in soil and litter for movement; predators of other soil and surface-dwelling organisms	Food quality and storage, plant physiology, crop productivity, creative landscape, leisure activities	Microfood web; microflora: bacteria, fungi; microfauna: protozoa, nematodes, microarthopods; litter transformers: megafauna—moles, little rodents, reptiles	Rodriguez et al. (2006), Macadam and Stockan (2015), Sehnal and Sutherland (2008) and Dzerefos and Witkowski (2014)
Cultural services	Physical or experimental use; intellectual representations		Beautiful landscape, participation in leisure activities, vital actor awareness		Vidal et al. (2014) and Morley et al. (2014)

capacities (regular procedures happening in soil). In the case of soil, water, and air, air compartments are interdependent, and their quality and maintainability are reliant on one another. Soil environment structure is in charge of the adjustments in individual life forms, and hence their role and capacity in soil modification and consequently in the biological community assembly. Soil biodiversity in this way manages soil structure and capacities (Morgado et al. 2018).

5.5 Conclusions and Future Studies

Earth environmentalists and soil modelers have generally depicted the occupants of soil as a black box named as "soil fauna" or "decomposers or detritivores," recognizing the fact that they reuse the stored dead material. The soil is a standout among the various living spaces on earth and contains a diverse collection of living organisms; but, the opaqueness of this world has extremely constrained our comprehension of their functional commitments to soil forms and ecosystem flexibility. Conventional scientific categorization, given morphological and anatomical angles, is getting to be supplanted by molecular techniques (e.g., with marker gene-based methodologies). Nonetheless, this might be impracticable in numerous natural or ecological situations, and therefore the larger part of the present learning still contributes little to our comprehension of their roles in ecosystem functioning.

The present chapter has investigated the reasons why soil fauna is considered for their inherent, useful, and functional qualities. Inherent reasons alone legitimize examination into these different and complex networks and their preservation through natural surroundings and suitable land management practices to address the issues for forthcoming generations. It is recommended that the significance of soil fauna remains an oddity in light of the fact that on one side broad research has demonstrated that they have significant consequences for soil biophysical forms at the scales at which the organisms are active, and on the other side these impacts are once in a while obvious at plot, biological community, and ecosystem scales. Three components are proposed to clarify this mystery. Firstly, that in very differing networks the "signal" of specific soil fauna impacts is concealed by the "noise" from other biophysical occasions that add to similar properties and procedures (e.g., carbon and nitrogen mineralization, soil structure). Secondly, that numerous procedures made by soil fauna have "sink" and "source" elements that can cancel the signal of these nearby impacts at bigger spatial scales. Thirdly, that at enormous temporal and spatial scales, biophysical parameters are utilized as rate determinants of biological system forms and the structure of above or subterranean networks is once in a while raised. However, there are conditions in which the environments are in transitional states, or in which natural occasions synchronize with activities of soil fauna when roles of soil fauna end up clear at the plot and environment level. Further examination into these procedures could define circumstances in which soil fauna are key determinants of biological system functions or ecosystem processes and increase acknowledgment of the soil fauna by different disciplines.

References

A report of the Millennium Ecosystem Assessment (2005) Ecosystems and human well-being: wetlands and water. World Resources Institute, Washington, DC

Albert C, Bonn A, Burkhard B, Daube S, Dietrich K, Engels B et al (2016) Towards a national set of ecosystem service indicators: insights from Germany. Ecol Indic 61(1):38–48

Bardgett RD (2002) Causes and consequences of biological diversity in soil. Zoology 105(4):367–375

Bardgett RD, van der Putten WH (2014) Belowground biodiversity and ecosystem functioning. Nature 515(7528):505–511. https://doi.org/10.1038/nature13855

Baron GL, Raine NE, Brown MJ (2014) Impact of chronic exposure to a pyrethroid pesticide on bumblebees and interactions with a trypanosome parasite. J Appl Ecol 51(2):460–469

Barrios E (2007) Soil biota, ecosystem services and land productivity. Ecol Econ 64(2):269–285

Beare M, Coleman D, Crossley D, Hendrix P, Odum E (1995) A hierarchical approach to evaluating the significance of soil biodiversity to biogeochemical cycling the significance and regulation of soil biodiversity. Plant Soil 170(1):5–22. Springer

Bongers T, van de Haar J (1990) On the potential of basing an ecological typology of aquatic sediments on the nematode fauna: an example from the river Rhine. Hydrol Bull 24(1):37–45

Bragazza L, Parisod J, Buttler A, Bardgett RD (2013) Biogeochemical plant–soil microbe feedback in response to climate warming in peatlands. Nat Clim Chang 3(3):273

Brevik EC, Pereg L, Pereira P, Steffan JJ, Burgess LC, Gedeon CI (2019) Shelter, clothing, and fuel: often overlooked links between soils, ecosystem services, and human health. Sci Total Environ 651:134–142

Briones MJI (2014) Soil fauna and soil functions: a jigsaw puzzle. Front Environ Sci 2:7

Briones MJ (2018) The serendipitous value of soil fauna in ecosystem functioning: the unexplained explained. Front Environ Sci 6:149

Brose U, Scheu S (2014) Into darkness: unraveling the structure of soil food webs. Oikos 123(10):1153–1156

Byers JE, Cuddington K, Jones CG, Talley TS, Hastings A, Lambrinos JG et al (2006) Using ecosystem engineers to restore ecological systems. Trends Ecol Evol 21(9):493–500

Cardoso EJBN, Vasconcellos RLF, Bini D, Miyauchi MYH, dos Santos CA, Alves PRL et al (2013) Soil health: looking for suitable indicators. What should be considered to assess the effects of use and management on soil health? Sci Agric 70(4):274–289

Carson JK, Gonzalez-Quiñones V, Murphy DV, Hinz C, Shaw JA, Gleeson DB (2010) Low pore connectivity increases bacterial diversity in soil. Appl Environ Microbiol 76(12):3936–3942

Cheeke TE, Coleman DC, Wall DH (2012) Microbial ecology in sustainable agroecosystems. CRC Press

Coleman DC (1996) Energetics of detritivory and microbivory in the soil in theory and practice. Food Webs:39–50. Springer

Coleman DC, Crossley DA Jr (1996) Fundamentals of soil ecology, vol 3. Academic Press, San Diego

Coleman DC, Whitman WB (2005) Linking species richness, biodiversity and ecosystem function in soil systems. Pedobiologia 49(6):479–497

Coyle DR, Nagendra UJ, Taylor MK, Campbell JH, Cunard CE, Joslin AH et al (2017) Soil fauna responses to natural disturbances, invasive species, and global climate change: current state of the science and a call to action. Soil Biol Biochem 110:116–133

Crawford JW, Harris JA, Ritz K, Young IM (2005) Towards an evolutionary ecology of life in the soil. Trends Ecol Evol 20(2):81–87

Davies G, Kirkpatrick J, Cameron E, Carver S, Johnson CN (2019) Ecosystem engineering by digging mammals: effects on soil fertility and condition in Tasmanian temperate woodland. R Soc Open Sci 6(1):180621

Decaëns TJ (2010) Macroecological patterns in soil communities. Glob Ecol Biogeogr 19(3):287–302

Delgado-Baquerizo, M., Bardgett, R. D., Vitousek, P. M., Maestre, F. T., Williams, M. A., Eldridge, D. J., ... & Sala, O. E. (2019). Changes in belowground biodiversity during ecosystem development. Proc Nat Acad Sci 116(14):6891–6896. 201818400
Digel C, Curtsdotter A, Riede J, Klarner B, Brose U (2014) Unraveling the complex structure of forest soil food webs: higher omnivory and more trophic levels. Oikos 123(10):1157–1172
Dzerefos CM, Witkowski ETF (2014) The potential of entomophagy and the use of the stinkbug, Encosternum delegorguei Spinola (Hemipera: Tessaratomidae), in sub-Saharan Africa. Afr Entomol 22(3):461–472
Eggleton P, Vanbergen AJ, Jones DT, Lambert MC, Rockett C, Hammond PM et al (2005) Assemblages of soil macrofauna across a Scottish land-use intensification gradient: influences of habitat quality, heterogeneity and area. J Appl Ecol 42(6):1153–1164
Evans DM, Pocock MJ, Brooks J, Memmott J (2011) Seeds in farmland food-webs: resource importance, distribution and the impacts of farm management. Biol Conserv 144(12):2941–2950
Faber JH, Van Wensem J (2012) Elaborations on the use of the ecosystem services concept for application in ecological risk assessment for soils. Sci Total Environ 415:3–8
FitzPatrick EA (1984) Micromorphoiogy of soils. Chapman and Hall, London
Fontaine C, Guimarães PR Jr, Kéfi S, Loeuille N, Memmott J, van Der Putten WH et al (2011) The ecological and evolutionary implications of merging different types of networks. Ecol Lett 14(11):1170–1181
Fusaro S, Gavinelli F, Lazzarini F, Paoletti MG (2018) Soil biological quality Index based on earthworms (QBS-e). A new way to use earthworms as bioindicators in agroecosystems. Ecol Indic 93:1276–1292
GarcíaPalacios P, Maestre FT, Kattge J, Wall DH (2013) Climate and litter quality differently modulate the effects of soil fauna on litter decomposition across biomes. Ecol Lett 16(8):1045–1053
Giller P (1996) The diversity of soil communities, the 'poor man's tropical rainforest'. Biodivers Conserv 5(2):135–168
Griffiths B, Faber J, Bloem J (2018) Applying soil health indicators to encourage sustainable soil use: the transition from scientific study to practical application. Sustainability 10(9):3021
Gupta S, Malik V (1996) Soil ecology and sustainability. Trop Ecol 37(1):43–55
Ha KV, Marschner P, Bünemann EK (2008) Dynamics of C, N, P and microbial community composition in particulate soil organic matter during residue decomposition. Plant Soil 303(1–2):253–264
Hågvar S (1998) The relevance of the Rio-convention on biodiversity to conserving the biodiversity of soils. Appl Soil Ecol 9(1–3):1–7
Hammer TJ, Fierer N, Hardwick B, Simojoki A, Slade E, Taponen J et al (2016) Treating cattle with antibiotics affects greenhouse gas emissions, and microbiota in dung and dung beetles. Proc R Soc B Biol Sci 283(1831):20160150
Haynes RJ (2014) Nature of the belowground ecosystem and its development during pedogenesis. Adv Agron 127:43–109
Hope PR, Bohmann K, Gilbert MTP, Zepeda-Mendoza ML, Razgour O, Jones G (2014) Second generation sequencing and morphological faecal analysis reveal unexpected foraging behaviour by Myotis nattereri (Chiroptera, Vespertilionidae) in winter. Front Zool 11(1):39
Jeffery S, Gardi CJ (2010) Soil biodiversity under threat– a review. Acta Soc Zoo Bohem 74(1–2):7–12
Jordán F (2009) Keystone species and food webs. Philos Trans Royal Soc B 364(1524):1733–1741
Jouquet P, Blanchart E, Capowiez Y (2014) Utilization of earthworms and termites for the restoration of ecosystem functioning. Appl Soil Ecol 73:34–40
Kardol P, Bezemer TM, van der Putten WH (2006) Temporal variation in plant-soil feedback controls succession. Ecol Lett 9(9):1080–1088. https://doi.org/10.1111/j.1461-0248.2006.00953.x
Kennedy AC, Gewin VL (1997) Soil microbial diversity: present and future considerations. Soil Sci 162(9):607–617
Koch A, McBratney A, Adams M, Field D, Hill R, Crawford J et al (2013) Soil security: solving the global soil crisis. Glob Policy 4(4):434–441
Lal R (2016) Soil health and carbon management. Food Energy Sec 5(4):212–222

Lavelle P (1997) Faunal activities and soil processes: adaptive strategies that determine ecosystem function. Adv Ecol Res 27:93–132
Lavelle P (2012) Soil as a habitat soil ecology and ecosystem services. Oxford University Press, Oxford, pp 1–27
Lavelle P, Spain AV (2001) Soil ecology. Kluwer Scientific, Amsterdam
Leal LC, Andersen AN, Leal IR (2014) Anthropogenic disturbance reduces seed-dispersal services for myrmecochorous plants in the Brazilian Caatinga. Oecologia 174(1):173–181
Ma C, Yin X, Wang HJAB (2019) Soil fauna effect on Dryas octopetala litter decomposition in an Alpine tundra of the Changbai Mountains, China. Alpine Bot 129(1):1–10
Macadam CR, Stockan JA (2015) More than just fish food: ecosystem services provided by freshwater insects. Ecol Entomol 40:113–123
Maleque MA, Maeto K, Ishii HT (2009) Arthropods as bioindicators of sustainable forest management, with a focus on plantation forests. Appl Entomol Zool 44(1):1–11
Maltby L, Duke C, Van Wensem J (2017) Ecosystem services, environmental stressors, and decision making: how far have we got? Integr Environ Assess Manag 13(1):38–40
Menta C (2012) Soil fauna diversity-function, soil degradation, biological indices, soil restoration. In: Biodiversity conservation and utilization in a diverse world. IntechOpen, London
Metcalfe DB, Asner GP, Martin RE, Silva Espejo JE, Huasco WH, Farfán Amézquita FF et al (2014) Herbivory makes major contributions to ecosystem carbon and nutrient cycling in tropical forests. Ecol Lett 17(3):324–332
Miethling R, Wieland G, Backhaus H, Tebbe C (2000) Variation of microbial rhizosphere communities in response to crop species, soil origin, and inoculation with Sinorhizobium meliloti L33. Microb Ecol 40(1):43–56. https://doi.org/10.1007/s002480000021
Morgado RG, Loureiro S, González-Alcaraz MN (2018) Changes in soil ecosystem structure and functions due to soil contamination. Soil Pollut:59–87. Elsevier
Morley EL, Jones G, Radford AN (2014) The importance of invertebrates when considering the impacts of anthropogenic noise. Proc R Soc B Biol Sci 281(1776):20132683
Orgiazzi A, Bardgett RD, Barrios E (2016) Global soil biodiversity atlas. European Commission Publication Office
Parker SS (2010) Buried treasure: soil biodiversity and conservation. Biodivers Conserv 19(13):3743–3756
Pringle RM, Doak DF, Brody AK, Jocqué R, Palmer TM (2010) Spatial pattern enhances ecosystem functioning in an African savanna. PLoS Biol 8(5):e1000377
Ramsden J, Kervalishvili PJ (2008) Soil as a paradigm of a complex system. In: Complexity and security, vol 37. Washington, DC, IOS Press, p 103
Ritz K, Six J, Strong DR, van der Putten WH (2012) Soil ecology and ecosystem services. Oxford University Press, Oxford
Robinson D, Hockley N, Cooper D, Emmett B, Keith A, Lebron I et al (2013) Natural capital and ecosystem services, developing an appropriate soils framework as a basis for valuation. Soil Biol Biochem 57:1023–1033
Rodríguez LC, Pascual U, Niemeyer HM (2006) Local identification and valuation of ecosystem goods and services from Opuntia scrublands of Ayacucho, Peru. Ecol Econ 57(1):30–44
Ruggiero MA, Gordon DP, Orrell TM, Bailly N, Bourgoin T, Brusca RC et al (2015) A higher level classification of all living organisms. PLoS One 10(4):e0119248
Scheu SJ (2002) The soil food web: structure and perspectives. Eur J Soil Biol 38(1):11–20
Sehnal F, Sutherland T (2008) Silks produced by insect labial glands. Prion 2(4):145–153
Shi E, Shofler D (2014) Maggot debridement therapy: a systematic review. Br J Community Nurs 19(Sup12):S6–S13
Thompson RM, Brose U, Dunne JA, Hall RO Jr, Hladyz S, Kitching RL, Tylianakis JM (2012) Food webs: reconciling the structure and function of biodiversity. Trends Ecol Evol 27(12):689–697
Turbé A, De Toni A, Benito P, Lavelle P, Lavelle P, Camacho NR et al (2010) Soil biodiversity: functions, threats, and tools for policy makers. Bio Intelligence Service; IRD, Paris/Marseille, 250 pp

Trubl G (2018) Pioneering soil viromics to elucidate viral impacts on soil ecosystem services. Doctoral dissertation, The Ohio State University

Vidal O, López-García J, Rendón-Salinas E (2014) Trends in deforestation and forest degradation after a decade of monitoring in the Monarch Butterfly Biosphere Reserve in Mexico. Conserv Biol 28(1):177–186

Wall DH, Snelgrove PVR, Covich AP (2001) Conservation priorities for soil and sediment invertebrates. In: Conservation biology: research priorities for the next decade. Island Press, Washington, DC, pp 99–123

Wall DH, Bradford MA, St. John MG, Trofymow JA, Behan-Pelletier V, Bignell DE et al (2008) Global decomposition experiment shows soil animal impacts on decomposition are climate-dependent. Global Change Biol 14(11):2661–2677

Wall DH, Bardgett RD, Kelly EJ (2010) Biodiversity in the dark. Nat Geosci 3(5):297

Wardle DA (2006) The influence of biotic interactions on soil biodiversity. Ecol Lett 9(7):870–886

Wardle DA, Bardgett RD, Klironomos JN, Setälä H, Van Der Putten WH, Wall DH (2004) Ecological linkages between aboveground and belowground biota. Science 304(5677):1629–1633

Williamson KE, Fuhrmann JJ, Wommack KE, Radosevich M (2017) Viruses in soil ecosystems: an unknown quantity within an unexplored territory. Annu Rev Virol 4:201–219

Wright JP, Jones CG, Flecker AS (2002) An ecosystem engineer, the beaver, increases species richness at the landscape scale. Oecologia 132(1):96–101

Wu P, Wang C (2019) Differences in spatiotemporal dynamics between soil macrofauna and mesofauna communities in forest ecosystems: the significance for soil fauna diversity monitoring. Geoderma 337:266–272

Wu H, Lu X, Wu D, Yin X (2010) Biogenic structures of two ant species Formica sanguinea and Lasius flavus altered soil C, N and P distribution in a meadow wetland of the Sanjiang Plain, China. Appl Soil Ecol 46(3):321–328

Wurst S, De Deyn GB, Orwin K (2012) Soil biodiversity and functions. In: Oxford handbook of soil ecology and ecosystem services. Oxford University Press, Oxford, pp 28–45

Yang G, Wagg C, Veresoglou SD, Hempel S, Rillig MC (2018) How soil biota drive ecosystem stability. Trends Plant Sci 23(12):1057–1067

Zhang P, Li B, Wu J, Hu S (2019) Invasive plants differentially affect soil biota through litter and rhizosphere pathways: a meta-analysis. Ecol Lett 22(1):200–210

An Insight into Current Trends of Pathogen Identification in Plants

6

Vinay Kumar, Vinukonda Rakesh Sharma, Himani Patel, and Nisha Dinkar

Abstract

Worldwide, plant diseases have caused major economic losses in the agricultural industry. Food security is a major issue to the rising global population. Plant health monitoring and the early detection of pathogens are important in decreasing the propagation of diseases and providing the best strategy. To reduce the damage caused by old and new pathogens, and to speed up the management and reduction of crop loss, a fast and reliable detection method in combination with decision support systems is essential. New tools and technologies were developed for both detection and diagnosis, often substituting old methods to make them quicker, more accurate, and precise. With the rising global population, there is need to boost crop production and protect the crop from seed to market. There is also need to use the latest technology for rapid and precise diagnosis of plant diseases to minimize losses for sustainable production. In addition to the traditional visual screening for symptoms, nucleic acid and serological-based tools will provide precise and rapid detection and diagnosis. Remote sensing technology, along with spectroscopy, will help in the capture of high spatialization of results, and can be very useful to identify primary infections quickly. Handheld antibody-based immuno-biosensor also helps the rapid detection of plant diseases. This chapter enlightens on the various recent technologies in plant

V. Kumar (✉) · V. R. Sharma
Genetics and Plant Molecular Biology, CSIR-National Botanical Research Institute, Lucknow, Uttar Pradesh, India

Academy of Scientific and Innovative Research (AcSIR), Ghaziabad, Uttar Pradesh, India

H. Patel
Birbal Sahni Institute of Palaeosciences, Lucknow, Uttar Pradesh, India

N. Dinkar
Department of Biotechnology, Khandelwal College of Management Science and Technology, Kalapur, Bareilly, Uttar Pradesh, India

© Springer Nature Singapore Pte Ltd. 2020

M. K. Solanki et al. (eds.), *Phytobiomes: Current Insights and Future Vistas*,
https://doi.org/10.1007/978-981-15-3151-4_6

pathogen identification for sustainable and effective crop production, and the safeguard of the crop produce.

Keywords
Pathogen identification · Molecular diagnostics · PCR · Plant pathology · Plant diseases

6.1 Introduction

Due to increasing population and globalization, food security, and prevention of the spread of invasive pests/pathogens, accurate identification and diagnosis of plants are very important to protect available crop produce. Plant pathogens causes significant losses in plant yield. Enormous economic losses due to wheat rusts and Fusarium head blight claimed US $5 and US $3 billion/year, respectively (Schumann and D'Arcy 2009). Precise, sensitive, and special diagnoses are necessary to manage plant diseases efficiently and economically. Identification of plant diseases have evolved from visual inspection to the use of high-profile serological tools such as the enzyme-linked immunosorbent assay (ELISA), and molecular methods like polymerase chain reaction (PCR). MLO-7 gene in the grapes against powdery mildew disease, which ultimately increase the precision for the latest technologies such as CRISPR (clustered regularly interspaced short palindromic repeats) in plant disease studies, has become possible through the application of biotechnology and bioinformatics in plant pathology (Malnoy et al. 2016). Although progress has been made about all aspects of plant disease diagnosis, there is a need for increased sensitivity and specificity of molecular studies for rapid detection and diagnosis. This chapter describes briefly the different techniques used for diagnosing plant diseases and their contribution to the current challenges for global food security.

6.1.1 Microbiome Diversity

The diversity of the plant microbes is defined as the number of microbe species present in the particular ecological area and relationship between them with their ecological, environmental conditions (Vos et al. 2016).

Nowadays, a different type of molecular methods was used in the detection of microbes present in the sample. Firstly we collect the sample and isolate the DNA/RNA/protein/lipids and then follow the different methods, viz., PCR, qPCR, MALDI-TOF MS, electrospray, and ionization mass spectrometry, for the assessment of diversity present in the sample (Table 6.1). Here, Table 6.1 discusses some molecular techniques for plant disease diagnosis.

Table 6.1 List of different techniques involved in the detection of plant microbes with their advantages and disadvantages

Detection methods		Merits	Demerits	References
Visual detection		Simplest methods	Lack of sensitivity in detecting that occur in low concentration, asymptomatic	Ali et al. (2019), Gomez et al. (2019) and Cooke and Cacciola (2007)
Polymerase chain reaction-based methods	Conventional PCR	Specific primer for particular species produces precise and rapid results	More costly and necessitate more labor	Milijasevic et al. (2006) and Walcott and Gitaitis (2000)
	Portable rapid cycler PCR	Identification of plant disease at on-site	Pricey and deficient in robustness	Love et al. (2012)
	Nested PCR	Utilization of two pairs of primers raises yield and accuracy of desired DNA amplification	More chances of contamination due to two times amplification reactions	Lee et al. (1997) and Bereswill et al. (1995)
	Multiplex PCR	Resources and time rescued by using the various set of primers in PCR reactions	Primer and probe interference reduces the sensitivity	Rico et al. (2003) and Fegan et al. (1998)
	Reverse transcriptase PCR	It is more accurate than standard PCR and produces the quantitative data	Information about every test is difficult and involves costly materials and chemicals	Liu et al. (2019a, b)
	Real-time qPCR	No requirement of post-amplification reaction and it is an automated process	Pricey and difficult in owing to concurrent detection of thermocycling with fluorescence	Schaad et al. (2007) and Salm and Geider (2004)
	Serial analysis of gene expression (SAGE)	Reference knowledge of the genome is not required	Accuracy of the tag sequences	Tarasov et al. (2007)

(continued)

Table 6.1 (continued)

Detection methods		Merits	Demerits	References
DNA/RNA probe-based methods	Northern blotting	Identification of size of RNA	It applies to the very specific sample of the gene sequences	Yin et al. (2019) and Boccardo et al. (2019)
	In situ hybridization	It required more supply of tissue	It is very complex to detect the target of sequences with low copies of DNA or RNA	Ellison et al. (2016) and Tanaka (2009)
	Fluorescence in situ hybridization (FISH)	It can be used in nondividing cells	It is essential to identifying the target DNA sequences	Sidra et al. (2017) and Ratan et al. (2017)
Post-amplification technique	Microarray Macroarray	It is high-throughput tool and allows identification of various pathogens	It does not differentiate the living and non-living cells analysis	Osmani et al. (2019), Leborgne and Bouhidel (2014) and Sato et al. (2010)
Isothermal amplification-based methods	Loop-mediated isothermal amplification (LAMP)	Fast, accurate, and extremely specific	Primer formation is more careful binding only on the specific pathogen	Almasi et al. (2013)
	Rolling circle amplification (RCA)	Simplicity, efficiency, tunability	–	Gomez et al. (2014)
	Nucleic acid sequence-based amplification (NASBA)	There is no need for expensive equipment, and also it is better than RT-PCR	Sample reaction temperature is 42 °C, and it cannot be increased because it depends on enzymes	Gabrielle et al. (1993)
RNA-sequence based next-gen. sequencing		Specificity and sensitivity higher	It requires more costly tools and information for the data analysis	Dawei and Peng (2014) and de Jonge et al. (2012)
The RNA interference (RNAi)		It is potential to investigate thousands of genes concurrently	Diversity and incompleteness of knockdowns and latent in nonredundancy of chemicals	Machado et al. (2017)

(continued)

Table 6.1 (continued)

Detection methods		Merits	Demerits	References
Spectroscopy	MALDI-TOF MS	Much faster, rapid turnaround time	Difficulty in distinguishing some organisms that are closely related genetically	Bruker (2016)
	Volatile compounds (GC-MS)	Highly indicative, accurate, high specificity	Capable of directly analyzing which are nonvolatile, polar, or thermally labile	Fang et al. (2014)
Antibody	Quartz crystal microbalance immunosensors (QCMI)	Real-time analysis possible	Better sensitivity required	Huang et al. (2014) and Zan et al. (2012)
	Agriculture nanobiosensors	Real-time analysis possible	Just beginning, trouble of reproducibility, and measurement errors	Etefagh et al. (2013) and Dubas and Pimpan (2008)
	Tissue blot immunoassay (TBIA)	Quick and simple to use	Costly	Chang et al. (2011)
	ELISA	It can manage a large amount of samples in the same time and produces more accurate results	Pre-enrichment is required for more surface antigens and more highly skilled manpower required	Kanakala and Kuria (2018)
	Flow cytometry	Concurrently identification and quantification of various pathogens in a consistent manner	More investment and very less information of this tools	Fang and Ramasamy (2015) and D'Hondt et al. (2011)
Quantum dots (QDs)		High sensitivity, real-time analysis possible	Limits for detection, trouble of reproducibility, and measurement errors	Syed and Ahmad (2013) and Kumar et al. (2007)

6.2 Toolbox for Diagnosing Plant Pathogens

The knowledge of the physiological method of interaction is lacking for many years between host and pathogen for many systems of pathogens (Table 6.2). Recently, quantitative high-performance imaging methods have been developed to phenotype plant growth and development (Mutka and Bart 2015). The following methods help to study the changing physiology of plants due to pathogens that further help in the development of mechanisms for disease symptoms.

Table 6.2 List of important fungal, bacterial, and viral pathogens affecting crop plants with their diseases, host, and percentage of crop loss

S.no.	Pathogen	Disease	Host	Crop loss	References
A	**Fungal**				
1	*Fusarium oxysporum*	Wilt	100 different hosts (mostly horticulture crops)	Nearly 80%	Debbi et al. (2018)
2	*Blumeria graminis*	Powdery mildews	Poaceae crops	Up to 45	Zulak et al. (2018)
3	*Botrytis cinerea*	Gray mold	200 crop hosts worldwide	50–100%	Rupp et al. (2017)
4	*Colletotrichum* spp.	Anthracnose	Economically important crops, especially fruits, vegetables, and ornamentals	Up to 100%	Han et al. (2016)
5	*Magnaporthe oryzae B.C. Couch*	Blast disease	Poaceae crops and their wild relatives	10–30%	Yoshida et al. (2016)
6	*Melampsora lini*	Rust	Flax	~80%	Nemri et al. (2014)
7	*Fusarium graminearum*	Head blight	Wheat and barley	Up to 70%	Yang et al. (2013)
8	*Mycosphaerella graminicola*	Blotch	Wheat	30–50%	Simón et al. (2012)
9	*Ustilago maydis*	Smut	Corn	<20%	Dean et al. (2012)
10	*Puccinia* spp.	Rust	Ferns, gymnosperms to highly evolved families of dicotyledons and monocotyledons	>80%	Dean et al. (2012)
B	**Bacterial**				
1	*Agrobacterium tumefaciens*	Crown gall tumor	140 species of eudicots	Considerable economic loss	Kerpen et al. (2019)
2	*Xylella fastidiosa*	Pierce's disease of grapevine (PD), citrus variegated chlorosis (CVC), and almond leaf scorch disease (ALSD)	Horticulture crops and especially grapes	Serious losses in grape production	Kyrkou et al. (2018)

(continued)

Table 6.2 (continued)

S.no.	Pathogen	Disease	Host	Crop loss	References
3	*Erwinia amylovora*	Fire blight	Apple, pear, quince, blackberry, raspberry, and many wild and cultivated rosaceous ornamentals	Serious losses in pome production	Aćimović et al. (2015)
4	*Xanthomonas axonopodis* pv. *manihotis*	Cassava bacterial blight	Tubers	12–100%	López and Bernal (2012)
5	*Pseudomonas syringae* pathovars	Blight, canker, brown spot, etc.	Mostly horticultural crops	40%	Mansfield et al. (2012)
6	*Pectobacterium carotovorum* (and *P. atrosepticum*)	Soft rot, blackleg	Mostly potato	Serious	Nykyri et al. (2012)
7	*Xanthomonas oryzae* pv. *oryzae*	Bacterial leaf blight	Poaceae crops	Up to 22.5%	Soto-Suárez et al. (2010)
8	*Ralstonia solanacearum*	Bacterial wilt	Very wide range of potential plants	25–75%	Artal et al. (2012)
9	*Xanthomonas campestris* pathovars	Black rot	57 families of dicotyledons and mostly cruciferous crops	20–45%	Hayward (1993)
C	**Virus**				
1	Tomato yellow leaf curl virus (TYLCV)	Leaf curl	Wide host range (>800 plant species)	Up to 60%	Ding et al. (2019)
2	Tomato spotted wilt virus (TSWV)	Spotted wilt	Around 500 species of crops	Up to 100%	Gupta et al. (2018)
3	Brome mosaic virus (BMV)	Mosaic of grasses	Poaceae crops	Up to 20%	Ding et al. (2018)
4	Potato virus Y (PVY)	Necrotic lesions	Solanaceae	50–80%	Hussain et al. (2016)
5	Cauliflower mosaic virus (CaMV)	Cauliflower mosaic virus	Brassicaceae family	25–59% loss	Schoelz et al. (2015)
6	Cucumber mosaic virus (CMV)	Mosaic	More than 40 families	Up to 60%	Phan et al. (2014)
7	Tobacco mosaic virus (TMV)	Necrotic local lesions	Many economically important crops	20–80%	Scholthof et al. (2011)
8	Plum pox virus (PPV)	Plum pox	Pome fruits	20–40%	Cambra et al. (2006)
9	African cassava mosaic virus (ACMV)	Cassava mosaic disease	Tubers	30–100%	Fargette et al. (1988)

6.2.1 Visual Detection

One of the simplest methods for plant pathogen detection was a plant germplasm visual inspection and subsequent selection of healthy matter. Disease scaling based on visual symptoms is being acceptable with accuracy since the last 80 years (Martinelli et al. 2015). Different methods can be used for the visual inspection including visible illumination, chlorophyll fluorescence imagery, infrared imaging, and electromagnetic spectrum imaging (Cooke and Cacciola 2007; Bock et al. 2010; Li et al. 2014a, b; Odilbekov et al. 2018). The environmental and biological factors have a strong influence on these methods. Traditional methods for measuring the severity of the disease are not reliable for population estimating diseases lack standardization and procedural (Nilsson et al. 2011). In asymptomatic plants and less virulent pathogens, visual inspections may not be helpful; many hosts remain symptomatic (Mutka and Bart 2015; Ali et al. 2019; Liu et al. 2019a, b; Gomez et al. 2019).

6.2.2 Serology-Based Diagnostics

Phenotypic observation of plants with pathogenic symptoms would require extensive experience in the diagnosis of plant diseases and isolation of the disease. But for all attempts, the exactness of the same cannot necessarily be the same. Furthermore, the detection of the pathogens in asymptomatic samples or seeds is very likely to be diagnosed properly. The precise identification and an adequate diagnosis procedure would contribute more promisingly to plant health management. Through inventions of precise methods of detection, better management strategies are developed, and then, the rapid detection tool, in particular, an onsite handy one, definitely helps to achieve management success. A handy detection kit for potato viruses has been developed recently, which growers can use easily in the field (Ansar and Singh 2016). No artificial cultivation of plant pathogens like viruses and obligate pathogens causing rusts and powdery mildew and downy mildew diseases was possible. To address this problem, there were developed serologic assays, which are also used to identify other plant pathogens (Caruso et al. 2002). Serological pathogens of plants involve identifying disease based upon color-changing antibodies of the assay. The antibodies consist of immunoglobulin (Ig) proteins produced within the animal body (mammalian), which in general include foreign proteins, complex carbon dioxide, multiple nucleotides, or lipopolysaccharides. As we know, each antigen binds to a particular antibody. The serological method most commonly used is ELISA (enzyme-linked immunosorbent assay) as previously developed for plant virus detection (Clark 1981).

6.2.2.1 Enzyme-Linked Immunosorbent Assay (ELISA)

ELISA is done with polystyrene plates capable of binding enzyme-substrate reaction antibodies or proteins (Corning Life Science 2001; Luminex 2010). The enzyme-substrate response must be optimized in the timing and development conditions to produce accurate and replicable results. ELISA is a popular test for the detection of plant pathogens present in plant resources and insect vectors (Clark and Adamas 1977; Naidua and Hughes 2001; Webster et al. 2004). Disease levels are observed on the bases of optical density in ELISA (Corning Life Science 2001; Webster et al. 2004). ELISA has benefits because it is more accurate; a big amount of tested materials can also be evaluated using a small number of antibodies to detect illnesses and a semiautomated process (Vemulapati et al. 2014; Naidua and Hughes 2001). Specific antibodies against the target pathogen have been developed (Torrance 1998). This technique is utilized for the identification of the various types of virus, i.e., Citrus tristeza virus (CTV), potato leaf roll virus (PLRV), potato virus X (PVX), and potato virus Y (PVY) (El-Araby et al. 2009; Sun et al. 2001). Since ELISA is a test based on antibody antigens, the availability of antibodies that react properly to the target agent is considered very important. ELISA often provides erroneous diagnostic because of false positive, which results primarily from unspecific or cross-reactive reactions with certain sample factors (Kfir and Genthe 1993). Several strains with a clear different symptom may be responded to by the antibody used in ELISA due to the absence of great specific places to bind.

Consequently, much related to different types of virus are not properly distinguished by ELISA (Boonham et al. 2014). Different type of additives were includes in extraction buffer for increased in ELISA sensitivity (Fegla and Kawanna 2013). When compared to molecular methods, ELISA is generally less sensitive. For these reasons, the utilization of ELISA in the way of diagnosis appears to progressively fall down, even when ELISAs have been used in the most up-to-date diagnostic purposes.

6.2.2.2 Tissue Blot Immunoassay (TBIA)

Tissue Blot Immunoassay (TBIA) is similar to the principle of ELISA in which antibodies are functional; TBIA has a similar consistency to ELISA to detect plant pathogen like that virus (Hančević et al. 2012). The main difference is that polystyrene plate is working as a platform for ELISA, while nitrocellulose and nylon membranes are treated with TBIA; that's why this assay is also known as TIBA or TBIA (Webster et al. 2004). Similar to ELISA, TBIA is also required for the particular antibody to find clear false positive and false negative. Therefore, TBIA has huge advantages over ELISA in different conditions such as time, money, sensitivity, and expediency of detection. In TIBA, we have detected a various number of viral diseases in which some are here, i.e., bamboo mosaic virus (BoMV), bean yellow mosaic virus (BYMV), CTV, *Cymbidium* mosaic virus (CyMV), papaya ringspot virus (PRSV), sweet potato feathery mottle virus (SPFMV), and tomato spotted wilt virus (TSWV) (Bove et al. 1988; Eid et al. 2008; Hančević et al. 2012; Lin et al. 1990; Makkouk and Kumari 2006; Shang et al. 2011; Webster et al. 2004).

6.2.2.3 Flow Cytometry (FCM)

Flow cytometry is a laser-based optical method. The technique is widely used in the fields of cell count, cell sorting, biomarker, and protein detection (O'Donnell et al. 2013). Many samples can be processed simultaneously via electronic detection devices, and the technique simultaneously measures many parameters (Fang and Ramasamy 2015). Flow cytometry is widely used to evaluate microorganisms in the processing of food in drinking and marine water. FCM is not currently widely used as an instrument for research and detection in plant pathology; however, it can be used to test the total DNA content in bacterial, oomycete, and fungal disease agents and check pathogen viability; FCM can be used in the field of plant pathology for multiplexed path detection (D'Hondt et al. 2011). Flow cytometer rating is highly precise and purifies little or complex populations, but some of them are not fast enough to achieve the desired results even for a high-speed rating, and it costs more than the alternatives such as radioimmunoassay and ELISA.

6.3 Molecular Methods

Molecular methods of diagnosis of diseases are one of the remarkable sciences that are fast-growing and have the potential to revolutionize numerous scientific research, innovations, healthcare, and agriculture disciplines.

6.4 Genomics-Driven Diagnostics

6.4.1 Conventional PCR

The innovation of PCR got an incredible boom in the field of plant pathology. This tool permits amplification of specific DNA sequences in lots of duplicates by utilizing specific primers (Cha and Thilly 1993). At first, PCR was very explicit for the detection of infections brought about by microscopic organisms, i.e., bacteria, fungi, and virus. Presently, this tool is broadly utilized for the detection of the pathogen from the infected plants (Fig. 6.1). Occasionally, efficiency is influenced by sampled inhibitors (Fang and Ramasamy 2015). To resolve this problem in plants, cetyltrimethylammonium bromide (CTAB) can be used to add particular treatments with different chemicals and enzymes. A few different techniques are available to decrease PCR inhibitor effects. In the case of DNA isolation through the phenol-chloroform method, as a substitute for ethanol precipitation and silica bed, recovery of DNA can be done by purification (Mancini et al. 2016). Also, it can detect plant pathogens; PCR technology requires first to start the DNA replication procedure that might decrease the convenient application for disease field sample technology. Sometimes a single pair of primer does not give accurate results so that its limiting effect is overcome by DNA samples and nested primers. (Compton 1991). The detection of *P. granati* was established using nested PCR (Yang et al. 2017). Presumably, this methodology gives exceptionally explicit outcomes. However, it is a lot very expensive and required much effort (Sidra et al. 2017).

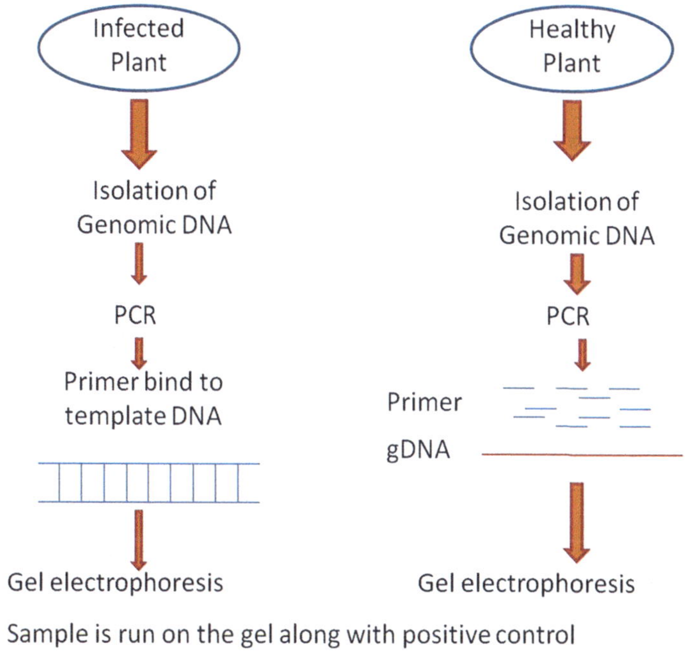

Fig. 6.1 PCR-based method for identification of microbes

6.4.2 Nested PCR

A high degree of specialty and sensitivity can be found using nested PCR (Sidra et al. 2017). For example, "Grote et al. (2002) reported that relative study on sensitivity and specificity of *phytophthora nicotianae* using both simple and nested PCR" revealed that the nested PCR sensitivity is 1000 times the detection of a single PCR. In this procedure, primers can be utilized for the amplification of large DNA sequence through the two continuous cycles, the afterthat sequence of amplified product works act as a specific sequence for second times by the use of two internal primers. In nested PCR, tremendous risk of contamination is assessed even though, in separate tubes, two amplification cycles should be performed. There is, therefore, the probability of false adverse population outcomes, and heavy labor is a significant defect (Rahman et al. 2013). Laboratories must take certain tough measures for the use of different instruments and space for each PCR cycle to conquer such borders (Trtkova and Raclavsky 2006). Two *Phytophthora* species *P. palmivora* and *P. parasitica* have been identified using this PCR (Tsai et al. 2006). Likewise, *P. cactorum* was particularly identified in strawberry infected plants through utilizing this technique (Bhat and Browne 2010), as well as rapid detection of *Cylindrocladium scoparium* on eucalyptus (Qiao et al. 2016). For the identification of *Candidates Liberibacter asiaticus* attached with citrus Huanglongbing (Hong et al. 2019), another report says that nested PCR can be used for detection of Lyme disease spirochete, *Borrelia burgdorferi*, in ticks (Wills et al. 2018).

6.4.3 Multiplex PCR

This method is utilized to demonstrate in less period as well as less money by utilizing few sets of preliminaries for a similar response to permits synchronous recognition of various focused on successions of DNA (Hyun et al. 2017). The method has such extraordinary significance of enormity in plant disease when crops are contaminated with many pathogens. Distinctive parts are clearly to target pathogenic parasites where at the same time enhanced and identified based on their molecular weight on agarose gel electrophoresis. DNA combining accuracy was unequivocally influenced by amplicon measure. To maintain a strategic distance from this entanglement, groundworks must be structured cautiously along with the comparative center and strengthening (Dasmahapatra and Mallet 2006). At present, lock tests (padlock primers) are utilized for the detection of pathogenic growths. This system has been utilized to the concurrent ID of parasitic pathogens, for example, *F. oxysporum*, *B. cactivora*, *P. nicotinae*, and *P. cactorum*, that key illness in joined cacti (Cho et al. 2016). Detection of various citrus disease-causing viruses in Jeju Island (Hyun et al. 2017), multiplex reverse transcription PCR assay for simultaneous detection of six main RNA viruses in tomato plants (Wu et al. 2018).

6.4.4 Portable PCR Systems

Fast PCR can manufacture a fully mobile configuration that will not just facilitate plant pathologists to carry out great efficiency with consistent experiment testing, but it might be significantly similar to downstream methods, i.e., those mandating in situ genomics detection tools. Palm PCR, manufactured in Korea by Ahram Biosystems, is easy to use, accurate, and high-competence thermal cycler (Love et al. 2012). Although it's small in size, this control tool provides tremendously well-defined and rapid amplification for different types of DNA materials (Monis and Giglio 2006). The amplification of the DNA to a sufficient amount is less than 25 minutes for the optimal detection in the agarose gel electrophoresis. The transferable way provides a greatly efficient easy-to-handle way for the learner and experienced academician to perform a different type of PCR tests. Koo et al. (2013) reported the successful identification of the six diverse fungal and bacterial plant pathogens in the sample.

Twista quantitative and portable real-time fluorometer are personalized tools produced to examine reactions to recombinase polymerase amplification (RPA) comparable with a different type of detection systems and the latest technology in DNA identifications, very high speed, easy to carry, and user-friendly with super quality. Twista RPA fluorometer provides urgently rapid diagnosis, enabling appropriate on-time therapy compared to usual microbiological test requiring at the minimum an hour and molecular assays typically requiring consolidated tools (Surrette et al. 2018).

6.4.5 *Real-Time* Polymerase Chain Reaction (RT-PCR)

Real-time PCR invention has developed more amount of modification in its protocol. A portion of maximum changes has extended the utility and detection quality of this PCR in numerous natural as well as therapeutic fields (Tang et al. 1997). These tools are insufficient to differentiate the living and dead parasites. PCR cannot differentiate between dead and living parasites. The DNA can be obtained from the liver and circulating DNA from other cells or bodies. The addition of a PCR test aimed at Leishmania-specific mRNA in a householder gene should make a difference between living and dead parasites (Bretagne et al. 2001). The development of RT-PCR overpowers this limitation. Contamination in mRNA is present in the dead cell; therefore, RT-PCR can detect the presence of mRNA in the cell (Capote et al. 2012). In this regard, RNA is conversely deciphered into cDNA by irregular groundworks and real-time compound and after that intensified by various PCR methods. In this way, RT-PCR is utilized to recognize and analyze the RNA-containing (*retroviruses*) contaminations. The finding of RNA-containing infections can help make or check the practicality of antimicrobial inoculations or treatment. For example, it can be utilized for evaluation of *Fusarium graminearum* growths to reason *Fusarium* early curse sickness within grains, for example, wheat, grain, and oat (Brown et al. 2011). *Cryphonectria parasitica* can be detected early using RT-PCR (Chandelier et al. 2019). Concurrently, it can detect the various pathogens of potato (Nikitin et al. 2018).

Recently, quantitative real-time PCR is a technique which affects microbial ecology greatly. The qPCR is a very sensitive technique to quantify the microbial population within the ecological sample (Trung et al. 2011). In short, qPCR amplified the specific sequence of nucleic acids and also calculates the sample amount of DNA. The fluorescent marker (SYBR Green) present at the end of the reaction detected by the PCR instrument. SYBR Green I and high-resolution melt dyes (LC Green, EvaGreen, BEBO, and SYTO9) can link between the dsDNA and emit the fluorescence at 494 to 521 nm wavelength.

Similarly, another way used in the qPCR analysis, i.e., TaqMan probes, works on a third oligonucleotide during annealing. The probe is made up of two molecules: (1) reporter and (2) quencher. When these two come together, then they repress the fluorescence of reporter in the reaction of DNA amplification. Also it generates the signal intensity with the amount of target DNA amplification. Detection of the pathogen such as *Phytophthora infestans* from the potato (Hassain et al. 2014; Li et al. 2014a, b; Clement et al. 2013) and *Verticillium dahliae*, *Colletotrichum acutatum*, and *C. gloeosporioides* from the strawberry (Bilodeau et al. 2012; Garrido et al. 2009) is through qPCR.

6.4.6 *Serial Analysis of Gene Expression* (SAGE)

Sequential examination for quality articulation was thorough based on grouping strategy for the quantitative quality articulation data that permits recognizable proof in numerous transcripts (Moreno 2003). SAGE is based on sequencing data of 15 bp or more nucleotides and comparability of successions beside accessibility of genome groupings to locate the relating communicated qualities (Velculescu 1995). It can utilize two examples that are bound and signed with the different preliminaries, and it's intensified. At that point, concatemers are shaped by the sticky finishes through the expulsion of groundwork. In sequencing of the cloned vector using the computational investigation, SAGE has a few downsides. In the first place, it needs mRNA in an extensive amount. Second, now and then the 15 bp tag isn't sufficient to explicitly recognize the quality of birthplaces with the more mind-boggling genomes. Thomas et al. (2002) reported that in the profiling of transcript of mildew pathogen *Blumeria graminis* in barley, SAGE is used, and another scientist also reported that gene expression of *Nicotiana tabacum* in *Rhizoctonia solani* is through SAGE (Portieles et al. 2017).

6.4.7 Sequencing-Independent Methods

The probe of DNA/RNA strategies is utilized to analyze plant infections caused by the microorganisms, for example, fungi with extraordinary sensitivity and speed (Sidra et al. 2017). This innovation is considered as the spine to a large portion of the present information. In these techniques, the probe shall be used without its amplification for the testing of nucleic acid. Samples are the shortest single-stranded DNA sequence that is named with chemiluminescent molecules or radio isotopes, for example, ^{32}P, ^{33}P, and ^{35}S (Sidra et al. 2017). For instance, engineered zinc finger proteins do not require targeted amplification for the detection of specific pathogen DNA (Kim and Kim 2016). It is utilized for the recognition of similar sequencing present on specific DNA. In traditional techniques, DNA probes is utilized for the most part as an option to PCR for the recognizable proof of microorganisms such as fungi. Be that as it may, in ongoing techniques, these are for the most part utilized related to PCR (McCartney et al. 2003; Khater et al. 2019). Engineered zinc finger proteins and pathogen targeted DNA immobilized on a chip of polymer (Ha et al. 2018; Guixia et al. 2019).

6.4.8 Northern Blotting or Northern Blot

This technique is also called as RNA smear technology and is generally used for the exchange of RNA onto a bearer for the recognizable proof of pathogenic growths utilizing the quality articulation (Zimmers-Koniaris 2001). The northern blot is equivalent to the southern smear apart from RNA material is utilized rather than DNA (Qi and Yang 2002). Right off the bat, the RNA from each tissue ought to be purged be inspected the statement of the quality of intrigue. RNA molecule was separated on the agarose gel electrophoresis. In this gel, litter molecule moves faster than a higher molecule; in this way, higher molecules present well as compared with the litter molecule (Kim et al. 2010). Hence every RNA molecule keeps its position regarding every single another particle (Berg 2007). The way is present to do a radioactive test to hybridize its related grouping molecule. After that, the channel is put for autoradiography to build up the film. At last, a band ought to show up on it, if both tests have hybridized a section of RNA atom on the channel. To start with, the situation of groups at smudge gives RNA approximation, but a measure of RNA is known; then they will give a guess to coding limit of transcripts as well as the extent of protein to which it recognized. Right from the beginning, the danger mRNA corruption amid electrophoresis is the primary impediment in the smudging system. Besides, recognition with numerous tests is hazardous. The affectability of the northern smudging strategy is low contrasted with RT-PCR. *Magnaporthe grisea* was perceived in rice plants through utilizing ongoing PCR and northern blot technique (Qi and Yang 2002). *Agrobacterium*-infiltrated place can be detected with the presence of more amount of GFP mRNA and siRNA through northern blot techniques (Yin et al. 2019; Boccardo et al. 2019).

6.4.9 In Situ Hybridization

In 1969, exploration of nucleic acids by in situ hybridization was first revealed (Gall and Pardue 1969). In situ hybridization is likewise called the hybridization histochemistry. It gives precious data to recognizing and listing the parasites. In these method, the ssRNA test is utilized known as riboprobes and 35S. It has an extraordinary closeness to northern blotting. These two methods rely upon hybridization of marked tests of DNA/RNA to the comparative groupings of mRNA molecule. These strategies vary in the utilization of beginning material. On account for northern smudge, the part of tissue digest is utilized as beginning material, while in situ hybridization histological area is utilized. Notwithstanding utilizing the immediate hybridization, the recognizable pieces of proof utilizing signal hybridization are most productively acquired after the organism's development or its natural intensification (Jensen 2014).

The preferred primary standpoint of this innovation is the most extreme utilization of tissue, for example, clinical biopsies and refined cells. Several unique hybridizations can be done in a similar tissue. Tissue libraries can be arranged by putting away in cooler for future use. There are various techniques for performing in-situ hybridization, for instance, testing with counterfeit oligonucleotides copy of DNA and RNA. These strategies give the most powerless and efficient outcomes. Tests may be named according to radioactive or nonradioactive nucleotides; after these tests, 35S riboprobes are a very delicate strategy for the ID of mRNA (Hayden et al. 2002). ISH was used to locate specific chromosome DNA sequences using radioisotope-labeled samples (Gall and Pardue 1971; Zeller et al. 2001). It was later used for detection of viral particles and mRNA high copy in cultivated cells and sectional tissue, so that gene expression patterns could then be located (Brahic and Haase 1978; Zeller et al. 2001). It is used for pathogen visualization in host plant tissue, using ISH technology (Tanaka 2009).

In situ hybridization has a few entanglements. Right off the bat, radiolabel tests are in all respects expensive and risky materials. It must be dealt with, transported, and discarded in all respects cautiously. A drawback of utilizing in situ hybridization procedure is the trouble in distinguishing focuses on with contains a low copy of DNA and RNA molecules (Qian and Lloyd 2003). It can be used for identification of various parasites, i.e., *Blastomyces dermatitidis*, *Coccidioides immitis*, *Cryptococcus neoformans*, *Histoplasma capsulatum*, and *Sporothrix schenckii* (Hayden et al. 2001). This type of hybridization can be used to imagine the difficulty in plant tissue of the rust organism, as well as it can be utilized for confinement region that is present in plant partition in rust fungi (Ellison et al. 2016).

6.4.10 Fluorescence In Situ Hybridization (FISH)

Because of the disadvantage of radiolabeled probe-based hybridization, the FISH technique has emerged, which is used for the speedy characterization of microorganism such as fungi (Sidra et al. 2017). This technology offers superior speed, resolution, and security and smoothens the way for simultaneous and quantitative phylogenetic analysis of multiple targets (Tsui et al. 2011). Besides this, FISH is also an essential type to identify break down in the spatial association of fungal networks.

Specific DNA sequence on chromosomes is detected using this cytogenetic technique (Ratan et al. 2017). The fluorescent probe is used in FISH which single binds those parts of the chromosomes, which shows a higher quantity of similarity. Fluorescent samples are established throughout the sample length by an enzymatic fuse of altered fluorophore base (Baschien et al. 2001). Conventional systems for detection of microorganism necessitates that cells should be effectively isolating (Ainsworth et al. 2006). However, for non-diving cells, FISH can be used, making it a highly resourceful technique. It is possible to recognize non-dividing cells by their low fluorescence intensity level. It is possible to use a different type of samples, for instance, an allele-specific sample, a centromeric repeat sample, and an entire chromosome sample. Many of them have various implementations. In FISH probe, preparing ready procedure is confounded because it is important to tailor probe to recognize the particular sequence of DNA (Volpi and Bridger 2008).

6.4.11 Microarray

A microarray is a technique that is used simultaneously to detect the expression of thousands of genes. It is a microscopic slide that is printed in defined positions with thousands of tiny spots with each spot referring to the known DNA sequence or gene, and these slides are known as DNA chips. It is used to detect potato viruses, cucurbit-infecting tobamoviruses, and grapevine viruses (Bystricka et al. 2003; Lee et al.

2003; Nicolaisen 2011). It is not quite the same as the above method in this point of view because it gives expression estimations of characterized genes of the set (Eshaque and Dixon 2006). Within a small surface area, thousands of DNA probes are displayed on a matrix that consists of glass chip or nylon filters in this technology. Sample position is called spot on the chip (Robinson et al. 2000). On the support matrix, these probes are immobilized, and focused cDNAs are useful as a hybridization chip. The Quantification encircled cDNA quantification is quantified using radio-labeled probes only after that signal is generated by hybridizing the probe to the focused mRNA that the specific software can identify and integrate. The Keen Software gives gene expression trend for every natural living sample (Russo et al. 2003). The microarray can detect a very large range of fungal cell. In the microarray techniques, large-scale DNA sequencing is not used. Even though this method can detect any changes in gene expression, noteworthy issues are noted. Initially, microarray required a lot of mRNA. Besides, this investigation is constrained by the expense and right to use (Singh and Kumar 2013). Because of the abundant error steps in microarray technique, it is minimized by the repetition of the experiments. Microarrays are referred to as critical testing because it requires physical cell trouble to gain accesses to its patterns of gene expression; it is remembered that false data can be generated by the mRNA degradation (Singh and Kumar 2013). This technique can detect bacteria, viruses, parasites, fungi, viroids, and phytoplasmas (Hadidi and Barba 2006). Postnikova and Nemchinov (2012) used microarray-based analysis for the comparative study of the various viruses in *Arabidopsis*. Leborgne and Bouhidel (2014) studied plant-microbe interaction using microarray. Osmani et al. (2019) used microarray analysis to detect responses of various viruses of potato.

6.4.12 Macroarray

It is also known as the hybridization of the DNA array or the plot of the reverse dot. It utilizes the feasibility of DNA amplification while radioisotopes are not required (Singh and Kumar 2013). In a growing number of labs around the world, macroarray is a quick molecular technique for the diagnostic of greenhouse plants (Le Floch et al. 2007; Lievens et al. 2003), ginseng (Punja et al. 2007), and potatoes (Fessehaie et al. 2003). It's working according to concurrent amplification of the linked species by the PCR. Also, in one hybridization reaction, it simultaneously analyzes several amplified sequence. It is a technique that is more sensitive and sophisticated than other PCR (Taoufik et al. 2004). In this evaluation amplification of PCR is joined with hybridization that builds affectability up to thousand folds or higher than PCR only technique, and it has a quick rotation time 1–2 days in contrast with radioactive culture techniques that involve 2 months for their conclusion (Taoufik et al. 2004). In this technique, the inward probe is considered to separate the variety. These probes are attached to the hold of the nylon membrane. By ultraviolet cross-linking, oligos are eternally bound to the membrane. The amplified PCR products are firstly hybridized and then spotted on the strips where species-specific interaction takes place (Tsui et al. 2011; Leinberger et al. 2005). To identify the species level of

Mycobacterium, the combination of the PCR and microarray technique is used (Leinberger et al. 2005). It is also used to detect *Alternaria alternata*, *Fusarium solani*, *Candida albicans*, *Aspergillus fumigates*, and *Cladosporium herbarum* (Sato et al. 2010) and identification of fungal as well as oomycete of the pathogen that causes the illness in solanaceous plants (Zhang et al. 2008).

6.4.13 *Loop-Mediated Isothermal Amplification* (LAMP)

LAMP has a vigorous and new approach for amplifying nucleic acid as a substitute for PCR. LAMP amplified specific nucleic acid with high specificity under the isothermal situation. It never requires PCR to produce changes in temperature but requires a single Tm for amplification of DNA molecule (Tsui et al. 2011). This technique is also based on auto-cycling and strand dislocations. LAMP utilizes polymerase of Bst DNA and two sets of internal and external primers. This reaction works at 650C for an hour while being put on either a dry or water bath, and after that, SYBR Green can be used for the detection of the amplified product. The end product of the LAMP has numerous reverse repeat with several loops which show cauliflower-like structure. LAMP is ten times greater exact and accurate than standard PCR (Ren et al. 2009). In LAMP, thermal changes are not needed; as compared with PCR, LAMP reaction requires only one tube (Fakruddin 2011). The sensitivity of LAMP reaction is affected by various factors, i.e., utilization of the DNA polymerase. This is significant for LAMP effectiveness. For recognition and diagnosis, LAMP is useful but it cannot be utilized for cloning purposes. While, the major disadvantage of LAM is, it cannot used for the direct assessment different dyes, for example Mn^{2+} (dye), SYBR Green dye, hydroxyl naphthol blue dye etc. that cannot be distinguished among the required specific amplified product size and non specific product size, So, lead to false positives result. It can be resolved due to the utilization of molecular beacons (MBs) by producing the fluorescence signals while it combines with specific DNA sequences. Therefore, MBs can detect the amplified product. The LAMP finds a suitable situation for MBs (25–45 bp beacon length, 60–65 °C reaction temperature, and 0.6–1 pmol/μL beacon concentration) for the evaluation technique. An original MB-LAMP-based method has proven direct identification of the LAMP product. MBs are the nucleic acid fluorescent sample with hairpin structure. The structure of the hairpin left the fluorescence as quencher that is near to a fluorophore (Liu et al. 2017). LAMP detects *Ascochyta rabiei* fungi that cause *Ascochyta* blight disease in chickpeas. The contrast of standard PCR and LAMP not only reveals superior accuracy, sensitivity, and specificity to detect *A. rabiei* but also utilizes simple apparatus and takes less time to operate (Chen et al. 2016). LAMP was successfully used to identify *Paracoccidioides brasiliensis*, a thermal-reliant dimorphic fungus (Endo et al. 2004). This method also detects the pathogenic fungus, i.e., *Ochroconis gallopava* (Tsui et al. 2011). The use of lamp technology has also efficiently diagnosed *Penicillium marneffei* (Sun et al. 2010). Recently, LAMP is used to identify *Ophiostoma clavatum*, main blue stain fungus (Villari et al. 2013), *Colletotrichum gloeosporioides* (Wang et al. 2017), and whitefly *Bemisia tabaci* (Blaser et al. 2018).

6.4.14 Sequencing of the Next Generation Based on RNA-Seq

RNA sequencing is a recently developed technique for deep sequencing. A huge populace of RNA are generally transformed into a library of cDNA with adapters that bind to single or both ends. After that, every sequence having with or without amplification is sequenced for obtaining minute sequence from one end as in single-end sequencing or both ends as in pair-end sequencing in a high-throughput way. Based on DNA, sequencing tools can be used for reads which are commonly up to 30,400 bp. Prepared library to observe how strongly the results of the RNA sequencing unveil that most of the unique RNA transcripts are identified in the step of preparation of the library. To build up an RNA sequence library, it is necessary to fragment RNA or DNA moreover to permit processing by sequencing of the next generation. In the real-time reaction, mRNA can be prepared using either oligo or random primers. The benefit for the utilization of oligonucleotide is the significant production of polyadenylated mRNA proportion of cDNA. Therefore most of the acquired sequence can be insightful (Mortazavi et al. 2008). The three very generally RNA-Seq technologies of next generation are SOLiD and Ion Torrent, which are produced by Life Technologies, and HiSeq by Illumina (Dawei and Peng 2014).

After that sequencing, we found the maximum reads that are either aligned with reference genome whose sequences are known or constructed with de novo sequence without genomics sequence to develop a transcriptional map consisting of both expression level of every gene and transcription structure. Although RNA sequencing data obtained that can be used for alignment with the referred genome data, Annotator and Trinity assembler that assemble the RNA sequencing data without reference genome data using the assemble of adjacent identification of shorts reads. These methods enable new transcripts to be discovered and numerical transcript to be detected fairly. They are allowing more efficient use of the RNA sequencing to detect transcripts and classify the transcriptome in a nonmodel organism (GrabHerr et al. 2011). The *Magnaporthe oryzae* fungus is detected using this technology, which induces rice blast disease (Soanes et al. 2012). *Verticillium dahliae* fungus in tomatoes that induces vascular wilt disease is also identified using this technique (de Jonge et al. 2012). RNA sequencing would be used to detect pathogens in plants. The prospective suitability of mRNA sequencing data to recognize nucleotide divergences can disclose fungal pathogenicity genes of the plant in their protein-coding transcriptome that is mutant. The technique allows us to diagnose plant disease using RNA measurements (Metzker 2009).

6.4.15 Lateral Flow Microarrays

Nucleic acid is rapidly detected using lateral flow microarrays (LFM) based on hybridization technique with colorimetric signals that can be easily visualized (Carter and Cary 2007). LFM based on lateral flow nitrocellulose membrane chromatography, hybridized in very less time, have detection limits and might decrease the requirement for high-price lab tools. Skill is dependent for accessibility of useful and reliable biomarkers of host and pathogen revealed by transcriptomics perspective (Martinelli et al. 2012a, 2013a). While we can use metabolomics for the detection of primary and secondary metabolites; it can be utilized as a biochemical marker for a variety of pathogenic diseases (Rizzini et al. 2010; Tosetti et al. 2012; Martinelli et al. 2012b, 2013b, 2014; Ibanez et al. 2014). Early pathogenic infections like Huanglongbing disease in citrus can be identified by an included omics approach (Dandekar et al. 2010). More highly synergistic proteins are necessary markers for plant health status, i.e., heat shock protein (HSP) or dehydrins, upregulated by various ecological ways (Natali et al. 2007).

6.5 Remote Sensing (RS)-Based Diagnostics

Spectroscopy is one of the main useful RS methods, including sensors such as VIS, NIR, SWIR, and imaging or nonimaging. Due to their possibility as a functioning instrument, elasticity, efficacy, and cost-effectiveness, these tools hold exciting promise for crop disease monitoring. Below are discussed the most applicable and recent developments in spectroscopy depending on techniques.

6.5.1 Nonimaging Spectroscopy

It is based on the natural things of leaf pigments, chemical factor, properties, and structural characteristics (Jacquemoud and Ustin 2001). Laboratory or field-collected leaf spectra determined spectral area of visibility for identification of diseases, i.e., leaf gall disease in sugarcane through remote sensing (Purcell et al. 2009), powdery mildew disease in wheat (Graeff et al. 2006), curl mite (Stilwell et al. 2013), yellow leaf virus in sugarcane (Grisham et al. 2010), also yellow rust in wheat during winter (Zhang et al. 2014), and grapevine viruses in grapes (Naidu et al. 2009). Yuan et al. (2014) experimented with the separation of winter contamination from pathogens, insects such as wheat aphids, and virus, i.e., yellow rust and powdery mildew. Methods for early identification of disease are of faithful interest (Malthus and Madeira 1993; Delalieux et al. 2007; Rumpf et al. 2010), while their actual use is incoherent across crops for crop management. Studies available are crop definite, and results cannot be comprehensive with comparable accuracy to other crops and places. When Huang et al. (2012) compared both, firstly canopy scale detection of leaf and severity determination of rice leaf folder disease, the best concurrence of recognition rates using VIS and NIR reflection by linear regression method and the NIR plateau has a very high negative correlation (737–1000 nm) and severity of infestation.

Many authors are working with radiometry to determine the harshness of the damage of crop (Nutter 1989). Yang et al. (2007) reported that rice plants are infested with brown plant hopper and leaf folder. Mirik et al. (2006) studied spoil from green bugs to winter wheat, Chen et al. (2008) Accessed loss of cotton from *Verticillium* wilt and studied leafhopper disease from Prabhakar et al. (2011). Also utilized for fruit value evaluation was spectroscopy, often connected with additional resources, i.e., e-nose data; booming integration of remote sensing depends on the way with VOC analysis (Costa et al. 2007).

6.5.2 Imaging Spectroscopy

Hyperspectral imaging tools have recently been integrated to evaluate and observe plant illness. Lab experiments detect various diseases, i.e., head blight disease in wheat, a fungal disease with *Fusarium* (Bauriegel et al. 2011), sugar beet disease (Mahlein et al. 2012) on leaves, rust, and powdery mildew (Mahlein et al. 2013). Diverse infections and its stage of maturity are of particular interest for effective intervention (Mahlein et al. 2012). This report utilized a wide range of statistical instruments for image investigation, such as linear regression, principal component analysis (PCA), spectral angle mapper (SAM), and support vector machine (SVM) categorization with elevated accuracy of detection of disease. However, these trials concentrated on one or a few plants with little probability of generalization. On the basis of field survey of wheat plant yellow rust disease (Bravo et al.) and attempts to distinguished among wheat disease and abiotic stress (Moshou et al. 2004), Reynolds et al. (2012) and Huang et al. (2007) used the hyperspectral field and aerial information for evaluating the seriousness of *Rhizoctonia* crown and root rot disease in sugar beet and yellow wheat rust, respectively. Airborne hyperspectral data is best suited for farm and regional-scale remote sensing (RS) applications.

6.6 Protein-Based Diagnostics

6.6.1 MALDI-TOF MS

Bacterial product isolates were subcultured on LB agar media (Thermo Fisher Scientific, UK) and incubated at 37 °C for a 24–48 h and recognized with a Bruker MALDI biotyper system (microflex LT, Bruker) as directed by the manufacturer. The calibration of the mass spectrometer was performed by an automatic calibration operation with Bruker bacterial test standard (BTS), which is a modification of *E. coli* BH5α, spiked using two extra enzymes to allow calibration with 4–17 kDa mass ranges. After 6x40 laser shots have been implemented, a sum spectrum is automatically acquired from separate locations on BTS control. The minimum amount of peaks reverted to seven after the calibration phase (Fig. 6.2).

Using a sterile toothpick, an individual colony from a new culture of each isolate was selected and smeared on the specified places on normal MALDI target plate, and the drying plates were put on RT (25 °C) for 5 min after loading the sample. 1 μl HCCA matrix (saturated solution of α-cyano-40hydroxycinnamic acid in 50% acetonitrile and 2.5% trifluoroacetic acid) was put on the top of a spot not more than 10 min of drying the samples. Before loading plate into a mass spectrometer, sample was air-dried again for 5 min. Every bacterial colony has been screened in duplicate. MALDI Biotyper Real-time Classification (MBT-RTC) software (Bruker) acquired data set. It compares unknown peak from the reference database (MBT

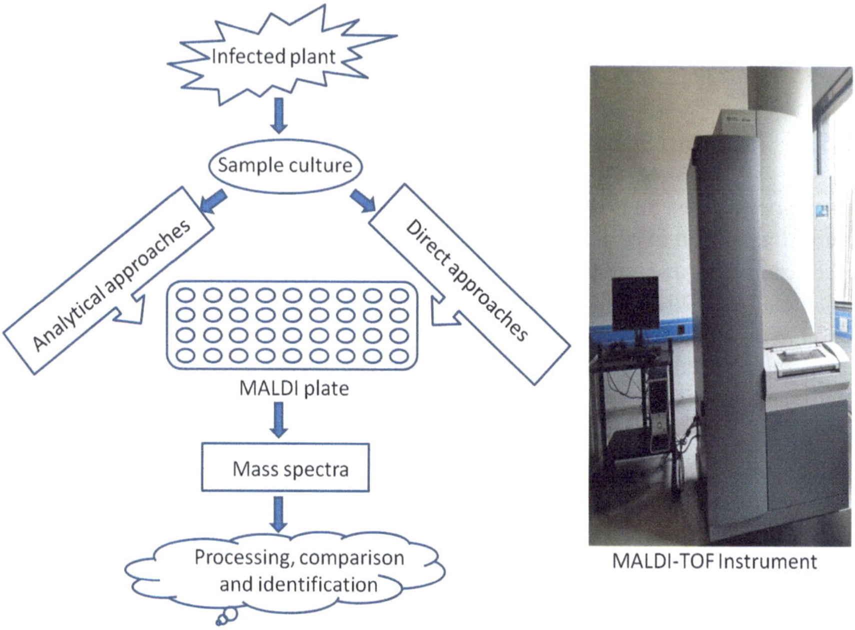

Fig. 6.2 MALDI-TOF method for the identification of the microbes from the infected plant

Compass 4.1, Bruker) as well as used statistical algorithm to produce a log score ranging from 0.000 to 3.000. The classification score was considered to be suitable for species-level recognition (>=2.000 confident species identification), although recognition of the genus level (1.700–1.999 confident genus identification), below <1.700 no reliable identification (Bruker 2016).

Bacterial proteins were obtained by the process of ethanol/formic acid extraction outlined by the manufacturer for isolates, which could not be identified directly by transfer of colonies to the MALDI target plates. A bacteria-packed loop was briefly moved and resuspended with a 10 μl inoculation loop in 0.3 mL of deionized water and then added with 0.9 mL 100% ethanol, through blending and centrifugation at 16,000 × g for 2 min. Bacterial pellets were shortly drained, blended sequentially with 70% formic acid and 50 μl acetonitrile with 50 μl (20 μl for tiny pellets), and centrifuged to 16,000 g for 2 min. 1 μl supernatant was put on the spot on a plate of MALDI, dried, and overlaid using 1 μl HCCA matrix solution; this can be detected using Bruker MALDI Biotyper System (Bruker 2011).

6.6.2 Quartz Crystal Microbalance Immunosensors (QCMI)

QCMI analysis is based on concurrent mass vibration and its frequency variations and extensively utilized to conclude little vacuum, gas, and liquid mass (Kurosawa et al. 2006; Mecea 2005, 2006). Transducer device has a mass sensitive, the immunological mixture with QCM resultant in QCMI (Owen et al. 2007).

Both the antigen and antibody compatibility reaction in a positive reaction affects the reduced frequency of quartz crystal oscillation. QCMI, which offers certain advantages, high accuracy, real-time output, portability, label-free entity, and low operating, manufacturing, and protection costs, is an attractive option to a conservative tool of analysis (Chen and Tang 2007; Lee and Chang 2005; Tang et al. 2006). The diagnostic signal is very less to identify specific material; sensitivity to identification can be bigger through the introduction of the signal enhancement step (Kurosawa et al. 2006). QCMI is highly sensitive to detection of pathogen sample with viruses (Bachelder et al. 2005; Eun et al. 2002; Kleo et al. 2011; Lee and Chang 2005; Owen et al. 2007; Su et al. 2003; Uttenthaler et al. 2001). Since a QCM detection tool is potable, and QCM covered with virus-specific antibodies to identify plant viruses has a finite life span, it may also be utilized for the identification of plant viruses on-site (Becker and Cooper 2011; Eun et al. 2002). QCMI has been clearly detecting plant viruses, i.e., *Cymbidium* mosaic virus (CMV), TMV, turnip yellow mosaic virus (TYMV), *Odontoglossum* ringspot tobamovirus (ORSV), and maize chlorotic mottle virus (MCMV) (Dickert et al. 2004; Eun et al. 2002; Huang et al. 2014; Zan et al. 2012).

6.7 Nanotechnology-Driven Diagnostic

6.7.1 Agriculture Nanosensors

The nano sensors are results in collective advances to biology and nanotechnology (Yang et al. 2008). These nanosensors have the ability for improved sensitivity and also increased sensitivity with those can hold the potential in sensors. Therefore they considerably decrease the reaction time to detect prospective illness challenges in plants (Small et al. 2001), and therefore they can contribute to improving efficiency and food safety in agriculture. Hashimoto et al. (2008) developed the latest biosensor system that contains two types of biosensors for the speedy diagnosis of soilborne diseases. The system was built utilizing two different microbes in equal quantities, each being separately immobilized on an electrode. The optical properties of the silver nanoparticles can be differentiated or evaluated between sulphuration ethyl herbicide and silver nano particles, taking into account the specific optical properties of silver nanoparticles (Dubas and Pimpan 2008). They found that silver nanoparticles in a solution are sensitive to improved herbicide concentration and induced a difference in nanoparticle color converted from yellow to orange and then finally giving the violet color. It is a useful identification of contamination in water bodies and the environment, like organic pollutants and microbial pathogens (Dubertret et al. 2001). Combined with antibody molecules, fluorescent silica nanoparticles (FSNP) effectively identify plant pathogens like *Xanthomonas axonopodis* pv. vesicatoria in tomatoes and peppers that cause bacterial spot disease (Yao et al. 2009). Solgel and spray pyrolysis strategies were used to analyze nanoparticles and nanolayers of copper oxide (CuO). The detection of *A. niger* was conducted using both CuO nanoparticles and nanostructure layer biosensors

(Etefagh et al. 2013). In contrast, silver nanoparticles (AgNPs) are frequently used in soil and water bodies to detect toxins and microbial pathogens. The utilization of nanosensors has enabled plant disease prediction and disease control in agriculture (Bogue 2008).

6.8 Other New Technologies

6.8.1 Quantum Dots (QDs)

QDs are nanoparticles with a semiconductor that fluoresce while it stimulates through a cause of excitation light. In addition, QDs are inorganic fluorophore with foremost advantages compared to conventional organic fluorophore (Alexa Fluor 488, Alexa Fluor 514, BODIPY FL, carboxyrhodamine 6G, Cy 5.5, Cy 7, fluorescein, etc.) used as sign-on nucleic acids or visual proteins (Wang et al. 2006; Arya et al. 2005). Ferrari and Bergquist (2007) compared QDs and organic fluorophore for the detection of *Cryptosporidium* parasites. Different microbes make the semiconductor nonmaterials through mycosynthesis in single cell yeast (Dameron et al. 1989) and are also used for Cadmium sulfide (CdS) biosynthesis; through, some reports reported its luminescent values. The fungus *F. oxysporum* produced QDs when combined with $Cdcl_2$ and $Secl_4$ at normal temperature (Kumar et al. 2007). A proficient myco-mediated synthesis of highly fluorescent CdTe QDs was accomplished by the F. oxysporum isolates when reacted with a combination of $CdCl_2$ and $TeCl_2$ at room temperature (Jain 2003). Biosynthesized CdTe nanoparticles were described using electron microscopy and electron diffraction technique (Syed and Ahmad 2013).

The CdTe QD units were widely present in yeast cells, particularly in the cytoplasm and nucleus, while nothing was present in the cell membrane. The CdTe QD units are only spread into the cytoplasm of the yeast cells, and none in the cell membrane are found. The organic semi-compatible crystals are formed by a core and a shell which enables the ligands to bind and thus the fluorescent marker to be attached to the pathogen. A biosensor, also known as quartz microbalance crystal biosensor, is the best example. It can vibrate under electrical stimulation, and a mass change with the connection of every compound to its surface can be detected by reducing the vibration rate. A modified shape and distributed resonant frequency of quartz and crystal biosensor significantly decrease when a nucleic acid sample is attached to the surface of a quartz crystal biosensor and instead subjected to a supplementary PCR product. The hybrid forms and significantly reduces the resonance frequency of QC biosensor even if a nucleic acid probe is inserted on the surface of a quartz crystal microbalance and subjected to existing PCR product. This scheme decreases the time needed for particular environmental pathogens in conjunction with quick PCR methods (Sharon et al. 2010).

6.9 Conclusion

An alternative to traditional methods may be molecular methods such as PCR and qPCR as well as serological methods and flow cytometry. These methods are very sensitive and specific and take very little time, while simultaneously, it is necessary to optimize this method to fulfill the specificity and sensitivity requirements. Furthermore, standardized protocols are needed as a standard protocol for quarantine purposes for the global acceptance of the method. The portable diagnostic instrument, nanoparticle-based, bio-barcoded DNA sensor, and QD have prospective utilization in numerous identifications of various plant pathogens and toxic fungi. It can serve as an analysis to establish identification of plant illness easily and it can be used to stop epidemics. These diagnostic kits depend on nanomaterials and not just enhance the speed of pathogen detection as well as increase diagnostic accuracy. These tools have experienced notable technological changes since their innovation, resulting in them remaining important characteristics of research methods devoted to exploring diversity and all of its microorganism variations. Microscopy is progressively augmented with molecular technologies to show data on relative species availability and existence of noncultural microbes in certain environmental fields. Researchers and business will expand and take up these techniques as expenses are lowered in a manner comparable to desktop or cell phones. Harshness and robustness of specifications and precise attributes are key to the production of any current diagnostic as well as detection of various technologies. Ideally, fresh tools should be resilient to detect distinct entities using air sampling devices merely by removing product such as primers, antibodies, and biosensors. Ultimately, the use of sophisticated but inexpensive diagnostic methods is expanded to automated mobile collecting instruments. Further, in the future automated collecting instruments will also assist plant health inspection monitoring the development of fresh pathogens. This could threaten both agricultural and environmental systems.

References

Aćimović SG, Zeng Q, McGhee GC, Sundin GW, Wise JC (2015) Control of fire blight (*Erwinia amylovora*) on apple trees with trunk-injected plant resistance inducers and antibiotics and assessment of induction of pathogenesis related protein genes. Front Plant Sci 6:16

Ainsworth TD, Fine M, Blackall LL, Hoegh-Guldberg O (2006) Fluorescence in situ hybridization and spectral imaging of coral-associated bacterial communities. Appl Environ Microbiol 72:3016–3020

Ali GS, Hu X, Reddy ASN (2019) Overexpression of the Arabidopsis thaumatin-like protein 1 in transgenic potato plants enhances resistance against early and late blights. bioRxiv 1:621–649

Almasi MA, Aghapour-Ojaghkandi M, Aghaei S (2013) Visual detection of curly top virus by the colorimetric loop-mediated isothermal amplification. J Plant Pathol Microb 4:198

Ansar M, Singh RP (2016) Field diagnosis and temporal progress of *Potato Virus X* and A in Tarai region of Uttarakhand. J Mycol Pl Pathol 46:368–373

Artal RB, Gopalakrishnan C, Thippeswamy B (2012) An efficient inoculation method to screen tomato, brinjal and chilli entries for bacterial wilt resistance. Pest Manag Hortic Ecosyst 18(1):70–73

Arya H, Kaul Z, Wadhwa R, Taira K, Hirano T, Kaul SC (2005) Quantum dots in bio-imaging: revolution by the small. Biochem Biophys Res Commun 329:1173–1177

Bachelder E, Ainslie K, Pishko M (2005) Utilizing a quartz crystal microbalance for quantifying CD4+ Tcell counts. Sens Lett 3:211–215

Baschien C, Manz W, Neu TR, Szewzyk U (2001) Fluorescence in situ hybridization of freshwater fungi. Int Rev Hydrobiol 86:371–381

Bauriegel E, Giebel A, Geyer M, Schmidt U, Herppich WB (2011) Early detection of *Fusarium* infection in wheat using hyper-spectral imaging. Comput Electron Agric 75:304–312

Becker B, Cooper MA (2011) A survey of the 2006–2009 quartz crystal microbalance biosensor literature. J Mol Recognit 24:754–787

Bereswill S, Burget P, Bruchmuller I, Geider K (1995) Identification of the fire blight pathogen, Erwinia amylovora, by PCR assays with chromosomal DNA. Appl Environ Microbiol 61:2636–2642

Berg JM (2007) Biochemistry, 6th edn. W. H. Freeman and Company, New York

Bhat R, Browne G (2010) Specific detection of *Phytophthora cactorum* in diseased strawberry plants using nested polymerase chain reaction. Plant Pathol 59:121–129

Bilodeau GJ, Koike ST, Uribe P, Martin FN (2012) Development of an assay for rapid detection and quantification of *Verticillium dahliae* in soil. Phytopathology 102:331–343

Blaser S, Diem H, Von Felten A, Gueuning M, Andreou M, Boonham N, Frey JE (2018) A loop-mediated isothermal amplification (LAMP) assay for rapid identification of Bemisia tabaci. J Vis Exp 140:58502

Boccardo NA, Segretin ME, Hernandez I, Mirkin FG, Chacón O, Lopez Y et al (2019) Expression of pathogenesis-related proteins in transplastomic tobacco plants confers resistance to filamentous pathogens under field trials. Sci Rep 9:2791

Bock C, Poole G, Parker P, Gottwald T (2010) Plant disease severity estimated visually, by digital photography and image analysis, and by hyperspectral imaging. Crit Rev Plant Sci 29:59–107

Bogue R (2008) Nanosensors: a review of recent progress. Sens Rev 28(1):12–17

Boonham N, Kreuze J, Winter S, Van der Vlugt R, Bergervoet J, Tomilinson J, Mumford R (2014) Methods in virus diagnostics: from ELISA to next generation sequencing. Virus Res 186:20–31

Bove JM, Vogel R, Albertini D, Bove JM (1988) Discovery of a strain of *Tristeza* virus (K) inducing no symptoms in Mexican lime. Proceedings of the 10th Conference of IOCV. Spain 1988. International Organization of Citrus Virologists Riverside, CA. pp 14–16

Brahic M, Haase AT (1978) Detection of viral sequences of low reiteration frequency by in situ hybridization. Proc Natl Acad Sci U S A 75:6125–6129

Bretagne S, Durand R, Olivi M, Garin JF, Sulahian A, Rivollet D, Deniau M (2001) Real-time PCR as a new tool for quantifying *Leishmania infantum* in liver in infected mice. Clin Diagn Lab Immunol 8:828–831

Brown NA, Bass C, Baldwin TK, Chen H, Massot F, Carion PW, Urban M, Van De Meene AM, Hammond-Kosack KE (2011) Characterisation of the *Fusarium graminearum*-wheat floral inter-action. J Pathog 2011:626345

Bruker (2011) Instructions for use MALDI biotarget 48 revision 1. Bruker Daltonics, Inc., Bremen

Bruker (2016) MBT compass user manual revision C. Bruker Daltonics, Inc., Bremen

Bystricka D, Lenz O, Mraz I, Dedic P, Sıp M (2003) DNA microarray: parallel detection of potato viruses. Acta Virol 47:41–44

Cambra M, Capote N, Myrta A, Llácer G (2006) Plum pox virus and the estimated costs associated with sharka disease. EPPO Bull 36:202–204

Capote N, Aguado A, Pastrana AM, Sánchez-Torres P (2012) Molecular tools for detection of plant pathogenic fungi and fungicide resistance. INTECH Open Access Publisher, Valencia

Carter DJ, Cary RB (2007) Lateral flow microarrays: a novel platform for rapid nucleic acid detection based on miniaturized lateral flow chromatography. Nucleic Acids Res 35:74

Caruso P, Gorris MT, Cambra M, Palomo JL, Collar J, López MM (2002) Enrichment double-antibody sandwich indirect enzyme-linked immunosorbent assay that uses a specific monoclonal antibody for sensitive detection of Ralstonia solanacearum in asymptomatic potato tubers. Appl Environ Microbiol 68:3634–3638

Cha RS, Thilly WG (1993) Specificity, efficiency, and fidelity of PCR. PCR Methods Appl 3:18–29
Chandelier A, Massot M, Fabreguettes O, Gischer F, Teng F, Robin C (2019) Early detection of Cryphonectria parasitica by real-time PCR. Eur J Plant Pathol 153:135–152
Chang PG, McLaughlin WA, Tolin SA (2011) Tissue blot immunoassay and direct RT-PCR of cucumoviruses and potyviruses from the same NitroPure nitrocellulose membrane. J Virol Methods 171:345–351
Chen ZG, Tang DY (2007) Antigen–antibody interaction from quartz crystal microbalance immunosensors based on magnetic $CoFe_2O_4/SiO_2$ composite nanoparticles functionalized biomimetic interface. Bioprocess Biosyst Eng 30:243–249
Chen B, Wang K, Li S, Wang J, Bai J, Xiao C, Lai J (2008) Spectrum characteristics of cotton canopy infected with Verticillium wilt and inversion of severity level. In: Li D (ed) Computer and computing technologies in agriculture, volume II, vol 259. Springer, Boston, pp 1169–1180
Chen X, Ma L, Qiang S, Ma D (2016) Development of a loop-mediated isothermal amplification method for the rapid diagnosis of *Ascochyta rabiei* L. in chickpeas. Sci Rep 6:25688
Cho HJ, Hong SW, Kim HJ, Kwak YS (2016) Development of a multiplex PCR method to detect fungal pathogens for quarantine on exported cacti. Plant Pathol J 32:53–54
Clark MF (1981) Immunosorbent assays in plant pathology. Annu Rev Phytopathol 19:83–106
Clark MF, Adamas AN (1977) Characteristic of the microplate of enzyme-linked immunosorbent assay for the detection of plant viruses. J Gen Virol 34:475–483
Clement JAJ, Baldwin TK, Magalon H, Glais I, Gracianne C, Andrivon D, Jacquot E (2013) Specific detection and quantification of virulent/avirulent *Phytophthora infestans* isolates using a real-time PCR assay that targets polymorphisms of the *Avr3a* gene. Lett Appl Microbiol 56:322–332
Compton J (1991) Nucleic acid sequence-based amplification. Nature 350:91–92
Cooke LS, Cacciola SO (2007) Tools to detect, identify and monitor Phytophthora species in natural ecosystems D.E.L. J Plant Pathol 89:13–28
Corning Life Science (2001) Selecting the detection system: Colori- metric, fluorescent, luminescent methods. ELISA Technical Bulletin 5:14
Costa G, Noferini M, Fiori G, Spinelli F (2007) Innovative application of non-destructive techniques for fruit quality and disease diagnosis. Acta Hortic 753:275
D'hondt L, Hofte M, Van Bockstaele E, Leus L (2011) Applications of flow cytometry in plant pathology for genome size determination, detection and physiological status. Mol Plant Pathol 12:815–828
Dameron CT, Reeser RN, Mehra RK, Kortan AR, Carroll PJ, Steigerwaldm ML, Brus LE, Winge DR (1989) Biosynthesis of cadmium sulphide quantum semiconductor crystallites. Nature 338:596–597
Dandekar AM, Martinelli F, Davis CE, Bhushan A, Zhao W, Fiehn O, Skogerson K, Wohlgemuth G, D'Souza R, Roy S, Reagan RL, Lin D, Bruce Cary R, Pardington P, Gupta G (2010) Analysis of early host responses for asymptomatic disease detection and management of specialty crops. Crit Rev Immunol 30:277–289
Dasmahapatra K, Mallet J (2006) DNA barcodes: recent successes and future prospects. Heredity 97:254–255
Dawei W, Peng Y (2014) Assessing the impact of dominant sequencing base gene expression profiling techniques (SGEPTs) on phytopathogenic fungi. Chiang Mai J Sci 41:922–9444
De Jonge R, van Esse HP, Maruthachalam K, Bolton MD, Santhanam P, Saber MK, Zhang Z, Usami T, Lievens B, Subbarao KV (2012) Tomato immune receptor Ve1 recognizes effector of multiple fungal pathogens uncovered by genome and RNA sequencing. Proc Natl Acad Sci USA 109:5110–5115
Dean R, Van Kan JA, Pretorius ZA, Hammond-Kosack KE, Di Pietro A, Spanu PD, Rudd JJ, Dickman M, Kahmann R, Ellis J, Foster GD (2012) The top 10 fungal pathogens in molecular plant pathology. Mol Plant Pathol 13:414–430
Debbi A, Boureghda H, Monte E, Hermosa R (2018) Distribution and genetic variability of *Fusarium oxysporum* associated with tomato diseases in Algeria and a biocontrol strategy with indigenous Trichoderma spp. Front Microbiol 9:282

Delalieux S, Van Aardt JAN, Keulemans W, Schrevens E, Coppin P (2007) Detection of biotic stress (*Venturia inaequalis*) in apple trees using hyperspectral data: non-parametric statistical approaches and physiological implications. Eur J Agron 27:130–143

Dickert FL, Hayden O, Bindeus R, Mann K, Blaas D, Waigmann E (2004) Bioimprinted QCM sensors for virus detection screening of plant sap. Anal Bioanal Chem 378:1929–1934

Ding XS, Mannas SW, Bishop BA, Rao X, Lecoultre M, Kwon S, Nelson RS (2018) An improved brome mosaic virus silencing vector: greater insert stability and more extensive VIGS. Plant Physiol 176:496–510

Ding TB, Li J, Chen EH, Niu JZ, Chu D (2019) Transcriptome profiling of the whitefly Bemisia tabaci MED in response to single infection of tomato yellow leaf curl virus, tomato chlorosis virus, and their co-infection. Front Physiol 10:302

Dubas ST, Pimpan VG (2008) Synthesis of silver nanoparticles for ammonia sensing. Talanta 76:29–33

Dubertret B, Calame M, Libchaber AJ (2001) Single mismatch detection using gold-quenched fluorescent oligonucleotides. Nat Biotechnol 19:365–370

Eid S, Atamian HS, Abou-Jawdah Y, Havey MJ (2008) Assessing the movement of cucurbit yellow stunting disorder virus in susceptible and tolerant cucumber germplasms using serological and nucleic acid based methods. J Phytopathol 156:438–445

El-Araby SW, Ibrahin AI, Hemeida AA, Mahmo A, Soliman MA, El-Attar KA, Mazyad MH (2009) Biological, serological and molecular diagnosis of three major potato viruses in Egypt. Int J Virol 5:77–88

Ellison MA, McMahon MB, Bonde MR, Palmer CL, Luster DG (2016) In situ hybridization for the detection of rust fungi in paraffin embedded plant tissue sections. Plant Methods 12:37

Endo S, Komori T, Ricci G, Sano A, Yokoyama K, Ohori A, Kamei K, Franco M, Miyaji M, Nishimura K (2004) Detection of gp43 of *Paracoccidioides brasiliensis* by the loop mediated isothermal amplification (LAMP) method. FEMS Microbiol Lett 234:93–97

Eshaque B, Dixon B (2006) Technology platforms for molecular diagnosis of cystic fibrosis. Biotechnol Adv 24:86–93

Etefagh R, Azhir E, Shahtahmasebi N (2013) Synthesis of CuO nanoparticles and fabrication of nanostructural layer biosensors for detecting *Aspergillus niger* fungi. Sci Iran 20:1055–1058

Eun AJC, Huang L, Chew FT, Li SFY, Wong SM (2002) Detection of two orchid viruses using quartz crystal microbalance (QCM) immunosensors. J Virol Methods 99:71–79

Fakruddin M (2011) Loop mediated isothermal amplification (LAMP)–an alternative to polymerase chain reaction (PCR). Bangladesh Res Pub J 5:425–439

Fang Y, Ramasamy RP (2015) Current and prospective methods for plant disease detection. Biosensors 5:537–561

Fang Y, Umasankar Y, Ramasamy RP (2014) Electrochemical detection of p-ethylguaiacol, a fungi infected fruit volatile using metal oxide nanoparticles. Analyst 139:3804–3810

Fargette D, Fauquet C, Thouvenel JC (1988) Yield losses induced by African cassava mosaic virus in relation to the mode and the date of infection. Int J Pest Manage 34:89–91

Fegan M, Croft BJ, Teakle DS, Hayward AC, Smith GR (1998) Sensitive and specific detection of Clavibacter xyli subsp. xyli, causal agent of ratoon stunting disease of sugarcane, with a polymerase chain reaction-based assay. Plant Pathol 47:495–504

Fegla G, Kawanna M (2013) Improved indirect ELISA for detection of some plant viruses. Int J Agric Biol 15:939–944

Ferrari BC, Bergquist PL (2007) Quantum dots as alternatives to organic fluorophore for Cryptosporidium detection using conventional flow cytometry and specific monoclonal antibodies: lessons learned. Cytometry Part A 71:265–271

Fessehaie A, De Boer SH, Lévesque CA (2003) An oligonucleotide array for the identification and differentiation of bacteria pathogenic on potato. Phytopathology 93:262–269

Gabrielle M, Van der V, Schukkink R, Van Gemen B, Schepers Y, Klatser Y (1993) Nucleic acid sequence-based amplification (NASBA) for the identification of mycobacteria. Microbiology 139:2423–2429

Gall JG, Pardue ML (1969) Formation and detection of RNA–DNA hybrid molecules in cytological preparation. Proc Natl Acad Sci U S A 63:378–383

Gall JG, Pardue ML (1971) Nucleic acid hybridization in cytological preparations. In: Grossman L, Moldave K (eds) Methods in enzymology, vol 21. Academic, New York, pp 470–480

Garrido C, Carbu M, Acreo FJ, Boonham N, Coyler A, Cantoral JM, Budge G (2009) Development of protocols for detection of *Colletotrichum acutatum* and monitoring of strawberry anthracnose using real-time PCR. Plant Pathol 58:43–51

Gomez A, Miller SN, Smolina I (2014) Visual detection of bacterial pathogens via PNA-based padlock probe assembly and isothermal amplification of DNAzymes. Anal Chem 86:1992–11998

Gomez MA, Lin ZD, Moll T, Chauhan RD, Hayden L, Renninger K, Beyene G, Taylor NJ, Carrington JC, Staskawicz BJ, Bart RS (2019) Simultaneous CRISPR/Cas9-mediated editing of cassava eIF4E isoforms nCBP-1 and nCBP-2 reduces cassava brown streak disease symptom severity and incidence. Plant Biotechnol J 17:421–434

Grabherr MG, Haas BJ, Yassour M, Levin JZ, Thompson DA, Amit I, Adiconis X, Fan L, Raychowdhury R, Zeng Q (2011) Full length transcriptome assembly from RNA-Seq data without a reference genome. Nat Biotechnol 29:644–652

Graeff S, Link J, Claupein W (2006) Identification of powdery mildew (*Erysiphe graminis* sp. *tritici*) and take all disease (*Gaeumannomyces graminis* sp. *tritici*) in wheat (*Triticum aestivum* L.) by means of leaf reflectance measurements. Cent Eur J Biol 1:275–288

Grisham MP, Johnson RM, Zimba PV (2010) Detecting sugarcane yellow leaf virus infection in asymptomatic leaves with hyperspectral remote sensing and associated leaf pigment changes. J Virol Method 167:140–145

Grote D, Olmos A, Kofoet A, Tuset JJ, Bertolini E, Cambra M (2002) Specific and sensitive detection of Phytophthora nicotianae by simple and nested-PCR. Eur J Plant Pathol 108:197–207

Guixia H, Desouky A, Yongping D, Ed S (2019) Transgenic citrus plants expressing a 'Candidatus Liberibacter asiaticus' prophage protein LasP235 display Huanglongbing-like symptoms. Agri Gene 12:100085

Gupta R, Kwon SY, Kim ST (2018) An insight into the tomato spotted wilt virus (TSWV), tomato and thrips interaction. Plant Biotechnol Rep 12:157–163

Ha DT, Ghosh S, Ahn CH, Segal DJ, Kim MS (2018) Pathogen-specific DNA sensing with engineered zinc finger proteins immobilized on a polymer chip. Analyst 143:4009–4016

Hadidi A, Barba M (2006) DNA microarrays: 21st century pathogen detection. In XX international symposium on virus and virus-like diseases of temperate fruit crops-fruit tree diseases. 781:331–340

Han YC, Zeng XG, Xiang FY, Ren L, Chen FY, Gu YC (2016) Distribution and characteristics of *Colletotrichum* spp. associated with anthracnose of strawberry in Hubei, China. Plant Dis 100:996–1006

Hančević K, Cerni S, Radić T, Škorić D (2012) Comparison of different methods for *Citrus tristeza* virus detection in Satsuma mandarins. J Plant Dis Protect 119:2–7

Hashimoto Y, Nakamura H, Koichi AK, Karube I (2008) A new diagnostic method for soil borne disease using a microbial biosensor. Microbes Environ 23:359

Hassain T, Singh BP, Anwar F (2014) A quantitative real-time PCR based method for the detection of *Phytophthora infestans* causing late blight of potato, in infested soil. Saudi J Biol Sci 21:380–386

Hayden RT, Qian X, Roberts GD, Lloyd RV (2001) In situ hybridization for the identification of yeast like organisms in tissue section. Diagn Mol Pathol 10:15–23

Hayden R, Qian X, Procop G, Roberts G, Lloyd R (2002) In situ hybridization for the identification of filamentous fungi in tissue section. Diagn Mol Pathol 11:119–126

Hayward AC (1993) The hosts of Xanthomonas. In: Swings JG, Civerolo EL (eds) Xanthomonas. Springer, Dordrecht

Hong Y, Luo Y, Yi J, He L, Dai L, Yi T (2019) Screening nested-PCR primer for 'Candidatus Liberibacter asiaticus' associated with citrus Huanglongbing and application in Hunan, China. PLoS One 14:e0212020

Huang W, Lamb DW, Niu Z, Zhang Y, Liu L, Wang J (2007) Identification of yellow rust in wheat using in-situ spectral reflectance measurements and airborne hyperspectral imaging. PrecisAgric 8:187–197

Huang J, Liao H, Zhu Y, Sun J, Sun Q, Liu X (2012) Hyperspectral detection of rice damaged by rice leaf folder (*Cnaphalocrocis medinalis*). Comput Electron Agric 82:100–107

Huang X, Xu J, Ji HF, Li G, Chen H (2014) Quartz crystal microbalance based biosensor for rapid and sensitive detection of maize chlorotic mottle virus. Anal Methods 6:4530–4536

Hussain A, Arif M, Abbas A, Hussain B, Ali M, Jaffar S (2016) A review on aphid-borne virus (potato virus Y). J Entomol Zool Stud 4:189–192

Hyun JW, Hwang RY, Jung KE (2017) Development of multiplex PCR for simultaneous detection of citrus viruses and the incidence of citrus viral diseases in late-maturity citrus trees in Jeju Island. Plant Pathol J 33:307–317

Ibanez AM, Martinelli F, Uratsu SL, Vo A, Tinoco MA, Phu ML, Chen Y, Rocke DM, Dandekar AM (2014) Transcriptome and metabolome analysis of Citrus fruit to elucidate puffing disorder. Plant Sci 217:87–98

Jacquemoud S, Ustin SL (2001) Leaf optical properties: a state of the art. In: Proceedings of international symposium physical measurements and signature in remote sensing, pp 223–232

Jain K (2003) Nanodiagnostics: application of nanotechnology (NT) in molecular diagnostics. Expert Rev Mol Diagn 2:153–161

Jensen E (2014) Technical review: in situ hybridization. Anat Rec 297:1349–1353

Kanakala S, Kuria P (2018) Chickpea chlorotic dwarf virus: an emerging monopartite dicot infecting mastrevirus. Viruses 11:5

Kerpen L, Niccolini L, Licausi F, Dongen JT, Weits DA (2019) Hypoxic conditions in crown galls induce plant anaerobic responses that support tumor proliferation. Front Plant Sci 10:56

Kfir R, Genthe B (1993) Advantages and disadvantages of the use of immunodetection techniques for the enumeration of microorganisms and toxins in water. Water Sci Technol 27:243–252

Khater M, de la Escosura-Muñiz A, Laura A, Arben M (2019) In situ plant virus nucleic acid isothermal amplification detection on gold nanoparticle-modified electrodes. Anal Chem 91:4790–4796

Kim MS, Kim J (2016) Multiplexed detection of pathogen-specific DNA using engineered zinc finger proteins without target amplification. Anal Methods 8:6696–6700

Kim SW, Li Z, Moore PS, Monaghan AP, Chang Y, Nichols M, John B (2010) A sensitive non-radioactive northern blot method to detect small RNAs. Nucleic Acids Res 38:e98

Kleo K, Kapp A, Ascher L, Lisdat F (2011) Detection of *vaccinia* virus DNA by quartz crystal microbalance. Anal Biochem 418:260–266

Koo C, Martha M-W, Hyun SK, Cifci OS, Vanessa LV-D, Bo M, Sungman K (2013) Development of a real-time microchip PCR system for portable plant disease diagnosis. PLoS One 8:12–e82704

Kumar S, Ansary A, Ayoobul A, Absar A, Khan MI (2007) Extra cellular biosynthesis of CdSe quantum dots by the fungus, *Fusarium oxysporum*. J Biomed Nanotechnol 3:190–194

Kurosawa S, Park JW, Aizawa H, Wakida SI, Tao H, Ishihara K (2006) Quartz crystal microbalance immunosensors for environmental monitoring. Biosens Bioelectron 22:473–481

Kyrkou I, Pusa T, Ellegaard-Jensen L, Sagot MF, Hansen LH (2018) Pierce's disease of grapevines: a review of control strategies and an outline of an epidemiological model. Front Microbiol 9:21–41

Le Floch G, Tambong J, Vallance J, Tirilly Y, Lévesque A, Rey P (2007) Rhizosphere persistence of three *Pythium oligandrum* strains in tomato soilless culture assessed by DNA macroarray and real-time PCR. FEMS Microbiol Ecol 61:317–326

Leborgne CN, Bouhidel K (2014) Plasma membrane protein trafficking in plant–microbe interactions: a plant cell point of view. Front Plant Sci 5:735

Lee YG, Chang KS (2005) Application of a flow type quartz crystal microbalance immunosensor for real time determination of cattle bovine ephemeral fever virus in liquid. Talanta 65:1335–1342

Lee IM, Bartoszyk IM, Gundersen-Rindal DE, Davis RE (1997) Phylogeny and classification of bacteria in the genera Clavibacter and Rathayibacter on the basis of 16S rRNA sequence analysis. Appl Environ Microbiol 63:2631–2636

Lee GP, Min BE, Kim CS, Choi SH, Harn CH, Kim SU (2003) Plant virus cDNA chip hybridization for detection and differentiation of four cucurbit-infecting Tobamoviruses. J Virol Methods 110:19–24

Leinberger DM, Schumacher U, Autenrieth IB, Bachmann TT (2005) Development of a DNA microarray for detection and identification of fungal pathogens involved in invasive mycoses. J Clin Microbiol 43:4943–4953

Li M, Ishiguro Y, Otsubo K, Suzuki H, Tsuji T, Miyake N, Nagai H, Suga H, Kageyama K (2014a) Monitoring by real-time PCR of three water-borne zoosporic *Pythium* species in potted flower and tomato greenhouses under hydroponic culture systems. Eur J Plant Pathol 140:229–242

Li L, Zhang Q, Huang D (2014b) A review of imaging techniques for plant phenotyping. Sensors 14:20078–20111

Lievens B, Brouwer M, Vanachter ACRC, Lévesque CA, Cammue BPA, Thomma BPHJ (2003) Design and development of a DNA array for rapid detection and identification of multiple tomato vascular wilt pathogens. FEMS Microbiol Lett 223:113–122

Lin NS, Hsu YH, Hsu HT (1990) Immunological detection of plant viruses and a mycoplasma like organism by direct tissue blotting on nitrocellulose membranes. Phytopathology 80:824–828

Liu W, Huang S, Liu N, Dong D, Yang Z, Tang Y, Ma W, He X, Ao D, Xu Y (2017) Establishment of an accurate and fast detection method using molecular beacons in loop-mediated isothermal amplification assay. Sci Rep 7:40125

Liu H, Wu K, Wu W, Mi W, Hao X, Wu Y (2019a) A multiplex reverse transcription PCR assay for simultaneous detection of six main RNA viruses in tomato plants. J Virol Methods 265:53–58

Liu D, Zhao Q, Cheng Y, Li D, Jiang C, Cheng L, Wang Y, Yang A (2019b) Transcriptome analysis of two cultivars of tobacco in response to cucumber mosaic virus infection. Sci Rep 9:3124

López CE, Bernal AJ (2012) Cassava bacterial blight: using genomics for the elucidation and management of an old problem. Trop Plant Biol 5(1):117–126

Love JM, Marquis-Nicholson R, Love RC, Love DR (2012) Portable battery-operated rapid PCR amplification of the CAG repeat region of the Huntington disease locus. Res J Biol 2:191–196

Luminex (2010) xMAP technology technical note: overcoming the cost and performance limitations of ELISA with xMAP technology. http://www.luminexcorp.com/prod/groups/public/documents/lmnxcorp/308-xmap-vs.-elisa-white-paper.pdf

Machado AK, Brown NA, Urban M, Kanyuka K, Hammond-Kosack KE (2017) RNAi as an emerging approach to control *Fusarium* head blight disease and mycotoxin contamination in cereals. Pest Manag Sci 74:790–799

Mahlein AK, Oerke EC, Steiner U, Dehne HW (2012) Recent advances in sensing plant diseases for precision crop protection. Eur J Plant Pathol 133:197–209

Mahlein AK, Rumpf T, Welke P, Dehne HW, Plümer L, Steiner U, Oerke EC (2013) Development of spectral indices for detecting and identifying plant diseases. Remote Sens Environ 128:21–30

Makkouk KM, Kumari SG (2006) Molecular diagnosis of plant viruses. Arab J Plant Protect 24:135–138

Malnoy M, Viola R, Jung MH, Koo OJ, Kim S, Kim JS, Nagamangala KC (2016) DNA-free genetically edited grapevine and apple protoplast using CRISPR/Cas9 ribonucleoproteins. Front Plant Sci 7:1904

Malthus TJ, Madeira AC (1993) High resolution spectroradiometry: spectral reflectance of field bean leaves infected by Botrytis fabae. Remote Sens Environ 45:107–116

Mancini V, Murolo S, Romanazzi G (2016) Diagnostic methods for detecting fungal pathogens on vegetable seeds. Plant Pathol 65:691–703

Mansfield J, Genin S, Magori S, Citovsky V, Sriariyanum M, Ronald P, Dow M, Verdier V, Beer SV, Machado MA, Toth I, Salmond G, Foster GD (2012) Top 10 plant pathogenic bacteria in molecular plant pathology. Mol Plant Pathol 13:614–629

Martinelli F, Basile B, Morelli G, D'Andria R, Tonutti P (2012a) Effects of irrigation on fruit ripening behavior and metabolic changes in olive. Sci Hortic 144:201–207

Martinelli F, Uratsu SL, Albrecht U, Reagan RL, Phu ML (2012b) Transcriptome profiling of citrus fruit response to Huanglongbing disease. PLoS One 7:e38039
Martinelli F, Reagan RL, Uratsu SL, Phu ML, Albrecht U (2013a) Gene regulatory networks elucidating Huanglongbing disease mechanisms. PLoS One 8:e74256
Martinelli F, Remorini D, Saia S, Massai R, Tonutti P (2013b) Metabolic profiling of ripe olive fruit in response to moderate water stress. Sci Hortic 159:52–58
Martinelli F, Scalenghe R, Giovino A, Pasquale M, Aksenov AA, Pasamontes A, Peirano DJ, Davis CE, Dandekar AM (2014) Proposal of a Citrus translational genomic approach for early and infield detection of Flavescence dorée in Vitis. Plant Biosyst 150:43–53
Martinelli F, Scalenghe R, Davino S, Panno S, Scuderi G, Ruisi P, Villa P, Stroppiana D, Boschetti M, Goulart LR, Davis CE (2015) Advanced methods of plant disease detection. A review. Agron Sustain Dev 35:1–25
McCartney HA, Foster SJ, Fraaije BA, Ward E (2003) Molecular diagnostics for fungal plant pathogens. Pest Manag Sci 59:129–142
Mecea VM (2005) From quartz crystal microbalance to fundamental principles of mass measurements. Anal Lett 38:753–767
Mecea VM (2006) Is quartz crystal microbalance really a mass sensor? Sens Actuators A 128:270–277
Metzker ML (2009) Sequencing technologies the next generation. Nat Rev Genet 11:31–46
Milijasevic S, Todorovic B, Balaz J (2006) Clavibacter michiganensis subsp. michiganensis, bacterial canker of tomato: 1. Conventional and molecular identification. Pestic Phytomed 21:185–192
Mirik M, Michels GJ Jr, Kassymzhanova-Mirik S, Elliott NC, Catana V, Jones DB, Bowling R (2006) Using digital image analysis and spectral reflectance data to quantify damage by greenbug (*Hemiptera: Aphididae*) in winter wheat. Comput Electron Agric 51:86–98
Monis PT, Giglio S (2006) Nucleic acid amplification based techniques for pathogen detection and identification. Infect Genet Evol 6:2–12
Moreno JC (2003) Identification of novel genes involved in congenital hypothyroidism using serial analysis of gene expression. Horm Res Paedia 60:96–102
Mortazavi A, Williams BA, McCue K, Schaeffer L, Wold B (2008) Mapping and quantifying mammalian transcriptomes by RNA-Seq. Nat Methods 5:621–628
Moshou D, Bravo C, West J, Wahlen S, McCartney A, Ramon H (2004) Automatic detection of 'yellow rust' in wheat using reflectance measurements and neural networks. Comput Electron Agric 44:173–188
Mutka AM, Bart RS (2015) Image based phenotyping of plant disease symptoms. Front Plant Sci 5:734
Naidu RA, Perry EM, Pierce FJ, Mekuria T (2009) The potential of spectral reflectance technique for the detection of grapevine leafroll associated virus-3 in two red berried wine grape cultivars. Comput Electron Agric 66:38–45
Naidua RA, Hughes JA (2001) Methods for the detection of plant virus diseases. In: Proceedings of a conference organized by IITA, plant virology in sub Saharan Africa. International Institute of Tropical Agriculture, Oyo State, pp 233–260
Natali L, Giordani T, Lercari B, Maestrini P, Cozza R, Pangaro T, Vernieri P, Martinelli F, Cavallini A (2007) Light induces expression of a dehydrin encoding gene during seedling de-etiolation in sunflower (*Helianthus annuus* L.). J Plant Physiol 164:263–273
Nemri A, Saunders DGO, Anderson C, Upadhyaya NM, Win J, Lawrence GJ, Jones DA, Kamoun S, Ellis JG, Dodds PN (2014) The genome sequence and effector complement of the flax rust pathogen *Melampsora lini*. Front Plant Sci 5:98
Nicolaisen M (2011) An oligonucleotide-based microarray for detection of plant RNA viruses. J Virol Methods 173:137–143
Nikitin M, Statsyuk N, Frantsuzov P, Dzhavakhiya V, Golikov A (2018) Matrix approach to the simultaneous detection of multiple potato pathogens by real-time PCR. J Appl Microbiol 124:797–809

Nilsson RH, Abarenkov K, Larsson KH, Kõljalg U (2011) Molecular identification of fungi: rationale, philosophical concerns, and the UNITE database. Open Appl Inf J 5:81–86

Nutter FW Jr (1989) Detection and measurement of plant disease gradients in peanut with multispectral radiometer. Phytopathology 79:958–963

Nykyri J, Niemi O, Koskinen P, Nokso-Koivisto J, Pasanen M, Broberg M, Palva ET (2012) Revised phylogeny and novel horizontally acquired virulence determinants of the model soft rot phytopathogen *Pectobacterium wasabiae* SCC3193. PLoS Pathog 8:e1003013

O'Donnell EA, Ernst DN, Hingorani R (2013) Multiparameter flow cytometry: advances in high resolution analysis. Immune Netw 13:43–54

Odilbekov F, Armonienė R, Henriksson T, Chawade A (2018) Proximal phenotyping and machine learning methods to identify *Septoria Tritici* blotch disease symptoms in wheat. Front Plant Sci 9:685

Osmani Z, Sabet MS, Shams-Bakhsh M, Moieni A, Vahabi K (2019) Virus-specific and common transcriptomic responses of potato (*Solanum tuberosum*) against PVY, PVA and PLRV using microarray meta-analysis. Plant Breed 138:216–228

Owen TW, Al-Kaysi RO, Bardeen CJ, Cheng Q (2007) Micro-gravimetric immunosensor for direct detection of aerosolized influenza a virus particles. Sens Actuators B 126:691–699

Phan MS, Seo JK, Choi HS, Lee SH, Kim KH (2014) Molecular and biological characterization of an isolate of cucumber mosaic virus from Glycine soja by generating its infectious full-genome cDNA clones. Plant Pathol J 30:159–167

Portieles R, Ochagavia ME, Canales E, Silva Y, Chacón O, Hernández I et al (2017) High-throughput SuperSAGE for gene expression analysis of Nicotiana tabacum-Rhizoctonia solani interaction. BMC Res Notes 10:603

Postnikova OA, Nemchinov LG (2012) Comparative analysis of microarray data in Arabidopsis transcriptome during compatible interactions with plant viruses. Virol J 9:101

Prabhakar M, Prasad YG, Thirupathi M, Sreedevi G, Dharajothi B, Venkateswarlu B (2011) Use of ground based hyperspectral remote sensing for detection of stress in cotton caused by leafhopper (Hemiptera: Cicadellidae). Comput Electron Agric 79:189–198

Punja ZK, Wan A, Goswami RS, Verma N, Rahman M, Barasubiye T, Seifert KA, Lévesque CA (2007) Diversity of *Fusarium* species associated with discolored ginseng roots in British Columbia. Can J Plant Pathol 29:340–353

Purcell DE, O'Shea MG, Johnson RA, Kokot S (2009) Near-infrared spectroscopy for the prediction of disease ratings for Fiji leaf gall in sugarcane clones. Appl Spectrosc 63:450–457

Qi M, Yang Y (2002) Quantification of Magnaporthe grisea during infection of rice plants using real-time polymerase chain reaction and northern blot/phosphor imaging analyses. Phytopathology 92:870–876

Qian X, Lloyd RV (2003) Recent developments in signal amplification methods for in situ hybridization. Diagn Mol Pathol 12:1–13

Qiao TM, Zhang J, Li SJ, Han S, Zhu TH (2016) Development of nested PCR, multiplex PCR, and loop-mediated isothermal amplification assays for rapid detection of Cylindrocladium scoparium on eucalyptus. Plant Pathol J 32:414–422

Rahman MT, Uddin MS, Sultana R, Moue A, Setu M (2013) Polymerase chain reaction (PCR): a short review. Anwer Khan Mod Med Coll J 4:30–36

Ratan ZA, Zaman SB, Mehta V, Haidere MF, Runa NJ, Akter N (2017) Application of fluorescence in situ hybridization (FISH) technique for the detection of genetic aberration in medical science. Cureus 9:e1325

Ren CH, Hu CQ, Luo P, Wang QB (2009) Sensitive and rapid identification of *Vibrio vulnificus* by loop mediated isothermal amplification. Microbiol Res 164:514–521

Reynolds GJ, Windels CE, MacRae IV, Laguette S (2012) Remote sensing for assessing *Rhizoctonia* crown and root rot severity in sugar beet. Plant Dis 96:497–505

Rico A, López R, Asensio C, Aizpún MT, Asensio-S-Manzanera MC, Murillo J (2003) Nontoxigenic strains of Pseudomonas syringae pv. phaseolicola are a main cause of halo blight of beans in Spain and escape current detection methods. Phytopathology 93:1553–1559

Rizzini FM, Bonghi C, Chkaiban L, Martinelli F, Tonutti P (2010) Effects of postharvest partial dehydration and prolonged treatments with ethylene on transcript profiling in skins of wine grape berries. Acta Hortic 877:1099–1104
Robinson B, Erle D, Jones D, Shapiro S, Metzger W, Albelda S, Parks W, Boylan A (2000) Recent advances in molecular biological techniques and their relevance to pulmonary research. Thorax 55:329–339
Rumpf T, Mahlein AK, Steiner U, Oerke EC, Dehne HW, Plümer L (2010) Early detection and classification of plant diseases with support vector machines based on hyperspectral reflectance. Comput Electron Agric 74:91–99
Rupp S, Weber RW, Rieger D, Detzel P, Hahn M (2017) Spread of *Botrytis cinerea* strains with multiple fungicide resistance in German horticulture. Front Microbiol 7:2075
Russo G, Zegar C, Giordano A (2003) Advantages and limitations of microarray technology in human cancer. Oncogene 22:6497–6507
Salm H, Geider K (2004) Real-time PCR for detection and quantification of Erwinia amylovora, the causal agent of fireblight. Plant Pathol 53:602–610
Sato T, Takayanagi A, Nagao K, Tomatsu N, Fukui T, Kawaguchi M, Kudoh J, Amagai M, Yamamoto N, Shimizu N (2010) Simple PCR- based DNA microarray system to identify human pathogenic fungi in skin. J Clin Microbiol 48:2357–2364
Schaad NW, Berthier-Schaad Y, Knorr D (2007) A high throughput membrane BIO-PCR technique for ultra-sensitive detection of Pseudomonas syringae pv. Phaseolicola. Plant Pathol 56:1–8
Schoelz JE, Angel CA, Nelson RS, Leisner SM (2015) A model for intracellular movement of cauliflower mosaic virus: the concept of the mobile virion factory. J Exp Bot 67:2039–2048
Scholthof KG, Adkins S, Czosnek H, Palukaitis P, Jacquot E, Hohn T, Hohn B, Saunders K, Candresse T, Ahlquist P, Hemenway C, Foster GD (2011) Top 10 plant viruses in molecular plant pathology. Mol Plant Pathol 12:938–954
Schumann GL, D'Arcy CJ (2009) Essential plant pathology, 2nd edn. APS Press, St Paul
Shang H, Xie Y, Zhou X, Qian Y, Wu J (2011) Monoclonal antibody-based serological methods for detection of cucumber green mottle mosaic virus. Virol J 8:228–234
Sharon M, Choudhary AK, Kumar RJ (2010) Nanotechnology in agricultural diseases and food safety. J Phytology 2:83–92
Sidra S, Aisha T, Muhammad FS, Muhammad Waqar A, Arshad AS, Sehrish S (2017) Recent advances in molecular techniques for the identification of phytopathogenic fungi – a mini review. J Plant Interact 12:493–504
Simón MR, Cordo CA, Castillo NS, Struik PC, Börner A (2012) Population structure of *Mycosphaerella graminicola* and location of genes for resistance to the pathogen: recent advances in Argentina. Int J Agron 2012:680275
Singh A, Kumar N (2013) A review on DNA microarray technology. Int J Cur Res Rev 5:01–05
Small J, Call DR, Brockman FJ, Straub TM, Chandler DP (2001) Direct detection of 16S rRNA in soil extracts by using oligonucleotide microarrays. Appl Environ Microbiol 67:4708–4716
Soanes DM, Chakrabarti A, Paszkiewicz KH, Dawe AL, Talbot NJ (2012) Genome-wide transcriptional profiling of appressorium development by the rice blast fungus *Magnaporthe oryzae*. PLoS Pathog 8:e1002514
Soto-Suárez M, González C, Piégu B, Tohme J, Verdier V (2010) Genomic comparison between *Xanthomonas oryzae pv. oryzae* and *Xanthomonas oryzae pv. oryzicola*, using suppression-subtractive hybridization. FEMS Microbiol Lett 308:16–23
Stilwell AR, Hein GL, Zygielbaum AI, Rundquist DC (2013) Proximal sensing to detect symptoms associated with wheat curl mite-vectored viruses. Int J Remote Sens 34:4951–4966
Su CC, Wu TZ, Chen LK, Yang HH, Tai DF (2003) Development of immunochips for the detection of dengue viral antigens. Anal Chim Acta 479:117–123
Sun W, Jiao K, Zhang S, Zhang C, Zhang Z (2001) Electro-chemical detection for horseradish peroxidase-based enzyme immunoassay using p-aminophenol as substrate and its application in detection of plant virus. Anal Chim Acta 434:43–50

Sun J, Li X, Zeng H, Xie Z, Lu C, Xi L, De Hoog GS (2010) Development and evaluation of loop mediated isothermal amplification (LAMP) for the rapid diagnosis of *Penicillium marneffei* in archived tissue samples. FEMS Immunol Med Microbiol 58:381–388

Surrette C, Scherer B, Corwin A, Grossmann G, Kaushik AM, Hsieh K, Puleo CM (2018) Rapid microbiology screening in pharmaceutical workflows. SLAS Technol 23:387–394

Syed A, Ahmad A (2013) Extracellular biosynthesis of CdTe quantum dots by the fungus *Fusarium oxysporum* and their antibacterial activity. Spectrochim Acta Part A: Mol Biomol Spectrosc 106:4147

Tanaka F (2009) Specific in situ visualization of the pathogenic endophytic fungus *Aciculosporium take*, the cause of witches' broom in bamboo. Appl Environ Microbiol 75:4829–4834

Tang YW, Procop GW, Persing DH (1997) Molecular diagnostics of infectious diseases. Clin Chem 43:2021–2038

Tang DQ, Zhang DJ, Tang DY, Ai H (2006) Amplification of the antigen–antibody interaction from quartz crystal microbal-anceimmunosensors via back filling immobilization of nanogold on biorecognition surface. J Immunol Methods 316:144–152

Taoufik A, Nijhof A, Hamidjaja R, Jongejan F, Pillay V, Sonnevelt M, de Boer M (2004) Reverse line blot hybridisation in the detection of tick borne diseases. BT*i* September. http://www.bti.org.in

Tarasov KV, Brugh SA, Tarasova YS, Boheler KR (2007) Serial analysis of gene expression (SAGE). In: Zhang J, Rokosh G (eds) Cardiac gene expression. Methods in molecular biology, vol 366. Humana Press

Thomas SW, Glaring MA, Rasmussen SW, Kinane JT, Oliver RP (2002) Transcript profiling in the barley mildew pathogen *Blumeria graminis* by serial analysis of gene expression (SAGE). Mol Plant-Microbe Interact 15:847–856

Torrance L (1998) Developments in serological methods to detect and identify plant viruses. Plant Cell Tissue Organ Cult 52:27–32

Tosetti R, Martinelli F, Tonutti P (2012) Metabolomics approach to studying minimally processed peach (*Prunuspersica*) fruit. Acta Hortic 934:1017–1022

Trtkova J, Raclavsky V (2006) Molecular genetic approaches to identification and typing of pathogenic *Candida* yeasts. Biomed Pap 150:51–61

Trung TT, Hetzer A, Göhler A, Topfstedt E, Wuthiekanun V, Limmathurotsakul D, Steinmetz I (2011) Highly sensitive direct detection and quantification of Burkholderia pseudomallei bacteria in environmental soil samples by using real-time PCR. Appl Environ Microbiol 77:6486–6494

Tsai HL, Huang LC, Ann PJ, Liou RF (2006) Detection of orchid Phytophthora disease by nested PCR. Bot Stud 47(4):379–387

Tsui CK, Woodhall J, Chen W, Lévesque CA, Lau A, Schoen CD, Baschien C, Najafzadeh MJ, de Hoog GS (2011) Molecular techniques for pathogen identification and fungus detection in the environment. IMA Fungus 2:177

Uttenthaler E, Schraml M, Mandel J, Drost S (2001) Ultrasensitive quartz crystal microbalance sensors for detection of M13-Phages in liquids. Biosens Bioelectron 16:735–743

Velculescu VE (1995) Serial analysis of gene expression. Science 270:484–487

Vemulapati B, Druffel KL, Husebye D, Eigenbrode SD, Pappu HR (2014) Development and application of ELISA assays for the detection of two members of the family Luteoviridae infecting legumes: pea enation mosaic virus (genus Enamovirus) and bean leafroll virus (genus Luteovirus). Ann Appl Biol 165:130–136

Villari C, Tomlinson JA, Battisti A, Boonham N, Capretti P, Faccoli M (2013) Use of loop-mediated isothermal amplification for detection of Ophiostoma clavatum, the primary blue stain fungus associated with Ips acuminatus. Appl Environ Microbiol 79:2527–2533

Volpi EV, Bridger JM (2008) Fish glossary: an overview of the fluorescence in situ hybridization technique. BioTechniques 45:385–386

Vos De A, Cumming GS, Cumming D, Ament JM, Baum J, Clements H, Grewar J, Maciejewski K, Moore C (2016) Pathogens, disease, and the social-ecological resilience of protected areas. Ecol Soc 21:20

Walcott RR, Gitaitis RD (2000) Detection of Acidovorax avenae subsp. citrulli in watermelon seed using immunomagnetic separation and the polymerase chain reaction. Plant Dis 84:470–474

Wang L, O'Donoghue MB, Tan W (2006) Nanoparticles for multiplex diagnostics and imaging. Nanomedicine 1:413–426

Wang S, Ye W, Tian Q, Dong S, Zheng X (2017) Rapid detection of Colletotrichum gloeosporioides using a loop-mediated isothermal amplification assay. Australas Plant Pathol 46:493–498

Webster CG, Wylie JS, Jones MGK (2004) Diagnosis of plant viral pathogens. Curr Sci 86:1604–1607

Wills MK, Kirby AM, Lloyd VK (2018) Detecting the Lyme disease spirochete, Borrelia burgdorferi, in ticks using nested PCR. J Vis Exp 132:e56471

Wu H, Yang C, Li Q, Gao Y, Zhang X, Tong J, Zhang Z, Hu J, Wang D, Zeng S, Li Z (2018) Intervention effect of Qi-Yu-San-long decoction on Lewis lung carcinoma in C57BL/6 mice: insights from UPLC–QTOF/MS-based metabolic profiling. J Chromatogr B 1102-1103:23–33

Yang CM, Cheng CH, Chen RK (2007) Changes in spectral characteristics of rice canopy infested with brown planthopper and leaffolder. Crop Sci 47:329–335

Yang H, Li H, Jiang X (2008) Detection of foodborne pathogens using bio conjugated nano materials. Microfluid Nanofluid 5:571–583

Yang F, Jacobsen S, Jørgensen HJ, Collinge DB, Svensson B, Finnie C (2013) *Fusarium graminearum* and its interactions with cereal heads: studies in the proteomics era. Front Plant Sci 4:37

Yang X, Hameed U, Zhang AF, Zang HY, Gu CY, Chen Y, Xu YL (2017) Development of a nested-PCR assay for the rapid detection of Pilidiella granati in pomegranate fruit. Sci Rep 7:40954

Yao KS, Li SJ, Tzeng KC, Cheng TC, Chang CY, Chiu CY, Liao CY, Hsu JJ, Lin ZP (2009) Fluorescence silica nanoprobe as a biomarker for rapid detection of plant pathogens. Multi-Funct Mater Struct II Parts 1 and 2 79–82:513–516

Yin C, Ramachandran SR, Zhai Y, Bu C, Pappu HR, Hulbert SH (2019) A novel fungal effector from Puccinia graminis suppressing RNA silencing and plant defense responses. New Phytol 222:1561–1572

Yoshida K, Saunders DG, Mitsuoka C, Natsume S, Kosugi S, Saitoh H, Terauchi R (2016) Host specialization of the blast fungus Magnaporthe oryzae is associated with dynamic gain and loss of genes linked to transposable elements. BMC Genomics 17:370

Yuan L, Huang Y, Loraamm RW, Nie C, Wang J, Zhang J (2014) Spectral analysis of winter wheat leaves for detection and differentiation of diseases and insects. Field Crop Res 156:199–207

Zan X, Sitasuwana P, Powellb J, Dreherb TW, Wang Q (2012) Polyvalent display of RGD motifs on turnip yellow mosaic virus for enhanced stem cell adhesion and spreading. Acta Biomater 8:2978–2985

Zeller R, Rogers M, Haramis AG (2001) In situ hybridization to cellular RNA. Curr Protoc Mole Biol 55:14–13

Zhang N, McCarthy ML, Smart CD (2008) A macroarray system for the detection of fungal and oomycete pathogens of solanaceous crops. Plant Dis 92:953–960

Zhang J, Pu R, Loraamm RW, Yang G, Wang J (2014) Comparison between wavelet spectral features and conventional spectral features in detecting yellow rust for winter wheat. Comput Electron Agric 100:79–87

Zimmers-Koniaris T (2001). Northern blotting and RNA detection. In: eLS (ed). https://doi.org/10.1038/npg.els.0000997

Zulak KG, Cox BA, Tucker MA, Oliver RP, Lopez-Ruiz FJ (2018) Improved detection and monitoring of fungicide resistance in *Blumeria graminis* f. sp. *hordei* with high-throughput genotype quantification by digital PCR. Front Microbiol 9:706

7 Linkages of Microbial Plant Growth Promoters Toward Profitable Farming

Priyanka Verma, Anjali Chandrol Solanki,
Manoj Kumar Solanki, and Baby Kumari

Abstract

Soil and plant health are linked with each other, and both are badly affected by the excess application of inorganic fertilizers and pesticides. The negative impact of chemical fertilizers toward the environment forced the scientific community to find out an alternative strategy that can improve crop yield and quality in an eco-friendly manner. The modern agriculture system is well furnished with microbial inoculants and plant defense elicitors. However, the application of microbes to manage plant growth and fitness needs to improve. Microbial inoculants play an important role in soil mineralization, energy mobilization and channelization, and also nitrogen fixation. This chapter aims to review the microbial plant helpers and their interlinks toward plant and human health. Microbial inoculums improve crop quality and yield, plant and soil health, and profit to farmers and reduce pollution. Proper utilization of microbial inoculants could help to improve the economic condition of the farmers and the country.

Keywords

Economic development · Human health · Microbial linkage · Sustainable agriculture

---

P. Verma (✉)
Department of Microbiology, Eternal University, Sirmaur, Himachal Pradesh, India

A. C. Solanki
Soil Science and Agriculture Chemistry, Jawaharlal Nehru Agricultural University, Jabalpur, Madhya Pradesh, India

M. K. Solanki
Department of Food Quality & Safety, Institute for Post-harvest and Food Sciences, The Volcani Center, Agricultural Research Organization, Rishon LeZion, Israel

B. Kumari
University Department of Biotechnology, Vinoba Bhave University, Hazaribag, Jharkhand, India

© Springer Nature Singapore Pte Ltd. 2020

M. K. Solanki et al. (eds.), *Phytobiomes: Current Insights and Future Vistas*,
https://doi.org/10.1007/978-981-15-3151-4_7

7.1 Introduction

Population growth and food demand both are expanding with time, and alleviation of fertile land and polluted natural resources place distinctive stress on current agriculture system (Lotze-Campen 2011). Moreover, increased demand for quality food without harming the environment and human health has become a global challenge for the farmers and agriculture researchers (Amundson et al. 2015). A smart agriculture system needs to develop for the country—as the country's economic growth is particularly based on agriculture—and this system can be helpful to solve the problem of malnutrition and hunger. The modern agriculture system needs to project well now to balance higher crop production and environmental safety. Global food demand might be increased by more than 9 billion in 2050 as the projected population will be increased more than the present time (Cole et al. 2018; Hunter et al. 2017).

Currently, eco-friendly farming is receiving attention globally because it offers the likelihood to organize our agricultural requirements with environmental and human concern (Amundson et al. 2015; Hunter et al. 2017; Morawicki and Díaz González 2018). This system accustoms a special farming technique whereby the environmental resources will be completely consumed without damaging the substratum that provides a healthy agricultural product. Agro-ecosystems are significantly managed by microorganism populations and soil fauna that play a critical role in soil fertility. Microbes are the major gadgets of topsoil and plant that drive stability and productivity by the multiple microbial processes such as nutrient cycling, biodegradation, biostimulation, restoration, and disease management (Jacoby et al. 2017; Kumari et al. 2019).

Biofertilizers are often delineating as constituents that incorporate with living microorganisms that are within the rhizosphere, or inside or outside of the plant (Patil and Solanki 2016). These microbes stimulate plant growth through fixing nitrogen, improve nutrient uptake, and work as a growth promoter when applied to seed, plant surfaces, or soil (Muraleedharan et al. 2010; Kumari et al. 2019). Presently, researchers are interested in eco-friendly and safe farming practices (Harish et al. 2009a; Harish et al. 2009b; Kavino et al. 2010; Schütz et al. 2018), and most of the scientists have characterized rhizospheric and endospheric microbes that cause beneficial impact to the plant are mostly utilized as microbial inoculants (Glick 1995; Harish et al. 2009b; Gouda et al. 2018; Patil and Solanki 2016). Phytobiome-associated microbes are synergistically interacting with plants and their community. These interactions regulate the microbiome structure, and it is not only influenced by the environmental factors where they reside but also have widespread effects of the host metabolism and the location of the host. For example, Grice and Segre (2012) proved that the composition of the human gut microbiome is directly linked with physiological and psychological health of the host.

Similarly, Chaparro et al. (2012) reported that the soil microbiomes are interconnected with plant health. Therefore, due to the great influence of the microbiome on ecosystems, proper attention is needed to look out for the beneficial insights into those different elements that contribute to plant soil health. Herein, this chapter presents the current insights of microbial linkages toward the soil and plant fitness and deliberations about the valuable microbes.

7.2 Effect of Micronutrients to Sustain Plant Growth

Many Asian countries are facing the problem of micronutrient deficiency, and to solve this problem they are utilizing chemical fertilizers or pesticides in an excess amount that is damaging the native beneficial soil activity that poses major constraints such as low or high pH, fewer minerals, less organic matter, contaminated water with high bicarbonate, drought, and salinity (Malakouti 2008; Fageria 2012). Deficiency of micronutrients, increases plant stress that destructively affects the crop yield and quality by causing hindrance to internal plant tissues such as xylem vessels, causing the spread of the pest and disease faster than a healthy plant, reducing the fertilizer use efficiency, and deactivating the phytosiderophores and phytostimulation that helps plants grow. Malakouti (2008) proved that Calcareous soil has Zn deficiency that is responsible for the yield loss and other important micronutrients such as Fe, B, Mn, Cu, and Mo that plays a significant role in the plant growth. Application of micronutrients can intensify grain yield up to 50% of durum wheat (*Triticum durum* L.), and mostly micronutrients are applied as a soil application, foliar spray, or seed treatment that improves the crop yield and quality, as well as macronutrient uptake efficiency (Malakouti 2008). The micronutrients play an essential role in crop production and quality that directly influences the crop and human health by providing sufficient nutrients (Table 7.1). Zohaib et al. (2018) reported that application of micronutrient mepiquat chloride enhanced the cotton yield and nutritional quality under boron-deficient and adequate boron conditions by nutrient accumulation in the seed tissues. Recently, a study proved the positive interaction between potassium and nitrogen advising that higher application of K helped to alleviate ammonium stress although growth vigor enhanced during the application of nitrate nutrition that improved the nutrient uptake and enhanced the growth of wheat plants (Guo et al. 2019).

7.3 Microbial Linkage for Sustainable Agriculture and Environment

Poor coordination of energy conversion is directly linked with the low agricultural production that frequently gets affected by physiological factors such as pH, temperature, and beneficial and pathogenic microbes (Souza et al. 2015). Plant-associated microbes that exist in either soil or plant rhizosphere improve the plant growth and accelerate plant defense against the plant pathogens. Beneficial plant growth-promoting (PGP) microbes mostly produce the secondary metabolites and solubilize the minerals that play a significant role in the plant rhizosphere (Olanrewaju et al. 2017). These kinds of beneficial microbes have a primary effect

Table 7.1 Essential plant nutrient elements and their primary forms utilized by plants

S. No.	Essential plant element		Symbol	Primary form
1.	Non-mineral elements			
		Carbon	C	CO_2 (g)
		Hydrogen	H	H_2O (l), H^+
		Oxygen	O	H_2O (l), O_2 (g)
2.	Mineral elements			
I	Primary macronutrients			
		Nitrogen	N	NH_4^+, NO_3^-
		Phosphorus	P	HPO_4^{2-}, $H_2PO_4^-$
		Potassium	K	K^+
		Calcium	Ca	Ca_2^+
ii	Secondary macronutrients			
		Magnesium	Mg	Mg_2^+
		Sulfur	S	SO_4^{2-}
		Iron	Fe	Fe_3^+, Fe_2^+
		Manganese	Mn	Mn_2^+
		Zinc	Zn	Zn_2^+
iii	Micronutrients			
		Copper	Cu	Cu_2^+
		Boron	B	$B(OH)_3$
		Molybdenum	Mo	MoO_4^{2-}
		Chlorine	Cl	Cl^-
		Nickel	Ni	Ni_2^+

Source: Parikh and James (2012)

on the soil and plant quality (Kumari et al. 2019). Beneficial microbes promote the plant growth through two different kinds of mechanism such as indirect and direct. In indirect growth promotion, microbes inhibit plant pathogen by different ways such as by antagonism or mycoparasitism, while direct mechanism projected by microbes involves the nutrient solubilization that improves the plant nutrient uptake from the environment (Glick 1995; Kafle et al. 2019; Kashyap et al. 2019). Moreover, microbes possess an antibiosis character, that is, antagonism for food and space, which induces systemic resistance of plants against the pathogens (Verma et al. 2017b). Microbes regulate various growth parameters/yields of crop/fruit plants that have been listed in Table 7.2.

7.4 Potential Microbes Act As Biofertilizers

Microbial plant growth helpers enhance the plant growth attributes and crop yield considerably. Ribeiro et al. (2018) reported that endophytic *Bacillus* could solubilize iron phosphate (Fe-P), produce siderophores and indole-acetic acid (IAA), and enhance the pearl millet plant biomass and nitrogen and phosphorus content of plant under a no-phosphorus-added condition. Bargaz et al. (2018) reported that

Table 7.2 Various crop diseases regulated by plant growth promoting microorganisms (PGPM) as biocontrol agents

S. No.	Disease	PGPM	References
1.	*Cucumber mosaic virus* (CMV; of genus *Cucumovirus* of tomato (*Lycopersicon esculentum*)	*Bacillus pumilus* strain SE34; *Kluyvera cryocrescens* strain IN114; *Bacillus amyloliquefaciens* strain IN937a; *Bacillus subtilus* strain IN937b	Rabie et al. (2017); Rendina et al. (2019); Wang et al. (2018); Zehnder et al. (2001)
2.	*Tomato mottle virus*	*Bacillus amyloliquefaciens* 937a; *B. subtilis* 937b; *B. pumilus* SE34	Ambros et al. (2017); Gill et al. (2019); Gong (2018); Murphy et al. (2000)
3.	Bacterial wilt disease in cucumber (*Cucumis sativus*)	*Bacillus pumilus* strain INR7 F	Park et al. (2013); Rojas et al. (2011); Zehnder et al. (2001)
4.	Sheath blight disease and leaf folder insect in rice (*Oryza sativa)*	*Pseudomonas fluorescens* based bioformulation	Commare et al. (2002); Kumar et al. (2016); Seenivasan et al. (2012)
5.	Blue mold disease of tobacco (*Nicotiana*)	*Bacillus pumilus* strain SE34	Sahoo et al. (2014); Wu et al. (2015); Zhang et al. (2002)
6.	Downy mildew in pearl millet (*Pennisetum glaucum*)	*Bacillus subtilis* strain GBO3; *Bacillus pumilus* strain INR7; *Bacillus* pumilus strain T	Jogaiah et al. (2014); Mahatma et al. (2011); Raj et al. (2003)
7.	CMV in cucumber	*Bacillus subtilis* strain IN937a	Borriss (2011); Jetiyanon et al. (2003); Zhang et al. (2010)
8.	Foliar diseases of tomato	*Bacillus cereus* strains B101R, B212R, and A068R	Koné et al. (2010); Silva et al. (2004); Zodape et al. (2011)
9.	Blight of bell pepper (*Capsicum annuum*)	*Bacillus* strains BB11 and FH17	Díaz-Pérez (2014); Jiang et al. (2006); Oh et al. (2011)
10.	Saline resistance in groundnut (*Arachis hypogea*)	*Pseudomonas fluorescens*	Asif et al. (2011); Saravanakumar and Samiyappan (2007)
11.	Maize (*Zea mays*) rot	*Burkholderia* strains MBf21 and MBf15	Hernández-Rodríguez et al. (2008); Löffler et al. (2010)
12.	Soil-borne pathogens of cucumber and pepper (*Piper*)	*Bacillus subtilis* ME488	Chung et al. (2008); Pliego et al. (2011)
13.	Significantly reduce the Banana bunchy top virus (BBTV) incidence	*P. fluorescens* strain CHA0 + chitin bioformulations	Kavino et al. (2008); Niyongere et al. (2013)
14.	Rice blast	*Bacillus* sp.; *Azospirillum* strains SPS2, WBPS1, and Z2–7	Shan et al. (2013); Zakira (2009)
15.	Rice sheath rot (*Sarocladium oryzae*)	Fluorescent *Pseudomonas* spp.	Hittalmani et al. (2016); Saravanakumar et al. (2009)
16.	Blight of squash	*Bacillus* strain	Ji et al. (2012); Zhang et al. (2010)

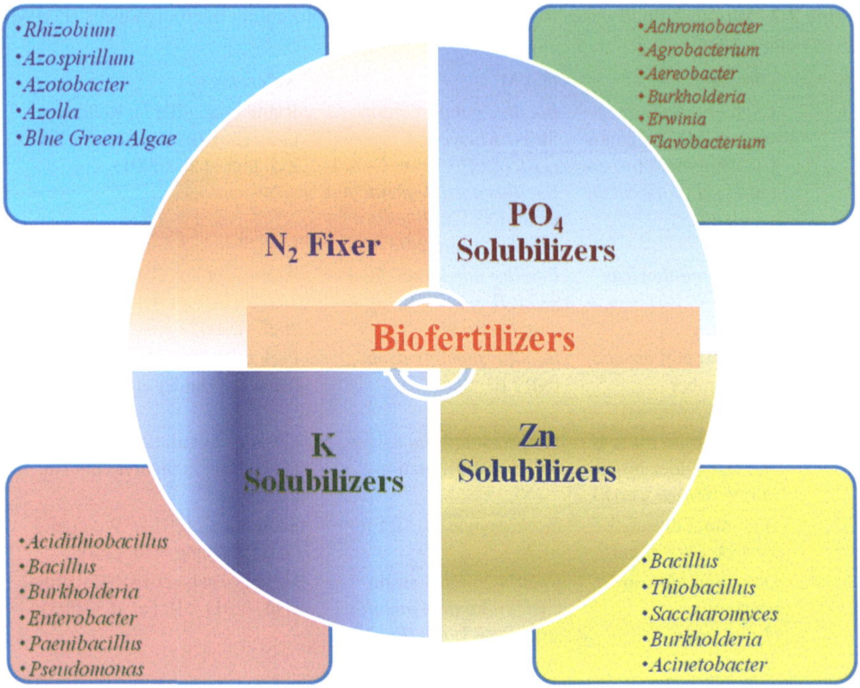

Fig. 7.1 Potential microbial linkage to improve agriculture and human health

nitrogen-fixing and P-solubilizing/mobilizing microbes are an especial element of soil fertilization for improving crop productivity and fertilizer efficiency. Moreover, these groups of microbes possess multitrophic interaction that helps in plant mineral uptake. The plants that are inoculated with appropriate microbial strains have been found to be more proficient in acquiring nutrient in even nutrient-deficient conditions (Ribeiro et al. 2018). Several beneficial plant growth-promoting microbes used as biofertilizers are represented in Fig. 7.1.

7.4.1 Nitrogen Fixers

7.4.1.1 *Rhizobium*

Among all plant growth-promoting bacteria, *Rhizobium* bacteria are heterotrophic soil bacteria that are known to be the most prominent nitrogen fixers. *Rhizobium* fixes approximately 50–100 kg/ha nitrogen in the leguminous plants (Checcucci et al. 2017). It belongs to family Rhizobiaceae, for example, the well-known symbiotic bacteria of the legumes that are present in the nodules and fix the atmospheric nitrogen for the plant. It is mostly detected from useful pulses, like black gram, red-gram, pea, chickpea, and lentils, and also isolated from oil-seed legumes like soybean and peanut, and some important forage crops like berseem and lucerne. Recently, Clúa et al. (2018) discussed the importance of legume and *Rhizobium* strain compatibility that largely depends on the availability of compatible strain for a particular legume. Moreover, rhizobia get signal molecules through the plant to infect the root tissue and colonize the nodule and depend on the plant growth stages and plant metabolism to get successful nodulation (Clúa et al. 2018; Zhao et al. 2018).The nodule is a bulb-like structure that acts as a nitrogen fixation factory (Geetha and Joshi 2013). It assists in converting atmospheric nitrogen by symbiosis with leguminous plants, and in certain cases also with the nonleguminous plants like *Parasponia* (Santi et al. 2013). Leguminous plants improved the *Rhizobium* population in the soil and to maintain the *rhizobium* population crop rotation and artificial inoculation methods used mostly that enhance the rate of nitrogen in the soil.

7.4.1.2 *Azospirillum*

After *Rhizobium*, *Azospirillum* (family Spirilaceae) is the next diazotrophs that is globally recommended as important nitrogen-fixing microbes, and it can fix about 20–40 kg/ha nitrogen by biological nitrogen fixation (BNF) process in legumes and nonleguminous plants (Pankievicz et al. 2015; Zeffa et al. 2019). It is capable of improving plant growth by various ways like amino acids formation and release, improving root growth that helps plants to uptake more water and nutrient, and working as plant growth stimulant via biosynthesis of phytohormones such as indole-acetic acid, gibberellins, cytokinins, and different forms of polyamines (Pereg et al. 2016; Mehnaz 2015; Vejan et al. 2016). The *Azospirillum* forms symbiosis in several plants species mainly with the C_4-dicarboxylic pathway of photosynthesis (Hatch and Slack pathway), because they mostly grow and fix nitrogen on salts of organic acids such as malic or aspartic acid (Arun 2007). Bashan and Levanony (1990) reported "Multiple Mechanism Theory" that mostly worked with the plant system, and it varied as per the host and growth stages (Bashan and de-Bashan 2010). Recently, Zeffa et al. (2019) proved that *Azospirillum brasilense* Ab-V5 inoculated maize plants improved the biomass and yield under nitrogen deficit condition. Thus, it is mainly endorsed for these cereal crops such as maize, sugarcane, sorghum, and pearl millet. Among all species of this genera, *Azospirillum lipoferum* and *A. brasilense* are extensively used as biofertilizer throughout the world (Zeffa et al. 2019).

7.4.1.3 *Azotobacter*

Azotobacter is a free-living aerobic soil bacteria that belongs to the family Azotobacteriaceae and is a known plant growth-promoting rhizobacteria (PGPR; Wani et al. 2016; Viscardi et al. 2016; Van Oosten et al. 2018). These bacteria play an eminent role in the BNF and fix an average of 20 kg N/ha per year. *Azotobacter* is easily colonized in the soil and plant tissues and improves the plant growth through the production of auxins, gibberellins, cytokinins, ammonia, vitamins, and beneficial metabolites that improve seed germination, provide protection to the plant root against pathogens, and work as growth stimulator of beneficial rhizospheric microbiomes that improve the crop yield. It can secrete 1-aminocyclopropa ne-1-carboxylate (ACC) deaminase enzyme that helps the plant to grow in extreme environments, which makes these bacteria most vital for the stress management of the *plant (*Kukreja et al. 2004; Van Oosten et al. 2018). However, the real mechanism that exactly helps the plant is not fully understood. *Azotobacter* population is influenced or regulated by various factors of soil such as soil physicochemical (e.g., organic matter, pH, temperature, soil moisture) and microbiological properties. Major *Azotobacter* species are *A. vinelandii*, *A. beijerinckii*, *A. insignis*, and *A. macrocytogenes* (Subba Roa 2001; Tippannavar and Reddy 1993), and most of the strains can restrict several plant pathogens such as *Alternaria*, *Fusarium*, and *Helminthosporium*. Higher abundance of *Azotobacter* has been reported from the rhizosphere of various crop plants such as rice, maize, sugarcane, bajra, vegetables, and plantation crops (Arun 2007; Wani et al. 2016; Viscardi et al. 2016; Van Oosten et al. 2018).

7.4.1.4 Blue-Green Algae (Cyanobacteria)

Photosynthetic prokaryotes that are mostly associated with both marine and freshwater environments and fix the environment nitrogen are known as blue-green algae (BGA). BGA are filamentous vegetative cell chains that are heterocyst, which are mostly symbiotically associated with different hosts such as fungi, liverworts, ferns, and flowering plants, but the most common symbiotic association has been found between a free-floating aquatic fern and rice paddy. BGA can fix 20–30 kg N/ha and are able to produce plant growth substances such as indole acetic acid, auxins, cytokinins, gibberellic betaines, amino acids, vitamins, and polyamines (Ronga et al. 2019). Several microalgae families such as Chlorophyceae, Trebouxiophyceae, Ulvophyceae, and Charophyceae are able to produce auxin and cytokinin that are utilized as plant growth stimulants (Stirk et al. 2013). Moreover, the major species of BGA available commercially are: *Isochrysis* spp., *Chaetoceros* spp., *Chlorella* spp., *Arthrospira* spp., and *Dunaliella* spp. (Priyadarshani and Rath 2012). *Arthrospira* spp. and *Chlorella* spp. are the most common BGA species that are cultivated and utilized commercially all over the world (Ronga et al. 2019). Several past reports underline a beneficial impact on plant nutrient uptake, biomass, and crop yields when BGA are applied as biofertilizers (Shaaban 2001a, b; Hall and Williams 2003), and due to these multiple characteristics, BGA could play a significant role in sustainable agriculture practices.

7.4.1.5 *Azolla*

Genus *Azolla* contains several species of aquatic ferns also known as floating plants. They are native to the tropics, subtropics, and warm temperate regions of Africa, Asia, and America (Costa et al. 2009). *Azolla* is habitually detected in different water resources such as stagnant waters, ponds, ditches, canals, or paddy fields. *Azolla* mats cover these water resources in a suitable condition. Generally, *Azolla* is associated with other free-floating plant species such as water lettuce (*Pistia stratiotes* L.), watermeal (*Wolffia* Horkel ex Schleid), water caltrop (*Trapa natans* L.), duckweed (*Lemna minor* L.), water purslane (*Ludwigia palustris* L.), knotweed (*Polygonum arenastrum*), and mud-rooting species such as hornwort (*Ceratophyllum demersum* L. [Kannaiyan and Kumar 2006; Mosha 2018]). *Azolla* is known as a potential source of organic manure and nitrogen source for the crops, particularly for rice. The dry biomass of *Azolla* contains 4–5% N and 0.2–0.4% wet biomass. *Azolla* biofertilizer is applied on rice crop as a nitrogen source; after application, it decomposes very fast and provides the available nitrogen to rice plants. Moreover, it also contains substantial quantities of P, K, S, Zn, Fe, Mb, and other micronutrients (Mosha 2018). *Azolla* is recommended as a green manure that is incorporated to the rice field before showing. *Azolla pinnata* is most commonly used in India on a commercial scale. However, *Azolla caroliniana*, *A. microphylla*, *A. filiculoides*, and *A. mexicana* are also utilized for biomass production and used as biofertilizer.

7.5 Phosphate Solubilizers

Soil is a reservoir of numerous biochemical reactions performed by microbes to transport minerals from the soil to the plant. Microbes that are involved in the solubilization of insoluble inorganic phosphate compounds, such as tricalcium phosphate, dicalcium phosphate, hydroxyapatite, and rock phosphate, into available phosphorus are known as phosphate (P) solubilizers. These microbes improve the microbial activity of soil and enhance plant growth (Alori et al. 2017; Raj 2014). Among all bacterial genera, *Achromobacter*, *Aereobacter*, *Agrobacterium*, *Bacillus*, *Burkholderia*, *Flavobacterium*, *Erwinia*, *Microccocus*, *Pseudomonas*, and *Rhizobium* are known as prominent P solubilizers and provide the available phosphorus to the plants (Rodríguez and Fraga 1999). Rhizosphere soil of plant mostly contains considerable amounts of P solubilizers that play an indispensable role in the P mineralization (Kasiamdari et al. 2002). Phosphate-solubilizing bacteria (PSB) are aerobic and anaerobic strains that exist in the soil or plant, with a prevalence of aerobic strains in submerged soils. *Pseudomonas* and *Bacillus* are widely accepted as major P solubilizers, and they mobilize the insoluble P compounds by the production of organic acids that enhance the acidic nature of substratum (Chen et al. 2016). The organic and inorganic acids help to convert tricalcium phosphate to di- and monobasic phosphates so that plants can easily absorb the available P. Moreover, different microbes produce different kinds of organic acid to solubilize P. Tri- and dicarboxylic acids are able to solubilize P more effectively as compared to the monobasic and aromatic acids. Similarly, aliphatic acids have better ability to solubilize P as compared to phenolic, citric, and fumaric acids (Chen et al. 2016; Khan et al. 2010; Manzoor et al. 2017). Panhwar et al. (2011) assessed that Christmas Island Rock Phosphate application with a phosphate-solubilizing bacteria of *Bacillus* spp. improved P uptake and plant biomass in aerobic rice (*Oryza sativa* L.). Oteino et al. (2015) proved the ability of an endophytic *Pseudomonas* bacteria to solubilize P by releasing gluconic acid (GA) and stimulate the growth of *Pisum sativum* L. plants. Chen et al. (2016) reported that phosphate-solubilizing activity of *Pseudomonas* sp. PSB12 was associated with the release of organic acids, specially gluconic acid, formic acid, butyrate, and propanedioic acid, and it varied as per the different phosphate forms, and phenol could be used as the carbon source to dissolve insoluble phosphorus. Manzoor et al. (2017) revealed that the integrated use of P solubilizers with insoluble rock phosphate enhanced the P availability in soil pool and improved the maize plant growth significantly. Ahmad et al. (2018) concluded that two P-solubilizer bacteria (*Bacillus subtilis* strain Q3 and *Paenibacillus* sp. strain Q6) could promote cotton growth under alkaline conditions by supplying P nutrients. Prakash and Arora (2019) discussed the importance of an integrated application of Tricalcium phosphate with P-solubilizer *Bacillus* sp. that could be used to increase menthol production and oil yield of *Mentha arvensis*.

7.6 Phosphate Absorbers

Phosphate is a major element of plant nutrients. Plant root associated symbiotic microbes play a significant role to absorb phosphate from the soil. These symbiotic fungal microbes are known as arbuscular mycorrhiza (AM; fungal roots) that basically colonize in two kinds of environment, host plant (inside the root, or in rhizoids or thalli) and in surrounding soil of the plant (Jansa and Gryndler 2010). Mycorrhizal hyphae helps to uptake plant nutrients such as phosphorus, zinc, copper, and sulfur. AM fungi enhance the plant phosphorus uptake by increasing the absorbing surface area of root that helps to mobilize available phosphorus. AM-associated plants display a greater volume of phosphorus invasion than the non-mycorrhizal controls. Yang et al. (2012) assessed that rice + AM improved 70% phosphorus uptake. Moreover, colonized mycelia of vesicular-arbuscular mycorrhiza (VAM) protects the host root from the fungal pathogens and nematodes. Moreover, the fungal mycelia that colonize inside the host plant give a stable environment to the plant even in extreme conditions, through the extended root system that provides greater volume of water and nutrients to the host (Pavithra and Yapa 2018; Rask et al. 2019; Wang et al. 2017). VAM is ubiquitous and widely colonized in the plants that even exist in the arctic, temperate, and tropical regions, which shows the broad ecological range of this microbe and its ability to proliferate from aquatic to desert environments (Jansa et al. 2013; van der Heijden et al. 2015). However, only a few fungi have the ability to form a mycorrhizal association with the plant, for example, order *Glomales* of class *Zygomycetes* have 150 fungal species, and, among them, a small number of fungi are presumed to be mycorrhizal. Two genera, *Glomus* and *Sclerocytis*, are able to produce chlamydospores, whereas four genera can form zygospores: they are similar to *Gigaspora*, *Scutellospora*, *Acaulospora*, and *Entrophospora* (Maia et al. 1994). Mycorrhizal fungi have a wide range of applications in the agriculture system such as in seedling growth enhancement (van der Heijden and Horton 2009), decomposition of plant and crop residues (Lindahl et al. 2007), soil aggregation, and soil formation (Rillig and Mummey 2006).

7.7 Potassium Solubilizers

Potassium (K) is recommended as the third most important macronutrient for plant fitness and is involved in several biochemical and physiological processes. Low level of soluble potassium (up to 90% exists as insoluble rocks and silicate minerals) in the soil impacts negatively on plant metabolism and plant development and grain quality; it plays a significant role in the synthesis of cells, proteins, enzymes, cellulose, starch, and vitamins in normal or in stress condition (Shahzad et al. 2019; Wakeel et al. 2017). Macronutrient K not only regulates nutrient by transportation and absorption but also modulates the plant defense system under abiotic and biotic stresses and thus improves the crop production and quality (Oosterhuis et al. 2013; Zahoor et al. 2017). Sheng and He (2006) reported that Nanjing feldspar and Suzhou illite are two potassium (K)-bearing minerals that are used with *Bacillus edaphicus* through soil application and bacterial inoculation, which enhances the K uptake significantly in plant components.

Verma et al. (2015a) described a gram-positive plant growth-promoting bacterial strain *Bacillus amyloliquefaciens* that showed a significant level of K solubilization in in-vitro and in-vivo conditions. Verma et al. (2016) reported that most of the bacilli-solubilized potassium, such as *B. aerophilus*, *B. atrophaeus*, *B. cereus*, *B. circulans*, *B. horikoshii*, *B. licheniformis*, *B. megaterium*, *B. mojavensis*, *B. pumilus*, *Lysinibacillus sphaericus*, *Exiguobacterium antarcticum*, *Paenibacillus amylolyticus*, *P. dendritiformis*, *P. polymyxa*, *Planococcus citreus*, and *Planococcus salinarum*. However, K-solubilizing bacteria may have been used in the amelioration of K-deficient soil in agriculture. The most important K-solubilizing microbes (KSM) *Acidithiobacillus ferrooxidans*, *B. circulans*, *B. edaphicus*, *B. globisporus*, *B. mucilaginous*, *B. subtilis*, *Burkholderia cepacia*, *Enterobacter hormaechei*, *Paenibacillus kribensis*, *P. mucilaginous*, and *Pseudomonas putida* are used as K solubilizers and solubilize K in an eco-friendly manner. Therefore, these efficient K-solubilizing microbes (KSM) should be applied for solubilization of a fixed form of K to an available form of K in the soils (Verma et al. 2017a).

7.8 Zinc Solubilizers

Zinc (Zn) is an essential element of soil and Zn deficiency is widely reported in Indian soils (>50%) that exhibit the critical level of 1.5 ppm of available zinc (Katyal et al. 1994). The plant alone cannot absorb Zn from the soil, and hence farmers apply a huge amount of zinc sulfate ($ZnSO_4$) as an external fertilizer application. However, external fertilizer application also does not work well because the plant utilizes a very low percentage (1–4%) of the total available Zn and 75% of the applied Zn is transformed into different mineral fractions (Goteti et al. 2013). Two core mechanisms are applied to fix Zn: one is cation exchange, which mostly operates in acidic soils, and the other operates in alkaline conditions, where fixation occurs using chemisorption (chemisorption of zinc on calcium carbonate [$CaCO_3$] forms a solid-solution of Zn by complexation of organic ligands; Alloway 2008). Microbe-based solubilization of Zn plays a significant role in the soil, and major bacterial inoculants that fix Zn are *Bacillus subtilis*, *Thiobacillus thiooxidans*, and *Saccharomyces* sp., *Burkholderia* sp., and *Acinetobacter* sp., and these microbial inoculants are used as biofertilizers (Azooz and Ahmad 2013; Vaid et al. 2014). Mahdi et al. (2010) reported that Zn-solubilizing microbes are mostly applied in soils where native Zn is higher or in conjunction with insoluble cheaper zinc compounds like zinc oxide (ZnO), zinc carbonate ($ZnCO_3$), and zinc sulfide (ZnS) instead of costly zinc sulfate ($ZnSO_4$)). Kamran et al. (2017) reported that Zn-solubilizing microbes include *Rhizobium* sp. (LHRW1), while *Enterobacter cloacae* (PBS 2) improved Zn uptake in wheat plants and enhanced the root and shoot biomass. Dinesh et al. (2018) concluded that Zn-solubilizing *Bacillus megaterium* ZnSB2 strain played a significant role in Zn dissolution in soil, and it would allow reducing the utilization of inorganic Zn application rates. Gontia-Mishra et al. (2017) assessed that Zn-solubilizing bacteria such as *Pseudomonas aeruginosa*, *Ralstonia picketti*, *Burkholderia cepacia*, and *Klebsiella pneumoniae* improve the growth of rice seedlings and these bacteria could be used as Zn mobilizers for profitable farming.

7.9 Microbes Play a Role in Biofertilization

In the past several decades, chemical fertilizers have been applied as a source of major plant nutrients such as nitrogen, phosphorus, and potassium. These fertilizers improved the crop production but enhanced the farming cost and also polluted the natural resources that increased the environmental risks (Gong et al. 2011; Machado et al. 2017; Mullin et al. 2010; Vitousek et al. 2009). The plants themselves cannot utilize the insoluble forms of P and N_2. Therefore, plant growth-promoting microbes that can fix N_2 and solubilize the P are used as biofertilizers (Zahir et al. 2004; Schütz et al. 2018; Zaidi and Khan 2006; Alori et al. 2017). These microbe-based biofertilizers are environment-friendly and less expensive as compared to the chemical fertilizers. These biofertilizers improve the nutrient accumulation in the plant root by a different mechanism, such as mobilization, solubilization, and absorption, and enhance P uptake and N_2-fixation. Moreover, biofertilizers also play a significant role in the uptake of micronutrients like Fe, Cu, Zn, B, Mn, Co, and Mo (Bargaz et al. 2018). Several reports also concluded that mixed inoculation of N_2-fixing and phosphate-solubilizing bacteria had provided more balanced nutrition to different agriculture crops such as sorghum, barley, black gram, soybean, and wheat (Abd-Alla et al. 2001; Alagawadi and Gaur 1992; Galal 2003; Tanwar et al. 2003)

Rhizobium is one of the prominent diazotrophs that is mostly used as a biofertilizer alone in India, but the use of *Rhizobium* in combination with phosphate-solubilizing bacteria (PSB) needs to be explored more extensively. Nevertheless, application of combined microbe inoculum to enhance the soil fertility is getting the attention of researchers (Basak and Biswas 2010; Dash et al. 2018; He et al. 2019; Kant et al. 2016; Sharma et al. 2019). Recent findings specified that combined inoculation of plant growth-promoting rhizobacteria (PGPR) in arable soils enhanced the crop yield significantly (Harish et al. 2009a; Harish et al. 2009b). The PGPR strains *Pseudomonas alcaligenes* PsA15, *Bacillus polymyxa* BcP26, and *Mycobacterium phlei* MbP18 had positive impact on maize plant health and the uptake of N, P, and K in nutrient-deficient Calcisol soils (Egamberdiyeva 2007; Verma et al. 2017a). These PGPRs have a strong ability to produce growth-promoting phytohormones and play a significant role in phosphate mobilization, production of siderophore and antibiotics, inhibition of plant ethylene synthesis, and induction of plant systemic resistances to pathogens that helps to enhance the crop yield (Cakmakçi et al. 2006; Han and Lee 2006; Kohler et al. 2006; Turan et al. 2005; Zahir et al. 2004; Zaidi and Khan 2006). PGP endophytes also play a significant role in plant growth regulation (Suman et al. 2016b). *Rhizobium*, *Azotobacter*, *Azospirillum*, *Cyanobacteria*, *Azolla*, Phosphate, and potassium-solubilizing microorganisms are the most applied microorganisms that are considered as beneficial for sustainable agriculture and used as biofertilizers. Silicate-solubilizing bacteria also help to promote plant growth and are available as liquid biofertilizers (Parewa et al. 2014; Suman et al. 2016a; Verma et al. 2014a; Zhang and Kong 2014). Microbes and their beneficial association with crop plants are listed in Table 7.3.

7.10 Economic Development in Agriculture

Microbe-based biofertilizer development is accepted as a relevant approach to increase food production (Schütz et al. 2018). Global population pressure as well as climate change have become serious constraints for the economic growth of several countries and food security (Cole et al. 2018). Fundamentally, to improve the soil and plant health, two major strategies need to be followed: (1) extensive application of microbe-based biofertilizer, and (2) the manipulation of naturally existing microbial populations (Patil and Solanki 2016; Schütz et al. 2018). Mostly, PGPRs that are grown as either saprophytic or endophytic symbionts are protagonists of applied microbial biotechnology in agriculture (Backer et al. 2018). Especially, microbial formulation, quality control, and modes of application of microbial inoculants are getting the attention of researchers (Singh et al. 2014; Velivelli et al. 2014). Several mechanisms underlying the plant–microbe interactions in the rhizosphere and plant still need to be explored in depth. Researchers are facing difficulties, mostly regarding how to utilize the large microbial structure, especially unculturable microbes, to boost the soil and plant health. Secondary, a plethora of culture-independent molecular techniques is becoming accessible and is presently being applied either to interpret the hidden networks of microorganisms inhabiting soil and rhizosphere microenvironments or to outline the molecular bases of the plant–microbiome interactions (Patil and Solanki 2016). Multitrophic factors, including nutrient deficits, salt, low water, air contamination, diseases, and pests, cause negative impacts on the functionality/productivity of agricultural products and plant systems (Fig. 7.2).

Agriculture plays an indispensable role in the Indian economy, and biofertilizers are essential to improve crop yield in an eco-friendly manner (Kesavan and

Table 7.3 Microbial linkage promoting various types of growth in plants to improve crop yield

S. No.	Types of growth promotion in plants	Microbial inoculants	References
1.	Growth enhancement of canola and lettuce	*Rhizobium leguminosarum*	García-Fraile et al. (2012); Noel et al. (1996); Singh (2013)
2.	Early developments of canola seedlings	*Pseudomonas putida* G 12-2	Glick et al. (1997); Hwang et al. (2011); Zhang et al. (2015)
3.	Growth enhancement of wheat and maize plants	*Azospirillum brasilense* and *Azospirillum irakense* strains	Couillerot et al. (2013); Dobbelaere et al. (2002); Fukami et al. (2016)
4.	Growth enhancement of pearl millet	*Pseudomonas fluorescens* strain	Fukami et al. (2016); Karnwal (2012); Raj et al. (2003)
5.	Growth stimulation of tomato plant	*Pseudomonas putida* strain	Gravel et al. (2007); Mariutto et al. (2011); Pastor et al. (2014)
6.	Growth and productivity enhancement of canola	*Azotobacter* and *Azospirillum* strains	Naderifar and Daneshian (2012); Naseri et al. (2013); Yasari and Patwardhan (2007)
7.	Enhanced uptake of N, P, and K by maize crop in nutrient-deficient Calcisol soil	*Pseudomonas alcaligenes* PsA15; *Bacillus polymyxa* BcP26; *Mycobacterium phlei* MbP18	Egamberdiyeva (2007); Saharan and Nehra (2011); Shrivastava and Kumar (2015)
8.	Growth and yield enhancement of chick pea (*Cicer arietinum*)	*Pseudomonas*, *Azotobacter*, and *Azospirillum* strains	Joseph et al. (2012); Rokhzadi et al. (2008); Rokhzadi and Toashih (2011)
9.	Improvement in the yield and phosphorus uptake in wheat	*R. leguminosarum* (Thal-8/ SK8); *Pseudomonas sp.* strain 54RB	Afzal and Bano (2008); Mehboob et al. (2012); Shaikh et al. (2016)
10.	Improvement in seed germination, seedling growth, and yield of maize	*P. putida* strains R-168 and DSM-291; *P. fluorescens* strains R-98 and DSM-50090; *A. brasilense* DSM-1691; *Azospirillum lipoferum* DSM-1690	Gholami et al. (2009); Karunakaran et al. (2013); Shen et al. (2010)
11.	Improvement in seed germination, growth parameters of maize seedling in greenhouse, and also grain yield of field-grown maize	*P. putida* strain R-168	Gholami et al. (2009); Noumavo et al. (2013); Singh et al. (2011)

(continued)

Table 7.3 (continued)

S. No.	Types of growth promotion in plants	Microbial inoculants	References
12.	Increase in growth, leaf nutrient contents, and yield of banana cv. Virupakshi (Musa spp. AAB) plants	*P. fluorescens* strain R-93; *P. fluorescens* DSM 50090; *P. putida* DSM291; *A. lipoferum* DSM 1691; *A. brasilense* SM 1690; *P. fluorescens* strains CHA0 and Pf1	Kavino et al. (2010); Patil (2013); Selvarajan and Balasubramanian (2014)
13.	Improvement in seed germination, growth parameters of wheat seedling in pots, and also grain yield of field-grown wheat	*Bacillus amyloliquefaciens* IARI-HHS2-30	Verma (2013, 2015, 2016); Verma et al. (2014a, b, 2016)

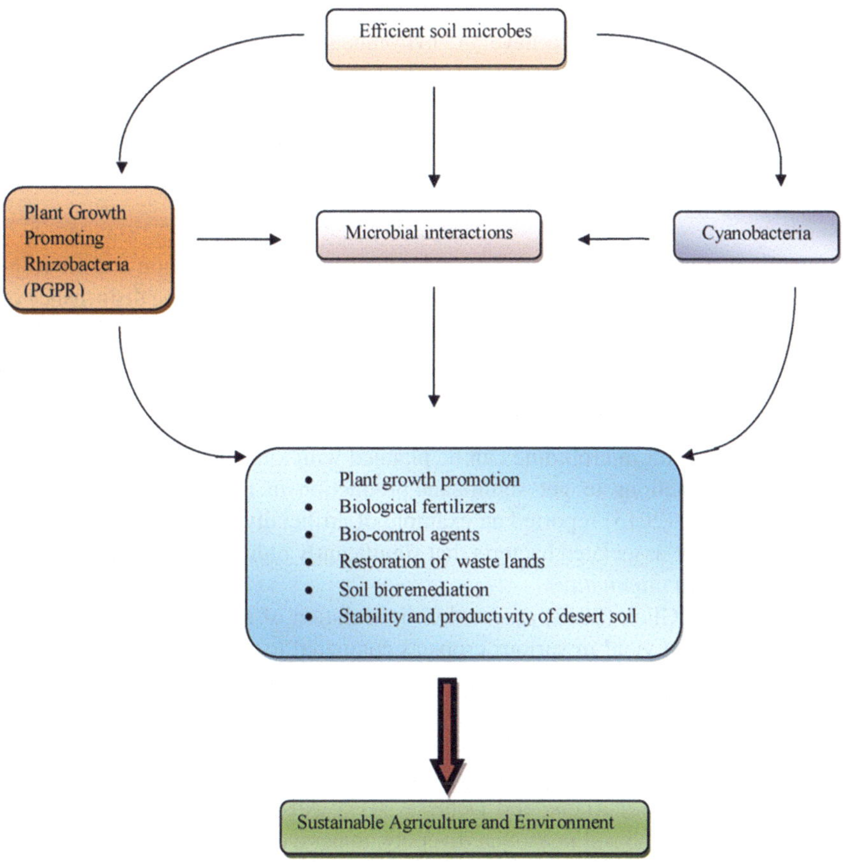

Fig. 7.2 Microbes and their role in the intensification of sustainable agriculture

Swaminathan 2008). Out of India's total geographical area (329 million hectares), about 114 million hectares are under farming (Raghuwanshi 2012). To gain better yield, farmers vaccinate the soil with fertilizers. These fertilizers belong to two classes. They are either inorganic chemical or organic biofertilizers. Increasingly high inputs of chemical fertilizers during the last 150 years have not only negatively influenced the fertility of the soil but also polluted the water sources. Soil fertility degraded significantly, and polluted resources have also posed severe health and environmental hazards (Schütz et al. 2018).

In contrast, microbe-based biofertilizers not only give a better yield, but are also harmless to the ecosystem (Patil and Solanki 2016). Biofertilizer-based organic farming methods would help solve these concerns and boost the ecosystem (Shukla et al. 2016). In the present time, organic biofertilizer market has reached around US$ 30 billion globally and with an excellent growth rate (8%). Approximately, 22 million hectares of agricultural land is now cultivated organically by the use of microbe-based fertilizers. However, organic farming represents less than 1% of the world's conventional agricultural production and about 9% of the total agricultural area (Verma et al. 2014b). To have an environmentally safe technology, microbial product-based agriculture system needs to be used efficiently.

7.11 Microbial Engineering Gets Better Agriculture and Human Health

A microbial structure is composed of and influenced by host location and environmental factors that generate undesirable phenotypes in the hosts. The disturbance of microbiome negatively affects the associated ecosystems that cause diseases and disorders in the host. Microbiome engineering can be used to modify or restore the microbial community to improve the plant and human health. Foo et al. (2017) discussed the vitality of microbiome engineering toward agriculture prospects and human health. Plant microbiome can be planned with known microbial inoculants with desired functions to get significant production in an eco-friendly manner. Santhanam et al. (2015) reported an example of artificially developed microbiome by using five root-associated bacteria that significantly reduced the sudden-wilt disease in *Nicotiana attenuata*.

Additionally, Glick (2012) assessed co-inoculation of rhizobacteria had shown higher growth and yield in various crops as compared to the single microbes. Soil microbiome engineering is usually flourished by hiring different agricultural practices such as intercropping, crop rotation, and tillage (Fageria 2012; Solanki et al. 2017, 2019). Application of green organic manure, farmyard manure, is also one of the important agriculture practices that enhance the soil microbiomes significantly (Chaparro et al. 2012; Abebe and Deressa 2017). All these agricultural practices targeted to modify the soil microbial diversity, mineral cycles, and biological activity, and these significantly helps to reduce the off-farm inputs, for example, chemical fertilizers and herbicides.

In the present scenario, microbial engineering is most extensively utilized to modify the human microbiome to improve immunity and disease resistance. Advanced biotechnological tools help to manipulate human microbiome to treat diseases such as diabetes and cancer; modified microbiota influence the host metabolism that helps to regulate diseases (Grice and Segre 2012). Meat industry badly suffered from antibiotic-resistant bacteria, and it has become the most common problem for livestock farming. Thus, microbiome alteration of livestock by using feed enzymes, prebiotics, and probiotics is a suitable alternative as compared to the overdose of antibiotics that also negatively affect human health. To promote gut health in swine and poultry, feed enzymes like phytase, amylase, non-starch polysaccharide (NSP)-degrading enzymes (e.g., xylanase, β-glucanase, and β-mannanase), proteases, and lysozyme are used that enhance the substrate digestion, improve the production of prebiotics from dietary NSPs, and reduce the antinutritive factors (Archna et al. 2015; Kiarie et al. 2013; Verma et al. 2015b).

7.12 Conclusion and Future Prospects

During the past several decades, microbial research has made significant progress in sustainable agriculture. Subsequently, advanced microbiological tools offer to use microbial engineering to alter microbiota of human and plant to improve health and increase production. However, plant microbiome and different environmental stress need to be explored to reveal a unique signaling network among microbes that help the plant to survive. Characterization of these signaling molecules can help to design advanced biotechnological strategies that unlock the plant adaptation mechanisms. Similarly, soil health can be flourished by using the crosstalk of soil microbes. Moreover, different microbial tools and agriculture technologies are being established, providing deeper insights into the plant/soil microbiomes, and these tools widely help to accelerate the microbial engineering efforts. With nonstop hard work in understanding and manipulating microbes, the microbial link will be a mainstream approach for refining human health and agriculture output.

Acknowledgments We are grateful to the Department of Microbiology, Akal College of Basic Science, Eternal University, Himachal Pradesh, and Department of Biotechnology (DBT), Ministry of Science and Technology, for providing financial support to carry out field studies.

References

Abd-Alla MH, Omar SA, Omar S (2001) Survival of rhizobia/bradyrhizobia and a rock-phosphate-solubilizing fungus *Aspergillus niger* on various carriers from some agro-industrial wastes and their effects on nodulation and growth of faba bean and soybean. J Plant Nutr 24:261–272

Abebe Z, Deressa H (2017) The effect of organic and inorganic fertilizers on the yield of two contrasting soybean varieties and residual nutrient effects on a subsequent finger millet crop. Agronomy 7:42. https://doi.org/10.3390/agronomy7020042

Afzal A, Bano A (2008) *Rhizobium* and phosphate solubilizing bacteria improve the yield and phosphorus uptake in wheat (*Triticum aestivum*). Int J Agric Biol 10:85–88

Ahmad M, Ahmad I, Hilger TH, Nadeem SM, Akhtar MF, Jamil M et al (2018) Preliminary study on phosphate solubilizing Bacillus subtilis strain Q3 and *Paenibacillus* sp. strain Q6 for improving cotton growth under alkaline conditions. PeerJ 6:e5122. https://doi.org/10.7717/peerj.5122

Alagawadi AR, Gaur A (1992) Inoculation of *Azospirillum brasilense* and phosphate-solubilizing bacteria on yield of sorghum [*Sorghum bicolor* (L.) Moench] in dry land tropical agriculture (Trinidad and Tobago). Trop Agric 69:347–350

Alloway B (2008) Zinc in soils and crop nutrition. International Zinc Association, Brussels

Alori ET, Glick BR, Babalola OO (2017) Microbial phosphorus solubilization and its potential for use in sustainable agriculture. Front Microbiol 8:971. https://doi.org/10.3389/fmicb.2017.00971

Ambros S, Martinez F, Ivars P, Hernandez C, de la Iglesia F, Elena SF (2017) Molecular and biological characterization of an isolate of tomato mottle mosaic virus (ToMMV) infecting tomato and other experimental hosts in eastern Spain. Eur J Plant Pathol 149:261–268

Amundson R, Berhe AA, Hopmans JW, Olson C, Sztein AE, Sparks DL (2015) Soil and human security in the 21st century. Science (80-) 348:1261071. https://doi.org/10.1126/science.1261071

Archna S, Priyank V, Nath YA, Kumar SA (2015) Bioprospecting for extracellular hydrolytic enzymes from culturable thermotolerant bacteria isolated from Manikaran thermal springs. Res J Biotechnol 10:33–42

Arun K (2007) Bio-fertilizers for sustainable agriculture. Mechanism of P solubilization. Agribios publishers, Jodhpur, pp 196–197

Asif MA et al (2011) Enhanced expression of AtNHX1, in transgenic groundnut (*Arachis hypogaea* L.) improves salt and drought tolerance. Mol Biotechnol 49:250–256

Azooz MM, Ahmad P (2013) Role of bio-fertilizers in crop improvement. In: Crop improvement: new approaches and modern techniques. Springer, New York, p 189

Backer R, Rokem JS, Ilangumaran G, Lamont J, Praslickova D, Ricci E et al (2018) Plant growth-promoting Rhizobacteria: context, mechanisms of action, and roadmap to commercialization of biostimulants for sustainable agriculture. Front Plant Sci 9:1473. https://doi.org/10.3389/fpls.2018.01473

Bargaz A, Lyamlouli K, Chtouki M, Zeroual Y, Dhiba D (2018) Soil microbial resources for improving fertilizers efficiency in an integrated plant nutrient management system. Front Microbiol 9:1606. https://doi.org/10.3389/fmicb.2018.01606

Basak BB, Biswas DR (2010) Co-inoculation of potassium solubilizing and nitrogen fixing bacteria on solubilization of waste mica and their effect on growth promotion and nutrient acquisition by a forage crop. Biol Fertil Soils 46:641–648. https://doi.org/10.1007/s00374-010-0456-x

Bashan Y, de-Bashan LE. (2010) How the plant growth-promoting bacterium *Azospirillum* promotes plant growth–a critical assessment. In: Advances in agronomy. Elsevier, San Diego, pp 77–136

Bashan Y, Levanony H (1990) Current status of *Azospirillum* inoculation technology: *Azospirillum* as a challenge for agriculture. Can J Microbiol 36:591–608

Borriss R (2011) Use of plant-associated *Bacillus* strains as biofertilizers and biocontrol agents in agriculture. In: Bacteria in agrobiology: plant growth responses. Springer, Berlin, pp 41–76

Cakmakçi R, Dönmez F, Aydın A, Şahin F (2006) Growth promotion of plants by plant growth-promoting rhizobacteria under greenhouse and two different field soil conditions. Soil Biol Biochem 38:1482–1487

Chaparro JM, Sheflin AM, Manter DK, Vivanco JM (2012) Manipulating the soil microbiome to increase soil health and plant fertility. Biol Fertil Soils 48:489–499

Checcucci A, DiCenzo GC, Bazzicalupo M, Mengoni A (2017) Trade, diplomacy, and warfare: the quest for elite rhizobia inoculant strains. Front Microbiol 8:2207. https://doi.org/10.3389/fmicb.2017.02207

Chen W, Yang F, Zhang L, Wang J (2016) Organic acid secretion and phosphate solubilizing efficiency of *Pseudomonas* sp . PSB12: effects of phosphorus forms and carbon sources. Geomicrobiol J 33:870–877. https://doi.org/10.1080/01490451.2015.1123329

Chung S, Kong H, Buyer JS, Lakshman DK, Lydon J, Kim S-D, Roberts DP (2008) Isolation and partial characterization of *Bacillus subtilis* ME488 for suppression of soilborne pathogens of cucumber and pepper. Appl Microbiol Biotechnol 80:115–123

Clúa J, Roda C, Zanetti ME, Blanco FA (2018) Compatibility between legumes and rhizobia for the establishment of a successful nitrogen-fixing symbiosis. Genes (Basel) 9. https://doi.org/10.3390/genes9030125

Cole MB, Augustin MA, Robertson MJ, Manners JM (2018) The science of food security. npj Sci Food 2:14. https://doi.org/10.1038/s41538-018-0021-9

Commare RR, Nandakumar R, Kandan A, Suresh S, Bharathi M, Raguchander T, Samiyappan R (2002) *Pseudomonas fluorescens* based bio-formulation for the management of sheath blight disease and leaf folder insect in rice. Crop Prot 21:671–677

Costa ML, Santos MCR, Carrapico F, Pereirac AL (2009) *Azolla*-Anabaena's behavior in urban wastewater and artificial media-influence of combined nitrogen. Water Res 43:3743–3750

Couillerot O et al (2013) Comparison of prominent *Azospirillum* strains in *Azospirillum–Pseudomonas–Glomus* consortia for promotion of maize growth. Appl Microbiol Biotechnol 97:4639–4649

Dash D, Gupta S, Deole S (2018) Study of characterization of tamarind associated rhizobium spp. and phosphate solubilizing bacteria and their potency for germination of tamarind seeds. Pharm Innov J 7:282–286. Available at: www.thepharmajournal.com. Accessed May 21, 2019

de Souza R, Ambrosini A, Passaglia LMP (2015) Plant growth-promoting bacteria as inoculants in agricultural soils. Genet Mol Biol 38:401–419. https://doi.org/10.1590/S1415-475738420150053

Díaz-Pérez JC (2014) Bell pepper (*Capsicum annum* L.) crop as affected by shade level: fruit yield, quality, and postharvest attributes, and incidence of phytophthora blight (caused by Phytophthora capsici Leon.). HortScience 49:891–900

Dinesh R, Srinivasan V, Hamza S, Sarathambal C, Anke Gowda SJ, Ganeshamurthy AN et al (2018) Isolation and characterization of potential Zn solubilizing bacteria from soil and its effects on soil Zn release rates, soil available Zn and plant Zn content. Geoderma 321:173–186. https://doi.org/10.1016/J.GEODERMA.2018.02.013

Dobbelaere S, Croonenborghs A, Thys A, Ptacek D, Okon Y, Vanderleyden J (2002) Effect of inoculation with wild type *Azospirillum brasilense* and *A. irakense* strains on development and nitrogen uptake of spring wheat and grain maize. Biol Fertil Soils 36:284–297

Egamberdiyeva D (2007) The effect of plant growth promoting bacteria on growth and nutrient uptake of maize in two different soils. Appl Soil Ecol 36:184–189

Fageria NK (2012) Role of soil organic matter in maintaining sustainability of cropping systems. Commun Soil Sci Plant Anal 43:2063–2113. https://doi.org/10.1080/00103624.2012.697234

Foo JL, Ling H, Lee YS, Chang MW (2017) Microbiome engineering: current applications and its future. Biotechnol J 12:1600099

Fukami J, Nogueira MA, Araujo RS, Hungria M (2016) Accessing inoculation methods of maize and wheat with *Azospirillum brasilense*. AMB Express 6:3

Galal Y (2003) Assessment of nitrogen availability to wheat (*Triticum aestivum* L.) from inorganic and organic N sources as affected by *Azospirillum brasilense* and *Rhizobium leguminosarum* inoculation. Egypt J Microbiol 38:57–73

García-Fraile P et al (2012) *Rhizobium* promotes non-legumes growth and quality in several production steps: towards a biofertilization of edible raw vegetables healthy for humans. PLoS One 7:e38122

Geetha SJ, Joshi SJ (2013) Engineering rhizobial bioinoculants: a strategy to improve iron nutrition. Sci World J 2013:315890. https://doi.org/10.1155/2013/315890

Gholami A, Shahsavani S, Nezarat S (2009) The effect of plant growth promoting rhizobacteria (PGPR) on germination, seedling growth and yield of maize. Int J Biol Life Sci 5:35–40

Gill U et al (2019) Ty-6, a major begomovirus resistance gene on chromosome 10, is effective against tomato yellow leaf curl virus and tomato mottle virus. Theor Appl Genet 132:1–12

Glick BR (1995) The enhancement of plant growth by free-living bacteria. Can J Microbiol 41:109–117

Glick BR (2012) Plant growth-promoting bacteria: mechanisms and applications. Scientifica 2012:963401

Glick BR, Liu C, Ghosh S, Dumbroff EB (1997) Early development of canola seedlings in the presence of the plant growth-promoting rhizobacterium *Pseudomonas putida* GR12-2. Soil Biol Biochem 29:1233–1239

Gong S (2018) Investigating Vector-Virus-Plant interactions influencing transmission efficiency of Tomato yellow leaf curl virus and Tomato mottle virus by Bemisia tabaci. Auburn Univ In Press

Gong P, Liang L, Zhang Q (2011) China must reduce fertilizer use too. Nature 473:284–285. https://doi.org/10.1038/473284e

Gontia-Mishra I, Sapre S, Tiwari S (2017) Zinc solubilizing bacteria from the rhizosphere of rice as prospective modulator of zinc biofortification in rice. Rhizosphere 3:185–190. https://doi.org/10.1016/J.RHISPH.2017.04.013

Goteti PK, Emmanuel LDA, Desai S, Shaik MHA (2013) Prospective zinc solubilising bacteria for enhanced nutrient uptake and growth promotion in maize (Zea mays L.). Int J Microbiol 2013:869697. https://doi.org/10.1155/2013/869697

Gouda S, Kerry RG, Das G, Paramithiotis S, Shin HS, Patra JK (2018) Revitalization of plant growth promoting rhizobacteria for sustainable development in agriculture. Microbiol Res 206:131–140. https://doi.org/10.1016/j.micres.2017.08.016

Gravel V, Antoun H, Tweddell RJ (2007) Growth stimulation and fruit yield improvement of greenhouse tomato plants by inoculation with *Pseudomonas putida* or *Trichoderma atroviride*: possible role of indole acetic acid (IAA). Soil Biol Biochem 39:1968–1977

Grice EA, Segre JA (2012) The human microbiome: our second genome. Annu Rev Genomics Hum Genet 13:151–170

Guo J, Jia Y, Chen H, Zhang L, Yang J, Zhang J et al (2019) Growth, photosynthesis, and nutrient uptake in wheat are affected by differences in nitrogen levels and forms and potassium supply. Sci Rep 9:1248. https://doi.org/10.1038/s41598-018-37838-3

Hall JL, Williams LE (2003) Transition metal transporters in plants. J Exp Bot 54:2601–2613

Han H-S, Lee K (2006) Effect of co-inoculation with phosphate and potassium solubilizing bacteria on mineral uptake and growth of pepper and cucumber. Plant Soil Environ 52:130

Harish S, Kavino M, Kumar N, Balasubramanian P, Samiyappan R (2009a) Induction of defense-related proteins by mixtures of plant growth promoting endophytic bacteria against Banana bunchy top virus. Biol Control 51:16–25

Harish S, Kavino M, Kumar N, Samiyappan R (2009b) Differential expression of pathogenesis-related proteins and defense enzymes in banana: interaction between endophytic bacteria, banana bunchy top virus and *Pentalonia nigronervosa*. Biocon Sci Technol 19:843–857

He Y, Pantigoso HA, Wu Z, Vivanco JM (2019) Co-inoculation of Bacillus sp. and Pseudomonas putida at different development stages acts as a biostimulant to promote growth, yield and nutrient uptake of tomato. J Appl Microbiol. https://doi.org/10.1111/jam.14273. jam.14273

Hernández-Rodríguez A, Heydrich-Pérez M, Acebo-Guerrero Y, Velazquez-Del Valle MG, Hernandez-Lauzardo AN (2008) Antagonistic activity of Cuban native rhizobacteria against *Fusarium verticillioides* (Sacc.) Nirenb. in maize (*Zea mays* L.). Appl Soil Ecol 39:180–186

Hittalmani S, Mahesh H, Mahadevaiah C, Prasannakumar MK (2016) De novo genome assembly and annotation of rice sheath rot fungus *Sarocladium oryzae* reveals genes involved in Helvolic acid and Cerulenin biosynthesis pathways. BMC Genomics 17:271

Hunter MC, Smith RG, Schipanski ME, Atwood LW, Mortensen DA (2017) Agriculture in 2050: recalibrating targets for sustainable intensification. Bioscience 67:386–391. https://doi.org/10.1093/biosci/bix010

Hwang S, Ahmed H, Strelkov S, Gossen B, Turnbull G, Peng G, Howard R (2011) Seedling age and inoculum density affect clubroot severity and seed yield in canola. Can J Plant Sci 91:183–190

Jacoby R, Peukert M, Succurro A, Koprivova A, Kopriva S (2017) The role of soil microorganisms in plant mineral nutrition-current knowledge and future directions. Front Plant Sci 8:1617. https://doi.org/10.3389/fpls.2017.01617

Jansa J, Gryndler M (2010) Biotic environment of the arbuscular mycorrhizal fungi in soil. In: Koltai H, Kapulnik Y (eds) Arbuscular mycorrhizas: physiology and function. Springer, Heidelberg, pp 209–236

Jansa J, Bukovská P, Gryndler M (2013) Mycorrhizal hyphae as ecological niche for highly specialized hypersymbionts – or just soil free-riders? Front Plant Sci 4:134. https://doi.org/10.3389/fpls.2013.00134

Jetiyanon K, Fowler WD, Kloepper JW (2003) Broad-spectrum protection against several pathogens by PGPR mixtures under field conditions in Thailand. *Plant Dis* 87:1390–1394

Ji P, Koné D, Yin J, Jackson KL, Csinos AS (2012) Soil amendments with *Brassica* cover crops for management of Phytophthora blight on squash. Pest Manag Sci 68:639–644

Jiang Z-Q, Guo Y-H, Li S-M, Qi H-Y, Guo J-H (2006) Evaluation of biocontrol efficiency of different *Bacillus* preparations and field application methods against Phytophthora blight of bell pepper. Biol Control 36:216–223

Jogaiah S, Sharathchandra R, Raj N, Vedamurthy A, Shetty HS (2014) Development of SCAR marker associated with downy mildew disease resistance in pearl millet (*Pennisetum glaucum* L.). Mol Biol Rep 41:7815–7824

Joseph B, Ranjan Patra R, Lawrence R (2012) Characterization of plant growth promoting rhizobacteria associated with chickpea (*Cicer arietinum* L.). Int J Plant Prod 1:141–152

Kafle A, Cope K, Raths R, Krishna Yakha J, Subramanian S, Bücking H et al (2019) Harnessing soil microbes to improve plant phosphate efficiency in cropping systems. Agronomy 9:127. https://doi.org/10.3390/agronomy9030127

Kamran S, Shahid I, Baig DN, Rizwan M, Malik KA, Mehnaz S (2017) Contribution of zinc solubilizing Bacteria in growth promotion and zinc content of wheat. Front Microbiol 8:2593. https://doi.org/10.3389/fmicb.2017.02593

Kannaiyan S, Kumar K (2006) Biodiversity of *Azolla* and its algal symbiont, *Anabaena azollae*. NBA Scientific Bulletin Number-2, National Biodiversity Authority, Chennai, Tamil Nadu, India, pp 1–31

Kant S, Kumar A, Kumar S, Kumar V, Pal Y, Shukla AK (2016) Effect of rhizobium, PSB and p-levels on growth, yield attributes and yield of Urdbean (Vigna Mungo L.). J Pure Appl Microbiol 10:3093–3099

Karnwal A (2012) Screening of plant growth-promoting rhizobacteria from maize (*Zea mays*) and wheat (*Triticum aestivum*). Afr. J. Food Agric. Nutr. Dev. 12:6170–6185

Karunakaran G, Suriyaprabha R, Manivasakan P, Yuvakkumar R, Rajendran V, Prabu P, Kannan N (2013) Effect of nanosilica and silicon sources on plant growth promoting rhizobacteria, soil nutrients and maize seed germination. IET Nanobiotechnol 7:70–77

Kashyap BK, Solanki MK, Pandey AK, Prabha S, Kumar P, Kumari B (2019) Bacillus as plant growth promoting rhizobacteria (PGPR): a promising green agriculture technology. In: Plant health under biotic stress. Springer Singapore, Singapore, pp 219–236. https://doi.org/10.1007/978-981-13-6040-4_11

Kasiamdari R, Smith S, Smith F, Scott E (2002) Influence of the mycorrhizal fungus, Glomus coronatum, and soil phosphorus on infection and disease caused by binucleate Rhizoctonia and *Rhizoctonia solani* on mung bean (*Vigna radiata*). Plant Soil 238:235–244

Katyal J, Venkateswarlu B, Das S (1994) Biofertilisers for nutrient supplementation in Dryland agriculture potentials and problems. Fertil News 39:27–27

Kavino M, Harish S, Kumar N, Saravanakumar D, Samiyappan R (2008) Induction of systemic resistance in banana (*Musa* spp.) against Banana bunchy top virus (BBTV) by combining chitin with root-colonizing *Pseudomonas fluorescens* strain CHA0. Eur J Plant Pathol 120:353–362

Kavino M, Harish S, Kumar N, Saravanakumar D, Samiyappan R (2010) Effect of chitinolytic PGPR on growth, yield and physiological attributes of banana (Musa spp.) under field conditions. Appl Soil Ecol 45:71–77

Kesavan PC, Swaminathan MS (2008) Strategies and models for agricultural sustainability in developing Asian countries. Philos Trans R Soc Lond Ser B Biol Sci 363:877–891. https://doi.org/10.1098/rstb.2007.2189

Khan MS, Zaidi A, Ahemad M, Oves M, Wani PA (2010) Plant growth promotion by phosphate solubilizing fungi – current perspective. Arch Agron Soil Sci 56:73–98. https://doi.org/10.1080/03650340902806469
Kiarie E, Romero LF, Nyachoti CM (2013) The role of added feed enzymes in promoting gut health in swine and poultry. Nutr Res Rev 26:71–88
Kohler J, Caravaca F, Carrasco L, Roldán A (2006) Contribution of *Pseudomonas mendocina* and *Glomus intraradices* to aggregate stabilization and promotion of biological fertility in rhizosphere soil of lettuce plants under field conditions. Soil Use Manag 22:298–304
Koné SB, Dionne A, Tweddell RJ, Antoun H, Avis TJ (2010) Suppressive effect of non-aerated compost teas on foliar fungal pathogens of tomato. Biol Control 52:167–173
Kukreja K, Suneja S, Goyal S, Narula N (2004) Phytohormone production by Azotobacter-a review. Agric Rev Agric Res Commun Cent India 25(1):70–75
Kumar KVK, Reddy M, Kloepper J, Lawrence K, Groth D, Miller M (2016) Sheath blight disease of rice (Oryza sativa L.)–an overview. Biosci Biotechnol Res Asia 6:465–480
Kumari B, Mallick MA, Solanki MK, Solanki AC, Hora A, Guo W (2019) Plant growth-promoting rhizobacteria (PGPR): modern prospects for sustainable agriculture. In: Ansari RA, Mahmood I (eds) Plant health under biotic stress. Springer Netherlands, Dordrecht. doi.org/10.1007/978-981-13-6040-4_6
Lindahl BD, Ihrmark K, Boberg J, Trumbore SE, Hogberg P, Stenlid J, Finlay RD (2007) Spatial separation of litter decomposition and mycorrhizal nitrogen uptake in a boreal forest. New Phytol 173:611–620
Löffler M, Kessel B, Ouzunova M, Miedaner T (2010) Population parameters for resistance to *Fusarium graminearum* and *Fusarium verticillioides* ear rot among large sets of early, mid-late and late maturing European maize (*Zea mays* L.) inbred lines. Theor Appl Genet 120:1053–1062
Lotze-Campen H (2011) Climate change, population growth, and crop production: an overview. In: Crop adaptation to climate change. Wiley-Blackwell, Oxford, pp 1–11. https://doi.org/10.1002/9780470960929.ch1
Machado CS, Fregonesi BM, Alves RIS, Tonani KAA, Sierra J, Martinis BS et al (2017) Health risks of environmental exposure to metals and herbicides in the Pardo River. Brazil Environ Sci Pollut Res 24:20160–20172. https://doi.org/10.1007/s11356-017-9461-z
Mahatma M, Bhatnagar R, Solanki R, Mittal G, Shah R (2011) Characterisation of downy mildew resistant and susceptible pearl millet (*Pennisetum glaucum* (L.) R. Br.) genotypes using isozyme, protein, randomly amplified polymorphic DNA and inter-simple sequence repeat markers. Arch Phytopathol Plant Protect 44:1985–1998
Mahdi S, Dar S, Ahmad S, Hassan G (2010) Zinc availability–a major issue in agriculture. Res J Agric Sci 3:78–79
Maia LC, Kimbrough JW, Benny GL (1994) Ultrastructure of spore germination in Gigaspora albida (Glomales). Mycologia 86:343. https://doi.org/10.2307/3760564
Malakouti MJ (2008) The effect of micronutrients in ensuring efficient use of macronutrients. Turk J Agric For 32:215–220
Manzoor M, Abbasi MK, Sultan T (2017) Isolation of phosphate solubilizing bacteria from maize rhizosphere and their potential for rock phosphate solubilization–mineralization and plant growth promotion. Geomicrobiol J 34:81–95. https://doi.org/10.1080/01490451.2016.1146373
Mariutto M et al (2011) The elicitation of a systemic resistance by *Pseudomonas putida* BTP1 in tomato involves the stimulation of two lipoxygenase isoforms. BMC Plant Biol 11:29
Mehboob I, Naveed M, Zahir ZA, Ashraf M (2012) Potential of rhizobia for sustainable production of non-legumes. In: Crop production for agricultural improvement. Springer, Dordrecht, pp 659–704
Mehnaz S (2015) Azospirillum: a biofertilizer for every crop. In: Arora NK (ed) Plant microbes symbiosis: applied facets. Springer India, New Delhi, pp 297–314
Morawicki RO, Díaz González DJ (2018) Food sustainability in the context of human behavior. Yale J Biol Med 91:191–196

Mosha SS (2018) A review on significance of Azolla meal as a protein plant source in finfish culture. J Aquacult Res Dev 09:1–7. https://doi.org/10.4172/2155-9546.1000544

Mullin CA, Frazier M, Frazier JL, Ashcraft S, Simonds R, vanEngelsdorp D et al (2010) High levels of Miticides and agrochemicals in North American apiaries: implications for honey bee health. PLoS One 5:e9754. https://doi.org/10.1371/journal.pone.0009754

Muraleedharan H, Seshadri S, Perumal K (2010) Biofertilizer (Phosphobacteria) Shri AMM Murugappa Chettiar Research Centre, Taramani, Chennai 600113, pp 1–16

Murphy JF, Zehnder GW, Schuster DJ, Sikora EJ, Polston JE, Kloepper JW (2000) Plant growth-promoting rhizobacterial mediated protection in tomato against tomato mottle virus. Plant Dis 84:779–784

Naderifar M, Daneshian J (2012) Effect of seed inoculation with *Azotobacter* and *Azospirillum* and different nitrogen levels on yield and yield components of canola (*Brassica napus* L.). Iran J Plant Physiol 3(1):619–626

Naseri R, Moghadam A, Darabi F, Hatami A, Tahmasebei GR (2013) The effect of deficit irrigation and *Azotobacter chroococcum* and *Azospirillum brasilense* on grain yield, yield components of maize (SC 704) as a second cropping in western Iran. Bull Environ Pharmacol Life Sci 2:104–112

Niyongere C et al (2013) Understanding banana bunchy top disease epidemiology in Burundi for an enhanced and integrated management approach. Plant Pathol 62:562–570

Noel TC, Sheng C, Yost C, Pharis R, Hynes M (1996) *Rhizobium leguminosarum* as a plant growth-promoting rhizobacterium: direct growth promotion of canola and lettuce. Can J Microbiol 42:279–283

Noumavo PA et al (2013) Effect of different plant growth promoting rhizobacteria on maize seed germination and seedling development. Am J Plant Sci 4:1013

Oh B-T et al (2011) Suppression of Phytophthora blight on pepper (Capsicum annuum L.) by bacilli isolated from brackish environment. Biocontrol Sci Tech 21:1297–1311

Olanrewaju OS, Glick BR, Babalola OO (2017) Mechanisms of action of plant growth promoting bacteria. World J Microbiol Biotechnol 33:197. https://doi.org/10.1007/s11274-017-2364-9

Oosterhuis DM, Loka DA, Raper TB (2013) Potassium and stress alleviation: physiological functions and management of cotton. J Plant Nutr Soil Sci 176:331–343. https://doi.org/10.1002/jpln.201200414

Oteino N, Lally RD, Kiwanuka S, Lloyd A, Ryan D, Germaine KJ et al (2015) Plant growth promotion induced by phosphate solubilizing endophytic Pseudomonas isolates. Front Microbiol 6:745. https://doi.org/10.3389/fmicb.2015.00745

Panhwar QA, Radziah O, Zaharah AR, Sariah M, Razi IM (2011) Role of phosphate solubilizing bacteria on rock phosphate solubility and growth of aerobic rice. J Environ Biol 32:607–612

Pankievicz VCS, do Amaral FP, KFDN S, Agtuca B, Xu Y, Schueller MJ et al (2015) Robust biological nitrogen fixation in a model grass-bacterial association. Plant J 81:907–919. pmid:25645593

Parewa HP, Yadav J, Rakshit A, Meena V, Karthikeyan N (2014) Plant growth promoting rhizobacteria enhance growth and nutrient uptake of crops. Agric Sustain Dev 2:101–116

Parikh S, James B (2012) Soil: the foundation of agriculture. Nat Educ Knowledge 3:2

Park K, Park J-W, Lee S-W, Balaraju K (2013) Disease suppression and growth promotion in cucumbers induced by integrating PGPR agent *Bacillus subtilis* strain B4 and chemical elicitor. ASM Crop Prot 54:199–205

Pastor N, Rosas S, Luna V, Rovera M (2014) Inoculation with *Pseudomonas putida* PCI2, a phosphate solubilizing rhizobacterium, stimulates the growth of tomato plants. Symbiosis 62:157–167

Patil VK (2013) Studies on integrated nutrient management in banana Cv. Ardhapuri (*Musa* AAA). Vasantrao Naik Marathwada Krishi Vidyapeeth, Parbhani

Patil HJ, Solanki MK (2016) Microbial inoculant: modern era of fertilizers and pesticides. In: Singh D, Singh H, Prabha R (eds) Microbial inoculants in sustainable agricultural productivity: vol. 1: research perspectives. Springer, New Delhi, pp 319–343. https://doi.org/10.1007/978-81-322-2647-5_19

Pavithra D, Yapa N (2018) Arbuscular mycorrhizal fungi inoculation enhances drought stress tolerance of plants. Groundw Sustain Dev 7:490–494. https://doi.org/10.1016/J.GSD.2018.03.005
Pereg L, de Bashan LE, Bashan Y (2016) Assessment of affinity and specificity of Azospirillum for plants. Plant Soil 399:389–414
Pliego C, Ramos C, de Vicente A, Cazorla FM (2011) Screening for candidate bacterial biocontrol agents against soilborne fungal plant pathogens. Plant Soil 340:505–520
Priyadarshani I, Rath B (2012) Commercial and industrial applications of micro algae—a review. J Algal Biomass Util 3:89–100
Rabie M, Ratti C, Calassanzio M, Aleem EA, Fattouh FA (2017) Phylogeny of Egyptian isolates of cucumber mosaic virus (CMV) and tomato mosaic virus (ToMV) infecting *Solanum lycopersicu*. Eur J Plant Pathol 149:219–225
Raghuwanshi R (2012) Opportunities and challenges to sustainable agriculture in India. NeBIO 3:78–86
Raj KS (2014) Paripex – invitro screening of trichoderma in rockphosphate solubilization and mycoremediation of heavy metal contaminated water. Indian J Res III:1–4
Raj SN, Deepak S, Basavaraju P, Shetty HS, Reddy M, Kloepper JW (2003) Comparative performance of formulations of plant growth promoting rhizobacteria in growth promotion and suppression of downy mildew in pearl millet. Crop Prot 22:579–588
Rask KA, Johansen JL, Kjøller R, Ekelund F (2019) Differences in arbuscular mycorrhizal colonisation influence cadmium uptake in plants. Environ Exp Bot 162:223–229. https://doi.org/10.1016/J.ENVEXPBOT.2019.02.022
Rendina N, Nuzzaci M, Scopa A, Cuypers A, Sofo A (2019) Chitosan-elicited defense responses in cucumber mosaic virus (CMV)-infected tomato plants. J Plant Physiol 234:9–17
Ribeiro VP, Marriel IE, de Sousa SM, de Paul Lana UG, Mattos BB, de Oliveira CA et al (2018) Endophytic Bacillus strains enhance pearl millet growth and nutrient uptake under low-P. Braz J Microbiol 49:40–46. https://doi.org/10.1016/j.bjm.2018.06.005
Rillig MC, Mummey DL (2006) Mycorrhizas and soil structure. New Phytol 171:41–53
Rodríguez H, Fraga R (1999) Phosphate solubilizing bacteria and their role in plant growth promotion. Biotechnol Adv 17:319–339
Rojas ES, Gleason M, Batzer J, Duffy M (2011) Feasibility of delaying removal of row covers to suppress bacterial wilt of muskmelon (Cucumis melo). Plant Dis 95:729–734
Rokhzadi A, Toashih V (2011) Nutrient uptake and yield of chickpea (*Cicer arietinum* L.) inoculated with plant growth-promoting rhizobacteria. Aust J Crop Sci 5:44
Rokhzadi A, Asgharzadeh A, Darvish F, Nour-Mohammadi G, Majidi E (2008) Influence of plant growth-promoting rhizobacteria on dry matter accumulation and yield of chickpea (*Cicer arietinum* L.) under field conditions American-Eurasian. J Agric Environ Sci 3:253–257
Ronga D, Biazzi E, Parati K, Carminati D, Carminati E, Tava A (2019) Microalgal biostimulants and biofertilisers in crop productions. Agronomy 9:192. https://doi.org/10.3390/agronomy9040192
Saharan B, Nehra V (2011) Plant growth promoting rhizobacteria: a critical review. Life Sci Med Res 21:30
Sahoo DK, Raha S, Hall JT, Maiti IB (2014) Overexpression of the synthetic chimeric native-T-phylloplanin-GFP genes optimized for monocot and dicot plants renders enhanced resistance to blue mold disease in tobacco (*N. tabacum* L.). Sci World J 2014:601314
Santhanam R, Weinhold A, Goldberg J, Oh Y, Baldwin IT (2015) Native root-associated bacteria rescue a plant from a sudden-wilt disease that emerged during continuous cropping. Proc Natl Acad Sci 112:E5013–E5020
Santi C, Bogusz D, Franche C (2013) Biological nitrogen fixation in non-legume plants. Ann Bot 111:743–767. https://doi.org/10.1093/aob/mct048
Saravanakumar D, Samiyappan R (2007) ACC deaminase from Pseudomonas fluorescens mediated saline resistance in groundnut (*Arachis hypogea*) plants. J Appl Microbiol 102:1283–1292
Saravanakumar D, Lavanya N, Muthumeena K, Raguchander T, Samiyappan R (2009) Fluorescent pseudomonad mixtures mediate disease resistance in rice plants against sheath rot (*Sarocladium oryzae*) disease. BioControl 54:273

Schütz L, Gattinger A, Meier M, Müller A, Boller T, Mäder P et al (2018) Improving crop yield and nutrient use efficiency via biofertilization—a global meta-analysis. Front Plant Sci 8:2204. https://doi.org/10.3389/fpls.2017.02204

Seenivasan N, David P, Vivekanandan P, Samiyappan R (2012) Biological control of rice root-knot nematode, Meloidogyne graminicola through mixture of *Pseudomonas fluorescens* strains. Biocontrol Sci Technol 22:611–632

Selvarajan R, Balasubramanian V (2014) Host–virus interactions in banana-infecting viruses. In: Plant virus–host interaction. Elsevier, Amsterdam, pp 57–78

Shaaban MM (2001a) Nutritional status and growth of maize plants as affected by green microalgae as soil additives. J Biol Sci 6:475–479

Shaaban MM (2001b) Green microalgae water extract as foliar feeding to wheat plants. Pak J Biol Sci 4:628–632

Shahzad AN, Rizwan M, Asghar MG, Qureshi MK, Bukhari SAH, Kiran A et al (2019) Early maturing Bt cotton requires more potassium fertilizer under water deficiency to augment seed-cotton yield but not lint quality. Sci Rep 9:7378. https://doi.org/10.1038/s41598-019-43563-2

Shaikh S, Gang S, Saraf M (2016) Rhizosphere microbial community structure: a review. J Gujarat Univ 1–2:14

Shan H et al (2013) Biocontrol of rice blast by the phenaminomethylacetic acid producer of *Bacillus methylotrophicus* strain BC79. Crop Prot 44:29–37

Sharma V, Sharma S, Sharma S, Kumar V (2019) Synergistic effect of bio-inoculants on yield, nodulation and nutrient uptake of chickpea (Cicer arietinum L) under rainfed conditions. J Plant Nutr 42:374–383. https://doi.org/10.1080/01904167.2018.1555850

Shen B et al (2010) Expression of ZmLEC1 and ZmWRI1 increases seed oil production in maize. Plant Physiol 153:980–987

Sheng XF, He LY (2006) Solubilization of potassium-bearing minerals by a wild-type strain of *Bacillus edaphicus* and its mutants and increased potassium uptake by wheat. Can J Microbiol 52:66–72

Shrivastava P, Kumar R (2015) Soil salinity: a serious environmental issue and plant growth promoting bacteria as one of the tools for its alleviation. Saudi J Biol Sci 22:123–131

Shukla L, Suman A, Yadav A, Verma P, Saxena AK (2016) Syntrophic microbial system for ex-situ degradation of paddy straw at low temperature under controlled and natural environment. J App Biol Biotech 4:30–37

Silva HSA, da Silva RR, Macagnan D, de Almeida H-VB, Pereira MCB, Mounteer A (2004) Rhizobacterial induction of systemic resistance in tomato plants: non-specific protection and increase in enzyme activities. Biol Control 29:288–295

Singh JS (2013) Plant growth promoting rhizobacteria. Resonance 18:275–281

Singh JS, Pandey VC, Singh D (2011) Efficient soil microorganisms: a new dimension for sustainable agriculture and environmental development. Agric Ecosyst Environ 140:339–353

Singh S, Gupta G, Khare E, Behal KK, Arora NK (2014) Effect of enrichment material on the shelf life and field efficiency of bioformulation of rhizobium sp. and P-solubilizing Pseudomonas fluorescens. Sci Res Rep 4:44–50

Solanki MK, Wang Z, Wang FY, Li CN, Lan TJ, Singh RK et al (2017) Intercropping in sugarcane cultivation influenced the soil properties and enhanced the diversity of vital Diazotrophic Bacteria. Sugar Tech 19:136–147. https://doi.org/10.1007/s12355-016-0445-y

Solanki MK, Wang F-Y, Wang Z, Li C-N, Lan T-J, Singh RK et al (2019) Rhizospheric and endospheric diazotrophs mediated soil fertility intensification in sugarcane-legume intercropping systems. J Soil Sediment 19:1911–1927. https://doi.org/10.1007/s11368-018-2156-3

Stirk WA, Ördög V, Novák O, Rolèík J, Strnad M, Bálint P, Staden J (2013) Auxin and cytokinin relationships in 24 microalgal strains. J Phycol 49:459–467

Subba Roa N (2001) An appraisal of biofertilizers in India. In: Kannaiyan S (ed) The biotechnology of biofertilizers. Narosa Publishing House, New Delhi

Suman A, Verma P, Yadav AN, Srinivasamurthy R, Singh A, Prasanna R (2016a) Development of hydrogel based bio-inoculant formulations and their impact on plant biometric parameters of wheat (*Triticum aestivum* L.). Int J Curr Microbiol App Sci 5:890–901

Suman A, Yadav AN, Verma P (2016b) Endophytic microbes in crops: diversity and beneficial impact for sustainable agriculture. In: Microbial inoculants in sustainable agricultural productivity. Springer, New Delhi, pp 117–143
Tanwar S, Sharma G, Chahar M (2003) Effect of phosphorus and biofertilizers on yield, nutrient content and uptake by black gram [*Vigna mungo* (L.) Hepper]. Legume Res 26:39–41
Tippannavar C, Reddy T (1993) Seed treatment of wheat (*Triticum aesativum* L.) on the survival of seed borne *Azotobacter chroococcum* Karnataka. J Agric Sci 6:310–312
Turan A, Kaya G, Karamanlioğlu B, Pamukcu Z, Apfel C (2005) Effect of oral gabapentin on postoperative epidural analgesia. Br J Anaesth 96:242–246
Vaid SK, Kumar B, Sharma A, Shukla A, Srivastava P (2014) Effect of Zn solubilizing bacteria on growth promotion and Zn nutrition of rice. J Soil Sci Plant Nutr 14:889–910
van der Heijden MGA, Horton TR (2009) Socialism in soil? The importance of mycorrhizal fungal networks for facilitation in natural ecosystems. J Ecol 97:1139–1150
van der Heijden MGA, Martin FM, Selosse M-A, Sanders IR (2015) Mycorrhizal ecology and evolution: the past, the present, and the future. New Phytol 205:1406–1423. https://doi.org/10.1111/nph.13288
Van Oosten MJ, Di Stasio E, Cirillo V, Silletti S, Ventorino V, Pepe O et al (2018) Root inoculation with Azotobacter chroococcum 76A enhances tomato plants adaptation to salt stress under low N conditions. BMC Plant Biol 18:205. https://doi.org/10.1186/s12870-018-1411-5
Vejan P, Abdullah R, Khadiran T, Ismail S, Nasrulhaq BA (2016) Role of plant growth promoting rhizobacteria in agricultural sustainability–a review. Molecules 21:573. pmid:27136521
Velivelli SLS, De Vos P, Kromann P, Declerck S, Prestwich BD (2014) Biological control agents: from field to market, problems, and challenges. Trends Biotechnol 32:493–496. https://doi.org/10.1016/j.tibtech.2014.07.002
Verma P (2013) Elucidating the diversity and plant growth promoting attributes of wheat (*Triticum aestivum*) associated acidotolerant bacteria from southern hills zone of India. Natl J Life Sci 10:219–226
Verma P (2015) Assessment of genetic diversity and plant growth promoting attributes of psychrotolerant bacteria allied with wheat (*Triticum aestivum*) from the northern hills zone of India. Ann Microbiol 65:1885–1899
Verma P (2016) Appraisal of diversity and functional attributes of thermotolerant wheat associated bacteria from the peninsular zone of India. Saudi J Biol Sci-In Press
Verma JP, Yadav J, Tiwari KN, Jaiswal DK (2014a) Evaluation of plant growth promoting activities of microbial strains and their effect on growth and yield of chickpea (*Cicer arietinum* L.) in India. Soil Biol Biochem 70:33–37
Verma P, Yadav AN, Kazy SK, Saxena AK, Suman A (2014b) Evaluating the diversity and phylogeny of plant growth promoting bacteria associated with wheat (*Triticum aestivum*) growing in central zone of India. Int J Curr Microbiol App Sci 3:432–447
Verma P, Yadav AN, Shukla L, Saxena AK, Suman A (2015a) Alleviation of cold stress in wheat seedlings by *Bacillus amyloliquefaciens* IARI-HHS2-30, an endophytic psychrotolerant K-solubilizing bacterium from NW Indian Himalayas. Natl J Life Sci 12:105–110
Verma P, Yadav AN, Shukla L, Saxena AK, Suman A (2015b) Hydrolytic enzymes production by thermotolerant *Bacillus altitudinis* IARI-MB-9 and *Gulbenkiania mobilis* IARI-MB-18 isolated from Manikaran hot springs. Int J Adv Res 3:1241–1250
Verma P, Yadav AN, Khannam KS, Kumar S, Saxena AK, Suman A (2016) Molecular diversity and multifarious plant growth promoting attributes of bacilli associated with wheat (*Triticum aestivum* L.) rhizosphere from six diverse agro-ecological zones of India. J Basic Microbiol 56:44–58
Verma P, Yadav AN, Khannam KS, Saxena AK, Suman A (2017a) Potassium-solubilizing microbes: diversity, distribution, and role in plant growth promotion. In: Microorganisms for green revolution. Springer, Singapore, pp 125–149
Verma P, Yadav AN, Kumar V, Singh DP, Saxena AK (2017b) Beneficial plant-microbes interactions: biodiversity of microbes from diverse extreme environments and its impact for

crop improvement. In: Plant-microbe interactions in agro-ecological perspectives. Springer, Singapore, pp 543–580

Viscardi S, Ventorino V, Duran P, Maggio A, De Pascale S, Mora ML et al (2016) Assessment of plant growth promoting activities and abiotic stress tolerance of Azotobacter chroococcum strains for a potential use in sustainable agriculture. J Soil Sci Plant Nutr 16:848–863

Vitousek PM, Naylor R, Crews T, David MB, Drinkwater LE, Holland E et al (2009) Nutrient Imbalances in Agricultural Development. Science (80-) 324:1519–1520. https://doi.org/10.1126/science.1170261

Wakeel A, Rehman H, Magen H (2017) Potash use for sustainable crop production in Pakistan: a review. Int J Agric Biol 19:381–390. https://doi.org/10.17957/IJAB/15.0291

Wang W, Shi J, Xie Q, Jiang Y, Yu N, Wang E (2017) Nutrient exchange and regulation in arbuscular mycorrhizal symbiosis. Mol Plant 10:1147–1158. https://doi.org/10.1016/J.MOLP.2017.07.012

Wang R, Li Y, Dong J-L, Ding W-L (2018) First report of tomato yellow leaf curl virus and cucumber mosaic virus infecting Huoxiang (Agastache rugosa) in CHINA. J Plant Pathol 100:581–581

Wani SA, Chand S, Wani MA, Ramzan M, Hakeem KR (2016) *Azotobacter chroococcum* – a potential biofertilizer in agriculture: an overview. In: Hakeem KR, Akhtar J, Sabir M (eds) Soil science: agricultural and environmental prospectives. Springer, Cham, pp 333–348

Wu X, Li D, Bao Y, Zaitlin D, Miller R, Yang S (2015) Genetic dissection of disease resistance to the blue mold pathogen, Peronospora tabacina, in tobacco. Agronomy 5:555–568

Yang SY, Grønlund M, Jakobsen I, Grotemeyer MS, Rentsch D, Miyao A et al (2012) Nonredundant regulation of rice arbuscular mycorrhizal symbiosis by two members of the PHOSPHATE TRANSPORTER1 gene family. Plant Cell 24(10):4236–4251

Yasari E, Patwardhan A (2007) Effects of (Azotobacter and Azospirillum) inoculants and chemical fertilizers on growth and productivity of canola (Brassica napus L.). Asian J Plant Sci 6:77–82

Zahir ZA, Arshad M, Frankenberger WT (2004) Plant growth promoting rhizobacteria: applications and perspectives in agriculture. Adv Agron 81:98–169

Zahoor R, Dong H, Abid M, Zhao W, Wang Y, Zhou Z (2017) Potassium fertilizer improves drought stress alleviation potential in cotton by enhancing photosynthesis and carbohydrate metabolism. Environ Exp Bot 137:73–83. https://doi.org/10.1016/j.envexpbot.2017.02.002

Zaidi A, Khan MS (2006) Co-inoculation effects of phosphate solubilizing microorganisms and *Glomus fasciculatum* on green gram-*Bradyrhizobium* symbiosis. Turk J Agric For 30:223–230

Zakira F (2009) Africa's New Path: Paul Kagame Charts a Way Forward Newsweek, July 18

Zeffa DM, Perini LJ, Silva MB, de Sousa NV, Scapim CA, de Oliveira ALM et al (2019) Azospirillum brasilense promotes increases in growth and nitrogen use efficiency of maize genotypes. PLoS One 14:e0215332. https://doi.org/10.1371/journal.pone.0215332

Zehnder GW, Murphy JF, Sikora EJ, Kloepper JW (2001) Application of rhizobacteria for induced resistance. Eur J Plant Pathol 107:39–50

Zhang C, Kong F (2014) Isolation and identification of potassium-solubilizing bacteria from tobacco rhizospheric soil and their effect on tobacco plants. Appl Soil Ecol 82:18–25

Zhang S, Moyne A-L, Reddy M, Kloepper JW (2002) The role of salicylic acid in induced systemic resistance elicited by plant growth-promoting rhizobacteria against blue mold of tobacco. Biol Control 25:288–296

Zhang S, White TL, Martinez MC, McInroy JA, Kloepper JW, Klassen W (2010) Evaluation of plant growth-promoting rhizobacteria for control of *Phytophthora blight* on squash under greenhouse conditions. Biol Control 53:129–135

Zhang J, Jiang F, Yang P, Li J, Yan G, Hu L (2015) Responses of canola (*Brassica napus* L.) cultivars under contrasting temperature regimes during early seedling growth stage as revealed by multiple physiological criteria. Acta Physiol Plant 37:7

Zhao R, Liu LX, Zhang YZ, Jiao J, Cui WJ, Zhang B et al (2018) Adaptive evolution of rhizobial symbiotic compatibility mediated by co-evolved insertion sequences. ISME J 12:101–111. https://doi.org/10.1038/ismej.2017.136

Zodape S, Gupta A, Bhandari S, Rawat U, Chaudhary D, Eswaran K, Chikara J (2011) Foliar application of seaweed sap as biostimulant for enhancement of yield and quality of tomato (*Lycopersicon esculentum* Mill)

Zohaib A, Jabbar A, Ahmad R, Basra SMA, Zohaib A, Jabbar A et al (2018) Comparative productivity and seed nutrition of cotton by plant growth regulation under deficient and adequate boron conditions. Planta Daninha:36. https://doi.org/10.1590/s0100-83582018360100040

Wheat Microbiome: Present Status and Future Perspective

8

Sunita Mahapatra, Pravallikasree Rayanoothala, Manoj Kumar Solanki, and Srikanta Das

Abstract

Wheat microbiome harbors a diverse array of microbial communities and play a vital role in maintaining wheat physiology as well assist in offering protection from biotic and abiotic stresses. Several research findings indicated that wheat microbiome encompasses predominantly fungi, bacteria, viruses, actinomycetes, cyanobacteria, protozoa, archaea, etc. which performed myriads of advantageous activities including bio-management of crop pathogens, abiotic stress amelioration, as well as plant growth promotion under adverse conditions. In this chapter, attempts have been made to provide comprehensive and up-to-date insights on wheat microbiome research with major emphasis on emerging microbiome-based sustainable solutions for profitable and quality wheat production under every changing climate.

Keywords

Abiotic stress · Bio-antagonist · Microbiome · PGPR · Wheat

8.1 Introduction

Wheat (*Triticum aestivum*) is the largest cultivated crop in the world occupying an area of 222.35 million hectares with an estimated production of 753.89 million tons in 2016–2017 (Jasrotia et al. 2018). It is a staple food in more than 40 countries and

S. Mahapatra (✉) · P. Rayanoothala · S. Das
Department of Plant Pathology, Bidhan Chandra Krishi Viswavidyalaya, Mohanpur, Nadia, West Bengal, India

M. K. Solanki
Department of Food Quality & Safety, Institute for Post-harvest and Food Sciences, The Volcani Center, Agricultural Research Organization, Rishon LeZion, Israel

© Springer Nature Singapore Pte Ltd. 2020

M. K. Solanki et al. (eds.), *Phytobiomes: Current Insights and Future Vistas*,
https://doi.org/10.1007/978-981-15-3151-4_8

finds a significant share of about 35% in the consumption basket of millions across the world. It is estimated that global population depends on wheat for 85% and 82%, respectively, for calories and protein (Chaves et al. 2013). Moreover, wheat meets up 21% of the world's food demand and is grown on 200 M ha of farmland globally (Tsvetanov et al. 2016). By 2050, the world will have the challenge to meet the food demand of an estimated population of 9.6 billions. Feeding for growing population at the era of climate change requires the use of optimized, reliable and existing resources which have less impact on the environment (Busby et al. 2017). Now research topics mainly highlighted on the issue of not only quantity but also quality of food product with minimal uses of existing earth resources. Under developing countries like India, which has self-sufficiency in wheat production, there is a need to give emphasis on quality wheat production by understanding plant microbiome and its interactions with environment and stress regimes.

Moreover, the future global climate scenarios predicted an increase in the occurrence of exceptionally hot days, together with an increase in average global temperatures and its implications on global food production. Asseng et al. (2015) reported that the global wheat production is influenced by the increase of temperature and becomes more variable over space and time. In Indian perspective, temperature changes in the last 30 years had a bigger impact on national wheat production, where over 90% of wheat is irrigated (Singh and Mustard 2012), than changes in precipitation (Lobell et al. 2011). One of the most important cropping patterns in South Asia is rice-wheat cropping system, facing immense pressure because of heat stress and degraded soil health due to high cropping intensity and tillage for growing rice and over-exploitation of the natural resources (Joshi et al. 2007). The most affected locations of South Asia are eastern Gangetic plains, central and peninsular India and Bangladesh. Besides, a substantial proportion of cultivated land under wheat in South Asia is salt affected. The salt-affected land under wheat in India is 4.5 m ha, whereas 6.0 m ha in Pakistan amongst the other South Asian countries (Singh and Chatrath 2001). Although soil reclamation and provision of proper drainage may be more effective solution, it does not seem possible in the near future due to huge acreage affected by salt. Another constraint is deficiency of macro-nutrients like zinc, sulphur, iron, manganese and boron which are being observed in some pockets of northern India, Bangladesh and Nepal due to imbalanced fertilization, overmining of essential plant nutrients and burning of crop residues (Chatrath 2004). Water is also becoming scarce as the water table is going down due to overmining of groundwater in intensive rice-wheat cultivation and comparatively less water recharge from monsoon rains.

Amongst the biotic stresses, rusts continue to be the major threat (Khan et al. 2017; Savadi et al. 2018; Singh et al. 2006; Kumar et al. 2019; Kashyap et al. 2019). Out of three rusts prevalent in Indian subcontinent, leaf rust is the major disease which affects almost whole of India, parts of Bangladesh and Nepal. Spot blotch caused by *Bipolaris sorokiniana* (Sacc.) Shoem is also considered an important disease in the eastern part of South Asia (Joshi et al. 2007; Devi et al. 2018). In addition, other diseases, viz. Karnal bunt, powdery mildew and wheat blast also affect wheat crop to some extent (Singh 2017; Kashyap et al. 2011, 2018, 2019;

Nallathambi et al. 2019; Kumar et al. 2020). The use of microbiomes increases crop productivity, reduces the cost of cultivation and sustains the health of soil as well as environment. Use of microbiome in wheat cultivation has the ability to change the entire scenario of the current agricultural and food security.

The plant microbiome is a key determinant of plant health and productivity and has received substantial attention in recent years. There is an expanded version of the tree of life dominated by bacterial diversification due to their ability to perform lateral gene transfer across disparate phylogenetic groups (Mcdonald and Currie 2017; Hug et al. 2016). Recent advance researches in high-throughput sequencing technique and the increasing number of microbial culture libraries, they can quickly proliferate and have high mutation rates (Kibota and Lynch 1996; Boe et al. 2000; Denamur and Matic 2006). Individual microbes of the same species could potentially bear different genetic endowments and thus functional characteristics (Sergaki et al. 2018). Amplicon sequencing has been invaluable in determining general patterns of microbial diversity within the plant microbiome (Bulgarelli et al. 2012; Lundberg et al. 2012; Peiffer et al. 2013; Tkacz et al. 2015).

However, use of microbiome on wheat is very limited. Most of the previous studies mostly based on identifying microbes in the root's rhizosphere (Hartmann et al. 2014; Mahoney et al. 2017; Ofek et al. 2013; Yin et al. 2017) and limited research work is focused on aboveground organs (Granzow et al. 2017; Huang et al. 2016; Karlsson et al. 2017). To our knowledge, till date no detailed published works are available on entire wheat microbiome, including both above- and belowground plant organs, mode of actions, how they work under stress conditions and what are the easy and fast way of identification with high-throughput sequencing techniques. Here we classify the microbiome in a detailed way and also use fullness of those microbiomes especially for wheat crop. But still microbial communities were dependent on type of plant organ, growth stages of the plant, environmental conditions, acceptability amongst the farmers and community composition, etc. and therefore need to be checked before their successful intervention and execution.

8.2 Plant Microbiome: Concept and Definitions

Microbes are associated with majority of living species and are omnipresent in nature. However, majority of them are noticed in harsh environments. The microbial diversity consists of various types of microbes such as archaea, bacteria, cyanobacteria, fungi and protozoa (Vessey 2003; Verma and Suman 2018). The plant microbiomes include rhizospheric, endophytic and epiphytic microbes which help in plant growth and in adaptation even in extreme environments. In our planet, researchers have recorded the presence of microbiomes amongst 1.7 million living species and noticed that 1–10% bacterial species are present amongst 5000 species of prokaryotes (Federhen 2014; Moore 2014). The microbiomes are useful in industry and medical processes too in addition to agriculture, and they act as abundant reservoirs of bioresources. Thus in keeping good soil health, enhancing productivity and improving the growth of plants, these plant microbiomes play vital role and are found helpful in providing a sustainable biosphere.

The following definitions are found for plant microbiome in literature.

1. Bulgarelli and Schlaeppi (2015) have defined microbiota or microbiome as a set of genomes of the microorganisms in some habitat.
2. Several researchers have defined that microbial communities are the ones associated with any plant that can live, thrive and interact with various tissues such as roots, shoots, leaves, flowers and seeds (Turner et al. 2013; Haney and Ausubel 2015; Mueller and Sachs 2015; Haney et al. 2015; Nelson 2018). It is also found in literature that microbiome includes all the microbes of a community.

8.3 Wheat Microbiomes: Concept and Types

The wheat microbiome is mainly composed of various kinds of organisms such as archaea, protozoa, bacteria, fungi and virus (Mueller and Sachs 2015). Research done on the microorganisms so far found that they are present in healthy crops, and their findings are limited to species of wild *Triticum* and of wheat (Marshall et al. 1999).

It is a well known fact that nematodes, ants, moles and several fungi and bacteria take common shelter in soil. They play an essential role in protecting the plant from potential plant pathogens and also in improving plant growth, health, and production (Berg et al. 2014; Haney et al. 2015). There is a closer association between plants and microbial communities and normally found at phyllosphere (above the ground), in rhizosphere (below the ground especially on roots and surrounding area of roots) and in endophytes (inside the root intercellular spaces) (Hirsch and Mauchline 2014). Relationship of plants with microbes is beneficial in improving disease resistance, in increasing stress tolerance levels and also in nutrient uptake. It is not always beneficial; rather, sometimes interactions of plants with microbes may be harmful too. This is dependent on kinds of bacteria involved, their characteristics and finally the ways of interactions.

Mainly, the interactions between plants and microbes are classified as rhizospheric, endophytic and epiphytic (as shown in Fig. 8.1). Wheat plants interact mostly with either common microbes or niche-specific microbes (Fig. 8.2). Common microbes are *Arthrobacter nicotianae*, *Bacillus amyloliquefaciens*, *B. sphaericus*, *B. subtilis*, *Paenibacillus amylolyticus*, *P. polymyxa*, *Micrococcus luteus*, *Pseudomonas aeruginosa* and *P. azotoformans*, and most predominant species were reported from various parts of the plant such as phyllosphere, rhizosphere and internal tissues (Bhattacharyya and Jha 2012; Verma and Suman 2018). Further, the wheat bacterial microbiomes belong to different taxonomic positions of phyla, namely, *Actinobacteria*, *Bacteroidetes*, *Firmicutes*, *Proteobacteria* and *Gemmatimonadetes* (Verma and Suman 2018). Some novel microbiomes like acidophiles, alkaliphiles,

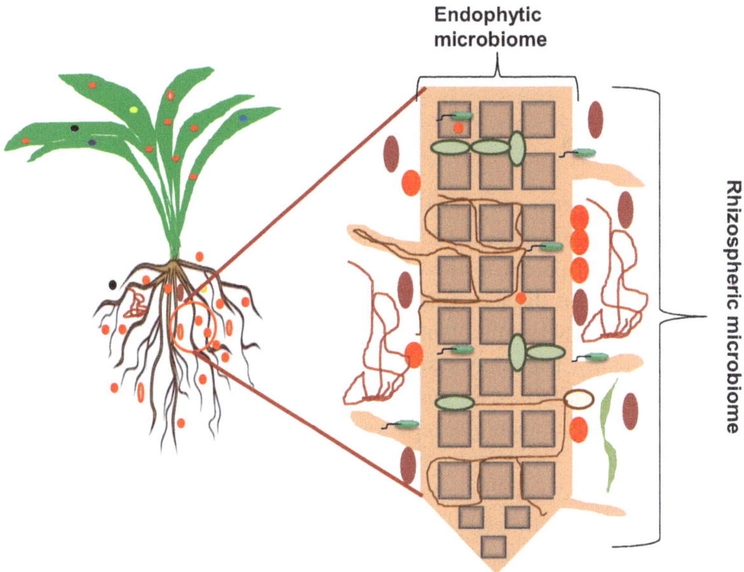

Fig. 8.1 Assemblage of microbes in and around wheat plant

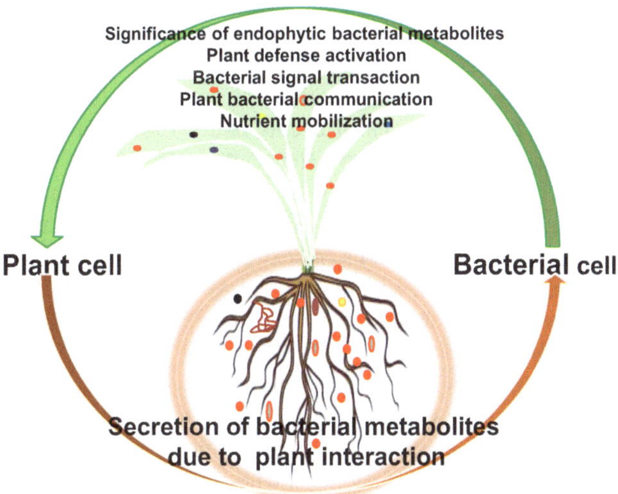

Fig. 8.2 Relative distribution amongst different microbes isolated from phyllospheric, endophytic and rhizospheric sample of wheat

halophiles, psychrophiles, thermophiles and xerophiles are also found in wheat (Narula et al. 2006). According to Verma and Suman (2018), bacterial microbiomes, viz. *Actinobacteria* (12%), *Bacteroidetes* (7%), *Firmicutes* (32) and *Proteobacteria* (49%), which occupy a major portion, and three classes of bacteria, namely,

Alphaproteobacteria (6%), *Betaproteobacteria* (8%) and *Gammaproteobacteria* (35%) are present in wheat.

8.3.1 Endophytes

Wilson (1995) and Brader et al. (2014) observed that the bacteria or fungi during their life cycle attack the plant tissues. This invasion causes infections in plant tissues which are not apparent to the naked eye. However, these infections do not cause any symptoms of disease. Endophytic microbiomes get into plants through wounds and root hairs to benefit plants by colonizing them and preventing effects of pathogenic organisms. They may systematically colonize the plants (Compant et al. 2010; Kushwaha et al. 2020a). Requisite chemicals are produced by them to prevent the growth of plant pathogenic competitor organisms. Bacterial endophytes help plants in their growth (Kushwaha et al. 2019, 2020b). Carroll (1988) has done research on endophyte associations and shown that they enhance the survival chances against fungal pathogens.

The specific endophytic species are *Achromobacter piechaudii*, *A. xylosoxidans*, *Delftia lacustris*, *D. acidovorans*, *Acinetobacter lwoffii*, *Ochrobactrum intermedium*, *Pantoea dispersa*, *P. eucalypti*, *Staphylococcus epidermis*, *Pseudomonas monteilii* and *Variovorax soli* and *Serratia* (Liu et al. 2010; Hallmann et al. 1997). Carroll (1988) has observed that host benefits is a common phenomenon related with fungal endophytes.

Endophytic fungi are used as genetic vectors and as abundant source of secondary metabolites (Fisher et al. 1986; Stierle et al. 1993; Strobel and Daisy 2003). Further they act as biological control agents (Clay 1989; Bacon 1990; Schardl et al. 1991; Dorworth and Callan 1996). Hence, biotechnological interest is shown very much on endophytic fungi (Murray et al. 1992). It is also studied that fungal endophytes cause water loss in leaves. Some kinds of fungal endophytes help plants to survive in drought conditions and extreme temperature climates (El-Daim et al. 2014; Naveed et al. 2014; Khalafallah and Abo-Ghalia 2008). Thus, focused research has been done on it in general and on grass endophytes in particular (Farooq et al. 2009).

Direct antagonism of microbial pathogens is one way that endophytic bacterial strains fuel plant growth. As biocontrol agents, endophytes induce resistance in plants to disease-causing organisms and allow plant to be healthy (Pleban et al. 1995). Figure 8.3 described the isolation of host-associated endophyte and their vital function.

The anti-microbial property of endophytic microbes generates secondary metabolites, and thus they act against pathogenic microbes (Tan and Zou 2001). There are many bioactive compounds different from endophytes, and such compounds belong to classes like alkaloids, flavones, peptides, phenols, quinines, steroids and terpenoids (Yu et al. 2010).

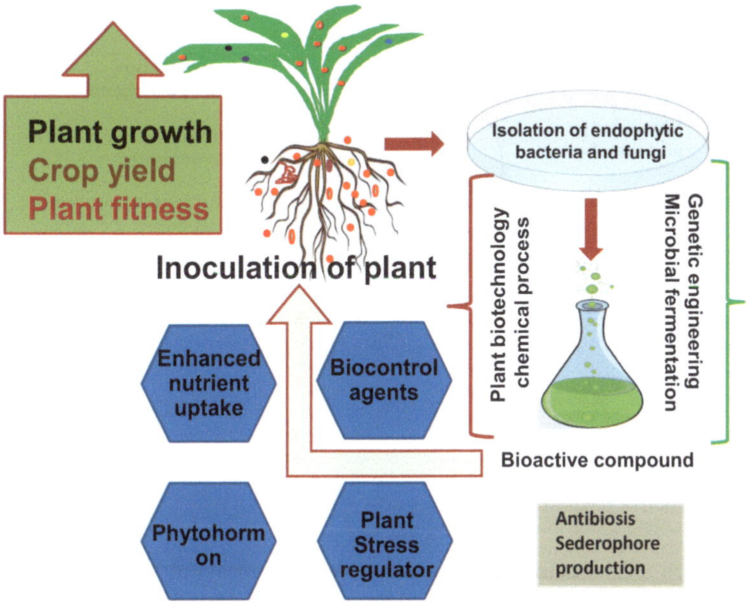

Fig. 8.3 Schematic representation of isolation, purification and function of endophytes in host plant

8.3.2 Epiphytic Microbiomes

Synergy between plant and bacteria is due to the presence of epiphytic microbes on phyllosphere. Microorganisms found on leaf surfaces are identified as extremophiles due to their capacity to survive even at high temperature (40–55 °C). Further they tolerate ultraviolet radiation during the day and even cool temperature (5–10 °C) during the night (Kushwaha et al. 2020b). The association between the microorganisms in the phyllosphere influences the plant growth in natural habitat and the productivity of agricultural and horticultural crops for human consumption.

The aerial parts of plants are normally exposed to air and dust. This helps typical microbiomes to get attached to the surface. Hence, survival and proliferation of the phyllospheric microbiomes on leaves depend on the leaf diffuses or exudates. Amino acids, glucose, fructose and sucrose are some nutrient factors found in the leaf exudates. Hence, the plant growth is accomplished by the essential process like nitrogen fixation (Iniguez et al. 2004; Venieraki et al. 2011).

In the phyllosphere of wheat, researchers have observed the presence of microbes, viz., *Achromobacter*, *Corynebacterium*, *Agrobacterium*, *Haemophilus*, *Alcaligenes*, *Arthrobacter*, *Bacillus*, *Azotobacter*, *Enterobacter*, *Lysinibacillus*, *Micrococcus*, *Brevundimonas*, *Paenibacillus*, *Methylobacterium*, *Micromonospora*, *Pseudomonas*, *Micrococcus*, *Streptomyces*, *Stenotrophomonas*, *Pantoea* and *Psychrobacter* (Sharma et al. 2011a, b; Verma et al. 2016).

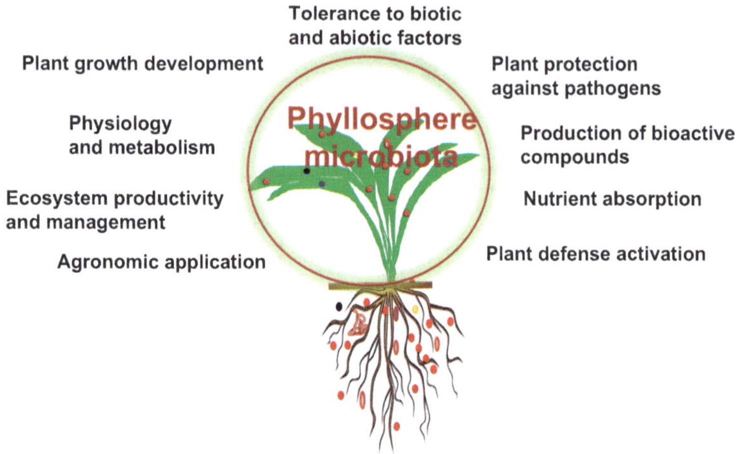

Fig. 8.4 Function of phyllosphere microbiota

Production of bioactive compounds, tolerance to biotic and abiotic stress, immune response, agronomic applications, nutrient acquisition, ecosystem productivity and management, physiology and metabolism, plant protection against pathogen are some of the benefits happen due to phyllosphere microbiota and responsible for improved plant growth (Roat and Saraf 2017) (Fig. 8.4).

8.3.3 Rhizospheric Microbiome

The microbiomes in rhizosphere predominantly gained importance due to attraction of different types of root exudate substrates, and this zone is considered a hotspot for microbial survival and its diversity compared to other plant parts (Arkhipova et al. 2005; Kuzmina and Melentev 2003). The rhizospheric microbiomes also play a vital role in growth of the plant. They are closely attached between roots and soils and help the plants in nutrient uptake. However, rhizospheric microbiome is greatly affected by several factors like soil types, pH values, soil moisture and presence of micronutrients in soil etc. (Ahemad and Kibret 2014; Khalid et al. 2004).

Rhizospheric microbes live in soil near roots, and due to gradient of root closeness they utilize metabolites from the surrounding roots as carbon and nitrogen sources and they inhabit spaces between cortical cells, they colonize rhizoplane and also they live in the specialized root structures like root nodules. Specific species of the rhizospheric are *Pseudomonas extremorientalis*, *P. rhizosphaerae*, *Arthrobacter nicotinovorans*, *Azotobacter tropicalis*, *Bacillus atrophaeus*, *B. horikoshii*, *B. mojavensis*, *B. siamensis*, *B. thuringiensis*, *Enterobacter asburiae*, *Exiguobacterium acetylicum*, *Serratia marcescens*, *Planomicrobium okeanokoites*, *Rhodobacter capsulatus* and *Rhodobacter sphaeroides* (Hassan and Bano 2015; Zahir et al. 2003).

Different genera related to the wheat rhizospheric microbiomes are *Arthrobacter*, *Alcaligenes*, *Acinetobacter*, *Azospirillum*, *Methylobacterium*, *Bacillus*, *Burkholderia*, *Erwinia*, *Flavobacterium*, *Enterobacter*, *Lysinibacillus*, *Paenibacillus*, *Rhizobium*, *Serratia* and *Pseudomonas* (Khalid et al. 2004; Verma and Suman 2018).

8.4 Wheat Microbiome Detection and Diagnosis

Both culturable and un-culturable techniques have been employed to wheat-associated microbes to study diversity of their association with wheat crops and also to know their distribution (Fig. 8.5). Particularly, surface sterilization and serial dilution techniques are used to isolate wheat endophytic and rhizospheric microbes (Forchetti et al. 2007; Suman et al. 2016a, b). The epiphytic microbes are isolated with the imprinting method and, alternatively, serial dilution, and then pour or spread plate methods are used in sequence to isolate the same (Akbari et al. 2007; Yadav et al. 2015a, b; Suman et al. 2016a, b). For isolation and enumeration of wheat microbiomes (eubacteria, archaea, fungi), the following specific methods are employed along with specific medium.

The specific medium for isolation of heterotrophic microbes is nutrient agar; for *Pseudomonads* isolation, King's B agar medium is used. *Arthrobacter* is isolated by growing it in trypticase soy agar media; soil extract agar is used for the isolation of

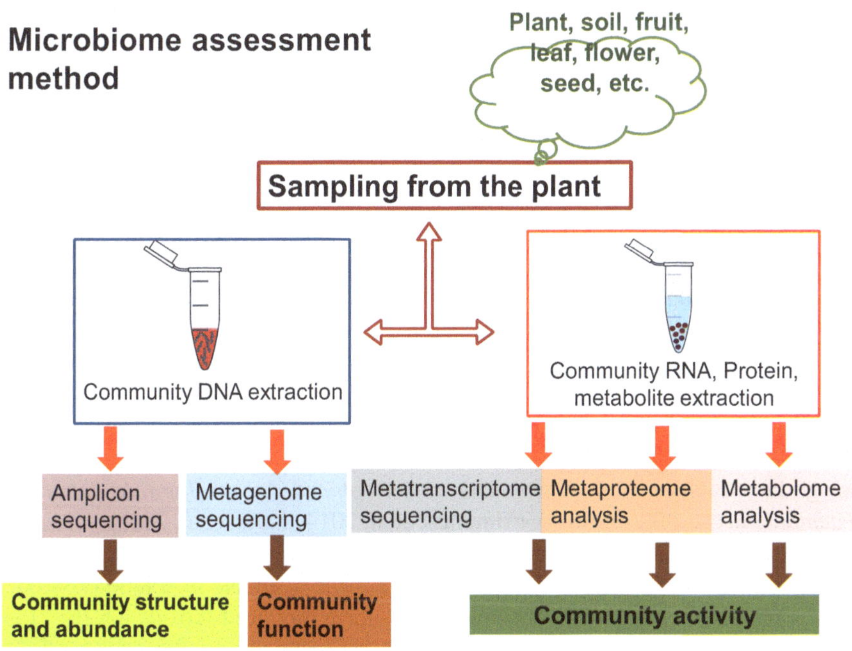

Fig. 8.5 The flowchart of the assessment of microbiomes

soil-specific microbes. *Bacillus and Bacillus*-derived genera can be isolated by T3A with heat treatment methods. Any chemically defined and complex medium is used for the isolation of archaea. Rose Bengal media and potato dextrose agar is used for isolation of fungi (Forchetti et al. 2007; Verma and Suman 2018).

Microbiome samples can be assessed either through community DNA extraction or community RNA extraction. Community DNA extraction is sequenced through PCR amplification (16S–18S rRNA, ITS, cpn60) or amplicon sequencing and metagenome sequencing through which species, number, abundance and its composition are analysed; and the function of the community is assessed, respectively. Community RNA, protein and metabolite extraction can be done through meta-transcriptome sequencing, meta-proteome analysis and metabolome analysis from which the activity of the community is known (Liu et al. 2014).

8.5 Role of Wheat Microbiomes

The wheat microbiomes stimulate plant growth, soil health and fertility (Canbolat et al. 2006; Singh et al. 2007; Khan et al. 2007). Further, diverse abiotic stresses are ameliorated directly through the following: N_2 fixation, production of siderophore and phytohormones (auxin, cytokinin and gibberellins) and solubilization of potassium, phosphorus and zinc (Sachdev et al. 2009; Kumar et al. 2001; Zaidi and Khan. 2005; Kudoyarova et al. 2014; Rroco et al. 2003). The same stresses are ameliorated indirectly through the following: production of ammonia, hydrogen cyanide, iron-chelating compounds, hydrolytic enzymes, antibiotics and antagonistic molecules for suppression of soilborne pathogens (Jankiewicz 2006; Scavino and Pedraza 2013; Upadhyay et al. 2012).

Plant growth-promoting microbes produce phytohormones or plant growth regulators (Glick 2015) for better growth of the plants. Phytohormones are metabolites derived from bacteria such as *Azotobacter*, *Bacillus*, *Azospirillum* and *Pseudomonas* which show benefits such as promotion and spread of root development; with these benefits, roots of the plants increase their ability to uptake water and nutrients from soil in an efficient manner (Mehnaz 2015).

Nitrogen-fixing microbe *Azospirillum brasilense* produces small amount of phytohormones like gibberellins, indole-3-acetic acid (IAA) and substances like cytokinin. PGP microbial strains like *Azospirillum*, *Pseudomonas* and *Bacillus* produce phytohormone like auxin which modulates shoot elongation and similar plant physiological processes and further promotes root development in plants (Verma et al. 2016). IAA and related compounds are also present, and these are demonstrated in many diazotrophs such as *Acetobacter diazotrophicus*, *Azospirillum*, *Azotobacter*, *Paenibacillus* and *Polymyxa* sp. (Timmusk et al. 2014; Aarab et al. 2015).

Plants produce soluble organic compounds such as chelators and phytosiderophores which help to bind iron (Fe^{3+}) and are made available in solution (Sarode et al. 2009; Rroco et al. 2003). Further, chelators' produced ferrous ion (Fe^{3+}) is reduced to ferric ion (Fe^{2+}) and absorbed immediately on the root surface. Phytosiderophores are low molecular weight and are ferric ion-specific ligands.

Thus, by production and discharge of phytosiderophore, PGPR are able to prevent the proliferation of phytopathogens which in turn facilitates plant growth (Mehnaz 2015; Sarode et al. 2013; Solanki et al. 2014; Jog et al. 2014).

8.6 Plant Growth-Promoting Bacteria (PGPB)

Plant growth-promoting substances such as rhizospheric or endophytic bacteria show beneficial effects in plant growth (Santoyo et al. 2016; Kushwaha et al. 2020b). Plant growth-promoting substances extend over plant roots and the soil in vicinity and show direct or indirect effect on plants (Glick 2012; Santoyo et al. 2012, Gupta et al. 2015).

The elements such as nitrogen, phosphorus and iron are available by plant growth-promoting bacteria, which are very much needed for plant growth and development (Calvo et al. 2014) (Fig. 8.6). Further, PGPB modulates level of hormones in the production of phytohormones like auxins, cytokinins and gibberellins (Yadav et al. 2017a, b). PGPB are also able to reduce the levels of the ethylene phytohormone by synthesizing an enzyme, 1-aminocyclopropane-1-carboxylate (ACC) deaminase, that slices the ACC compound. ACC is a precursor compound of ethylene in higher plants (Glick et al. 1998; Yadav et al. 2016).

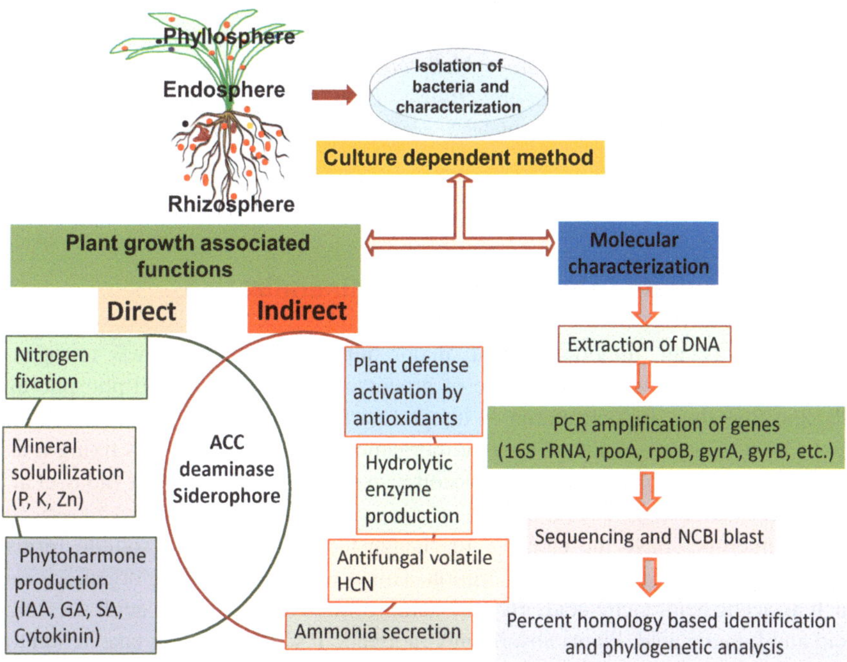

Fig. 8.6 A schematic representation of the isolation and characterization of plant growth-promoting bacteria

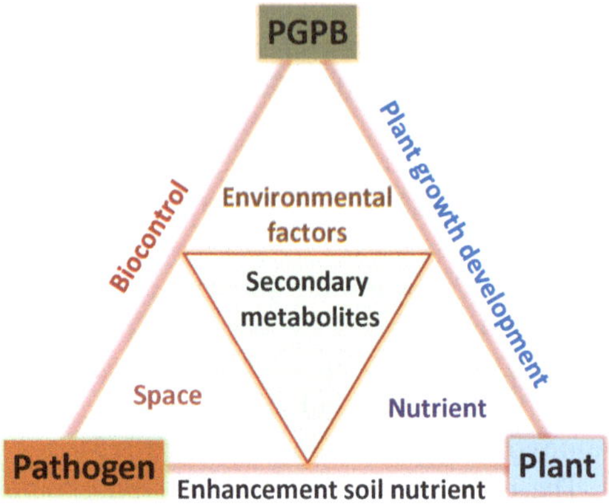

Fig. 8.7 Plant-microbe interactions and their action of plant growth promotion

After plant pathogen infection, PGPB decreases plant damage by indirect mechanism and thus promotes plant growth (Fig. 8.7). This is being demonstrated because of the inhibition of the pathogens which is induced by PGPB (Santoyo et al. 2012, 2016). Generally, the mechanisms are synthesis and release of antibiotics (such as 2,4-diacetylphloroglucinol); bacteriocins; chitinases; lipopeptides like iturin A, bacillomycin D and mycosubtilin; proteases; siderophores; and volatile organic compounds (Santoyo et al. 2012; Glick 2012) (Table 8.1).

As already stated earlier, wheat microbiomes (phyllospheric, endophytic and rhizospheric) such as *Azotobacter*, *Achromobacter*, *Alcaligenes*, *Azospirillum*, *Enterobacter*, *Herbaspirillum*, *Methanospirillum*, *Klebsiella*, *Pantoea*, *Burkholderia*, *Bacillus*, *Paenibacillus*, *Lysinibacillus*, *Methylobacterium*, *Pseudomonas*, *Rhodosporidium*, *Serratia*, *Staphylococcus*, *Penicillium*, *Streptomyces* and *Thermomonospora* are isolated and characterized for plant growth promotion (Cakmakci et al. 2017).

8.7 Phosphate Solubilization and Mineralization

Phosphorus is an important plant major nutrient which follows nitrogen. Microorganisms undergo a sequence of processes that transform soil phosphorus (P), and thus they are considered as integral part of soil phosphorous cycle. But, application of huge amount of soluble inorganic phosphate into soil is fixing it as insoluble phosphate, and hence they become unavailable to the plants (Yadav et al. 2015a, b; Sharma et al. 2011a, b).

Soil microbes have the capacity to change insoluble forms to soluble forms which are desired by plants for absorption. This is done by producing organic acids such as acetic acid, lactic acid, glycolic acid, formic acid, propionic acid, succinic acid and fumaric acid. Plants absorb only inorganic P so that bacteria take action to hydrolyse the organic P compounds with the help of the phosphatase enzyme that originates from their roots (Yadav et al. 2015a, b). This enzyme also helps in

Table 8.1 Microbes from different wheat niche involved in plant growth regulation

Wheat microbe	Source and its application towards plant	References
Bacillus, Azospirillum, Azotobacter	Rhizosphere Nutrient and water uptake from soil	Verma and Suman (2018)
Azospirillum brasilense *Klebsiella pneumoniae*	Nitrogen-fixing microbe Gibberellin and IAA production	El-Razek and El-Sheshtawy (2013)
Pseudomonas, Azospirillum, Bacillus	PGPB Cytokinin production	Verma and Suman (2018)
Paenibacillus, Polymyxa, Acetobacter	Diazotrophs IAA and related compounds	Timmusk et al. (2014) and Aarab et al. (2015)
Pseudomonas sps.	Rhizosphere Phosphorous solubilization, siderophore, IAA, DAPG	Roesti et al. (2006)
Providencia sp. PW5	Rhizosphere HCN, IAA, P solubilization, Zn solubilization	Rana et al. (2012)
Acinetobacter calcoaceticus	Rhizosphere P solubilization, siderophore, IAA	Prashant et al. (2009)
Pseudomonas putida	Produces several types of antibiotics, siderophores and slight quantity of hydrogen cyanide (HCN)	Flaishman et al. (1996)
Azotobacter chroococcum *Pantoea agglomerans*	Gibberellic acid (GA) IAA	Narula et al. (2006)
Azorhizobium caulinodans	N fixation	Sabry et al. (1997)
Bacillus sp. (AW1) *Providencia* sp. (AW5) *Brevundimonas diminuta* (AW7)	P solubilization, N2 fixation, ACC deaminase, siderophore, ammonia, HCN	Rana et al. (2011)
Bacillus thuringiensis *Azotobacter chroococcum* *Paenibacillus ehimensis* *Pseudomonas pseudoalcaligenes*	Higher heavy metal resistance Siderophore, indole acetic acid, HCN, P solubilization	Kumar et al. (2015)
Pseudomonas spp.	IAA, P solubilization, rhamnolipids, siderophores	Mishra et al. (2009)
Bacillus sp.	Indole-3-acetic acid Antioxidant defence system SOD shoots and roots Shoot POD and CAT	Wang et al. (2013)
Pseudomonas denitrificans *Pseudomonas rathonis* *Azotobacter chroococcum* *Pantoea agglomerans*	Gibberellic acid (GA), IAA, and auxin	Narula et al. (2006) and Egamberdiyeva and Höflich (2003)

mineralization of organic phosphorous compounds. Through production of organic acids, chelation and exchange reactions process, the phosphorous-solubilizing bacteria convert insoluble inorganic P to a form available to plants (Yadav et al. 2015a, b). Mostly in soils, one can find a range of percentage of required amount of phytates which are organic P forms roughly 10–50% of phosphorous, and phytases should be mineralized to make ready supply of P to the plants (Singh et al. 2014). Reports showed that P solubilization is positive during their study on *Alcaligenes* sp., *Providencia* sp., *Bacillus* sp. and *Brevundimonas* (Verma and Suman 2018). Role of PGP microbes is very much observed in producing phosphatase, β-gluconase, dehydrogenase and antibiotics. Phosphate and other nutrient solubilization improved soil structure with stabilized soil aggregates (Verma and Suman 2018).

8.8 Biological Nitrogen Fixation

Through biological nitrogen fixation, 60% of available nitrogen on this earth is fixed (Verma et al. 2010). This fixation is alternative to chemical fertilizers and hence an economically beneficial and environmentally friendly alternative (Islam et al. 2002, Venieraki et al. 2011). The above-said process of nitrogen fixation is done by *nif* genes, i.e. coded form of the nitrogenase enzyme. In the case of nitrogen-poor soils, *Azospirillum* sp. is isolated, and it is diazotroph for nitrogen fixation (Hegazi et al. 1998; Verma and Suman 2018). Members of these bacterial genera have the ability in fixing atmospheric nitrogen. Symbiotic N_2-fixing *Rhizobium* and others like *Azospirillum* are some microbial groups which fix nitrogen by colonization of root zones. *Actinomycetes* in non-leguminous trees and free living N_2 fixers as blue green algae, *Bacillus*, *Acetobacter*, *Klebsiella*, *Azotobacter* and *Pseudomonas* are also helping nitrogen fixation (Prasanna et al. 2012).

8.9 Biological Control

To prevent the proliferation of plant pathogens, a variety of antibiotics are identified and are having compounds such as amphisin, 2,4-diacetylphloroglucinol (DAPG), phenazine, oomycin A, pyoluteorin, tensin, pyrrolnitrin, tropolone, cyclic lipopeptides produced by *Pseudomonads* and oligomycin A, kanosamine, zwittermicin A and xanthobaccin produced by *Bacillus*, *Streptomyces* and *Stenotrophomonas* sp. (Verma and Suman 2018).

Many bacterial genera such as *Agrobacterium* (Hammami et al. 2009), *Arthrobacter* (Banerjee et al. 2010), *Azotobacter* (Kannan and Sureender 2009), *Bradyrhizobium* (Akhtar et al. 2010; Mishra et al. 2009), *Bacillus*, *Enterobacter* (Wang et al. 2012; Solanki et al. 2012; Kushwaha et al. 2019; Singh et al. 2014), *Burkholderia* (Santiago et al. 2014) and *Pseudomonas* (Solanki et al. 2015; Goswami et al. 2013) have shown their potential in biocontrol under in vitro and in vivo conditions, and they are able to resist soilborne fungal pathogens (Table 8.2).

Table 8.2 Microbes involved in biological control

Microbiome genera	Antagonist against	References
Bacillus, *Enterobacter* sps.	Rusts of wheat	Wang et al. (2012) and Li et al. (2013)
Verticillium lecanii, *Erwinia herbicola*, *Pseudomonas aurantiaca*		Alien (1982), Srivastava (1985), Kempf and Wolf (1989), and Wang (2011)
Verticillium lecanii + *Paecilomyces fumosoroseus*, *Beauveria bassiana*		Hall (1981)
Pseudomonas putida		Flaishman et al. (1996) and Peng et al. (2014)
Trichoderma harzianum, *Streptomyces viridosporus*, *Bacillus subtilis*, *Saccharomyces cerevisiae*		Eldoksch et al. (2001), Kalappanavar et al. (2008) and El-Sharkawy et al. (2015)
Combined application of arbuscular mycorrhizal fungi and *Azospirillum amazonense*		Ghoneem et al. (2015)
T. harzianum	Tan spot, spot blotch and *Helminthosporium* leaf blight, *Fusarium* head blight of wheat and septoria blotch	Perelló et al. (1997), Monte (2001), Hossain et al. (2015), and Mahmoud (2016)
B. subtilis NJ-18	Sharp eye spot	Peng et al. (2014)
Combined application of yeasts like *Rhodosporidium kratochvilovae* strain UM350, *Cryptococcus laurentii* strain UM108 and *Aureobasidium pullulans* strain LS30	Powdery mildew of wheat	Curtis et al. (2012)
Bacillus subtilis strain E1R-j	Fusarium head blight disease of wheat and powdery mildew of wheat	Gao et al. (2015) and Mahmoud (2016)
Bacillus subtilis BTS 3, *B. amyloliquefaciens* BTS 4, *Staphylococcus saprophyticus* BTS 5 and *B. amyloliquefaciens* BTLK6A	Wheat blast and black point complex of wheat	Surovy et al. (2017) and El-Gremi et al. (2017)
B. subtilis D1/2 (DAOM 231163) and *B. subtilis* RC 218, *Brevibacillus* sp. RC 263, *Bacillus amyloliquefaciens*	*Fusarium* head blight of wheat; take all disease of wheat	Chan et al. (2003), Nasraoui et al. (2007), Crane et al. (2014), Palazzini et al. (2016) and Liu et al. (2009)
Lysobacter enzymogenes strain C3	*Fusarium* head blight of wheat	Jochum et al. (2006)
Cryptococcus, *Brevibacillus* sp. RC 263		Schisler et al. (2002, 2006, 2014)
Clonostachys rosea strain ACM941 (CLO-1) FHB		Xue et al. (2014)

(continued)

Table 8.2 (continued)

Microbiome genera	Antagonist against	References
Pseudomonas fluorescens (strains MKB 158 and MKB 249) and *P. frederiksbergensis* (strain 202)	Karnal bunt of wheat	Khan and Doohan (2009) and Vajpayee et al. (2015)
Aureobasidium pullulans		Wachowska and Głowacka (2014)
Trichoderma viride, *T. harzianum* and *Gliocladium deliquescens*		Sharma and Basandrai (2000)
Trichoderma pseudokoningii, *T. lignorum*, *T. koningii*, *G. deliquescens* and *G. virens*, *Azotobacter chroococcum*		Amer et al. (2000)
Trichoderma viride, *Gliocladium deliquescence*, *T. harzianum*, *Pseudomonas fluorescence* and *Bacillus subtilis*	Loose smut of wheat and black point complex of wheat	Agarwal and Nagarajan (1992), Singh and Maheshwari (2001), Monaco et al. (2004) and El-Meleigi et al. (2007)
Bacillus megaterium B5, *B. amyloliquefaciens* B28, *T. harzianum* T37 and *Epicoccum* sp. E52	Black point complex of wheat	El-Meleigi et al. (2007)
Streptomyces, *Bacillus*, *Pseudomonas fluorescens*, *P. putida*, *Pseudomonas chlororaphis* MA 342, *Gliocladium* and *Trichoderma harzianum*	Common bunt disease of wheat	Borgen and Davanlou (2000), McManus et al. (1993) and Kollmorgen and Jones 1975
Phialophora radicicola var. *radicicola*	Take all disease of wheat	Wong and Southwell (1980), Wong et al. (1996) and Mathre et al. (1998)
Trichoderma koningii, *T. harzianum*, *T. viride*		Simon and Sivasithamparam (1989) and Zafari et al. (2008)
Pandoraea apista (S18 and S19) and *Cylindrocarpon destructans* S22	*Rhizoctonia* root rot disease of wheat	Barnett et al. (2017)

Bacillus amyloliquefaciens is understood for lipopeptide and polyketide production for biological management and plant growth promotion activity against soil-borne pathogens. Some bacteria also are capable of producing volatile compound referred to as hydrogen cyanide (HCN) for biocontrol of black root rot of tobacco, caused by *Thielaviopsis basicola*; also reported is the production of DAPG and HCN by *Pseudomonas* contributing to the biological management of bacterial canker of tomato (Sacherer et al. 1994; Lanteigne et al. 2012).

Wheat microbiome bacteria can reduce virulence of a plant pathogenic fungus by altering histone acetylation (Chen et al. 2018). The molecular mechanisms between *Pseudomonas piscium* bacterium isolated from the wheat head microbiome and the plant pathogenic fungus *Fusarium graminearum* have revealed that a compound secreted by the bacteria (phenazine-1-carboxamide) directly affects the activity of fungal protein FgGcn5, which is a histone acetyltransferase of the

Spt-Ada-Gcn5-acetyltransferase (SAGA) complex which led to deregulation of histone acetylation in *F. graminearum* and subsequently suppression of fungal growth and virulence and biosynthesis of mycotoxin. Thus, it has been proven that an antagonistic bacterium can inhibit growth and virulence of a plant pathogenic fungus by manipulating fungal histone regulation (Chen et al. 2018).

According to Liu et al. (2017), effects of jasmonic acid signalling on the wheat microbiome differ between body sites. The influence of jasmonic acid (JA) on the diversity and function of wheat microbiome and whether there is any specificity of the influence with respect to plant part were determined. By exogenous application of methyl jasmonate, JA pathway was activated and was confirmed by significant increases in the abundance of JA signalling-related gene transcripts. Phylogenetic marker gene sequencing revealed that JA signalling reduced the diversity and changed the composition of root endophytic, but the composition of shoot endophytic or rhizospheric bacterial communities was unchanged. But the total enzymatic activity and substrate utilization profiles of rhizosphere bacterial communities were not affected by JA signalling. Therefore, the conclusion drawn after this study was that JA signalling on the wheat microbiome is specific to individual plant parts.

8.10 Wheat Microbiome Under Different Stress Conditions and Advance Research

Till date, the status of wheat microbiome research is at budding stage. However the research in the following aspects has provided insights, and the work presented here provided a way forward to study fundamental microbiome research with the aim of better understanding of identification of potentially beneficial microbes, their dynamics and their role in reducing the pathogen pressure and improving plant yield. To successfully reach the goals in the microbial research, the composition of plant-associated community that can fulfil the required need in plant disease management and in sustainable management has to be determined.

8.10.1 Performance of Microbiome Under Nutrient Stress Status

Very recently Page et al. (2019) described the microbiome profiling of *Triticum aestivum* under nitrogen- and phosphorus-starved situations. As wheat farming consumes approximately 20% of the worldwide production of inorganic N and P fertilizers and keeping in view the fact that plants have natural partnerships with microbes that can enhance their nitrogen (N) and/or phosphorus (P) acquisition, it is essntial to evaluate the association between wheat and its microbes in boosting N/P availability. They have tested whether N/P-starved *Triticum aestivum* showed microbiome profiles that are similarly different from those of N/P-amended plants in their own bulk soils. The conclusions drawn that six N/P starvation and plant specific microbial communities that may represent the attraction of microbes towards *T. aestivum* when it experiences N/P starvation. However, additional research will be needed to validate this interpretation under different climatic conditions. But the work done helps to keep a step forward. Identification of the potential *T. aestivum* partners give us a target list for subsequent relationship-assessing microbioata for boosting yields of the crop and therefore need to be standardized in farmers field with common package of practices. Microbiome enhances N and P uptake and experimently proved by Panwar and Singh (2000). *A. chroococcum* and *P. agglomerans* like microbe not only enhance P and K uptake but also increase plant growth and plant dry matter (Narula et al. 2007, Islam et al. 2002). Application of biofertilizers also reported to enhance the accumulation of nutrient in wheat crop by different scientists (Khan and Zaidi 2007; Abbasi and Yousra 2012). Not only macronutrient, but microbiomes also provide positive impact on the availability of essential micronutrients of wheat (Jog et al. 2014; Mishra et al. 2011; Yasin et al. 2015).

8.10.2 Performance of Microbiome Under Drought Condition

In the introduction, it is already mentioned that areas in South Asian countries are suffering from drought. Here microbiomes could be a good supplement to overcome this stress as they counteract with damage due to water stress. Rhizosphere biology mainly influences on the productivity of plants (Watt et al. 2006; Richardson et al. 2009 and Berendsen et al. 2012). Plant growth-promoting (PGP) strains of *Azospirillum* and *Herbaspirillum* have been reported to colonize *Brachypodium* roots and enhance growth of some *Brachypodium* genotypes under low or no nitrogen conditions (Amaral et al. 2016). Another group of scientists also used the PGP strain *Bacillus subtilis* B26 to increase *Brachypodium* biomass and improve plant drought resistance (Gagne et al. 2015). Also some microbiome provided a better water status in osmotic stress condition on wheat plant (Chakraborty et al. 2013; Pereyra et al. 2012). Artificial inoculation of *Azospirillum brasilense* sp. 245-primed wheat seed can sustain under water stress condition by increasing the apoplastic water function in both shoot and root compared to non-inoculated wheat plant (Creus et al. 2004).

8.10.3 Performance of Microbiome Under Stress Cultivation System

Nowadays governments are gearing towards conservation agriculture. So another selection criteria for microbiome that they can survive under stress cultural practices or minimal management approaches. A long-term field experiment demonstrates the influence of tillage on the bacterial potential to produce soil structure-stabilizing agents such as exopolysaccharides and lipopolysaccharides (Cania et al. 2019). The non-metric multidimensional scaling (NMDS) ordination plot showed a difference between the composition of bacterial families from the deepest sampled soil layer and the uppermost soil layers. It was revealed that there is no clear separation of the tillage treatments like the effects of tillage, depth and their interaction on the general microbial community structure. Even salicylic acid signalling in wheat microbiome also depends on soil type and on strategic tillage (Liu et al. 2016).

Another study by Hartman et al. (2018) revealed that cropping practices manipulate abundance patterns of root and soil microbiome members. The soil and wheat root microbiomes under conventional and organic managements with different tillage intensities were conducted to analyse the effects of management type and tillage on microbial communities as well as their infulence on various ecosystem processes and regulations. The study revelaed that the microbial communities play an important roles in nutrient cycling via decomposition of organic matter, and transformation and fixation of soil nutrients like nitrogen and phosphorus. It was revealed that there was pronounced influence of cropping patterns on microbial community composition which were specific for the respective microbiomes. In roots, management type was the important factor for bacteria when compared with fungi, and this is generally determined by changes in tillage intensity. There were many taxonomically diverse cropping-sensitive microbes, in which their response practices are specific. Cropping practices may allow manipulation of influential community members in which members co-occurring with many other microbes in the community. Understanding the patterns of cropping-sensitive microbe abundance helps in developing microbiota management strategies for smart farming.

Even the soil microbiomes perform better in non-suppressive soil than suppressive soil (Hayden et al. 2018). The comparative metatranscriptomic approach was applied to assess the taxonomic and functional characteristics of the rhizosphere microbiome of wheat plants grown in adjacent fields which are suppressive and non-suppressive to the plant pathogen *R. solani* AG8. Soils collected prior to sowing showed similar inoculum levels of the pathogen *R. solani* AG8 in both the suppressive and non-suppressive fields, as determined by quantitative PCR. The inoculum was observed in both fields throughout the cropping season, in particular during the initial 8 weeks of crop growth. The inoculum potential was more in the non-suppressive soils compared to that in the suppressive soils, resulting in significantly higher *R. solani* AG8 DNA in the non-suppressive soils, and the disease index was higher in non-suppressive soil compared to suppressive soils.

The status of wheat microbiome under different management strategies and potentiality of endophytes in disease protection were studied by Gdanetz and Trail (2017) in detail. They analysed the fungal and bacterial communities of leaves, stems and roots of wheat throughout the growing season across growth stage and four crop management strategies like organic, low input, no till and conventional using 16S and fungal internal transcribed spacer region gene amplicon sequencing, and endophytes were isolated from plants and are tested for antagonistic activity against wheat pathogen *Fusarium graminearum*, and also antagonistic strains were assessed for plant protective activity in seedling assays. But contrary to the expectations based on the previous studies on different management strategies, management strategy does not show a strong influence on the plant microbial communities.

A meta-analysis was done to determine whether plant domestication affects the composition of the root-associated microbiome (Jaramillo et al. 2018). The significant and consistent differences in the abundance of *Bacteroidetes*, *Actinobacteria* and *Proteobacteria* for different plant species were observed. The higher relative abundance of different bacterial families on the roots of wild relatives of the different plant species was observed. For conducting this study, it is emphasized that the same computational pipeline was used by adopting a rarefaction of the operational taxonomic unit (OTU) table to the same sequencing depth and also found similar differences between the prokaryotic composition of the rhizosphere microbiome of wild and domesticated plant species with a significantly higher abundance of *Bacteroidetes* in the roots of wild plant relatives. However, the reasons behind the relative abundance of *Bacteroidetes* in the root and rhizosphere compartments of wild relatives of various crop plant species are yet unknown. Therefore, their prevalence in the root compartments of wild plant species may be a phylogenetic signal associated with the presence of complex biopolymers in the root exudates (Table 8.3).

8.10.4 Role of Microbiome to Trigger Flowering

Sometimes microbiomes have strong influence on the crop physiology like timing of flowering (Lu et al. 2018). The rhizosphere microbial communities can modulate the timing of flowering of *Arabidopsis thaliana*. The abundance of rhizosphere microorganisms and prolonged nitrogen bioavailability by nitrification, delayed flowering by converting tryptophan to the indole acetic acid (IAA), thus downregulating genes that trigger flowering, and stimulating plant growth. Lu et al. (2018) also observed a novel metabolic network in which soil microbiota influenced plant flowering time, thus providing a water font on the key role of soil microbiota on plant functioning and thus helping to mitigate the effects of climate change and environmental stress on plants using microbiomes.

8.11 Future Prospectives and Conclusion

Over the last 150 years, plant pathologists only dealt with individual microbes that have adapted to specific niches on their hosts. Now we need to think more to perform high-throughput sequencing of these niches to explore microbial communities that can affect disease outcomes. A fundamental understanding of the plant microbiome is necessary for their successful execution amongst the farmers.

The rhizospheric microbiomes are regulated by root exudates, selection of plant genotype and adoption of environment. So, plant microbiome must be a part of future breeding programmes so that next-generation plant genotypes have enhanced ability to interact with beneficial microbes either of natural soil microbiota or of microbial inoculants. Worldwide many wheat cultivars that are biofortified, for

Table 8.3 Microbes involved in amelioration of various stresses

Wheat microbes	Stress	References
Azospirillum lipoferum	Drought	Naveed et al. (2014)
Bacillus thuringiensis	Heavy metal	Kumar et al. (2015)
Bacillus safensis, Ochrobactrum pseudogregnonense	Drought	Chakraborty et al. (2013)
Bacillus sp.	Cold stress	Sezen et al. (2016)
Bacillus aerophilus, Lysinibacillus sphaericus	Acid and alkalinity stress	Verma et al. (2016)
Azospirillum brasilense INTA Az-39 wheat roots	Drought	Diaz-Zorita and Fernández-Canigia (2009)
Bacillus amyloliquefaciens and *Azospirillum brasilense*	Temperature	Abd-Alla et al. (2013)
Pseudomonas fluorescens, Pantoea agglomerans, Mycobacterium sp.		Egamberdiyeva and Höflich (2003)
Pseudomonas fluorescens 153, 169, *Pseudomonas putida* 108	Salinity	
Achromobacter xylosoxidans, Serratia marcescens		Barra et al. (2014)
Pseudomonas putida N21, *Pseudomonas aeruginosa* N39 and *Serratia proteamaculans* M35		Zahir et al. (2009)
Azospirillum sp.		Nabti et al. (2010)
Pseudomonas putida, Pseudomonas extremorientalis, Pseudomonas chlororaphis and *Pseudomonas aurantiaca*		Abbaspoor et al. (2009)
Bacillus thuringiensis, Azotobacter chroococcum, Paenibacillus ehimensis, Pseudomonas pseudoalcaligenes	Heavy metal	Kumar et al. (2015)
Bacillus megaterium M3, *Bacillus subtilis* OSU142, *Azospirillum brasilense* Sp245, *Raoultella terrigena*	Cold	Turan et al. (2012)

example, rich in iron, rich in zinc, rich in anthocyanin, etc., are already released and also popular. In this scenario not only the microbial but also the host part of the interactions need to take more attentions, since the associated microbiome is emerging as a fundamental trait for controlling plant biotic and abiotic stress and optimizing plant growth promotion (Bulgarelli et al. 2013; Sclaeppi and Bulgerelli 2015). In the case of managing pathogen burden, the combined deployment of beneficial services of the plant microbiome (agricultural probiotics) and innate immune functions (resistance genes) is expected to deliver durable and sustainable protection from disease (Dangl et al. 2013). Identification of the genetic components of the host-microbiome control will be key factor for its ultimate inclusion into breeding programs.

Developing the next-generation agricultural tools and practices will be dependent on integration of all covariates present in agro-ecosystem. Under a specific agro-climatic condition, the performance of native soil microbiome and inoculated microbiome sometime interact with one or both with individual plant species and genotype, and this interaction drastickly influence the performance of the crop. Hence, we can exploit the use of plant microbiome at work and next-generation agriculture in reality to solve the aforesaid multivariate equations.

The ultimate goals of the researchers for the fate of the farmers are to minimize the use of fertilizers, protect plants from chemicals as well as reduce the cost of cultivation in the era of sustainable agriculture. But to reach this goal, we need to know the fundamental relationships between microbes and microbes and wheat plants and microbes and how much longer this relation may work under changing climatic conditions. Knowledge gain from this chapter will help to understand the selection criteria of microbes for alternative use of microbiomes as fertilizers, biocontrol agent, growth promoters and in efficenet soil nutrient cycling. But need more depth of researches for long duration under different cropping systems to translate microbiome concept in agriculture. Cropping practice and microbiome engineering would help only in sustainable agriculture system. Due to intensive cultivation, soil just become empty of nutrients and the physical and biological health of the existing soil hamphered badly. The microbiome is a fundamental part of basic processes such as plant development or growth of essential organs such as the root and for improved acquisition of nutrients and water; the mentioned capabilities make the microbiome an important component for the plant to carry out physiological functions. Analysis of metagenome and comparison with plant-associated communities will lead to novel phylogenetic and functional insight. The application of microbial inoculants to the field presents one of the several conceivable next-generation agricultural tools or practices.

References

Aarab S, Ollero JF, Megias M, Laglaoui A, Bakkali M, Arakrak A (2015) Isolation and screening of bacteria from rhizospheric soils of rice fields in Northwestern Morocco for different plant growth promotion (PGP) activities: an in vitro study. Int J Curr Microbiol App Sci 4(1):260–269

Abbasi M, Yousra M (2012) Synergistic effects of biofertilizer with organic and chemical N sources in improving soil nutrient status and increasing growth and yield of wheat grown under greenhouse conditions. Plant Biosyst Int J Deal Asp Plant Biol 146(sup1):181–189

Abbaspoor A, Zabihi HR, Movafegh S, Asl MA (2009) The efficiency of plant growth promoting rhizobacteria (PGPR) on yield and yield components of two varieties of wheat in salinity condition. Am Eurasian J Sustain Agric 3(4):824–828

Abd-Alla MH, El-Sayed E-SA, Rasmey A-HM (2013) Indole-3-acetic acid (IAA) production by Streptomyces atrovirens isolated from rhizospheric soil in Egypt. J Biol Earth Sci 3(2):B182–BB93

Agarwal R, Nagarajan S (1992) Possible biocontrol of loose smut of wheat (*Ustilago segetum* var. *tritici*). J Biol Control 6(2):114–115

Ahemad M, Kibret M (2014) Mechanisms and applications of plant growth promoting rhizobacteria: current perspective. J King Saud Univ Sci 26(1):1–20

Akbari GA, Arab SM, Alikhani H, Allakdadi I, Arzanesh M (2007) Isolation and selection of indigenous Azospirillum spp. and the IAA of superior strains effects on wheat roots. World J Agric Sci 3(4):523–529

Akhtar MS, Shakeel U, Siddiqui ZA (2010) Biocontrol of *Fusarium* wilt by *Bacillus pumilus*, *Pseudomonas alcaligenes* and *Rhizobium* sp. on lentil. Turk J Biol 34(3):1–7

Alien DJ (1982) *Verticillium lecanii* on the bean rust fungus, *Uromyces appendiculatus*. Trans Brit Myco Soc 55:173

Amaral FP, Pankievicz VCS, Arisi ACM, Souza EM, Pedrosa F, Stacey G (2016) Differential growth responses of *Brachypodium distachyon* genotypes to inoculation with plant growth promoting rhizobacteria. Plant Mol Biol 90(6):689–697. https://doi.org/10.1007/s11103-016-0449-8. PMID: 26873699

Amer GAM, Singh DV, Aggarwal R, Prem-Dureja (2000) Microbial antagonism to Neovossia indica, causing Karnal bunt of wheat. International conference on integrated plant disease management for sustainable agriculture. New Delhi, India, pp 1281

Arkhipova T, Veselov S, Melentiev A, Martynenko E, Kudoyarova G (2005) Ability of bacterium *Bacillus subtilis* to produce cytokinins and to influence the growth and endogenous hormone content of lettuce plants. Plant Soil 272(1–2):201–209

Asseng S, Ewert F, Martre P, Rotter RP, Lobell DB, Cammarano D (2015) Rising temperatures reduce global wheat production. Nat Clim Chang 5(2):143–147

Bacon, C. W. (1990). Isolation, culture and maintenance of endophytic fungi of grasses in: isolation of biotechnological organisms from nature. Ed. by D. P. Labeda, McGraw-Hill, New York

Banerjee S, Palit R, Sengupta C, Standing D (2010) Stress induced phosphate solubilisation by *Arthrobacter* sp. and *Bacillus* sp. isolated from tomato rhizosphere. Aust J Crop Sci 4(6):378–383

Barnett S, Zhao S, Ballard R, Franco C (2017) Selection of microbes for control of *Rhizoctonia* root rot on wheat using a high throughput pathosystem. Biol Control. 113: 45–57, https://doi.org/10.1016/j.biocontrol.2017.07.003

Barra P, Inostroza NG, Sobarzo JA, Mora ML (2014) Formulation of bacterial consortia from avocado (persea americana mill.) and their effect on growth, biomass and superoxide dismutase activity of wheat seedlings under salt stress. Appl Soil Ecol 102(2016):80–91

Berendsen RL, Pieterse CMJ, Bakker P (2012) The rhizosphere microbiome and plant health. Trends Plant Sci 17(8):478–486. https://doi.org/10.1016/j.tplants.2012.04.001

Berg G, Grube M, Schloter M, Smalla K (2014) Unraveling the plant microbiome: looking back and future perspectives. Front Microbiol 5:148

Bhattacharyya P, Jha D (2012) Plant growth-promoting rhizobacteria (PGPR): emergence in agriculture. World J Microbiol Biotechnol 28(4):1327–1350

Boe L, Danielsen M, Knudsen S, Petersen JB, Maymann J, Jensen PR (2000) The frequency of mutators in populations of *Escherichia coli*. Mutat Res 448:47–55. https://doi.org/10.1016/S0027-5107(99)00239-0

Borgen A, Davanlou M (2000) Biological control of common bunt in organic agriculture. J Crop Prod 3:159–174

Brader G, Compant S, Mitter B, Trognitz F, Sessitsch A (2014) Metabolic potential of endophytic bacteria. Curr Opin Biotechnol 27:30–37

Bulgarelli D, Schlaeppi K (2015) The plant microbiome at work. MPMI 28(3):212–217. https://doi.org/10.1094/MPMI-10-14-0334-FI

Bulgarelli D, Schlaeppi K, Spaepen S, van Themaat EVL, Schulze-Lefert P (2013) Structure and functions of the bacterial microbiota of plants. Annu Rev Plant Biol 64(1):807–838

Bulgarelli D, Rott M, Schlaeppi K, Ver Loren van Themaat E, Ahmadinejad N, Assenza F et al (2012) Revealing structure and assembly cues for Arabidopsis root-inhabiting bacterial microbiota. Nature 488:91–95. https://doi.org/10.1038/nature11336

Busby PE, Soman C, Wagner MR et al (2017) Research priorities for harnessing plant microbiomes in sustainable agriculture. PLoS Biol 15(3):e2001793

Cakmakci R, Turan M, Kitir N, Gunes A, Nikerel E, Ozdemir BS, Yildirim E, Olgun M, Topcuoglu B, Tufenkci S, Karaman MR, Tarhan L, Mokhtari NEP (2017) The role of soil beneficial bacteria in wheat production: a review. Intech Open 7. https://doi.org/10.5772/67274

Calvo P, Nelson L, Kloepper JW (2014) Agricultural uses of plant biostimulants. Plant Soil 383(1–2):3–41

Canbolat MY, Bilen S, Çakmakçı R, Şahin F, Aydın A (2006) Effect of plant growth-promoting bacteria and soil compaction on barley seedling growth, nutrient uptake, soil properties and rhizosphere microflora. Biol Fertil Soils 42(4):350–357

Cania B, Vestergaard G, Krauss M, Fliessbach A, Schloter M, Schulz S (2019) A long-term field experiment demonstrates the influence of tillage on the bacterial potential to produce soil structure-stabilizing agents such as exopolysaccharides and lipopolysaccharides. Environ Microbiome 14:1. https://doi.org/10.1186/s40793-019-0341

Carroll GC (1988) Fungal endophytes in stems and leaves: from latent pathogens to mutualistic symbiont. Ecology 69(1):2–9. https://doi.org/10.2307/1943154

Chakraborty U, Chakraborty BN, Chakraborty AP, Dey PL (2013) Water stress amelioration and plant growth promotion in wheat plants by osmotic stress tolerant bacteria. World J Microbiol Biotechnol 29(5):789–803. https://doi.org/10.1007/s11274-012-1234-8

Chatrath R (2004) Breeding strategies for developing wheat varieties targeted for rice-wheat cropping system of Indo-Gangetic plains of Eastern India. In: Joshi AK et al (eds) A compendium of the training program on wheat improvement in eastern and warmer regions of India: conventional and non-conventional approaches, 26–30 December, 2003, NATP project, (ICAR). BHU, Varanasi

Chaves MS, Martinelli JA, Wesp-Guterres C, Graichen FAS, Brammer SP, Scagliusi SM, Consoli L (2013) The importance for food security of maintaining rust resistance in wheat. Food Secur 5(2):157–176

Chan YK, McCormick WA, Seifert KA (2003) Characterization of an antifungal soil bacterium and its antagonistic activities against *Fusarium* species. Can J Microbiol 49:253–262

Chen Y, Wang J, Yang N, Wen Z, Sun X (2018) Wheat microbiome bacteria can reduce virulence of a plant pathogenic fungus by altering histone acetylation. Nat Commun 9:3429. https://doi.org/10.1038/s41467-018-05683-7

Clay K (1989) Clavicipitaceous endophytes of grasses: their potential as biocontrol agents. Mycol Res 92:1–12

Compant S, Clement C, Sessitsch A (2010) Plant growth-promoting bacteria in the rhizo- and endosphere of plants: Their role, colonization, mechanisms involved and prospects for utilization. Soil Biol Biochem 42:669–678

Crane JM, Frodyma ME, Bergstrome JC (2014) Nutrient-induced spore germination of a *Bacillus amyloliquefaciens* biocontrol agent on wheat spikes. J Appl Microbiol 116(6):1572-1583, https://doi.org/10.1111/jam.12480

Creus CM, Sueldo RJ, Barassi CA (2004) Water relations and yield in Azospirillum-inoculated wheat exposed to drought in the field. Can J Bot 82(2):273–281. Microbiology 160(4):778–788

Curtis FD, Cicco VD, Lima G (2012) Efficacy of biocontrol yeasts combined with calcium silicate or sulphur for controlling durum wheat powdery mildew and increasing grain yield components. Field Crops Res 134:36–46

Dangl JL, Horvath DM, Staskawicz BJ (2013) Pivoting the plant immune system from dissection to deployment. Science 341:746–751

De Curtis F, De Cicco V, Lima G (2012) Efficacy of biocontrol yeasts combined with calcium silicate or sulphur for controlling durum wheat powdery mildew and increasing grain yield components. Field Crop Res 134:36–46

Diaz-Zorita M, Fernández-Canigia MV (2009) Field performance of a liquid formulation of Azospirillum brasilense on dryland wheat productivity. Eur J Soil Biol 45(1):3–11

Denamur E, Matic I (2006) Evolution of mutation rates in bacteria. Mol Microbiol 60(4):820–827

Devi HM, Mahapatra S, Das S (2018) Assessment of yield loss of wheat caused by spot blotch using regression model. Indian Phytopathol 71:291–294

Dorworth CE, Callan BE (1996) Manipulation of endophytic fungi to promote their utility as vegetation biocontrol agents. In: Reddin SC, Carris LM (eds) Endophytic fungi in grasses and woody plants. APS Press, St. Paul

Egamberdiyeva D, Höflich G (2003) Influence of growth-promoting bacteria on the growth of wheat in different soils and temperatures. Soil Biol Biochem 35(7):973–978

El-Daim IAA, Bejai S, Meijer J (2014) Improved heat stress tolerance of wheat seedlings by bacterial seed treatment. Plant Soil 379(1–2):337–350

Eldoksch HA, Atteia MF, Abdel-Moity SMH (2001) Management of brown leaf rust, *Puccinia recondita* of wheat using natural products and biocontrol agents. Pak J Biol Sci 4:550–553

El-Gremi SM, Draz IS, Youssef WA (2017) Biological control of pathogens associated with kernel black point disease of wheat. Crop Prot 91:13–19

El-Meleigi MA, Hassan ZM, Ibrahim GH (2007) Biological control of common root rot of spring wheat by coating seeds with *Bacillus* or *Trichoderma* spp. in Central Saudi Arabia. JKAU: Met Env Arid Land Agric Sci 18:3–12

El-Razek UA, El-Sheshtawy A (2013) Response of some wheat varieties to bio and mineral nitrogen fertilizers. Asian J Crop Sci 5(2):200

El-Sharkawy HHA, Tohamey S, Khalil AA (2015) Combined effects of *Streptomyces viridosporus* and *Trichoderma harzianum* on controlling wheat leaf rust caused by *Puccinia triticina*. Plant Pathol J 14:182–188. https://doi.org/10.3923/ppj.2015.182.188

Farooq M, Wahid A, Kobayashi N, Fujita D, Basra S (2009) Plant drought stress: effects, mechanisms and management. Sustainable agriculture. Springer, pp 153–88

Federhen S (2014) Type material in the Ncbi Taxonomy Database. Nucleic Acids Res 1127

Fisher PJ, Andoson AE, Petrini O (1986) Fungal endophytes in *Ulex europaeus* and *Ulex gallii*. Trans Br Mycol Soc 86:153–156

Flaishman MA, Eyal Z, Zilberstein A, Voisard C, Haas D (1996) Suppression of *Septoria tritici* blotch and leaf rust of wheat by recombinant cyanide-producing strains of *Pseudomonas putida*. Mol Plant-Microbe Interact 9:642–664

Forchetti G, Masciarelli O, Alemano S, Alvarez D, Abdala G (2007) Endophytic bacteria in sunflower (*Helianthus annuus* L.): isolation, characterization, and production of jasmonates and abscisic acid in culture medium. Appl Microbiol Biotechnol 76(5):1145–1152

GagneA-Bourque F, Mayer BF, Charron JB, Vali H, Bertrand A, Jabaji S (2015) Acceleration growth rate and increased drought stress resilience of the model grass *Brachypodium distachyon* colonized by *Bacillus subtilis* B26. PLoS One 10(6). https://doi.org/10.1371/journal.pone.0130456

Gao X, Gong Y, Huo Y, Han Q, Kang Z, Huang L (2015) Endophytic *Bacillus subtilis* strain E1R-J is a promising biocontrol agent for wheat powdery mildew. Biomed Res Int 2015:462645. https://doi.org/10.1155/2015/462645

Gdanetz K, Trail F (2017) The wheat microbiome under four management strategies and potential for endophytes in disease protection. Phytobiomes 1:158–168

Ghoneem KM, Saber WIA, Youssef IAM, Mohamed MR, Al-Askar AA (2015) Postulation and efficiency of leaf rust resistance genes of wheat and biological control of virulence formulae of Puccinia *triticina* races. Egypt J Biol Pest Control 25(1):23–31

Glick BR (2012) Plant growth-promoting bacteria: mechanisms and applications. Scientifica 2012:1–15

Glick BR (2015) Beneficial plant-bacterial interactions. Springer, Cham, p 243

Glick BR, Penrose DM, Li J (1998) A model for the lowering of plant ethylene concentrations by plant growth-promoting bacteria. J Theor Biol 190(1):63–68

Goswami D, Vaghela H, Parmar S, Dhandhukia P, Thakker JN (2013) Plant growth promoting potentials of *Pseudomonas* spp. strain OG isolated from marine water. J Plant Interact 8 (4):281–290

Granzow S, Kaiser K, Wemheuer B, Pfeiffer B, Daniel R, Vidal S, Wemheuer F (2017) The effects of cropping regimes on fungal and bacterial communities of wheat and faba bean in a greenhouse pot experiment differ between plant species and compartment. Front Microbiol 8:902

Gupta G, Parihar SS, Ahirwar NK, Snehi SK, Singh V (2015) Plant growth promoting rhizobacteria (pgpr): current and future prospects for development of sustainable agriculture. J Microbial Biochem Technol 7:096–102. https://doi.org/10.4172/1948-5948.1000188

Hall RA (1981) The fungus *Verticillium lecanii* as a microbial insecticide against aphids and scales. In: Microbial control of pests and plant diseases. Academic, London, pp 483–498

Hallmann J, Hallmann AQ, Mahaffee WF, Kloepper JW (1997) Bacterial endophytes in agricultural crops. Can J Microbiol 43(10):895–914

Hammami I, Rhouma A, Jaouadi B, Rebai A, Nesme X (2009) Optimization and biochemical characterization of a bacteriocin from a newly isolated bacillus subtilis strain 14b for biocontrol of agrobacterium spp. Strains. Lett Appl Microbiol 48(2):253–260

Haney CH, Ausubel FM (2015) Plant microbiome blueprints. Science 349(6250):788–789

Haney CH, Samuel BS, Bush J, Ausubel FM (2015) Associations with rhizosphere bacteria can confer an adaptive advantage to plants. Nat Plants 1(6):15051. https://doi.org/10.1038/nplants.2015.51

Hartman K, Marcel GA, Heijden VD, Wittwer RA, Banerjee S, Walser JC, Schlaeppi K (2018) Cropping practices manipulate abundance patterns of root and soil microbiome members paving the way to smart farming. Microbiome 6:14. https://doi.org/10.1186/s40168-017-0389-9

Hartmann M, Frey B, Mayer J, Mader P, Widmer F (2014) Distinct soil microbial diversity under long-term organic and conventional farming. ISME J 9:1177–1194

Hassan TU, Bano A (2015) Role of carrier-based biofertilizer in reclamation of saline soil and wheat growth. Arch Agron Soil Sci 61(12):1719–1731

Hayden HL, Savin KW, Wadeson J, Gupta VVSR, Mele PM (2018) Metatranscriptomics of wheat rhizosphere microbiomes in disease suppressive and non-suppressive soils for *Rhizoctonia solani* AG8. Front Microbiol 9:859. https://doi.org/10.3389/fmicb.2018.00859

Hegazi N, Fayez M, Amin G, Hamza M, Abbas M, Youssef H (1998) Diazotrophs associated with non-legumes grown in sandy soils. In: Nitrogen fixation with non-legumes. Springer, Dordercht, pp 209–222

Hirsch PR, Mauchline T (2014) Who's who in the plant root microbiome? Nat Biotechnol 30(10):961–962. https://doi.org/10.1038/nbt.2387

Hossain MM, Hossain I, Khalequzzaman KM (2015) Effect of seed treatment with biological control agent against *Bipolaris* leaf blight of wheat. Int J Sci Res Agric Sci 2(7):151–158

Huang Y, Kuang Z, Wang W, Cao L (2016) Exploring potential bacterial and fungal biocontrol agents transmitted from seeds to sprouts of wheat. Biol Control 98:27–33

Hug LA, Baker BJ, Anantharaman K, Brown CT, Probst AJ, Castelle CJ et al (2016) A new view of the tree of life. Nat Microbiol 1:16048. https://doi.org/10.1038/nmicrobiol.2016.48

Iniguez AL, Dong Y, Triplett EW (2004) Nitrogen fixation in wheat provided by *Klebsiella pneumoniae* 342. Mol Plant-Microbe Interact 17(10):1078–1085

Islam N, Rao C, Kennedy I (2002) Facilitating a N_2-fixing symbiosis between diazotrophs and wheat. In: Biofertilisers in action. Rural Industries Research and Development Corporation, Canberra, pp 84–93

Jankiewicz U (2006) Synthesis of siderophores by soil bacteria of the genus *Pseudomonas* under various culture conditions. Acta Sci Pol Agric 5(2):s. 33–s. 44

Jaramillo JEP, Carrion VJ, Hollander MD, Raaijmakers JM (2018) The wild side of plant microbiomes. Microbiome 6:143. https://doi.org/10.1186/s40168-018-0519-z260-269

Jasrotia P, Kashyap PL, Bhardwaj AK, Kumar S, Singh GP (2018) Scope and applications of nanotechnology for wheat production: a review of recent advances. Wheat Barley Res 10(1):1–14

Jochum CC, Osborne LE, Yuen GY (2006) Fusarium head blight biological control with *Lysobacter enzymogenes* strain C3. Biol Control 39:336–344

Jog R, Pandya M, Nareshkumar G, Rajkumar S (2014) Mechanism of phosphate solubilization and antifungal activity of *Streptomyces* spp. isolated from wheat roots and rhizosphere and their application in improving plant growth. Microbiology 160:778–788

Joshi AK, Chand R, Arun B, Singh RP, Ortiz R (2007) Breeding crops for reduced-tillage management in the intensive, rice-wheat systems of south Asia. Euphytica 153:135–151

Kalappanavar IK, Patidar RK, Kulkarni S (2008) Management strategies of leaf rust of wheat caused by *Puccinia recondita* f. sp. *tritici* Rob. ex. Desm. Karnataka J Agric Sci 21(1):61–64

Kannan V, Sureender R (2009) Synergistic effect of beneficial rhizosphere microflora in biocontrol and plant growth promotion. J Basic Microbiol 49(2):158–164

Karlsson I, Friberg H, Kolseth AK, Steinberg C (2017) Organic farming increases richness of fungal taxa in the wheat phyllosphere. Mol Ecol 2017:1–13

Kashyap PL, Jasrotia P, Kumar S, Singh DP, Singh GP (2018) Identification guide for major diseases and insect-pests of wheat. Technical bulletin 18. ICAR Indian Institute of Wheat and Barley Research, Karnal, India, p 38

Kashyap PL, Kaur S, Pannu PPS (2019) Expression analysis of pathogenesis related proteins induced by compatible and incompatible interactions of *Tilletia indica* in wheat plants. J Mycol Plant Pathol 49(1):39–47

Kashyap PL, Kaur S, Pannu PPS (2018) Induction of systemic tolerance to *Tilletia indica* in wheat by plant defence activators . Arch Phytopathol Plant Prot 51(1-2):1–13

Kashyap PL, Kaur S, Sanghera GS, Kang SS, Pannu PPS (2011) Novel methods for quarantine detection of Karnal bunt (*Tilletia indica*) of wheat. Elixir Agric 31:1873–1876

Kempf HJ, Wolf G (1989) *Erwinia herbicola* as a biocontrol agent of *Fusarium culmorum* and *Puccinia recondita* f. sp. *tritici* on wheat. Phytopathology 79(9):990–994

Khalafallah AA, Abo-Ghalia HH (2008) Effect of arbuscular mycorrhizal fungi on the metabolic products and activity of antioxidant system in wheat plants subjected to shorter water stress, followed by recovery at different growth stages. J Appl Sci Res 4(5):559–569

Khalid A, Arshad M, Zahir Z (2004) Screening plant growth-promoting rhizobacteria for improving growth and yield of wheat. J Appl Microbiol 96(3):473–480

Khan MR, Doohan FM (2009) Bacterium-mediated control of Fusarium head blight disease of wheat and barley and associated mycotoxin contamination of grain. Biol Control 48(1):42–47

Khan MS, Zaidi A (2007) Synergistic effects of the inoculation with plant growth-promoting rhizobacteria and an arbuscular mycorrhizal fungus on the performance of wheat. Turk J Agric For 31(6):355–362

Khan MS, Zaidi A, Wani PA (2007) Role of phosphate solubilizing microorganisms in sustainable agriculture-a review. Agron Sustain Dev 27:29–43

Khan H, Bhardwaj SC, Gangwar OP, Prasad P, Kashyap PL, Savadi PLS, Kumar S, Rathore R (2017) Identifying some additional rust resistance genes in Indian wheat varieties using robust markers. Cereal Res Commun 45(4):633–646

Kibota TT, Lynch M (1996) Estimate of the genomic mutation rate deleterious to overall fitness in *E. coli*. Nature 381(6584):694–696. https://doi.org/10.1038/381694a0

Kollmorgen JE, Jones LC (1975) The effects of soil-borne micro-organisms on the germination of chlamydospores of *Tilletia caries* and *T. foetida*. Soil Biol Biochem 7:407–410

Kudoyarova GR, Melentiev AI, Martynenko EV, Timergalina LN, Arkhipova TN, Shendel GV (2014) Cytokinin producing bacteria stimulate amino acid deposition by wheat roots. Plant Physiol Biochem 83:285–291

Kumar V, Behl RK, Narula N (2001) Establishment of phosphate-solubilizing strains of Azotobacter chroococcum in the rhizosphere and their effect on wheat cultivars under greenhouse conditions. Microbiol Res 156(1):87–93

Kumar V, Singh S, Singh J, Upadhyay N (2015) Potential of plant growth promoting traits by bacteria isolated from heavy metal contaminated soils. Bull Environ Contam Toxicol 94(6):807–814

Kumar S, Singroha G, Bhardwaj SC, Bala R, Saharan MS, Gupta V, Khan A, Mahapatra S, Sivasamy M, Rana V, Mishra CN, Prakash O, Verma A, Sharma P, Sharma I, Chatrath R, Singh GP (2019) Multienvironmental evaluation of wheat (*Triticum aestivum* L.) germplasm identifies donors with multiple fungal disease resistance. Genet Resour Crop Evol 66:797–808

Kumar S, Kashyap PL, Singh GP (2020) Wheat blast. CRC Press Boca Raton, pp 214, https://doi.org/10.1201/9780429470554

Kushwaha P, Kashyap PL, Srivastava AK, Tiwari RK (2020a) Plant growth promoting and antifungal activity in endophytic *Bacillus* strains from pearl millet (*Pennisetum glaucum*). Brazilian Journal of Microbiology 51(1):229–241

Kushwaha P, Kashyap PL, Bhardwaj AK, Kuppusamy P, Srivastava AK, Tiwari RK (2020b) Bacterial endophyte mediated plant tolerance to salinity: growth responses and mechanisms of action. World Journal of Microbiology and Biotechnology 36:26. https://doi.org/10.1080/11263504.2019.1651773

Kushwaha P, Kashyap PL, Kuppusamy P, Srivastava AK, Tiwari RK (2019) Functional characterization of endophytic bacilli from pearl millet (*Pennisetum glaucum*) and their possible role in multiple stress tolerance . Plant Biosystems - An International Journal Dealing with all Aspects of Plant Biology:1-12, https://doi.org/10.1080/11263504.2019.1651773

Kuzmina LY, Melentev AI (2003) The effect of seed bacterization by *Bacillus* Cohn bacteria on their colonization of the spring wheat rhizosphere. Microbiology 72:230–235

Lanteigne C, Gadkar VJ, Wallon T, Novinscak A, Filion M (2012) Production of DAPG and HCN by *Pseudomonas* sp. LBUM300 contributes to the biological control of bacterial canker of tomato. Phytopathology 102(10):967–973

Li H, Zhao J, Feng H, Huang L, Kang Z (2013) Biological control of wheat stripe rust by an endophytic *Bacillus subtilis* strain E1R-j in greenhouse and field trials. Crop Prot 43:201–206

Liu B, Qiao H, Huang L, Buchenauer H, Han Q, Kang Z, Gong Y (2009) Biological control of take-all in wheat by endophytic *Bacillus subtilis* E1R-j and potential mode of action. Biol Control 49(3):277–285

Liu X, Jia J, Atkinson S, Camara M, Gao K (2010) Biocontrol potential of an endophytic *Serratia* sp. G3 and its mode of action. World J Microbiol Biotechnol 26(8):1465–1471

Liu W, Liu J, Triplett L, Leach JE, Wang GL (2014) Novel insights into rice innate immunity against bacterial and fungal pathogens. Annu Rev Phytopathol 52:213–241. https://doi.org/10.1146/annurev-phyto-102313-045926

Liu H, Caravalhais LC, Schenk PM (2016) Effects of Salicylic acid signaling on Wheat microbiome are dependent on soil type, Effects of strategic tillage and plant hormone treatments on Wheat associated microbial communities. The University of Queensland, pp 148–168

Liu H, Carvalhais LC, Schenk PM, Dennis PG (2017) Effects of jasmonic acid signaling on the wheat microbiome differ between body sites. Sci Rep 7:41766. https://doi.org/10.1038/srep41766

Lobell DB, Schlenker W, Costa-Roberts J (2011) Climate trends and global crop production since 1980. Science 333:616–620

Lu T, Ke M, Lavoie M (2018) Rhizosphere microorganisms can influence the timing of plant flowering. Microbiome 6:231

Lundberg DS, Lebeis SL, Paredes S, Yourstone S (2012) Defining the core *Arabidopsis thaliana* root microbiome. Nature 488:86–90

Mahmoud AF (2016) Genetic variation and Biological control of *Fusarium graminearum* isolated from wheat in Assiut-Egypt. Plant Pathol J 32(2):145–156. https://doi.org/10.5423/PPJ.OA.09.2015.0201

Mahoney AK, Yin C, Hulbert SH (2017) Community structure, species variation, and potential functions of rhizosphere-associated bacteria of different winter wheat (*Triticum aestivum*) cultivars. Front Plant Sci 8:132

Marasco R, Rolli E, Ettoumi B, Vigani G, Mapelli F et al (2012) A drought resistance-promoting microbiome is selected by root system under desert farming. PLoS One 7(10):e48479. https://doi.org/10.1371/journal.pone.0048479

Marshall D, Tunali B, Nelson LR (1999) Occurrence of fungal endophytes in species of wild *Triticum*. Crop Sci 39(5):1507–1512

Mathre DE, Johnston RH, Grey WE (1998) Biological control of take-all disease of wheat caused by *Gaeumannomyces graminis* var. *tritici* under field conditions using a *Phialophora* sp. Biocontrol Sci Tech 8(3):449–457. https://doi.org/10.1080/09583159830243

McDonald BR, Currie CR (2017) Lateral gene transfer dynamics in the ancient bacterial genus Streptomyces. MBio 8:e00644-17. https://doi.org/10.1128/mBio.00644-17

McManus PS, Ravenscrofi AV, Fulbright DW (1993) Inhibition of *Tilletia laevis* teliospore germination and suppression of common bunt of wheat by *Pseudomonas fluorescens* 2–79. Plant Dis 77:1012–1015

Mehnaz S (2015) Azospirillum: a biofertilizer for every crop. In: Plant microbes symbiosis: applied facets. Springer, New Delhi, pp 297–314

Mishra PK, Mishra S, Bisht SC, Selvakumar G, Kundu S, Bisht J (2009) Isolation, molecular characterization and growth-promotion activities of a cold tolerant bacterium *Pseudomonas* sp. NARs9 (MTCC9002) from the Indian Himalayas. Biol Res 42(3):305–313

Mishra PK, Bisht SC, Ruwari P, Selvakumar G, Joshi GK, Bisht JK (2011) Alleviation of cold stress in inoculated wheat *(Triticum aestivum* L.) seedlings with psychrotolerant Pseudomonads from NW Himalayas. Arch Microbiol 193(7):497–513

Monaco C, Sisterna M, Perello A, Bello GD (2004) Preliminary studies on biological control of the blackpoint complex of wheat in Argentina. World J Microbiol Biotechnol 20:285–290

Monte E (2001) Underestanding *Trichoderma*: between biotechnology and microbial ecology. Int Microbiol 4:1–4

Moore B (2014) Life in our universe elephants in space. Springer, Dordrecht, pp 99–113

Mueller UG, Sachs JL (2015) Engineering microbiomes to improve plant and animal health. Trends Microbiol 23(10):606–617. https://doi.org/10.1016/j.tim.2015.07.009

Murray FR, Latch GCM, Scott DB (1992) Surrogate transformation of perennial rye grass, Lolium perenne, using genetically modified Acremonium endophyte. Mol Gen Genet 233(1–2):1–9

Nabti E, Sahnoune M, Ghoul M, Fischer D, Hofmann A, Rothballer M, Schmid M, Hartmann A (2010) Restoration of growth of durum wheat (*Triticum durum* var. waha) under saline conditions due to inoculation with the rhizosphere bacterium Azospirillum brasilense NH and extracts of the marine alga *Ulva lactuca*. J Plant Growth Regul 29:6–22

Nallathambi P, Maheswari CU, Singh DP, Watpade S , Kashyap PL, Aarthy B, Ravikumar P, Sharma A, Kumar R (2019) Rapid nass culturing method of wheat (*Triticum* species) powdery mildew pathogen [*Blumeria graminis* (DC) Speer f. sp. *tritici* Marchal under controlled conditions at Wellington. J Mycol Plant Pathol 49(2):211–216

Narula N, Deubel A, Gans W, Behl R, Merbach W (2006) Paranodules and colonization of wheat roots by phytohormone producing bacteria in soil. Plant Soil Environ 52(3):119

Narula N, Remus R, Deubel A, Granse A, Dudeja S, Behl R (2007) Comparison of the effectiveness of wheat roots colonization by Azotobacter chroococcum and Pantoea agglomerans using serological techniques. Plant Soil Environ 53(4):167

Nasraoui B, Hajlaoui MR, Aïssa AD, Kremer RJ (2007) Biological control of wheat take-all disease: I – characterization of antagonistic bacteria from diverse soils toward *Gaeumannomyces graminis* var. *tritici*. Tunis J Plant Prot 2:23–34

Naveed M, Hussain MB, Zahir ZA, Mitter B, Sessitsch A (2014) Drought stress amelioration in wheat through inoculation with *Burkholderia phytofirmans* strain PsJN. Plant Growth Regul 73(2):121–131

Nelson EB (2018) The seed microbiome: origins, interactions, and impacts. Plant Soil 422:7–34

Ofek M, Voronov-Goldman M, Hadar Y, Minz D (2013) Host signature effect on plant root-associated microbiomes revealed through analyses of resident vs. active communities. Environ Microbiol 16:2157–2167

Page AP, Tremblay J, Masson L, Greer CW (2019) Nitrogen- and phosphorus-starved *Triticum aestivum* show distinct belowground microbiome profile. PLoS One 14(2):e0210538. https://doi.org/10.1371/journal.pone.0210538

Palazzini JM, Alberione E, Torres A, Donat C, Köhl J, Chulze S (2016) Biological control of *Fusarium graminearum* sensu stricto, causal agent of *Fusarium* head blight of wheat, using formulated antagonists under field conditions in Argentina. Biol Control 94:56–61

Panwar J, Singh O (2000) Response of Azospirillum and Bacillus on growth and yield of wheat under field conditions. Indian J Plant Physiol 5(1):108–10.513

Peiffer JA, Spor A, Koren O (2013) Diversity and heritability of the maize rhizospheric microbiome under field conditions. Proc Natl Acad Sci 110(16):6548–6553

Peng D, Li S, Chen C, Zhou M (2014) Combined application of Bacillus subtilis NJ-18 with fungicides for control of sharp eyespot of wheat. Biol Control 70:28–34

Perelló A, Mónaco C, Cordo C (1997) Evaluation of *Trichoderma harzianum* and *Gliocladium roseum* in controlling leaf blotch of wheat (Septoria tritici) under *in vitro* and greenhouse conditions. Z Pflanzenkr Pflanzenschutz 104:588–598

Pereyra M, Garcia P, Colabelli M, Barassi C, Creus C (2012) A better water status in wheat seedlings induced by Azospirillum under osmotic stress is related to morphological changes in xylem vessels of the coleoptile. Appl Soil Ecol 53:94–97

Pleban S, Ingel F, Chet I (1995) Control of *Rhizoctonia solani* and *Sclerotium rolfsii* in the green house using endophytic *Bacillus* spp. Eur J Plant Pathol 101(6):665–667

Prasanna R, Nain L, Pandey AK (2012) Microbial diversity and multidimensional interactions in the rice ecosystem. Arch Agron Soil Sci 58(7):723–744

Prashant S, Makarand R, Bhushan C, Sudhir C (2009) Siderophoregenic *Acinetobacter calcoaceticus* isolated from wheat rhizosphere with strong PGPR activity. Malays J Microbiol 5(1):6–12

Rana A, Saharan B, Joshi M, Prasanna R, Kumar K, Nain L (2011) Identification of multi-trait PGPR isolates and evaluating their potential as inoculants for wheat. Ann Microbiol 61:893–900

Rana A, Joshi M, Prasanna R, Shivay YS, Nain L (2012) Biofortification of wheat through inoculation of plant growth promoting rhizobacteria and cyanobacteria. Eur J Soil Biol 50:118–126

Richardson AE, Barea JM, McNeill AM, Prigent-Combaret C (2009) Acquisition of phosphorus and nitrogen in the rhizosphere and plant growth promotion by microorganisms. Plant Soil 321(1–2):305–339. https://doi.org/10.1007/s11104-009-9895

Roat, C., Saraf, M. (2017) Isolation and screening of resveratrol producing endophytes from wild grape *Cayratia trifolia*. Int J Adv Agric Sci Technol 4(11):27–33

Roesti D, Gaur R, Johri B, Imfeld G, Sharma S, Kawaljeet K (2006) Plant growth stage, fertiliser management and bio-inoculation of arbuscular mycorrhizal fungi and plant growth promoting rhizobacteria affect the rhizobacterial community structure in rainfed wheat fields. Soil Biol Biochem 38(5):1111–1120

Rroco E, Kosegarten H, Harizaj F, Imani J, Mengel K (2003) The importance of soil microbial activity for the supply of iron to sorghum and rape. Eur J Agron 19(4):487–493

Sabry SR, Saleh SA, Batchelor CA, Jones J, Jotham J, Webster G (1997) Endophytic establishment of Azorhizobium caulinodans in wheat. Proc R Soc Lond B Biol Sci 264(1380):341–346

Sachdev DP, Chaudhari HG, Kasture VM, Dhavale DD, Chopade BA (2009) Isolation and characterization of indole acetic acid (IAA) producing *Klebsiella pneumoniae* strains from rhizosphere of wheat (*Triticum aestivum*) and their effect on plant growth. Indian J Exp Biol 47(12):993

Sacherer P, Défago G, Haas D (1994) Extracellular protease and phospholipase C are controlled by the global regulatory gene gacA in the biocontrol strain *Pseudomonas fluorescens* CHA0. FEMS Microbiol Lett 116:155–160

Santiago TR, Grabowski C, Rossato M (2014) Biological control of eucalyptus bacterial wilt with rhizobacteria. Biol Control 80:14–22

Santoyo G, Orozco-Mosqueda Mdel C, Govindappa M (2012) Mechanisms of biocontrol and plant growth-promoting activity in soil bacterial species of *Bacillus* and *Pseudomonas*: a review. Biocontrol Sci Tech 22(8):855–872

Santoyo G, Moreno-Hagelsieb G, Glick BR (2016) Plant growth-promoting bacterial endophytes. Microbiol Res 183:92–99

Sarode Prashant D, Rane Makarand R, Chaudhari Bhushan L, Chincholkar SB (2009) Siderophoregenic Acinetobacter calcoaceticus isolated from wheat rhizosphere with strong PGPR activity. Malays J Microbiol 5(1):6–12

Sarode P, Rane M, Kadam M, Chincholkar S (2013) Role of microbial siderophores in improving crop productivity in wheat. In: Bacteria in agrobiology: crop productivity. Springer, Berlin, pp 287–308

Savadi S, Prasad P, Kashyap PL, Bhardwaj SC (2018) Molecular breeding technologies and strategies for rust resistance in wheat (*Triticum aestivum*) for sustained food security. Plant Pathol 67(4):771–791

Scavino AF, Pedraza RO (2013) The role of siderophores in plant growth-promoting bacteria. In: Bacteria in agrobiology: crop productivity. Springer, Berlin, pp 265–285

Schardl CL, Liu J, White JK, Finkel RA, An Z, Siegel M (1991) Molecular phylogenetic relationship of nonpathogenic grass mycosymbionts and clavicipitaceous plant pathogens. Plant Syst 178(1–2):27–41

Schisler DA, Khan NI, Boehm MJ, Slininger PJ (2002) Greenhouse and field evaluation of biological control of *Fusarium* head blight on durum wheat. Plant Dis 86(12):1350–1356

Schisler DA, Khan NI, Boehm MJ, Lipps PE, Slininger PJ, Zhang S (2006) Selection and evaluation of the potential of choline-metabolizing microbial strains to reduce *Fusarium* head blight. Biol Control 39:497–506

Schisler DA, Core AB, Boehm MJ, Horst L, Krause C, Dunlap CA, Rooney AP (2014) Population dynamics of the *Fusarium* head blight biocontrol agent *Cryptococcus flavescens* OH 182.9 on wheat anthers and heads. Biol Control 70:17–27

Schlaeppi K, Bulgarelli D (2015) The plant microbiome at work. Mol Plant-Microbe Interact 28(3):212–217

Sergaki C, Lagunas B, Lidbury I, Gifford ML, Schafer P (2018) Challenges and approaches in microbiome research: from fundamental to applied. Front Plant Sci 9:1205. https://doi.org/10.3389/fpls.2018.01205

Sezen A, Ozdal M, Koc K, Algur OF (2016) Isolation and characterization of plant growth promoting rhizobacteria (PGPR) and their effects on improving growth of wheat. J Appl Biol Sci 10(1):41–46

Sharma BK, Basandrai AK (2000) Effectiveness of some fungicides and biocontrol agents for the management of Karnal bunt of wheat. Indian J Mycol Plant Pathol 30:76–78

Sharma SK, Johri BN, Ramesh A, Joshi OP, Prasad SS (2011a) Selection of plant growth-promoting *Pseudomonas* spp. that enhanced productivity of soybean-wheat cropping system in Central India. J Microbiol Biotechnol 21:1127–1142

Sharma S, Kumar V, Tripathi RB (2011b) Isolation of phosphate solubilizing microorganism (psms) from soil. J Microbiol Biotechnol Res 1(2):90–95

Simon A, Sivasithamparam K (1989) Pathogen suppression: a case study in biological suppression of *Gaeumannomyces graminis* var. *tritici* in soil. Soil Biol Biochem 21:331–337

Singh DP (2017) Management of wheat and barley diseases. Apple Academic Press, Waretown, p 682. ISBN 9781771885478

Singh D, Maheshwari VK (2001) Biological seed treatment for the control of loose smut of wheat. Indian Phytopathol 54(4):457–460

Singh KN, Chatrath R (2001) Salinity tolerance. In: Reynolds MP, Ortiz-Monasterio JI, McNab A (eds) Application of physiology in wheat breeding. CIMMYT, Mexico, pp 101–110

Singh S, Mustard (2012) India grain and feed annual grain report number IN2026. USDA Foreign Agricultural Services, Washington, DC

Singh RP, Hodson DP, Jin Y, Huerta-Espino J, Kinyua MG, Wanyera R, Njau P, Ward RW (2006) Current status, likely migration and strategies to mitigate the threat to wheat production from race *Ug99* (TTKS) of stem rust pathogen. CAB Rev 1(054):1–13. http://www.cababstractsplus.org/cabreviews

Singh R, Behl R, Jain P, Narula N, Singh K (2007) Performance and gene effects for root characters and micronutrient uptake in wheat inoculated with arbuscular mycorrhizal fungi and *Azotobacter chroococcum*. Acta Agronomica Hungarica 55(3):325–330

Singh RK, Kumar DP, Singh P, Solanki MK, Srivastava S, Kashyap PL, Kumar S, Srivastava AK, Singhal PK, Arora DK (2014) Multifarious plant growth promoting characteristics of chickpea rhizosphere associated Bacilli help to suppress soil-borne pathogens. Plant Growth Regul 73(1):91–101

Singh P, Kumar V, Agrawal S (2014) Evaluation of phytase producing bacteria for their plant growth promoting activities. Int J Microbiol. https://doi.org/10.1155/2014/426483

Solanki MK, Kumar S, Pandey AK, Srivastava S, Singh RK, Kashyap PL, Srivastava AK, Arora DK (2012) Diversity and antagonistic potential of *Bacillus* spp. associated to the rhizosphere of tomato for the management of *Rhizoctonia solani*. Biocontrol Sci Technol 22(2):203–217

Solanki MK, Singh RK, Srivastava S, Kumar S, Kashyap PL, Srivastava AK, Arora DK (2014) Isolation and characterization of siderophore producing antagonistic rhizobacteria against *Rhizoctonia solani*. J Basic Microbiol 54(6):585–597

Solanki MK, Singh RK, Srivastava S, Kumar S, Kashyap PL, Srivastava AK (2015) Characterization of antagonistic-potential of two *Bacillus* strains and their biocontrol activity against *Rhizoctonia solani* in tomato. J Basic Microbiol 55(1):82–90

Srivastava AK, Defago G, Kern H (1985) Hyperparasitism of *Puccinia horiana* and other microcyclic rusts. Phytopathol Z 114:73

Stierle A, Strobel G, Stierle D (1993) Taxol and taxane production by Taxomyces andreanae and endophytic fungus of Pacific Yew. Science 260:214–216

Strobel G, Daisy B (2003) Bioprocessing for microbial endophytes and their natural products. Microbiol Mol Biol Rev 67:491–502

Suman A, Verma P, Yadav AN et al (2016a) Development of hydrogel-based bio-inoculant formulations and their impact on plant biometric parameters of wheat (*Triticum aestivum* L.). Int J Curr Microbiol App Sci 5(3):890–901

Suman A, Verma P, Yadav AN et al (2016b) Endophytic microbes in crops: diversity and beneficial impact for sustainable agriculture. In: Singh D, Singh H, Prabha R (eds) Microbial inoculants in sustainable agricultural productivity. Springer, New Delhi, pp 117–143. https://doi.org/10.1007/978-81-322-2647-5-7

Surovy MZ, Gupta DR, Chanclud E, Win J, Kamoun S, Tofazzal I (2017) Plant probiotic bacteria suppress wheat blast fungus *Magnaporthe oryzae Triticum* pathotype. https://doi.org/10.6084/m9.figshare.5549278.v1

Tan RX, Zou WX (2001) Endophytes: a rich source of functional metabolites. Nat Prod Rep 18:448–459

Timmusk S, Abd El-Daim IA, Copolovici L, Tanilas T, Kannaste A (2014) Drought-tolerance of wheat improved by rhizosphere bacteria from harsh environments: enhanced biomass production and reduced emissions of stress volatiles. PLoS One 9(5):e96086. https://doi.org/10.1371/journal.pone.0096086

Tkacz A, Cheema J, Chandra G, Grant A, Poole PS (2015) Stability and succession of the rhizosphere microbiota depends upon plant type and soil composition. ISME J 9:2349–2359. https://doi.org/10.1038/ismej.2015.41

Tsvetanov T, Qi L, Mukherjee D, Shah F, Bravo-Ureta B (2016) Climate change and land use in Southeastern US: did the "dumb farmer" get it wrong? Clim Change Econ 7(03):1650005

Turan M, Gulluce M, Şahin F (2012) Effects of plant-growth-promoting rhizobacteria on yield, growth, and some physiological characteristics of wheat and barley plants. Commun Soil Sci Plant Anal 43(12):1658–1673

Turner TR, James EK, Poole PS, Gilbert J, Meyer F, Jansson J (2013) The plant microbiome. Genome Biol 14:209. https://doi.org/10.1186/gb-2013-14-6-209

Upadhyay SK, Singh JS, Saxena AK, Singh DP (2012) Impact of PGPR inoculation on growth and antioxidant status of wheat under saline conditions. Plant Biol 14(4):605–611

Vajpayee G, Asthana S, Purwar S, Sundaram S (2015) Screening of antagonistic activity of bacteria against *Tilletia indica*. Indian J Nat Sci 5(29):4302–4313

Venieraki A, Dimou M, Pergalis P, Kefalogianni I, Chatzipavlidis I, Katinakis P (2011) The genetic diversity of culturable nitrogen-fixing bacteria in the rhizosphere of wheat. Microb Ecol 61(2):277–285

Verma P, Suman A (2018) Wheat microbiomes: ecological significances, molecular diversity and potential Bioresources for sustainable agriculture. EC Microbiol 14(9):641–665

Verma JP, Yadav J, Tiwari KN (2010) Impact of plant growth promoting rhizobacteria on crop production. Int J Agric Res 5(11):954–983

Verma P, Yadav AN, Kazy SK, Kumar S (2016) Molecular diversity and multifarious plant growth promoting attributes of bacilli associated with wheat (*Triticum aestivum* L.) rhizosphere from six diverse agro-ecological zones of India. Int J Curr Microbiol App Sci 56(1):44–58

Vessey JK (2003) Plant growth promoting rhizobacteria as biofertilizers. Plant Soil 255(2):571–586

Wachowska U, Głowacka K (2014) Antagonistic interactions between *Aureobasidium pullulans* and *Fusarium culmorum*, a fungal pathogen of winter wheat. BioControl 59:635–645

Wang Y-F (2011) Biocontrol of wheat leaf rust caused by *Puccinia triticina* and induction of systemic acquired resistance by salicyclic acid. Agricultural University of Hebei (Master dissertation)

Wang YG, Xia QY, Gu WL (2012) Isolation of a strong promoter fragment from endophytic enterobacter cloacae and verification of its promoter activity when its host strain colonizes banana plants. Appl Microbiol Biotechnol 93(4):1585–1599
Wang H, Xu R, You L, Zhong G (2013) Characterization of Cu-tolerant bacteria and definition of their role in promotion of growth, Cu accumulation and reduction of Cu toxicity in *Triticum aestivum* L. Ecotoxicol Environ Saf 94:1–7
Watt M, Kirkegaard JA, Passioura JB (2006) Rhizosphere biology and crop productivity – a review. Aust J Soil Res 44(4):299–317. https://doi.org/10.1071/sr05142
Wilson D (1995) Endophyte: the evolution of a term and clarification of its use and definition. Oikos 73:274–276. https://doi.org/10.2307/3545919
Wong PTW, Southwell RJ (1980) Field control of take-all of wheat by avirulent fungi. Ann Appl Biol 94:41–49
Wong PTW, Mead JA, Holley MP (1996) Enhanced field control of wheat take-all using cold tolerant isolates of *Gaeumannomyces graminis* var. *graminis* and *Phialophora* sp. (lobed hyphopodia). Plant Pathol 45(2):285–293
Xue AG, Chen Y, Voldeng HD, Fedak G, Savard ME, Längle T (2014) Concentration and cultivar effects on efficacy of CLO-1 biofungicide in controlling Fusarium head blight of wheat. Biol Control 73:2–7. https://doi.org/10.1016/j.biocontrol.2014.02.010
Yadav AN, Verma P, Kumar M (2015a) Diversity and phylogenetic profiling of niche-specific bacilli from extreme environments of India. Ann Microbiol 65(2):611–629
Yadav AN, Verma P, Tyagi SP, Kaushik R, Saxena AK (2015b) Culturable diversity and functional annotation of psychrotrophic bacteria from cold desert of Leh ladakh (India). World J Microbiol Biotechnol 31(1):95–108
Yadav AN, Kazy SK, Kumar S (2016) Molecular diversity and multifarious plant growth promoting attributes of bacilli associated with wheat (*Triticum aestivum* L.) rhizosphere from six diverse agro-ecological zones of India. Int J Curr Microbiol Appl Sci 56(1):44–58
Yadav AN, Verma P, Kour D (2017a) Plant microbiomes and its beneficial multifunctional plant growth promoting attributes. Int J Environ Sci Nat Resour 3(1):1–8
Yadav AN, Verma P, Kumar R, Kumar V (2017b) Current applications and future prospects of eco-friendly microbes. EU Voice 3(1):21–22
Yasin M, El-Mehdawi AF, Anwar A, Pilon-Smits EA, Faisal M (2015) Microbial-enhanced selenium and iron biofortification of wheat (*Triticum aestivum* L.)-applications in phytoremediation and biofortification. Int J Phytoremediation 17(4):341–347
Yin C, Mueth N, Hulbert S, Schlatter D, Paulitz TC, Schroeder K, Prescott A, Dhingra A (2017) Bacterial communities on wheat grown under long-term conventional tillage and no-till in the Pacific Northwest of the United States. Phytobiomes 1:83–90
Yu H, Zhang L, Li L, Zheng C, Guo L, Li W, Sun P, Qin L (2010) Recent developments and future prospects of antimicrobial metabolites produced by endophytes. Microbiol Res 165(6):437–449
Zafari DM, Koushki M, Bazgir E (2008) Biocontrol evaluation of wheat take all disease by *Trichoderma* screened isolates. Afr J Biotechnol 7:3653–3659
Zahir ZA, Arshad M, Frankenberger WT (2003) Plant growth promoting rhizobacteria: applications and perspectives in agriculture. Adv Agron 81:97–168
Zahir ZA, Ghani U, Naveed M, Nadeem SM, Asghar HN (2009) Comparative effectiveness of Pseudomonas and Serratia sp. containing ACC-deaminase for improving growth and yield of wheat (*Triticum aestivum* L.) under salt-stressed conditions. Arch Microbiol 191(5):415–424
Zaidi A, Khan S (2005) Interactive effect of rhizotrophic microorganisms on growth, yield, and nutrient uptake of wheat. J Plant Nutr 28(12):2079–2092

9 Entomopathogenic Fungi: A Potential Source for Biological Control of Insect Pests

Anjney Sharma, Ankit Srivastava, Awadhesh K. Shukla, Kirti Srivastava, Alok Kumar Srivastava, and Anil Kumar Saxena

Abstract

Insect pests with high population level are inimical and cause enormous damage to agricultural crops and economy as well as decrease the food security for the growing human population. For controlling insect pests, the use of entomopathogenic fungi (EPF) is very much useful, and it was developed as eco-friendly mycopesticide that is useful in the regulation of insect pests. These entomopathogenic fungi have a unique mode of infection on different orders of insects. Recently, it was investigated that in addition to insect pest control, these entomopathogenic fungi also act as endophytes and biocontrol agents of plant pathogens and promote plant growth as rhizosphere fungi. Numerous environmental abiotic and biotic factors are recognized to inhibit or enhance the fungal efficacy against the insect pathogens. Advanced research in the genome biology of pathogenic insect fungi has shown that genetic features of these organisms are developed for fungal adaptation with various host insects. The efforts toward genetic engineering of entomopathogens, the knowledge of its virulence and tolerance to adverse conditions will potentiate cost-effective applications of mycoinsecticides for pest control in the agricultural fields. The studies suggest that the exploitation of ecological, genetic and functional diversity of these fungi increases our potential for integrated pest management. Consideration of the insect microbiome in fungal insect pathology represents a new frontier which may contribute in deciphering the obscure pathological aspects of the biology of entomopathogenic fungi and its ecology. Taking these into account, in the present chapter, we highlight the importance, classification, mode of action, factors affecting its virulence effi-

A. Sharma (✉) · A. Srivastava · K. Srivastava · A. K. Srivastava · A. K. Saxena
ICAR-National Bureau of Agriculturally Important Microorganisms,
Mau Nath Bhanjan, Uttar Pradesh, India

A. K. Shukla
K. S. Saket P. G. College (Dr. Rammanohar Lohia Avadh University),
Ayodhya, Uttar Pradesh, India

© Springer Nature Singapore Pte Ltd. 2020

M. K. Solanki et al. (eds.), *Phytobiomes: Current Insights and Future Vistas*,
https://doi.org/10.1007/978-981-15-3151-4_9

cacy, the role of genetic engineering in strain improvement, and ground level of entomopathogenic fungi for integrated pest management (IPM), and plant growth promotion with commercial available entomopathogenic products is critically discussed.

Keywords
Entomopathogenic fungi · Genetic engineering · Insect pests · Integrated pest management · Virulence

9.1 Introduction

Due to the dependency of the market at a global scale to agricultural merchandise, leading-edge technology in agriculture is increasingly used that is required to develop new agricultural practices and has potential to alleviate the adverse negative effects on the ecosystem and consequently results in the production of valuable safe products for human consumption. However, there is a major constraint in increasing agricultural production yield due to loss caused by insect pests, plant diseases, weeds, etc. (Oerke et al. 1994). In evolutionary agricultural practices, pest problems are unavoidable and arise largely due to simplified agroecosystems and additionally because of the creation of less stable natural ecosystems. It is investigated that insect pests such as locusts, grasshoppers, termites, and cattle ticks have caused huge economic and agricultural production losses, that is, about 40%, in many parts of the world (Thacker 2002; Chandler et al. 2011). Although the use of pesticides has increased marginally, crop losses have remained relatively stable (Oerke 2006). Various industrialized nations are trying to shift their strategy to using transgenic plants that express particular traits like resistance to insects, fungi, or viruses for pest management. Several agrochemicals pesticides are still being used by farmers for the control of insect pests and diseases in their agricultural practices. They have been responsible for maintaining and increasing the quality and quantity of food and fiber worldwide. The extensive use has resulted to adverse effect on nontarget organisms, deposition on edible food crops, groundwater pollution, and resistance of insects against the chemicals, and negative effect on human health has forced scientists to focus on the development of alternative environmentally safe strategies that are cost-effective and reliable. Integrated pest management (IPM) is a comprehensive approach to crop production (Chandler et al. 2011). This is a shift from the traditional individual pest-centered strategies that relied heavily on chemical pesticides to a more holistic approach of viewing the entire crop production system as a whole and managing rather than eradicating the pests. In line with this point, several microbial agents have been developed to manage insect pests. Entomopathogenic fungal biological plant protection plays a key role in the program of sustainable pest management. For integrated pest management, ready-made use of entomopathogenic fungus as biocontrol agents to control the insect density and increase the crop yield has several benefits compared with conventional insecticides. These encompass low expenses, high performance, safety for beneficial organisms, reduction of residues within the environment, and accelerated biodiversity (Asi et al. 2013; Gul et al. 2014). Entomopathogenic fungus has been known as important inhibitory factor of plant bugs for more than a century and is continuously used in pest biocontrol management throughout the world (Jaber and Enkerli 2017). Entomopathogenic fungi fill an extremely important niche and are well suited for development as microbial biopesticides for control of the harmful insect pests.

Almost all orders of insects are susceptible to fungal sicknesses. Entomopathogenic fungi are potential eco-friendly biocontrol agents especially because of their excessive reproductive abilities, goal-specific activity, minimal impacts on beneficial and other nontarget organisms, brief era time, and resting stage or saprobic phase—generating competencies which can make sure their survival for an extended time when no host is present (Mudgal et al. 2013). Entomopathogenic fungi can be found in a wide range of environmental conditions (including arid to tropical settings, terrestrial to aquatic habitats, and arctic to temperate climates), possess potential to colonize a wide variety of plant species, and infect a broad array of insects (Meyling et al. 2012; Mantzoukas et al. 2015). Researchers have stated that approximately a thousand species of fungi are identified as causal agents of disease in arthropods (Goettel et al. 2010; Vega et al. 2012). The majority of entomopathogenic fungi used for biological control of insects and mites are in the orders Hypocreales and Entomophthorales. Most of the biological control fungi produced commercially are included in Hypocreales. Various other fungi such as *Metarhizium* spp., *Beauveria bassiana*, *Lecanicillium* spp., and *Isaria fumosorosea* are being used against a large number of pests in different variety of crops. About 170 fungal formulations from at least 12 species of fungi have been developed (Faria and Wraight 2007). It is believed that there are about more than 750 species of fungi that have been identified to cause various infections in insects. Most of the entomopathogenic strains or species are obligate pathogens that show specific ecomorphological adaptations of their host's life cycles, such as the production of infective spores that are produced on insect cadavers during the night. These fungi often cause natural epizootics in insect and mite populations. Natural epizootics of entomophthoralean fungi are being used as a natural form of pest control. Pioneering work of Inglis et al. (2001) conducted with *Beauveria bassiana* (Balsamo) and *Vuilleminia* (Ascomycota: Hypocreales), as entomopathogenic endophytes and ubiquitous soilborne fungus and one of the most commercialized fungal biopesticides, reveals that they infect >700 insect species. A number of recent studies reveal that entomopathogenic fungi, often entirely taken into consideration as insect pathogens, play extra roles in nature, such as endophytism, plant disease antagonism, plant growth promotion, rhizosphere colonization, and management of various abiotic stresses (Vidal and Jaber 2015). Moreover, most studies investigating the interactions among plants and endophytic entomopathogens have up to now centered on the advantages of such interactions to the host plant through multiplied tolerance and resistance to biotic elements including pests and illnesses (Lopez and Sword 2015). Due to low virulence and inconsistencies of their performance within various field environmental stresses, currently, fungal pathogens have a small market share and small chances of killing insects in comparison to chemical insecticides (Fang et al. 2012). Low virulence may be incorporated as an evolutionary balance between microorganisms, and their hosts may have adapted to the pathogen to avoid rapid killing even at high doses, in which cost-effective biocontrol will require genetic modification of the fungi (Gressel 2007). Genetic engineering is an effective green device to improve the efficacy of mycoinsecticides by enhancing their virulence and tolerance to environmental stresses. Modern biotechnological tools that now enable genetic improvement of entomopathogens for myco-biocontrol of fungi are now feasible. Molecular biology techniques, coupled with the cloning of genes to express insect proteins and insecticidal proteins/peptides from insect pathogens, will create more effective strains for pest management. Further research is desirable under variable environmental conditions to monitor the environmental fate of recombinant fungal strains. Comparative genomics has facilitated the identification of fungal fitness characteristics and the selective forces acting on them to enhance our knowledge of why and how entomopathogenic fungi interact with insects and other additive components of their environment. So far, there may be plenty of extra-genomic facts on ascomycete insect pathogens, as sequences are available from *Metarhizium* strains, *Beauveria bassiana, Cordyceps militaris, Ophiocordyceps sinensis* (anamorph, *Hirsutella sinensis*), *Ophiocordyceps unilateralis, Tolypocladium inflatum*, and *Hirsutella thompsonii* (Zheng et al. 2011; Xiao et al. 2012; Bushley et al. 2013; Pattemore et al. 2014; de Bekker et al. 2015; Agrawal et al. 2015). Molecular studies provided valuable information on the phylogenetic relationships with various other fungi. Hence, sequence data can, therefore, provide crucial information on how these organisms reproduce and persist in different environments (Wang and St. Leger 2014). It is important to understand the ecology of fungal entomopathogens. Further, it is reported that various entomopathogenic fungi play important roles in nature, such as endophytes, useful rhizosphere associates, antagonists of plant pathogens, and possibly even plant growth promoters. A growing quantity of research has currently established the ability of several entomopathogenic fungi after endophytic establishment to promote plant growth (Sasan and Bidochka 2012; Lopez and Sword 2015). Increased plant growth, mediated through colonization by fungal endophytes, is the outcome in the suppression of numerous abiotic and biotic stresses, consisting of plant diseases (Kuldau and Bacon 2008). Hence, it is an urgent need to review and understand how entomopathogenic fungi solely act as biopesticides and also promote plant growth. Keeping the above facts, we are trying to summarize the recent information regarding the role of entomopathogenic fungi in biocontrol and plant growth promotion. Additionally, the factors affecting the activities of entomopathogenic fungi are also being discussed.

9.2 Classification and Type of Entomopathogenic Fungus

In eukaryotes, the kingdom Fungi is a major group with approximately 700 described entomopathogenic fungi species (Roberts and Humber 1981) which represent <1% of the total fungal species (McLaughlin et al. 2009). Entomopathogenic fungi are found in three main groups (Entomophthoromycota, Blastocladiomycota, and Microsporidia) and 12 different classes in six phyla of fungi. These pathogenic fungi are found in the divisions Ascomycota, Zygomycota, Deuteromycota, Oomycota, and Chytridiomycota (Humber 1997). Many of the potential entomopathogenic fungi genera known so far belong to either the order Entomophthorales in the Zygomycota or the class Hyphomycetes in Deuteromycota (Jaber and Enkerli 2017). The detailed descriptions of the classification are illustrated in Table 9.1. Among the different phyla and orders, the species that belong to genus *Metarhizium*, *Beauveria*, *Nomuraea*, *Verticillium*, and *Hirsutella* of different environmental habitats are most prominently agriculturally important entomopathogenic fungus which is commercially used successfully in the field level. The detail of this genus is illustrated in Table 9.2 and Fig. 9.1.

9.3 Entomopathogenic Fungus Infection

Entomopathogenic fungi have a unique mode of infection. The infection on different orders of insects (Lepidoptera, Coleoptera, Diptera, Hymenoptera, Homoptera) becomes started when insects come into contact with entomopathogenic fungal spores. The life cycle of insect pathogenic fungi starts with the spore germination on the cuticle of the host (Fig. 9.2). Attachment can be passive in spores that are covered with a sticky or slimy substance (e.g., *Lecanicillium* spp., *Entomophthorales* spp., and *Hirsutella* spp.). Adhesive processes involve physical, chemical, and enzymatic interactions. Different hydrolytic and detoxifying enzymes like protease, chitinase, lipase, catalyzes, cytochrome P450, and their secondary metabolite are generated during final insect adaptive responses. Two adhesion genes, namely, Mad1 and Mad2 in *M. anisopliae*, have been characterized (Wang and St. Leger 2007). Under favorable environmental condition such as optimum temperature, humidity, nutritional and chemical cue, and cuticular secreted material of the host, spore's germination occurs (Tanada and Kaya 1993). Through penetration hook, the germinating spores produce germ tubes which enzymatically and physically penetrate the cuticle and into the thinner softer part of the insect's body by sensory pores or wounds (Gabarty et al. 2014). Samuels and Paterson (1995) reported that a variety of cuticle-degrading proteases are produced by both insects and entomopathogenic fungi. An extensive range of proteases have been recognized, consisting of trypsin, chymotrypsin, esterase, collagenase, and chymoelastase (Sanchez-Perez et al. 2014). Once inside the body, the fungus spore starts multiplying by taking internal nutrient material of the insects, which create nutrient deficiency and disruption of biological functions, produce toxic compounds, invade insect's tissues, block the vessels, and ultimately cause host death (Goettel et al. 2000). After the insect host is killed and all nutrition has been consumed, hyphae grow out of the cadaver, particularly at the margins of the intersegmental regions, and produce resting or infective spores that promote the spread of the fungus. Finally, they must be transmitted to new hosts (Boomsma et al. 2014). Entomopathogenic fungi such as *Beauveria, Metarhizium Hirsutella, Paecilomyces, Isaria, Lecanicillium,* and

Table 9.1 Classification of the entomopathogenic fungus

Phylum	Order	Description	Genus/species	Infected insect host
Oomycota	Lagenidiales, Leptomitales	Cellulose in their coenocytic hyphae (without chitin) and they have biflagellate zoospores, reproduced by a thick-walled oospore, at the cellular level; they possess mitochondria with tubular cristae	*Lagenidium giganteum*, *Leptolegnia chapmanii*	Mosquitoes, crabs, and other aquatic crustaceans, arthropods
Chytridiomycota	Blastocladiales, Chytridiales, Blastocladiales	Coenocytic hyphae; chitin is a major constituent of cell wall, and zoospores are with a single flagellum	*Coelomomyces*, *Myriophagus*, *Coelomycidium*, *Myriophagus* (*Chytridiales*), *Coelomycidium* (*Blastocladiales*)	Hemipterans and dipterans, mosquitoes and flies
Zygomycota	Entomophthorales	Mycelium is multicellular, nonseptate gametangia that after fusion form zygospores	*Batkoa apiculata*, *Batkoa major*, *Entomophaga grylli*, *Entomophaga calopteri*, *Entomophaga maimaiga*, *Pandora neoaphidis*, *Pandora delphacis*, *Pandora blunckii*, *Pandora bullata*, *Zoophthora radicans*, *Conidiobolus thromboids*, *Neozygites parvispora*	Aphids, hemipterans, flies, lepidopterans, grasshopper, gypsy moth larvae, homoptera, lepidoptera, diptera, leafhoppers, psyllids, leaf rollers, clover leaf weevil
Ascomycota and Deuteromycota	Hypocreales	Septate mycelia, ascospores (sexual spores) formed in the fruiting body called ascus, haploid	*Aspergillus*, *Metarhizium*, *Hirsutella*, *Beauveria*, *Aschersonia*, *Culicinomyces*, *Lecanicillium*, *Paecilomyces*, *Tolypocladium*, etc.	Lepidoptera, hymenoptera (wasps), whitefly, pine caterpillar, potato Beetle, corn borer, coding moth, grasshoppers, chinch bug, boll weevil, cowpea curculio, pecan weevil, mosquitoes, lygus bug, granary weevil, brown plant hopper, termites, spider mite, fire ants, European cockchafer, sugarcane borer, brown planthopper, citrus rust mite

(continued)

Table 9.1 (continued)

Phylum	Order	Description	Genus/species	Infected insect host
Basidiomycota	Septobasidiales, Atheliales	Produce sexual spores called basidiospores, formed outside the specialized reproductive cells called basidia. These spores are forcibly discharged by specialized structures Unique traits for this group are clamp connections	*Septobasidium and Uredinella, Fibularhizoctonia*	Scale insects (Hemiptera, Diaspididae), termite eggs
Entomophthoromycota,	Entomophthorales	Somatic state consisting of a well-defined mycelium, coenocytic or septate Protoplasts either hyphoid or amoeboid and changeable in shape; cystidia or rhizoids formed by some taxa Conidiophores branched or unbranched. Primary spores true conidia, uni-, pluri-, or multinucleate	*Pandora neoaphidis and Entomophthora planchoniana, Entomophaga maimaiga*	Different aphid species, certain lepidopteran larvae

Verticillium are known as commercially used and sold biopesticides in multiple formulations around the world (Anchez-Rodríguez et al. 2015; Lacey et al. 2015).

9.4 Enzymatic Virulence Mechanisms of the Entomopathogenic Fungus

Virulence can be defined as a process involved in insect death during pathogenesis (Mondal et al. 2016). The production of cuticle-degrading enzymes has been proposed as an important attribute of the pathogenic fungi determining the virulence of the entomopathogenic fungi that cause insect death during pathogenesis (Samuels et al. 2011). The insect cuticle is a highly heterogeneous structure that can vary

Table 9.2 Some agriculturally important entomopathogenic fungi and their commercially available bioproducts against the insect pests

Genus	Spp.	Characteristics	Plants	Host insects/pests	Toxins against the insect pests	Commercial products manufacturer
Metarhizium	*M. anisopliae*, *M. hiemalis*, *M. acridum*, *M. album*, *M. flavoviride*, *M. brunneum*, *M. pingshaense*, *M. robertsii*, *M. guizhouense*, *M. lepidiotae*, *M. majus*, and *M. globosum*	Initially, the colony color is white which turns to green at later stage. The hyphae are hyaline, septate, branched. Conidiophores are short, erect, hyaline, septate, simple, or branched, terminating in single or a cluster of phialides. Synnemata absent. Conidia may be hyaline, lilac, brown or green, smooth and form chains. Size ranges from 4.0–14.5 × 0.0–5.0 μm; shape may be cylindrical, globose, ellipsoidal. Conidia are hydrophobic due to the presence of cysteine-rich proteins called hydrophobins in the cell wall	Rice, maize cotton, sugarcane, coconut	Brown plant hopper (*Nilaparvata lugens*), western corn rootworm (*Diabrotica virgifera*), *Dysdercus peruvianus*, Scirpophaga incertulas, green leafhopper (GLH), brown planthopper (BPH), coconut rhinoceros beetle, groundnut cutworm, rice brown planthopper, diamondback moth and early shoot borer, top shoot borer and internode borer of sugarcane white grub, red cotton bug, rhinoceros beetle	Destruxins (DTXs), cytochalasins, and ribotoxin anisoplin	**India:** M/s Pest Control (India) Pvt. Ltd., Bengaluru; M/s Sri Biotech; M/s Viswa Mithra Bio-Agro P. Ltd., Guntur; M/s Microplex (India), Wardha; M/s Varsha Bioscience & Technology, Hyderabad, M/s Agri Life, Secunderabad, AP; M/s Sri Venkateshwara Chemicals, Bangalore; M/s Biotech International Ltd, New Delhi; M/s DVS BioLife; M/s International Panacea Ltd; M/s T. Stanes & Co. Ltd; M/s Khodke Agro Products Pvt. Ltd.; M/s Ruchi Oyster Mushroom, Gondia (MS); M/s Advance Cropcare (India) Pvt. Ltd., Indore; M/s Adiraj Agro Industries, Pune (MS); M/s Nirmal Seeds Pvt. Ltd., Jalgaon (MS); M/s Arya Biotech & Research Laboratory, Amravati **USA:** Novozymes Biological Inc. **Spain:** Trichodex S.A. **Australia:** Becker Underwood Inc.

Beauveria	*B. bassiana*, *B. velata*, *B. brongniartii*, *B. amorpha*, and *B. caledonica*	A filamentous fungus. Grows naturally in soils. The colonies are white with cottony aerial mycelium. Conidiophores are single or branched, oblong, cylindrical, or flask shaped and are bearing laterally or at extremity and vesicles giving rise to sporogenous cells (phialides). Phialides generally are globose, sometimes cylindrical, flask-like, and curved or straight. Hydrophobic properties of conidia, due to the presence of cysteine-rich proteins called hydrophobins in the cell wall. Conidia are globose (1–4 μ) to oval shape (1.5–5.5 × 1–3 μ)	Rice, black gram, red gram, groundnut, sugarcane, mango, castor, maize, potato, pine	Pales weevils (*Hylobius pales*), black bug (*Scotinophara lurida*), zigzag leafhopper (*Recilia dorsalis*), spotted pod borer (*Maruca testulalis*), aphids, *H. armigera* (American boll worm), yellow hairy caterpillar, shoot borer, leaf webber, hairy caterpillar, stem borer, western corn rootworm, ladybird beetle, Colorado potato beetle, bark beetle	Beauvericin, bassianolide, isarolides, and beauverolides	**India:** M/s T. Stanes & Company Ltd.; M/s International Panacea, New Delhi; M/s Sri Biotech; M/s. Romvijay Bio-Tech (P) Ltd; M/s Multiplex Bio-Tech Pvt. Ltd; M/s Viswa Mithra Bio-Agro P. Ltd., Guntur; M/s Metroplex (India); M/s Pest Control (India) Pvt. Ltd; M/s Varsha Bioscience & Technology, Hyderabad; M/s Sri Venkateswara Chemicals, Secunderabad; M/s DVS BioLife Ltd; M/s Jai Biotech Industries; M/s Vaibhav Lakshmi Bio-control Laboratories; M/s Bio-Agro Ferticon, Pune; M/s R.B. Herbal; M/s Jai Kisan Agro; M/s Avishkar Bio-farm Pvt. Ltd.; M/s Khodke Agro Products Pvt. Ltd.; M/s Choudhary Agrotech (I); M/s Govinda Agro Tech. Ltd; M/s Om Agro Organics, Yavatmal, Maharashtra; M/s INORA, Pune; M/s Ellor Bio Tech & Agro Services, Aurangabad; M/s Super Pesticides & Agro (India) Pvt. Kolkata **USA:** Jabb of the Carolinas Inc., Laverlarn International Corporation, Mycotech Corp., Troy Bioscience Inc., **Spain:** Trichodex S.A. **South Africa**: Biological Control Products SA (Pty) Ltd. **Colombia:** Live System Technology S.A.

(continued)

Table 9.2 (continued)

Genus	Spp.	Characteristics	Plants	Host insects/pests	Toxins against the insect pests	Commercial products manufacturer
Verticillium or *Lecanicillium*	*V. lecanii*, *V. chlamydosporium*, *V. dahlia*, *V. albo-atrum* or *L. lecanii*, *L. attenuatum*, *L. longisporum*, *L. muscarium*, *L. nodulosum*	Conidiophores are formed in verticillate whorls, on which conidia are borne in slime or mucus balls. Conidiospores are ovoid/ellipsoid, single-celled; branching is verticillate, i.e., the branches arise in whorls on the upper portions of conidiophores. Hydrophilic conidia vary in shape and size, with a mean length of 5.0 ± 1.0 μm and width of 1.9 ± 0.5 μm. They have/form an ascus (and may, therefore, be described as asci fungi). They form resting structures, long phialides. The vegetative mycelium has the following characteristics: hyaline, multinucleate, and septet	Groundnut, guava, potato, castor, maize	*Aphis craccivora*, *Mylabris pustulata*, aphids, whiteflies, grasshoppers, *Icerya seychellarum*, *Myzus persicae*, *Pericallia ricini*, thrips, brown planthopper, scale insects, groundnut pests, hairy caterpillar, stem borer	VD toxins	**India:** M/s T. Stanes & Company, M/s. Romvijay Bio-Tech (P) Ltd, M/s Viswa MithraBio-Agro P. Ltd., M/s Indian Institute of Horticultural Research, M/s Microplex M/s Agri Life, M/s Varsha Bio Science & Technology, M/s Jai Biotech Industries, M/s T. Stanes & Co. Ltd., M/s Sri Biotech, M/s Avishkar Bio farm (P) Ltd., M/s DVS Bio Life Ltd., M/s Bio-Agro Ferticon, M/s Pravara Agro Biotech, M/s R.B. Herbal, M/s Institute of Natural Organic Agriculture, M/s Khodke Agro Products Pvt. Ltd., M/s Choudhary Agrotech (I), M/s Om Agro Organics, M/s Sai Agrotec, M/s Sri Venkateswara Chemicals, M/s Junna Life Sciences, M/s Shri Ram Solvent Extractions Pvt. Ltd., M/s Mitcon Biotechnology, M/s Pruthvi Fertilizers Organics, M/s Nirmal Organo Bio Tech, M/s Sun Agro Biosystem Pvt. Ltd., M/s Sri Dutta Gro-Tech Equipments **Russia**: Biodron **United Kingdom:** Koppert Biological Systems **Spain:** Trichodex **Brazil:** Natural Rural

Paecilomyces	*P. lilacinus, P. marquandii, P. farinosus, P. amoeneroseus, P. fumosoroseus, P. tenuipes, P. javanicus, P. variotii*	The colonies are white, red, or yellow. Well-developed synnematous conidiophores are found in many species, septate bearing whorls of divergent branches and phialide, the conidiogenous cells. The phialides are flask to oval shaped or subglobose with a distinct neck, borne singly or in whorls; conidia are unicellular, hyaline to light colored. Phialides occur as divergent loose group	Groundnut, castor	*Aproaerema modicella, Pericallia ricini*, Yellow and red mites, whiteflies, etc.	Leucinostatins	**India:** M/s Indian Institute of Horticultural Research, M/s T. Stanes and Co. Ltd., M/s RPC Balaji Crop Care Pvt. Ltd; M/s Nico Orgo Manures, M/s Shree Shiva BioTech, M/s Agriland Biotech Ltd., M/s Kanbiosys Pvt. Ltd., M/s. Devi Biotech (P) Ltd., M/s Varsha Bioscience & Technology, M/s Gujarat Life Sciences (P) Ltd., M/s. Chitra Agri Organics, M/s Agri Life, M/s Jyothiraditya Bio Solutions Ltd., M/s Viswa Mithra Bio-Agro (P) Ltd., M/s Excel Crop Care Ltd., M/s Romvijay Bio Tech Private Ltd., M/s Jhass Agro Industries, M/s Camson Bio Technologies Ltd., M/s Agriva Agro Tech Kappallur, M/s Poabs Biotech Pvt. Ltd. **USA:** Certis

(continued)

Table 9.2 (continued)

Genus	Spp.	Characteristics	Plants	Host insects/pests	Toxins against the insect pests	Commercial products manufacturer
Hirsutella	*H. thompsonii*, *H. gigantea*, *H. citriformis*	Fungal colonies show fluffy mycelial growth, gray with a brownish to grayish-green substratum color. The hyphae are 1.5–2.0 μm wide and smooth. The conidiogenous cells arise singly at intervals from the vegetative hyphae, mono- or polyphialide, unevenly verrucous, with a conical- to flask-shaped base and a narrow neck. The neck may be unbranched or branched, bearing enteroblasts conidia singly at the tip of the branch	Rice, citrus plants	*N. lugens*, *N. virescens*, and *S. furcifera*, citrus rust mite, (*Phyllocoptruta oleivora*) hoppers and bug pests, whiteflies, red mites BPH, GLH, and WBPH	Hirsutellin A (HtA)	**India:** M/s International Panaacea Ltd.

Nomuraea	*N. rileyi*, *N. atypicola*	Colonies of the fungus are white initially and later turn to green in color (malachite green). Hyphae are 2–3 μ in diameter, smooth, septate, hyaline, and slightly pigmented. Conidiophores are long (160 μ) and consist of dense, compacted clusters of phialides and branches in whorls on the upper section, short and swollen. Phialides are short and cylindrical to globose, with very swollen base tapering abruptly to a narrow neck. The conidia are 3–4 and 2.5 μ and pale green	Sorghum, soybean, groundnut, rice	Tobacco caterpillar (*S. litura*), American bollworm (*H. armigera*), *Helicoverpa*, *Trichoplusia ni*, *Heliothis zea*, *Plathypena scabra*, *Bombyx mori*, *Pseudoplusia includens*, Anticarsia gemmatalis, ear head caterpillar (*Helicoverpa armigera* Hubner), pod borers, cutworms, cabbage borers, leaf folder	Diketopiperazine	**India:** M/s Nirmal Seeds Pvt. Ltd.

Sources: Mathew et al. (1998), Sandhu and Vikrant (2006), Inglis and Tigano (2006), Sandhu et al. (2012), Sujeetha and Sahayaraj (2014), Lacey et al. (2015), Vega (2018), and Reddy et al. (2013), http://ppqs.gov.in/divisions/cib-rc/registered-products

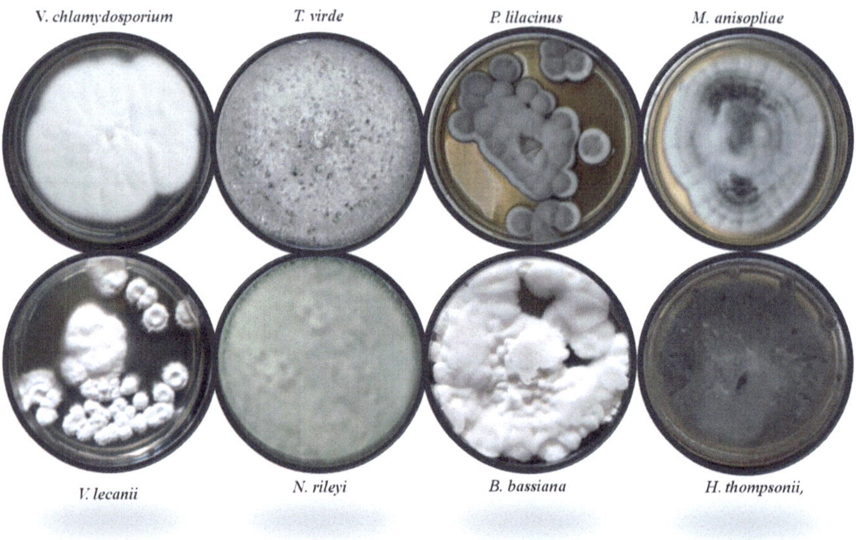

Fig. 9.1 Agriculturally important entomopathogenic fungi

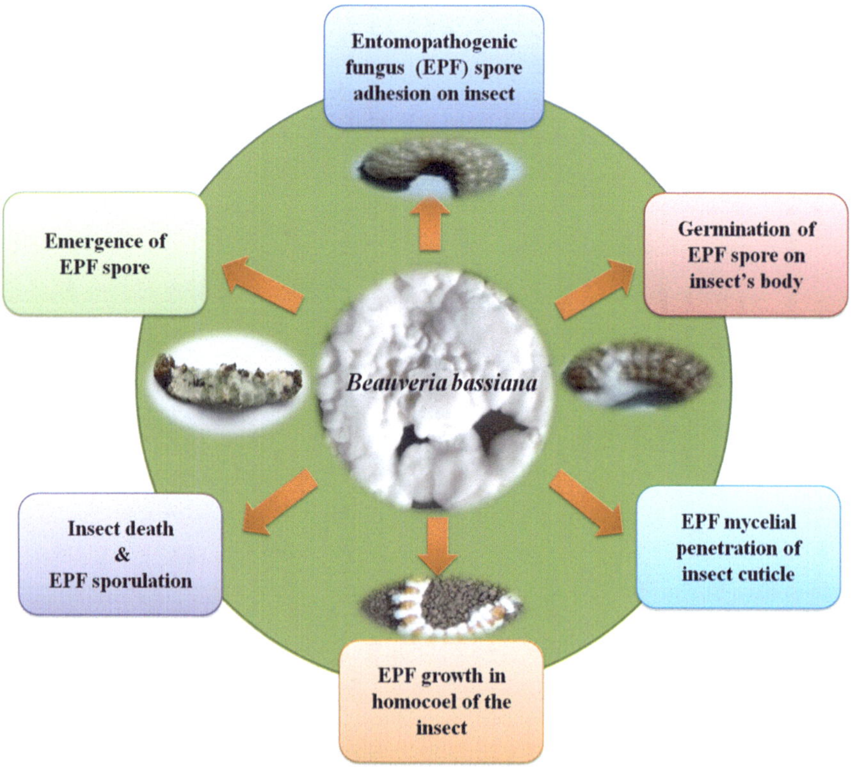

Fig. 9.2 Life cycle or mode of action of entomopathogenic fungi against the insect pests

greatly in composition and composed of wax, protein, and chitin associated with lipids and phenolic compounds; these represent a significant barrier to the invading fungus during the various life stages of a particular insect (Ortiz-Urquiza and Keyhani 2015. The entomopathogenic fungus can produce diverse enzymes like lipase, chitinase, and proteases as a virulence factor in response to different insects. During pathogenesis, degradation of the cuticular polymers in insect cell wall takes place due to these enzymes, which helps in penetrating the insect exoskeleton and getting nutrients for fungal growth; thus, they control insect pests and pathogens that attack productive crops and provide potential economic benefit to agribusiness (Petrisor and Stoian 2017). Most studies were focused on cuticle-degrading enzymes produced by entomopathogenic fungi and on extracellular activities of them (Charnley 2003; Ali et al. 2009). Several studies have carried out the biochemical characterization of cuticle-degrading enzymes such as chitinase, protease, and lipase, the key enzymes involved in the process of pathogenesis, and have been carried out to understand the host-pathogen interaction (Mondal et al. 2016; Cristina and Stoian 2017).

9.4.1 Proteases

Proteins and proteases that constitute the majority of insects' cuticle are a large group of hydrolytic enzymes that break the peptide bonds of cuticle proteins into small peptides and amino acids.

Proteases (endopeptidase, aminopeptidase, carboxypeptidase) attack insect cuticle before chitinases, because this masks cuticular chitin microfibers, followed by epicuticle that has been broken down through lipases. Entomopathogenic fungi produce a variety of endo- and exo-acting proteolytic enzymes in culture. Many of the proteases of entomopathogens are classified as collagenases or chymoelastases that show strong homology with the subtilisin family of proteases (Chrzanowska and Kolaczkowska 1998). The invading fungi produce great quantities of virulence indicator subtilisin-like serine protease Pr1 and trypsin-like protease Pr2, which degrade the proteinaceous material. It was validated that protease Pr1's extracellular involvement in cuticular penetration is initialized by cuticular infection (Mustafa and Kaur 2009). Pr1 and Pr2's extracellular activities were determined in *B. bassiana, M. anisopliae, Lecanicillium lecanii, Metarhizium flavoviride,* and *Nomuraea rileyi* (Liu et al. 2007). Also, degradation by amino peptidases and exopeptidases of solubilized proteins into amino acids is done to provide nutrients for entomopathogenic fungi (Mondal et al. 2016).

9.4.2 Lipase

Due to the great potential of lipases in commercial fields, studies on lipases, mainly of microbial origin, have increased in recent years. The epicuticle, the outer layer of the cuticle of the insect, is hydrophobic and acts as the first microbial attack barrier.

The main constituent of insect cuticles is a heterogeneous mixture of lipids, long-chain alkenes, esters, and fatty acids. Lipases are responsible for the hydrolysis of glycerol ester bonds, lipoproteins, and long-chain fatty acids found at the interior part of the insect integument (Mondal et al. 2016). Extracellular lipase activity is extra as compared to protease and chitinase; therefore, lipase can be taken into consideration as a crucial enzyme in the metabolic activities of *B. bassiana* (Hegedus and Khachatourians 1988). Secretion of pathogenic fungal lipases that significantly penetrate the cuticle may be linked to invasion of an insect host and then ease of nutrient absorption from host's nutrient source. The importance of lipases in the tegument penetration and breaking down process and defense mechanism in the insect has already been demonstrated (Silva et al. 2010). In *B. bassiana*, the Bbcyp52x1 gene and MrCYP52 gene cluster from *Metarhizium robertsii* encoded the lipase activity with an enzymatic complex (Zhang et al. 2012). Microorganism's lipase production varies not only from the source of the lipid but also from its concentration. In the initial stages of insect adhesion and penetration, the lipase secreted by entomopathogenic fungi was involved (Silva et al. 2005). Also, Dhawan and Joshi (2017) showed that the *B. bassiana* MTCC 4495 strain was more virulent to third instar larvae of *P. brassicae* that exhibited the highest levels of lipolytic activity. Apart from the role of fungal lipase in insect morality, fungal lipase also catalyzes numerous distinctive reactions which might be widely applied in more than one industry, which includes dairy and meals manufacture, leather-based and detergent industries, and manufacturing of cosmetics and pharmaceuticals, and natural synthesis reactions, mainly reactions in nonaqueous media (Akoh et al. 2007).

9.4.3 Chitinase

Chitin is a combined polymer of b-1,4 N-acetyl glucosamine and a principal structural component of cell walls of entomopathogen and invertebrate exoskeletons (Seidl 2008). Chitinases hydrolyze the b-1,4 bonds of chitin polymer, producing a foremost N,N′-diacetylchitobiose. That is accomplished via the breakdown of N-acetyl glucosamine (GlcNAc) monomer by using chitobiose. They collaborate with proteases to degrade the insect's cuticle (St Leger 1991) and are related with various stages of the life cycle of entomopathogenic fungi which includes germination, hyphal growth, morphogenesis, nutrition, and defense against competitors (Adams 2004). The genome of filamentous fungi consists of chitinases liable for diverse physiological functions along with (a) chitin degradation in the exoskeletons of arthropods used as nutrient assets; (b) remodeling of cell partitions at some stage in hyphae growth, branching, hyphae fusion, autolysis, and competence; and, (c) additionally, safety from different fungi placed within the equal ecological niche (Adams 2004; Yang et al. 2007). From *M. anisopliae*, the chit1 gene that encoded chitinase was first molecularly characterized (Bogo et al. 1998). Enzyme chitinases play a vital role in the process of insect cuticle degradation. Fungal virulence is often determined with the extracellular chitinases (Hegedus and Khachatourians 1996). From the growth media of *M. anisopliae*, *M. flavoviride*, and *B. bassiana* supplemented with insect cuticle, production of enzymes chitinolytic, N-acetyl-b-D-glucosaminidase, and endochitinases was reported (St Leger 1991).

9.5 Factors Affecting Entomopathogenic Fungus Activity

Numerous environmental elements are recognized to inhibit fungal efficacy. The extensive version in the susceptibility to individual abiotic elements has been determined among and within fungal species and genera. Abiotic stresses together with ultraviolet (UV) radiation, humidity, high temperature, pH, and low water pastime result in inconsistent performances, restricting the effectiveness of the entomopathogenic fungi, delay pest mortality, and contribute to their constrained use in agricultural manufacturing. Sunlight and ultraviolet (UV) rays are probably the most detrimental environmental factor that affects the viability of entomopathogenic fungi and can reduce the process of infection. Radiation harms entomopathogenic fungi performance (Fang et al. 2012). UV radiation causes DNA damage through the induction of chemical base modification. It has also been found that UV-A radiation inactivates and delays the germination of certain fungi's conidia (Braga et al. 2008), while the exposure of UV-B radiation damaged or killed the spores of many entomopathogenic fungi (Goettel et al. 2001). A high temperature is another adverse factor in reducing the performance of the entomopathogenic fungi by reducing virulence and persistence in field conditions. It has been found that generally most of the entomopathogenic fungi germinate, grow, and sporulated at optimal temperatures between 20 and 30 °C. Fluctuation in the temperature range may influence the germination, growth, and performance of the fungus (Goettel et al. 2000; de Crecy et al. 2009). Further, for developmental stages of entomopathogenic fungi, high humidity is required. Low humidity has been implicated in failures of germination or field trials, by timing the application of fungi when humidity levels are naturally higher; these low-moisture problems can be addressed. Similarly, soil moisture, organic matter in the soil, and pH also are essential in figuring out infection level. Rainfall also can cause speedy and full-size loss of inoculum. For instance, 15 min of mild rainfall in a simulator resulted in approximately 90% elimination of *B. bassiana* conidia from foliage (Inglis et al. 2001). The diversity of entomopathogenic fungi varies according to the nutritional mode of insects from biotrophy to necrotrophy.

Apart from abiotic factors, a range of biotic factors such as the host stage of the pest, competitive microbial organisms, and antagonistic enzymes and compounds on the plant or host surface can also impose negative impact on the efficacy of entomopathogenic fungus. Immature stages of the insects tend to be more sensitive to fungal infection than the mature stage (Brownbridge et al. 2010). Insect behavior also can reduce the effectiveness of a fungal infection. Some termites avoid coming in contact with infected individuals within the colony and thereby escape infection (Chouvenc et al. 2008). The termite *Coptotermes lacteus* showed an avoidance response, walling off tunnels with *M. anisopliae*, thus protecting the colony from infection. Aphids and mites are sometimes able to escape infection by molting before the fungus enters the body (Alavo et al. 2002).

9.6 Genetic Engineering to Improve Virulence of EFP

Due to low virulence and inconsistencies in their performance compared to the chemical insecticides, entomopathogenic fungus currently has a small market share (Fang et al. 2012). The inconsistencies in performance within the field are mainly due to the sensitivity to environmental stresses of entomopathogenic fungi (Lovett and St. Leger 2015). Barriers to the large-scale application of fungal biocontrol agents still exist due to their slow killing speed and environmental stability issues. With growing public concern about the continued use of synthetic chemical insecticides and increasing public acceptance of genetically modified organisms, new types of genetically engineered biological insecticides offer a range of environmentally pleasant alternatives for cost-effective insect pest control. Genetic engineering has been demonstrated to enable the virulence efficacy of mycoinsecticides to be substantially advanced by improving their tolerance to damaging environmental stress. To improve virulence, genetic engineering has focused on reducing both lethal conidial dosage and killing time. Virulence enhancement has been carried out by engineering fungi to express insect proteins and insecticidal peptides from insect predators and other insect pathogens or by overexpressing the pathogen's genes. For genetic improvement in fungal virulence, the first trial was performed by engineering *Metarhizium* to overexpress the gene encoding the endogenous cuticle-degrading protease Pr1, and the resulting transformant took 25% less time to kill insects (Hu and St. Leger 2002). In another similar study in constitutive overproduction of chitinase, CHIT1 improved virulence by 23% in *B. bassiana* (Fang et al. 2005). An insect-selective scorpion neurotoxin peptide gene (AaIT) was synthesized and used to transform *M. anisopliae* for targeted expression in insect hemolymph. The expression results of this gene decreased the killing time with the aid of 40% lethal spore dose in mosquitoes, caterpillars, and beetles by up to 22-fold (Pava-Ripoll et al. 2008). The current information of molecular mechanisms for *B. bassiana* and *Metarhizium* spp. pathogenesis has enabled many gene-related pathogenicity like Pr1A, Bbchit1, Mr-Npc2a, ATM1, Mr-Ste1, and BbBqrA responsible for subtilisin-like protease, chitinase, sterol carrier, trehalase, esterase, and benzoquinone oxidoreductase, respectively, to be characterized, and these genes may be used as a genetic enhancement resource for entomopathogenic fungi. The virulence of *B. bassiana* was increased through the expression of endogenous insect genes, such as those for a diuretic hormone, neuropeptide, and serpin, to disrupt or inhibit normal hormone levels, electrophoretic behavior, or phenoloxidase activation in insects (Fan et al. 2012; Yang et al. 2012). The events of horizontal gene transfer (HGT) revealed that *Metarhizium* species, like other fungi, acquired various genes from bacteria and archaea and even arthropods, plants, and vertebrates (Zhao et al. 2014). By transferring the sterol carrier gene into *B. bassiana*, this evolutionary event was reproduced, lacking an endogenous Mr-NPC2a homologue, and enhances its pathogenicity (Ortiz-Urquiza et al. 2013).

9.6.1 Improve Virulence Against the Abiotic Stresses

Among the abiotic stresses, UV radiation, high temperature, and low water activity cause prominent damage to molecular and metabolic functions of the pathogenic fungus that results in inconsistent limiting performances in the field. The range of genes and metabolic pathway genes like trxA (thioredoxin), Try (tyrosinase), MrPhr1 (CPD photolyase), BbSOD1 (superoxide dismutase), HsPHR2 (CPD photolyase), and heat shock protein 25 (HSP25) have been found suitable for improving virulence in *Metarhizium, Beauveria*, and other insect pathogens against the abiotic stresses (Zhao et al. 2016). The engineering of *Beauveria* or *Metarhizium* to overexpress a DNA repair photolyase also increased fungal resistance to solar radiation (Fang and Leger 2012). Expression of thioredoxin (trxA) also improved the tolerance of *B. bassiana* against heat, UV-B irradiation, and oxidation (Ying and Feng 2011). In another study, a look at genetically engineering *B. bassiana* with an exogenous tyrosinase gene expanded fungal production of melanins for stepped forward conidial tolerance to ultraviolet radiation and elevated virulence in opposition to diverse insects (Shang et al. 2012). The integration of a PKS gene cluster for melanin biosynthesis from a plant pathogen to *M. anisopliae* enabled the fungus to resist UV irradiation and improved fungal virulence. To improve the virulence against high temperature, *M. robertsii* was genetically transformed to overexpress an endogenous heat shock proteins (HSPs)-encoding gene (Hsp25), and the results showed that mutant gene not only survived extreme temperatures but also showed resistance to oxidative stress and osmotic agents (Liao et al. 2014). Thus, genetic engineering is an effective way to improve fungal environmental stability and, therefore, the efficacy of field applications. Combining the available genomes of *Bassiana* and several *Metarhizium* species in the future, with robust genetic manipulation technologies, will make it possible to characterize the full range of pathogenicity and host-specific genes by which novel combinations of insect specificity and virulence will be created (Gao et al. 2011; Xu et al. 2014). Depending on the regulations in different countries, developing these genetically modified strains will potentiate the cost-effective application of mycoinsecticides for the control of different insect pests or disease vectors.

9.7 Entomopathogenic Fungi and Integrated Pest Management Strategies

For plant protection from harmful pests and insects, chemical insecticides are commonly used which impose a negative impact on the ecosystem and increase the resistance of insects to various chemical substances. Therefore, this issue forces to seek an alternative new effective eco-friendly approach of reducing outbreaks of insects or pests. In current years, more interest is paid to the opportunity of the usage of natural insect enemies, which include entomopathogenic fungi, in biological control or inhibition of insect pests (Sahayaraj 2014). Entomopathogens may be a vital tool in IPM techniques in both natural and conventional manufacturing systems. There are numerous examples of entomopathogen-based bio-insecticides that have performed a critical role in pests control. EPF has been extensively investigated for the biological control of pests as well as human and animal disease-causing arthropod vectors (Blanford et al. 2011; Paula et al. 2013; Nana et al. 2015). Entomopathogens may be used alone or in combination with chemicals, botanical pesticides, or other entomopathogens depending on the crop, pest, and environmental conditions.

The most prominent instance of EPF inoculation biological control is the use of *Beauveria brongniartii* against *Melolontha melolontha*, European cockchafer beetles, that's a critical pest of pastureland, orchards, and wooded area and trees in Central Europe. The fungal application can provide useful control levels for up to 9 years (Keller et al. 1997). An IPM research in Californian strawberries validated the capacity function of entomopathogenic fungi for handling the western tarnished plant bug (*Lygus hesperus*) and other insect pests (Dara 2016). The combination of azadirachtin and *B. bassiana* reduced root aphid of rice and populations of honeysuckle aphid by 62% in natural celery in California (Dara 2015). Development of formulations can greatly improve the activity and performance of entomopathogenic fungi. Since the 1960s, worldwide, among the 171 products, a considerable number of potential formulation have been developed of which 33.9% is from *Metarhizium* spp., 33.9% from *Beauveria* spp., 5.8% from *Isaria fumosorosea*, and 4.1% from *B. brongniartii*, in which three-quarters are currently commercially available for the targeted Hemiptera, Coleoptera, Lepidoptera, Diptera, Orthoptera, and Acari pests. Some commercial formulations are illustrated in Table 9.2. These organic formulations used in augmentation, classic, and conservation biological control pose a minimum threat to useful organisms and hence are in all likelihood to be categorized as "low-danger" substances by authorities regulators.

Also, beneficial EPF is known to elicit induced systemic resistance (ISR), which emerged as another potential mechanism through which entire plant accelerates own defense system against a wide range of insects/pest pathogens (Pieterse et al. 2014). *Beauveria bassiana*-inoculated plant showed significantly reduced bacterial blight disease on the leaves compared with the noninoculated control plants. Similarly, in another study, *Lecanicillium*-pre-inoculated cucumber plants significantly induced systemic resistance against *P. ultimum* and powdery mildew *S. fuliginea* (Benhamou and Brodeur 2001; Hirano et al. 2008). Later *B. bassiana* and *Lecanicillium* spp. colonization of date palm plants showed upregulation of defense-related proteins for alleviation against the plant biotic stress (Gómez-Vidal et al. 2009). Hartley et al. (2015) and Khan et al. (2012) reported that fungal endophytes and *M. anisopliae*-inoculated plants secret isoflavonoid phytoalexin's bioactive secondary metabolites which play a major role in the adaptation of plants to different adverse environmental condition. In a recent study, colonization of *B. bassiana* in tomato plants has shown a significant positive effect on terpenoid accumulation in tomato plants compared with arbuscular mycorrhizal fungus *Rhizophagus intraradices* colonization (Shrivastava et al. 2015).

By enhancing plant growth, fungal entomopathogens can contribute to protecting their host plant against disease pathogens. A developing variety of research has recently proven the ability of several fungal entomopathogens to promote plant growth following endophytic establishment (Sasan and Bidochka 2012; Lopez and Sword 2015; Jaber and Enkerli 2016). Kuldau and Bacon (2008) reported that EPF-primed plants increased plant growth with suppression of various abiotic and biotic stresses. Jaber and Salem (2014) state that when plant is challenged with zucchini yellow mosaic virus (ZYMV), *B. bassiana*-inoculated plants not only reduced the severity of disease incidence but also significantly showed faster growth and development in plants compared to control. The same has been demonstrated in *M. robertsii*-inoculated plants that reduced disease indices and enhanced plant growth upon challenge inoculums of *F. solani*, through the production of plant growth-promoting attributes like phytohormones or siderophores (Sasan and Bidochka 2013). This evidence is supported by the study of Khan et al. (2012), where *M. anisopliae* significantly supports soybean plant's health and growth through the production of phytohormones in inoculated plants. Furthermore, Sánchez-Rodríguez et al. (2015) reported that entomopathogenic-fungi-induced plant growth promotion might be due to the increased uptake and transportation of nutrients (Behie and Bidochka 2014).

9.8 Conclusions, Future Opportunities, and Challenges

Due to increasing insect pest disease problems in agriculture and environmental concerns, myco-biocontrol of insects has been viewed as a substantial significant substitute for synthetic chemical pesticides and a key component of eco-friendly pest management. Understanding the ecological and environmental parameters associated with insect-pathogenic fungi is important in advancing our basic knowledge about the evolution and maintenance of pathogenesis of these organisms as well as in field applications for biological control of insect pests. Entomopathogenic fungi are a unique and enormously specialized group of microbial agents that possess several desirable traits favoring their development as biopesticides. However, the majority of the commercially produced fungi are handiest based on a few species of *Metarhizium, Beauveria, Lecanicillium, Isaria,* etc. In addition to regulatory issues relating to the premarket authorization of new active substances, the availability of entomopathogens bio-processed products is still restricted to certain crop pests. Indeed, a better understanding of the ecology of fungal entomopathogens would induce the development and uptake of more commercially available biopesticides based on these fungi in mainstream farming. Furthermore, to encourage the extensive use of fungal entomopathogen-based biopesticides, there is a need for microbial products with activity against multiple pests in addition to their genetically modified insecticidal efficacy, improved delivery methods, and increased persistence. Following these trends, as a result, researches and industrial interest on entomopathogenic fungi are required, and new discoveries with increased compatibility, maximized efficacy, increasing investments, and expansion of entomopathogens formulations for integrated pest management are expected in the near future.

References

Adams DJ (2004) Fungal cell wall chitinases and glucanases. Microbiology 150:2029–2035

Agrawal Y, Khatri I, Subramanian S, Shenoy BD (2015) Genome sequence, comparative analysis, and evolutionary insights into chitinases of entomopathogenic fungus *Hirsutella thompsonii*. Genome Biol Evol 7:916–930

Akoh CC, Chang SW, Lee GC, Shaw JF (2007) Enzymatic approach to biodiesel production. J Agric Food Chem 55:8995–9005

Alavo TBC, Sermann H, Bochow H (2002) Virulence of strains of the entomopathogenic fungus Verticillium lecanii to aphids: strain improvement. Arch Phytopathol Plant Prot 34:379–398

Ali S, Huang Z, Ren S (2009) Biocontrol Sci Technol:523–535

Anchez-Rodríguez AR, Del Campillo MC, Quesada-Moraga E (2015) Beauveria bassiana: an entomopathogenic fungus alleviates Fe chlorosis symptoms in plants grown on calcareous substrates. Sci Hortic 197:193–202

Asi MR, Bashir MH, Afzal M, Zia K, Akram M (2013) Potential of entomopathogenic fungi for biocontrol of Spodoptera litura Fabricius (Lepidoptera: Noctuidae). J Anim Plant Sci 23(3):913–918

Behie SW, Bidochka MJ (2014) An additional branch of the soil nitrogen cycle: ubiquity of insect-derived nitrogen transfer to plants by endophytic insect pathogenic fungi. Appl Environ Microb 80:1553–1560

Benhamou N, Brodeur J (2001) Pre-inoculation of Ri T-DNA transformed cucumber roots with the mycoparasite, Verticillium lecanii, induces host defense reactions against *Pythium ultimum* infection. Physiol Mol Plant Pathol 58:133–146

Blanford S, Shi WP, Christian R, Marden JH, Koekemoer LL, Brooke BD, Coetzee M, Read AF, Thomas MB (2011) Lethal and pre-lethal effects of a fungal biopesticide contribute to substantial and rapid control of malaria vectors. PLoS One 6:e23591. https://doi.org/10.1371/journal.pone.0023591

Bogo MR, Rota CA, Rota H, Pinto H Jr, Ocampos M, Correa CT, Vainstein MH, Schrank A (1998) A chitinase encoding gene (chit1 gene) from the entomopathogen *Metarhizium anisopliae*: isolation and characterization of genomic and full- length cDNA. Curr Microbiol 37:221–225

Boomsma JJ, Jensen AB, Meyling NV, Eilenberg J (2014) Evolutionary interaction networks of insect pathogenic fungi. Annu Rev Entomol 59:467–485

Braga-Neto R, Luiz~ao RCC, Magnusson WE, Zuquim G, Castilho VC (2008) Leaf litter fungi in a central Amazonian forest: the influence of rainfall, soil and topography on the distribution of fruiting bodies. Biodivers Conserv 17(11):2701–2712

Brownbridge M, Reay SD, Cummings NJ (2010) Association of entomopathogenic fungi with exotic bark beetles in New Zealand pine plantations. Mycopathologia 169(1):75–80

Bushley KE, Raja R, Jaiswal P, Cumbie JS, Nonogaki M, Boyd AE, Spatafora JW (2013) The genome of Tolypocladium inflatum: evolution, organization, and expression of the cyclosporin biosynthetic gene cluster. PLoS Genet 9:e1003496

Chandler D, Bailey AS, Tatchell GM, Davidson G, Greaves J, Grant WP (2011) The development, regulation and use of biopesticides for integrated pest management. Philos Trans R Soc B 366:1987–1998

Charnley AK (2003) Fungal pathogens of insects: cuticle degrading enzymes and toxins. Adv Bot Res 40:242–321

Chouvenc T, Su NY, Elliott ML (2008) Interaction between the subterranean termite Reticulitermes flavipes (Isoptera: Rhinotermitidae) and the entomopathogenic fungus Metarhizium anisopliae in foraging arenas. J Econ Entomol 101:885–893

Chrzanowska J, Kolaczkowska M (1998) Production of extracellular proteolytic enzymes by *Beauveria bassiana*. Acta Mycol 33(2):277–285

Cristina P, Stoian G (2017) The role of hydrolytic enzymes produced by entomopathogenic fungi in pathogenesis of insects. Rom J Plant Prot X

Dara SK (2015) Reporting the occurrence of rice root aphid and honeysuckle aphid and their management in organic celery. UCANR eJournal Strawberries and Vegetables, 21 August. http://ucanr.edu/blogs/blogcore/postdetail.cfm?postnum=18740

Dara SK (2016) IPM solutions for insect pests in California strawberries: efficacy of botanical, chemical, mechanical, and microbial options. CAPCA Advis 19(2):40–46

de Bekker C, Ohm RA, Loreto RG, Sebastian A, Albert I, Merrow M, Hughes DP (2015) Gene expression during zombie ant biting behavior reflects the complexity underlying fungal parasitic behavioral manipulation. BMC Genomics 16:620

de Crecy E, Jaronski S, Lyons B, Lyons TJ, Keyhani NO (2009) Directed evolution of a filamentous fungus for thermotolerance. BMC Biotechnol 9:74

Dhawan M, Joshi N (2017) Enzymatic comparison and mortality of *Beauveria bassiana* against cabbage caterpillar Pieris brassicae LINN. Braz J Microbiol 48:522–529

Fan Y, Pereira RM, Kilic E, Casella G, Keyhani NO (2012) Pyrokinin beta-neuropeptide affects necrophoretic behavior in fire ants (S. invicta), and expression of beta- NP in a mycoinsecticide increases its virulence. PLoS One 7:e26924

Fang W, St. Leger RJ (2012) Enhanced UV resistance and improved killing of malaria mosquitoes by photolyase transgenic entomopathogenic fungi. PLoS One 7:e43069

Fang W, Leng B, Xiao Y, Jin K, Ma J, Fan Y et al (2005) Cloning of Beauveria bassiana chitinase gene Bbchit1 and its application to improve fungal strain virulence. Appl Environ Microbiol 71:363–370

Fang W, Azimzadeh P, St. Leger RJ (2012) Strain improvement of fungal insecticides for controlling insect pests and vector-borne diseases. Curr Opin Microbiol 15:232–238

Faria MR, Wraight SP (2007) Mycoinsecticides and mycoacaricides: a comprehensive list with worldwide coverage and international classification of formulation types. Biol Control 43:237–256

Gabarty A, Salem HM, Fouda MA, Abas AA, Ibrahim AA (2014) Pathogenicity induced by the entomopathogenic fungi *Beauveria bassiana* and *Metarhizium anisopliae* in Agrotis ipsilon (Hufn.). J Radiat Res Appl Sci 7:95–100

Gao Q, Jin K, Ying SH, Zhang Y, Xiao G et al (2011) Genome sequencing and comparative transcriptomics of the model entomopathogenic fungi *Metarhizium anisopliae* and *M acridum*. PLOS Genet 7:e1001264

Goettel MS, Inglis GD, Wraight SP (2000) Fungi. In: Lacey LA, Kaya HK (eds) Field manual of techniques in invertebrate pathology. Kluwer Academic Publishers, Dordrecht, pp 255–282. 7985055235

Goettel MS, Hajek AE, Siegel JP, Evans HC (2001) Safety of fungal biocontrol agents. In: Butt T, Jackson C, Magan N (eds) Fungi as biocontrol agents-progress, problems and potential. CABI Press, Wallingford, pp 347–375

Goettel MS, Eilenberg J, Glare TR (2010) Entomopathogenic fungi and their role in regulation of insect populations. In: Gilbert LI, Gill DS (eds) Insect control: biological and synthetic agents. Academic, San Diego, pp 387–431

Gómez-Vidal S, Salinas J, Tena M, Lopez-Llorca LV (2009) Proteomic analysis of date palm (*Phoenix dactylifera* L.) responses to endophytic colonization by entomopathogenic fungi. Electrophoresis 30:2996–3005

Gressel J (2007) Failsafe mechanisms for preventing gene flow and organism dispersal of enhanced microbial biocontrol agents. In: Vurro M, Gressel J (eds) Novel biotechnologies for biocontrol agent enhancement and management. Springer, Dordrecht

Gul HT, Saeed S, Khan FZA (2014) Entomopathogenic fungi as effective insect pest management tactic: a review. Appl Sci Bus Econ 1:10–18

Hartley SE, Eschen R, Horwood JM, Gange AC, Hill EM (2015) Infection by a foliar endophyte elicits novel arabidopside-based plant defence reactions in its host, Cirsium arvense. New Phytol 205:816–827

Hegedus DD, Khachatourians GG (1988) Production of an extracellular lipase by Beauveria bassiana. Biotechnol Lett 10(9):637–642

Hegedus DD, Khachatourians GG (1996) Identification and differentiation of the entomopathogenic fungus Beauveria bassiana using polymerase chain reaction and single-strand conformation polymorphism analysis. J Invertebr Pathol 67:289–299

Hirano E, Koike M, Aiuchi D, Tani M (2008) Pre-inoculation of cucumber roots with Verticillium lecanii (Lecanicillium muscarium) induces resistance to powdery mildew. Res Bull Obihiro Univ 29:82–94

Hu G, St. Leger RJ (2002) Field studies using a recombinant mycoinsecticide (Metarhizium anisopliae) reveal that it is rhizosphere competent. Appl Environ Microbiol 68:6383–6387

Humber RA (1997) Fungi: identification. In: Lacey LA (ed) Manual of techniques in insect pathology. Academic, London, pp 153–185

Inglis GD, Goettel MS, Butt TM, Strasser H (2001) Use of hyphomycetous fungi for managing insect pests. In: Butt TM, Jackson C, Magan N (eds) Fungi as biocontrol agents: progress problems and potential. CABI Publishing, Wallingford, pp 23–70

Inglis PW, Tigano MS (2006) Identification and taxonomy of some entomopathogenic Paecilomyces spp. (Ascomycota) isolates using rDNA-ITS sequences. Genet Mol Biol 29(1):132–136

Jaber LR, Enkerli J (2016) Effect of seed treatment duration on growth and colonization of Vicia faba by endophytic *Beauveria bassiana* and *Metarhizium brunneum*. Biol Control 103:187–195

Jaber LR, Enkerli J (2017) Fungal entomopathogens as endophytes: can they promote plant growth? Biocontrol Sci Tech 27:28–41

Jaber LR, Salem NM (2014) Endophytic colonisation of squash by the fungal entomopathogen *Beauveria bassiana* (Ascomycota: Hypocreales) for managing Zucchini yellow mosaic virus in cucurbits. Biocontrol Sci Tech 24:1096–1109

Keller S, Schweizer C, Keller E, Brenner H (1997) Control of white grubs (Melolontha melolontha L.) by treating adults with the fungus *Beauveria brongniartii*. Biocontrol Sci Technol 7:105–116

Khan AL, Hamayun M, Khan SA, Kang SM, Shinwari ZK, Kamran M, Ur Rehman S, Kim JG, Lee IJ (2012) Pure culture of *Metarhizium anisopliae* LHL07 reprograms soybean to higher growth and mitigates salt stress. World J Microbiol Biotechnol 28:1483–1494

Kuldau G, Bacon C (2008) Clavicipitaceous endophytes: their ability to enhance resistance of grasses to multiple stresses. Biol Control 46:57–71

Lacey LA, Grzywacz D, Shapiro-Ilan DI, Frutos R, Brownbridge M, Goettel MS (2015) Insect pathogens as biological control agents: back to the future. J Invertebr Pathol 132:1–41. https://doi.org/10.1016/j.jip.2015.07.009

Liao X, Lu HL, Fang W, St. Leger RJ (2014) Overexpression of a *Metarhizium robertsii* HSP25 gene increases thermotolerance and survival in soil. Appl Microbiol Biotechnol 98:777–783

Liu SQ, Meng ZH, Yang JK, Fu YK, Zhang KQ (2007) Characterizing structural features of cuticle-degrading proteases form fungi by molecular modeling. BMC Struct Biol 7:33

Lopez DC, Sword GA (2015) The endophytic fungal entomopathogens Beauveria bassiana and Purpureocillium lilacinum enhance the growth of cultivated cotton (Gossypium hirsutum) and negatively affect survival of the cotton bollworm (Helicoverpa zea). Biol Control 89:53–60

Lovett B, St Leger RJ (2015) Stress is the rule rather than the exception for *Metarhizium*. Curr Genet 61:253–261

Mantzoukas S, Chondrogiannis C, Grammatikopoulos G (2015) Effect of three endophytic entomopathogens on sweet sorghum and on the larvae of the stalk borer Sesamia nonagrioides. Entomol Exp Appl 154:78–87

Mathew SO, Sandhu SS, Rajak RC (1998) Bioactivity of *Nomuraea rileyi* against Spilosoma obliqua: effect of dosage, temperature and relative humidity. J Indian Bot Soc 77:23–25

McLaughlin DJ, Hibbett DS, Lutzoni F, Spatafora JW, Vilgalys R (2009) The search for the fungal tree of life. Trends Microbiol 17:488–497

Meyling NV, Schmidt NM, Eilenberg J (2012) Occurrence and diversity of fungal entomopathogens in soils of low and high Arctic Greenland. Polar Biol 35:1439–1445

Mondal S, Baksi S, Koris A, Vatai G (2016) Journey of enzymes in entomopathogenic fungi. Pac Sci Rev 18:85–99

Mudgal S, De Toni A, Tostivint C, Hokkanen H, Chandler D (2013) Scientific support, literature review and data collection and analysis for risk assessment on microbial organisms used as active substance in plant protection products –lot 1 environmental risk

Mustafa U, Kaur G (2009) Extracellular enzyme production in *Metarhizium anisopliae* isolates. Folia Microbiol 54:499–504

Nana P, Nchu F, Ekesi S, Boga HI, Kamtchouing P, Maniania NK (2015) Efficacy of spot spray application of *Metarhizium anisopliae* formulated in emulsifiable extract of *Calpurnia aurea* in attracting and infecting adult *Rhipicephalus appendiculatus* ticks in semi-field experiments. J Pest Sci 88:613–619

Oerke EC (2006) Crop losses to pests. J Agric Sci 144:31–43

Oerke EC, Dehne HW, Schoenbeck F, Weber A (1994) Crop production and crop protection: estimated losses in major food and cash crops. Elsevier Science Publishers B.V, Amsterdam

Ortiz-Urquiza A, Keyhani NO (2015) Action on the surface: entomopathogenic fungi versus the insect cuticle. Insects 4:357–374

Ortiz-Urquiza A, Keyhani NO, Quesada-Moraga E (2013) Culture conditions affect virulence and production of insect toxic proteins in the entomopathogenic fungus *Metarhizium anisopliae*. Biocontrol Sci Tech. https://doi.org/10.1080/09583157.2013.822474

Pattemore JA, Hane JK, Williams AH, Wilson BA, Stodart BJ, Ash GJ (2014) The genome sequence of the biocontrol fungus *Metarhizium anisopliae* and comparative genomics of *Metarhizium* species. BMC Genomics 15:660

Paula AR, Carolino AT, Silva CP, Pereira CR, Samuels RI (2013) Testing fungus impregnated cloths for the control of adult Aedes aegypti under natural conditions. Parasit Vectors 6:1–6. https://doi.org/10.1186/1756-3305-6-256

Pava-Ripoll M, Posada FJ, Momen B, Wang C, St. Leger RJ (2008) Increased pathogenicity against coffee berry borer, Hypothenemus hampei (Coleoptera: Curculionidae) by *Metarhizium anisopliae* expressing the scorpion toxin (AaIT) gene. J Invertebr Pathol 99:220–226

Petrisor C, Stoian G (2017) The role of hydrolytic enzymes produced by entomopathogenic fungi in pathogenesis of insects. Rom J Plant Prot X

Pieterse CM, Zamioudis C, Berendsen RL, Weller DM, Van Wees SC, Bakker PA (2014) Induced systemic resistance by beneficial microbes. Annu Rev Phytopathol 52:34

Reddy KRK, Kuamr PD, Reddy KRN (2013) Entomopathogenic fungi: a potential bioinsecticide. KAVAKA 41:23–32

Roberts DW, Humber RA (1981) Entomogenous fungi. In: Cole GT, Kendrick B (eds) Biology of conidial fungi. Academic, New York, pp 201–236

Sahayaraj K (ed) (2014) Basic and applied aspects of biopesticides. Springer, New Delhi, pp 31–46

Samuels RI, Paterson IC (1995) Cuticle-degrading proteinase from insect moulting fluid and culture filtrates of entomopathogenic fungi. Comp Biochem Physiol B Biochem Mol Biol 110:661–669

Samuels RI, Santos AV, Silva CP (2011) Enzymology of entomopathogenic fungi. In: Borgio JF et al (eds) Microbial insecticides -principles and applications. Nova Science Publishers, Inc, pp 71–92

Sanchez-Perez LC, Barranco-Floriab JE, Rodriguez-Navarro S, Cervantes-Mayagoitia JF, Ramos-Lopez MA (2014) Enzymes and entomopathogenic fungi: advances and insights. Adv Enzym Res 2:65–76

Sánchez-Rodríguez AR, del Campillo MC, Quesada-Moraga E (2015) Beauveria bassiana: an entomopathogenic fungus alleviates Fe chlorosis symptoms in plants grown on calcareous substrates. Sci Hortic 197:193–202

Sandhu SS, Vikrant P (2006) Evaluation of mosquito larvicidal toxins in the extra cellular metabolites of two fungal genera *Beauveria* and *Trichoderma*. In: Bagyanarayana G, Bhadraiah B, Kunwar IK (eds) Emerging trends in mycology, plant pathology and microbial biotechnology. BS Publications, Hyderabad

Sandhu SS, Anil KS, Vikas B, Gunjan GPB, Anil Kumar Sundeep J, Sharma AK, Malhotra S (2012) Myco-biocontrol of insect pest factors involved mechanism, and regulation. J Pathog:1–10

Sasan RK, Bidochka MJ (2012) The insect-pathogenic fungus *Metarhizium robertsii* (Clavicipitaceae) is also an endophyte that stimulates plant root development. Am J Bot 99:101–107

Sasan RK, Bidochka MJ (2013) Antagonism of the insect- pathogenic fungus Metarhizium robertsii against the bean plant pathogen *Fusarium solani* f. sp. phaseoli. Can J Plant Pathol 35:288–293

Seidl V (2008) Chitinases of filamentous fungi: a large group of diverse proteins with multiple physiological functions. Fungal Biol Rev 22:36–42

Shang YF, Duan ZB et al (2012) Improving UV resistance and virulence of *Beauveria bassiana* by genetic engineering with an exogenous tyrosinase gene. J Invertebr Pathol 109:105–109
Shrivastava G, Ownley BH, Augé RM, Toler H, Dee M, Vu A, Köllner TG, Chen F (2015) Colonization by arbuscular mycorrhizal and endophytic fungi enhanced terpene production in tomato plants and their defense against a herbivorous insect. Symbiosis 65:65–74
Silva W, Mitidieri S, Schrank A, Vainstein MH (2005) Production and extraction of an extracellular lipase from the entomopathogenic fungus *Metarhizium anisopliae*. Process Biochem 40:321–326
Silva WOB, Santi L, Scharank A, Vainstein MH (2010) *Metarhizium anisopliae* lipolytic activity plays a pivotal role in Rhipicephalus (Boophilus) microplus infection. Fungal Biol 144:10–15
St Leger RJ (1991) A model to explain differentiation of appressoria by germlings of Metarhizium anisopliae. J Invertebr Pathol 57:299–310
Sujeetha A, Sahayaraj K (2014) Role of entomopathogenic fungus in pest management. In: Tanada Y, Kaya HK (eds) Insect pathology. Academic, New York. ISBN 0-12-683255-2
Tanada Y, Kaya HK (1993) Insect pathology. Academic Press, San Diego. 666 pp002E
Thacker JRM (2002) An introduction to arthropod pest control. Cambridge University Press, Cambridge
Vega FE (2018) The use of fungal entomopathogens as endophytes in biological control: a review. Mycologia 110(1):4–30
Vega FE, Meyling NV, Luangsa-Ard JJ, Blackwell M (2012) Fungal entomopathogens. In: Vega FE, Kaya HK (eds) Insect pathology, 2nd edn. Academic, San Diego, pp 171–220
Vidal S, Jaber LR (2015) Entomopathogenic fungi as endophytes: plant- endophyte-herbivore interactions and prospects for use in biological control. Curr Sci 109:46–54
Wang CS, St. Leger RJ (2007) The MAD1 adhesion of *Metarhizium anisopliae* links adhesion with blastospore production and virulence to insects, and the MAD2 adhesion enables attachment to plants. Eukaryot Cell 6:808–816
Wang CS, St. Leger RJ (2014) Genomics of entomopathogenic fungi. In: Martin F (ed) The ecological genomics of fungi. Wiley, pp 243–260
Xiao GH, Ying S-H, Zheng Z, Wang ZL, Zhang S, Xie XQ, Feng MG (2012) Genomic perspectives on the evolution of fungal entomopathogenicity in *Beauveria bassiana*. Sci Rep 2:483
Xu C, Zhang X, Qian Y, Chen X, Liu R, Zeng G et al (2014) A high-throughput gene disruption methodology for the entomopathogenic fungus *Metarhizium robertsii*. PLoS One 9:e107657
Yang J, Tian B, Liang L, Zhang KQ (2007) Extracellular enzymes and the pathogenesis of nematophagous fungi. Appl Microbiol Biotechnol 75:21–31
Yang MM, Meng LL, Zang YA, Wang YZ, Qu LJ, Wang QH, Ding JY (2012) Baculoviruses and insect pest control in China. Afr J Microbiol Res 6:214–218
Ying SH, Feng MG (2011) Integration of *Escherichia coli* thioredoxin (trxA) into Beauveria bassiana enhances the fungal tolerance to the stresses of oxidation, heat and UV-B irradiation. Biol Control 59:255–260
Zhang S, Wideman E, Bernard G, Lesot A, Pinot E, Pedrini N, Keyhani NO (2012) CYP52X1, representing new cytochrome P450 subfamily, displays fatty acid hydroxylase activity and contributes to virulence and growth on insect cuticular substrates in entomopathogenic fungus *Beauveria bassiana*. J Biol Chem 28:13477–13486
Zhao H, Xu C, Lu HL, Chen X, St. Leger RJ, Fang W (2014) Host- to-pathogen gene transfer facilitated infection of insects by a pathogenic fungus. PLoS Pathog 10:e1004009
Zhao H, Lovettx B, Fang W (2016) Genetically engineering entomopathogenic fungi. Adv Genet 94:137–163
Zheng P, Xia Y, Xiao G, Xiong C, Hu X, Zhang S et al (2011) Genome sequence of the insect pathogenic fungus Cordyceps militaris, a valued traditional Chinese medicine. Genome Biol 12:R116

Role of Microbiotic Factors Against the Soil-Borne Phytopathogens

10

Nasreen Musheer, Shabbir Ashraf, Anam Choudhary, Manish Kumar, and Sabiha Saeed

Abstract

Phytopathogenic association with beneficial microbiotic factors influences rhizospheric soil as well as plant growth parameters. Rhizospheric microbiotic factors check nutrients to supplement the lethal sensitivity against soilborne phytopathogens. All microbes besides harming plant growth are also able to reduce or check infection or disease caused by phytopathogens. Each microorganism showed specific antagonistic mechanisms against specific phytopathogens. This chapter discussed the importance of nematodes belonging to order Aphelenchida and Tylenchida which proved to be good management model organisms to inhibit or kill phytopathogens just like plant growth, promoting bacteria and fungi. Beneficial microbes protect plants from a greater extension of damage and induced plant vigor, growth, and development. Besides the beneficial role of microbiotic factors interacting with plants against soilborne phytopathogens present in soil ecology, it can also helps to develop products for agricultural biotechnology, biofertilizers, plant strengtheners, phytostimulation, and biopesticides. This chapter appraises the importance of microbiota factors and their mechanisms against soilborne phytopathogens.

Keywords

Arbuscular mycorrhiza · Biocontrol · Microbiome · Plant pathogens · Soil

N. Musheer (✉) · S. Ashraf · A. Choudhary · M. Kumar · S. Saeed
Department of Plant Protection, Aligarh Muslim University, Aligarh, Uttar Pradesh, India

© Springer Nature Singapore Pte Ltd. 2020

M. K. Solanki et al. (eds.), *Phytobiomes: Current Insights and Future Vistas*,
https://doi.org/10.1007/978-981-15-3151-4_10

10.1 Introduction

Soil ecosystem is full of microbiotic communities that help to improve plant sustainability and control soilborne phytopathogens through different kinds of mechanisms (Ahemad et al. 2009; Patil and Solanki 2016a, b; Solanki et al. 2016). Some of the beneficial effects observed are the following:

1. Stimulates plant growth
2. Increased nutrient mobilization through plant's root from surrounding soils
3. Induced plant growth regulators
4. Activated defense system of plants against phytopathogenic infections
5. Improved topsoil properties through bioremediation by sequestering soil pollutants such as toxic heavy metal and xenobiotic residues

Beneficial plant–microbe interaction was found to be of great importance for both plant- and soil-ecosystem functions over plant breeding (Smith et al. 1999). Soil ecosystem association with rhizo-microbiotic factors helps to reduce plant pathogenic disease stress over plants and promotes growth by enhancing nutrient uptake facilities (Thakore and Ehlers 2006; Solanki et al. 2019b). The improved soil ecosystem approach helps in the management of soilborne phytopathogens sustainably by increasing disease resistance, crop production, and quality (Ahemad et al. 2009; Solanki et al. 2012a, b; Singh et al. 2014). *Trichoderma* species were found to be used most widely as fungal biocontrol agents against many phytopathogens and played a significant role in bioremediation and soil restoration (Solanki et al. 2011, 2018, 2019a; Rai et al. 2019). Under field conditions, *Trichoderma* spp. applied as conidia and mycelia biomass showed great stability and viability (Rosane et al. 2008). Several plant growth-promoting rhizobacteria (PGPR) flourished rhizospheric soil and plant root and stimulate various mechanisms against soilborne pathogens to prevent plants from infection and diseases (Patil and Solanki 2016a; Kashyap et al. 2019; Kumari et al. 2019). It includes competitions, phytohormones, secondary active metabolites which increased nutrient solubilization and mobilization, availability, and activation of physical, chemical, and active defense responses in the host against their antagonists (Hammerschmidt and Kuc 1995; Vessey 2003; Solanki et al. 2012a, b, 2014, 2015, 2017; Kumar et al. 2012; Wang et al. 2017). The present chapter enlightens the importance and role of microbiotic factors against soilborne pathogens causing various plant diseases and the definition of the microbiome, biocontrol agents (bacterial, fungal, and nematodes), soilborne phytopathogens, and mode of action of biocontrol agents, mechanisms, and future approaches.

10.2 Microbiome

Microbiome or rhizosphere is the specific microenvironment that includes plant roots, soil niche, and plant microbes, viz., protozoans, filamentous fungi, viruses, bacteria, actinomycetes, and nematodes, which formed an association (Hiltner 1904; Saunders et al. 2010). Microbes compete for water, nutrients, and space and

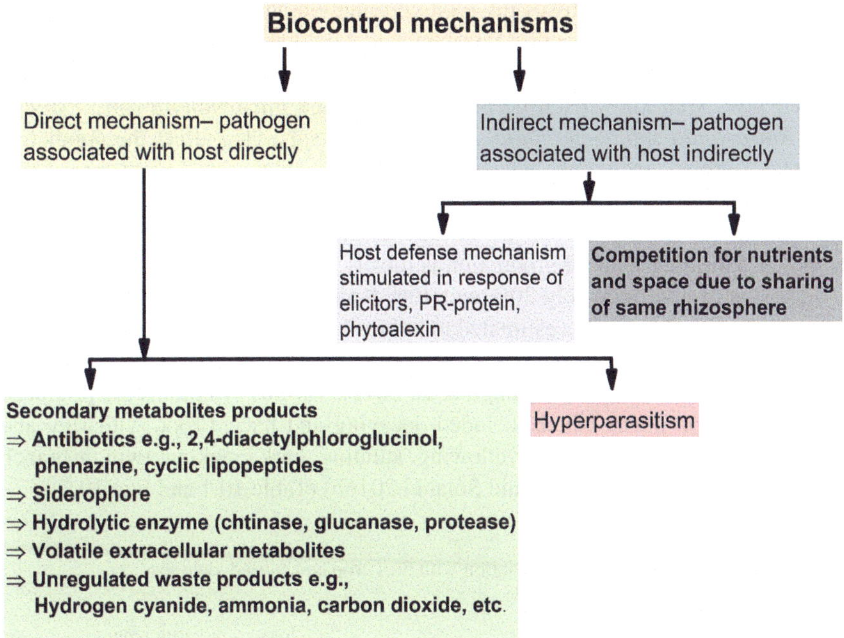

Fig. 10.1 Different mechanisms of biocontrol that are involved in plant disease management

sometimes improve microbial attractiveness toward plant roots more than phytopathogens (Cook and Baker 1983; Whipps 2001). Microbe colonization is established generously in all plant-associated microenvironments, especially rhizospheric region and root (Malviya et al. 2019; Solanki et al. 2019b). Berg et al. (2006) performed an in vitro assessment of plant-associated microbial isolates between 1% and 35% exhibited in antimicrobial properties. The narrow zone of soil around the plant's root is rich in nutrients due to the secretion of a variety of chemical constituents by plants. Root exudates may contain some protein and sugars. It promotes bacteria growth around the nutrient-rich zone (Gray and Smith 2005). Therefore, rhizospheric microorganisms are responsible for increasing plant vigor in an environmentally and eco-friendly manner (Leach et al. 2017; Igiehon and Bbaloal 2018). Solanki et al. (2018) assessed that tomato plant root extract played a significant role in the antagonism of *Bacillus*, *Streptomyces*, and *Trichoderma* against *Rhizoctonia solani* in the soil microecosystem (Fig. 10.1.).

10.3 Soilborne Phytopathogens and Their Causes in Plants

Soil ecosystem is a sophisticated resident of microflora, and their effect may be neutral, positive, or negative toward plants (Thrall et al. 2007). Complex microbial interactions form between (1) soilborne pathogens and their host and (2) pathogens and soil health. Mostly plant microbiota known as "phytobiota" survive in or near the soil and affect plant vigor, for example, cotton seed association with *Epicoccum nigrum* or its exudate improves the seed germination and seedling growth against *Pythium* damping off and root rot (Hashem and Ali 2004; Nelson 2017). Moreover, soilborne diseases are often much influenced by soil conditions, even in the absence of their host. Well-known soil-borne pathogens, e.g., *Phytophthora* spp., *Pythium* spp., *Rhizoctonia* spp., *Sclerotinia* spp., *Fusarium* spp., etc., affect the number of important crops, including wheat, cotton, vegetables, and temperate fruits (Mihajlovic et al. 2017; Patil and Solanki 2016b; Thaware et al. 2017). These pathogen inoculums survive in soil as microsclerotia, sclerotia, chlamydospores, or oospores. World agriculture production is mostly affected by microbes and caused yield losses of approximately 20% and 40% (Oerke 2006; Wild and Gong 2010). Fusarium wilt pathogen was estimated to induced 50–75% yield loss in many crops (Tiwari et al. 2018). Soilborne phytopathogens initiate specific symptoms when shown compatible association (Singh et al. 2019). Various symptoms are produced by pathogens on plants which included decaying and flaccid root, withering and drooping of plants, and flaccid, yellowing, stunting, bark cracking, twig or branch dieback, etc. (Horst 2001; Patil and Solanki 2016b) (Table 10.1 and Fig. 10.2).

Table 10.1 Soilborne pathogen association with host crop and caused yield loss

Host	Pathogen	Yield loss	References
Cereals (wheat barley, oats, rye, rice)	*Heterodera avenae* (cereal cyst nematode)	10%	Ali et al. (2019)
Ginger	*Ralstonia solanacearum* (bacterial wilt)	51.4–51.9%	Guji et al. (2019)
Wheat	*Gaeumannomyces graminis* var. *tritici* (take all disease of wheat)	2–92%	Ramanauskienė et al. (2019)
Soybean	*Rhizoctonia solani*	48–52%	Chang et al. (2018)
Chickpea	*F. oxysporum* f. sp. *ciceri*	34–57.33%	Murali et al. (2018)
Lentil	*Fusarium oxysporum* f. sp. *lentis* (wilt)	50%	Tiwari et al. (2018)
Mung bean	*Fusarium oxysporum*	80%	Kelly (2017)
Tomato	Root knot nematode (*Meloidogyne* spp.)	11–35%	Manjunatha et al. (2017)
Tomato	*Fusarium oxysporum* f. sp. *lycopersici*	10–50%	Ghazalibiglar et al. (2016)
Brinjal	*Sclerotium rolfsii* (foot and root rot)	30–50%	Siddique et al. (2016)
Chickpea	*F. oxysporum* f. sp. *ciceri*	48.29%	Thaware et al. (2015)
Turmeric	*Pythium* spp.	74.50%	Anoop et al. (2017)
	Fusarium spp.	30.51%	
	Rhizoctonia spp.	28.80%	
Tomato	*Rhizoctonia solani*	67–87%	Solanki et al. (2012a, 2014)
Bean	*Fusarium decline disease*	52%	Saremi et al. (2011)

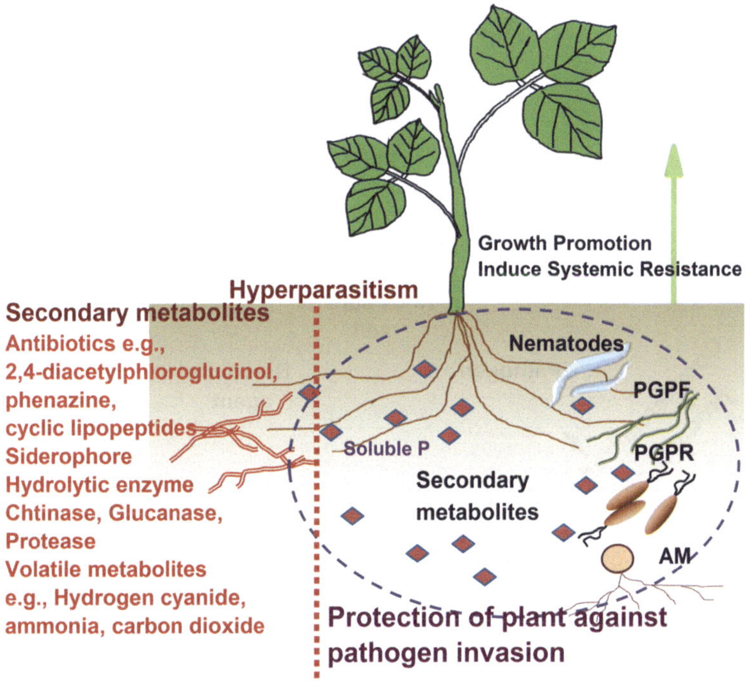

Fig. 10.2 Rhizosphere shows the interaction between beneficial microbes (plant growth-promoting fungi (PGPF), plant growth-promoting rhizobacteria (PGPR), and arbuscular mycorrhiza (AM)) and protection of plant against the pathogens, and under the rhizosphere microbes perform several activities that played a significant role in disease management such as nutrient competition, mycoparasitism, siderophores, antibiotics, and phytohormones

10.4 Biocontrol Agents Mechanisms Employed Against Soilborne Phytopathogens

Biological control agents (BCA) are natural enemies that are used to control the activity of virus, bacteria, fungi, and protozoans (Saunders et al. 2010). Functional interaction or association between bioagents and pathogenic soil microorganisms led to reduced root infection by destroying various infection propagules such as conidia, chlamydospores, zoospores, mycelium, and egg mass (Beneduzi et al. 2012; Hassani et al. 2018). Therefore, the additive effect of antagonistic biocontrol agents was found to be more efficient in declining of multiple soilborne diseases (Mazzola and Freilich 2016). Biocontrol agents are a nonpathogenic and an environmentally alternative safe method. Some bioagents show dominance over chemical for plant protection against their pathogens below or up to threshold level (Neshev 2008). However, commercialization of BCAs followed Central Insecticide Board (CIB) regulatory that provide registration and permit their commercial use after reviewing the submitted reports for registration. It is an important point about the bioagents that can reduce harmful effects of some pathogens below a certain

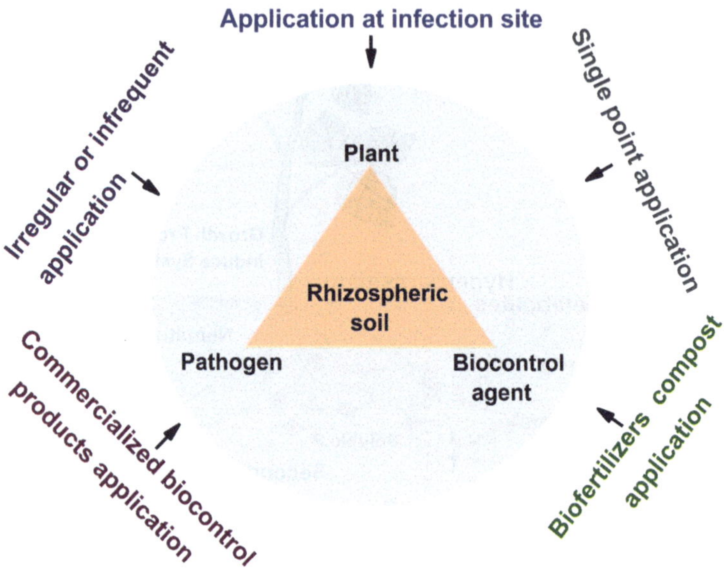

Fig. 10.3 Rhizospheric soil association with plant, pathogen, and biocontrol agents

threshold with no substantial changes in soil microbiological balance, something that does not occur when chemicals are applied (Neshev 2008) (Fig. 10.3 and Table 10.2).

10.5 Biocontrol Agent's Mechanisms and Their Mode of Application

Soil rhizosphere and the plant' root exudates attracts a group of the mixed microbial community including *Trichoderma* spp., yeast, PGRP, nematodes, etc. (Druzhinina et al. 2011; Lombardi et al. 2018). *Trichoderma* is a free-living ubiquitous antagonistic filamentous fungi that is mostly found in soil ecosystem (Harman et al. 2004). It is capable of producing different antimicrobial compounds that help in plant growth promotion and defense management. Biocontrol agents have strong survival ability and high reproductive rate than phytopathogens (Singh et al. 2016). *Trichoderma* shows bioefficacy against several fungal seeds or soilborne pathogenic fungi such as *Fusarium*, *Pythium*, *Macrophomina*, *Verticillium*, *Rhizoctonia*, etc. (Peteira et al. 2001; Smitha et al. 2014). Among various fungal bioagents, yeast is also found commonly in soil ecosystem with variable soil texture, composition, humidity, pH, and diverse climatic conditions (Yurkov 2018). Yeast was found in association with certain plant roots and exhibited as biostimulants against many soilborne pathogens, e.g., *Arabidopsis thaliana*, stimulate a defensive response against *Botrytis cinerea* and increase soil sporulation of arbuscular mycorrhiza

Table 10.2 Biocontrol agents and their agro-products applied to different crops to protect against the pathogens

Biocontrol agents	Commercial name	Target organism	Crop	Manufacturer company name	References
Bacillus subtilis FZB24	Rhizo-Plus Rhizo-Plus Konz	*Rhizoctonia*, *Fusarium*, *Verticillium* and *Streptomyces*	Vegetables and ornamental plants	FZB Biotechnik, GmbH	Shukla et al. (2019)
Pseudomonas chlororaphis	Cedomon	Soilborne pathogenic fungi	Barley and oat	BioAgri AB	Shukla et al. (2019)
Burkholderia cepacia	Deny	Soilborne pathogenic fungi	Vegetables	Stine microbial products	Shukla et al. (2019)
Pseudomonas aureofaciens strain TX-1	Bio -jet, spotless	*Rhizoctonia solani*, *Pythium*	Vegetables and ornamentals in greenhouses	EcoSoil system	Tsegaye et al. (2018)
Trichoderma viride	Trichostar	Collar/root/ stem rot, wilt, damping off	Vegetables and pulses	Super Agro India,	Kumar and Sarma (2016)
T. harzianum	Tricha	*Pythium* rot, *Rhizoctonia* rot, *Fusarium* wilt, *Botrytis* rot, and *Sclerotium*, *Sclerotinia*, *Ustilago*	Food and ornamental crop	Balaji Crop Care Pvt. Ltd., India	Kumar and Sarma (2016)
P. fluorescens	Greenmax	*Rhizoctonia solani*, *Pythium ultimum*	Paddy, oilseed, pulses, and vegetables	GreenMax AgroTech, India	Kumar and Sarma (2016)
B. subtilis strain GB34	GB34	*Rhizoctonia*, *Fusarium*	Soybean	Gustafon, USA	Junaid et al. (2013)
B. subtilis strain GB 03	Kodiac, companion	*Rhizoctonia*, *Aspergillus*	Wheat, barley, peas	Growth products, USA	Junaid et al. (2013)
P. aureofaciens strain TX-1	Bio -jet, spotless	*Rhizoctonia solani*, *Pythium*	Vegetables and ornamentals greenhouses	EcoSoil system	Junaid et al. (2013)
T. harzianum T-22	Root shield, plant shield	Soil-borne pathogens	Greenhouse nurseries	Bio works, USA	Junaid et al. (2013)

(continued)

Table 10.2 (continued)

Biocontrol agents	Commercial name	Target organism	Crop	Manufacturer company name	References
Gliocladium catenulatum strain JI446	Prima stop soil guard	Soilborne pathogens	Vegetables, herbs, spices	Kemira Agro Oy, Finland	Junaid et al. (2013)
B. pumilus GB34	Yield shield	Soil-borne fungal pathogens	Legumes, cereals grains, soybean, cotton, peanuts	(Gustafson); Bayer Crop Science	Berg (2009)
P. syringae	Bio-save	Yeast and bacteria	Apple, pear, potato, lemon	JET Harvest, Longwood, FL, USA	Janisiewicz and Peterson (2004)
T. harzianum T-22	Trianum-P	*Pythium* spp., *Rhizoctonia* spp., *Fusarium* spp., and *Sclerotinia* spp.	Vegetables, soft fruits, cereals, corn	Koppert Biological Systems partners with nature	

(Ferreira-Saab et al. 2018). *Yarrowia lipolytica* (yeast) is nonpathogenic and enables to clean the environment from oil spill contamination (Goncalves et al. 2014). Yeast helps in nitrogen, sulfur oxidation, and phosphorus solubilization near the growing plants. Oxidation and solubilization of macronutrients easily intake by plants that help to enhance the health and also protect from the soilborne pathogen infection (Fu et al. 2016; Alonso et al. 2008). El-Tarabily (2004) studied the sugar beet root rot caused by *Rhizoctonia solani* and documented that the disease is biologically controlled by using yeasts, like *Rhodotorula glutinis*, *Trichosporon asahii*, and *Candida valida*. El-Mehalawy et al. (2004) reported that various yeast species, e.g., *Candida*, *Trichosporon*, *Rhodotorula*, *Saccharomyces*, etc., establish antagonistic activity against soilborne fungal pathogens. These all were used either independently or in combination to reduced late-wilt disease occurrence on maize and bean caused by *Cephalosporium maydis*. *Kazachstania exigua*, *Saccharomyces cerevisiae*, *Clavispora*, and *Candida* strains exhibited antimicrobial activity against *Penicillium italicum* (Perez et al. 2016). Beneficial fungal microbiota shows specific mechanism against target pathogens present in or near the soil and in or on the infected host. Several mechanisms are shown by fungal antagonist microbes such as competitions, antibiosis, mycoparasitism or hyperparasitism, lytic enzymes, hydrogen cyanide, defense response, and plant growth hormone (Zaidi and Singh 2013; Al-Naemi et al. 2016). Protists were found to enhance plant growth by increasing plant root's nutrient absorption and maintaining the soil microbiomes by parasitizing soil bacteria, fungi, and other eukaryotic organisms (Geisen et al. 2018). Macias et al. (2018) demonstrated that sucrose-mediated tomato root association with *Trichoderma atroviride* promotes growth and stimulates antagonistic biota against *Phytophthora cinnamomi* (Table 10.3).

Table 10.3 Biocontrol agents producing secondary metabolites against pathogens

Biocontrol agents	Target organisms	Secondary metabolites	References
Bacillus velezensis	Soilborne phytopathogens	Lipopeptides (i.e., surfactin, bacillomycin-D, fengycin, and bacillibactin) and polyketides (i.e., macrolactin, bacillaene, and difficidin).	Rabbee et al. (2019)
Bacillus velezensis	*Ralstonia solanacearum and Fusarium oxysporum*	Lipopeptide compounds (surfactin, iturin, and fengycin)	Cao et al. (2018)
Trichoderma harzianum	*Rhizoctonia solani*	Harzianic acid	Manganiello et al. (2018)
Bacillus amyloliquefaciens	*Fusarium graminearum*	Bacillomycin D	Gu et al. (2017)
Trichoderma asperellum ZJSX5003	*Fusarium graminearum (corn stalk rot of maize)*	Cell wall-degrading enzymes (chitinase, protease b-glucanases, and peptaibols)	Li et al. (2016a, b)
Bacillus amyloliquefaciens SQR-9	*Ralstonia solanacearum*	Volatile organic compounds (VOCs)	Raza et al. (2016)
Trichoderma harzianum	*Aspergillus flavus*	2-Phenylethanol	Chang et al. (2015)
Streptomyces vinaceusdrappus S5 MW2	*Rhizoctonia* root rot in tomato	Chitinase	Yandigeri et al. (2015)
Streptomyces glauciniger	*F. oxysporum*	Chitinase	Awad et al. (2014)
Bacillus subtilis (B-CM191, B-CV235, B-CL-122)	*Fusarium oxysporum* f. sp. *ciceri* (wilt)	Indole acetic acid, siderophore, phosphate solubilization, and hydrolytic enzymes	Singh et al. (2014)
Pseudomonas fluorescens MPF47	*Rhizoctonia* root rot in tomato	Indole acetic acid, siderophore, phosphate solubilization, and hydrolytic enzymes	Solanki et al. (2014)
Arthrobacter agilis UMCV2	*Botrytis cinerea and Phytophthora cinnamomi*	Dimethyl hexadecylamine (antibiotics)	Velázquez-Becerra et al. (2013)
Bacillus alvei NRC 14	*Fusarium oxysporum*	Lytic enzymes	Abdel-Aziz (2013)
Bacillus subtilis strain BS-99-H	Gaeumannomyces graminis var. tritici	Phenazines	Recinosa et al. (2012)

(continued)

Table 10.3 (continued)

Biocontrol agents	Target organisms	Secondary metabolites	References
Lysinibacillus fusiformis B-CM18	Growth inhibition of *Fusarium oxysporum* f. sp. *ciceri*, *F. solani*, *F. oxysporum* f. sp. *Lycopersici*, and *Macrophomina phaseolina*	Chitinase	Singh et al. (2012)
Bacillus amyloliquefaciens MB101	*Rhizoctonia* root rot in tomato	Mycolytic enzyme production and plant defense activation	Solanki et al. (2012a, b)
Pseudomonas fluorescens F113	Soilborne phytopathogens	2, 4-Diacetyl phloroglucinol	Couillerot et al. (2011)
Trichoderma asperellum UTP-16	Growth inhibition of *Fusarium* spp.	Chitinase	Praveen Kumar et al. (2012)
Trichoderma/*Hypocrea* spp.	*Rhizoctonia* root rot in tomato	Hydrolytic enzymes	Solanki et al. (2011)
Trichoderma spp.	*Botrytis cinerea*, *Rhizoctonia solani*, *Phytophthora citrophthora*	Hydrolytic enzymes (β-1,6-Glucanases)	Druzhinina et al. (2011)
Bacillus subtilis strain BS-99-H	*Botrytis cinerea* and *Rhizoctonia solani*	Iturin A	Lin et al. (2010)
Rhizobacteria	*Fusarium* wilt	Chitinase	Siddiqui et al. (2009)
Bacillus licheniformis	*Gibberella saubinetii* and *Aspergillus niger*	Chitinase	Xiao et al. (2009)
Pseudomonas PGC2	*Rhizoctonia solani* and *Phytophthora capsici*	Chitinase, β-1,3-glucanase, protease, etc.	Arora et al. (2008)
Bacillus and *Pseudomonas*	Phytopathogens	Chitinase	Yu (2008)
Trichoderma viride	*Sclerotium rolfsii*	Antibiotic pyrone (Viridepyronone)	Evidente et al. (2003)

10.5.1 Biocontrol Agent Produces Active Secondary Metabolites Against Phytopathogens

Biocontrol agents produce active secondary metabolites against several soilborne phytopathogens such as *Botrytis cinerea* and *Rhizoctonia solani* (Kloepper et al. 2004). Plant root exudates are the main attractants that induced signals and later recognized by beneficial microbes to increase microbial colonization. Beneficial microbes antagonized directly to the soilborne phytopathogens by some mechanisms include (1) production of antibiotics, toxins, and cell wall-degrading enzymes such as chitinases and *β*-1,3-glucanase; (2) competition for establishment and supplement of macronutrients and micronutrients, e.g., increase siderophore iron-chelating microbes that help in iron uptake of phytosystems; and (3) toxin production degrades pathogenicity factors of pathogens (Thangavelu and Mustaffa 2012; Zaidi and Singh 2013; Hassani et al. 2018; Yandigeri et al. 2015). A recent study by Van Agtmaal et al. (2018) shows confirmation about the microbial community of rhizosphere-induced natural volatile organic compounds that cause suppression of phytopathogens. *Trichoderma* spp. are found to produce various metabolites: (a) volatile antibiotics and toxins like trichothecene and a sesquiterpene, trichodermin, (b) hydrophilic compounds heptelidic acid or koningic acid, and (c) peptaibols (Musoni et al. 2015; Marik et al. 2018; Gebarowska et al. 2019).

10.5.1.1 Antibiosis

Many microbes produce and secrete a simple or multiple forms of toxic compounds with antimicrobial properties at low concentrations (Homma et al. 1989). Lanteigne et al. (2012) reported that *Pseudomonas* sp. produced HCN and DAPG antibiotics which help to suppress the infection of black root rot bacterial canker disease of tobacco caused by *Thielaviopsis basicola* and *P. fluorescens;* BL915 inhibited the tomato bacterial canker caused by *Clavibacter michiganensis*. Iturin-A is a lipopeptide poisonous antibiotic and can kill or inhibit the growth of numerous pathogens at low concentration (Meena and Kanwar 2015). Biosynthesis of antibiotics demonstrated the antagonistic association between microbes and soilborne pathogens (Raaijmakers and Mazzola 2012; Schulz-Bohm et al. 2017). *Coniothyrium minitans* sclerotia are producing phenazines antibiotics attack on fungal hyphae of *Pythium oligandrum* (Thomashow et al. 1990; Perez et al. 2016). For instance, *Sporothrix flocculosa* and *Sporothrix rugulosa* liquid culture produced heptadecenoic and methyl-heptadecenoic acid antibiotics with antimycotic and antibacterial activity (Choudhury et al. 1994; Benyagoub et al. 1996). Both *Botrytis* sp. and *Fusarium oxysporum* sp. spore germination and biomass production were reduced in the presence of antibiotic produced by *Salvia flocculosa* (Hajlaoui et al. 1994). In vitro studies showed that yeasts *Saccharomyces* and *Zygosaccharomyces* were found to inhibit the growth of soilborne fungal plant pathogens such as *Rhizoctonia fragariae*, *Sclerotinia sclerotiorum*, and *Macrophomina phaseolina* due to the release of volatile antifungal constituents (Zakaria 2018).

10.5.1.2 Siderophores

Siderophores are iron carrier called iron-chelating compounds that are produced by rhizospheric microbes. Siderophore scavenging bound form of iron ions and solubilize around the host plants that facilitate the growth and vital physiological functions of plants, e.g., photosynthesis, respiration, translation, and transcription under stress conditions (Ahmed and Holmstrom 2014; Sah and Singh 2015). Fungal siderophores like fusigen, fusarine A, and ferricrocin stop the availability of iron for various pathogenic microorganisms but for plants scavenging more iron in a solubilized form that help to promote plant growth (Verma et al. 2011; Vinale et al. 2014). There are some reports which showed that most of the fungal species are also capable of producing iron-chelating siderophore under aerobic conditions except the *Saccharomyces* species (Andrey 2018). Fungi belonging to class zygomycetes produced hydroxamate-type (coprogens, fusigen, fusarine A, and ferricrocin) and carboxylate-type (rhizoferrin) siderophores (Lehner et al. 2013; Sah and Singh 2015). Moreover, some yeast species, *Candida* sp. and *Rhodotorula* sp., produced siderophores named as rhodotorulic acid that have an inhibitory effect on spore germination of *B. cinerea* that caused gray mold disease on apple wound (Sansone et al. 2005). Yeast antimicrobial compounds were tested against the *Corynespora cassiicola* and *Botrytis cinerea* fungal pathogens of *Theobroma cacao* fruits (Ferreira-Saab 2018).

10.5.2 The Enzymatic Action of Biocontrol Agents Against Phytopathogens

Transgenic plants carry genes for endochitinase enzyme activity. In such type of plants, resistance develops against attacking fungal phytopathogens as well as environmental stress (Smith and Osburn 2016; Toufiq et al. 2017). *Trichoderma* spp. and yeast species are capable of producing β-1,3-glucanase and chitinase enzymes that initiated lysis in the cell wall of pathogen and thus leads to cytoplasmic leakage and cause death (Srivastava et al. 2014). Yeast strains, e.g., *Pichia* sp., *Rhodotorula* sp., *Cryptococcus* sp., *Aureobasidium* sp., and *Tilletiopsis* sp., exhibited to produce β-1,3-glucanase enzyme against fungal test pathogens such as *Botrytis cinerea*, *Penicillium expansum*, *Rhizoctonia stolonifer*, *Aspergillus niger*, *Sphaerotheca fuliginea*, and *Puccinia xanthii* (El-Tarabily 2004; Hartmann et al. 2010). Chatterton and Punja (2009) stated that *Clonostachys rosea* f. sp. *catenulate* produced both chitinase and β-1,3-glucanase enzymes and showed a deterioration of mycelia of *Fusarium* and *Pythium* pathogens which caused root rot, stem rot, and damping-off diseases in cucumber plants. *Aureobasidium pullulans* 1WA1 stimulated secretion of both endo- and exo-β-1,3-glucanases in the presence of *Botrytis cinerea* which caused gray mold disease of grapevine (Bauermeister et al. 2015). *Bacillus subtilis*, *Bacillus pumilus*, *Bacillus megaterium*, and *Agrobacterium radiobacter* had showed antifungal activity by secreting chitinase, glucanase, and protease enzyme against citrus and blue mold incited by *Penicillium digitatum* (Mohammadi et al. 2017). Chitosan exhibited a suppressive effect on spore germination and hyphae growth of *Fusarium oxysporum* f. sp. *radicis lycopersici* and *Verticillium dahliae* (Palma-Guerrero et al. 2008). Chitosan shows antifungal activity against pathogenic yeasts, e.g., *Candida* spp. and *Cryptococcus* spp. (Camacho et al. 2017; Garcia et al. 2018).

10.5.2.1 Detoxification of Pathogen Toxin

Detoxification is a mechanism that reduced the virulence of pathogenic toxins by binding with some amino acids (Aliashkevich et al. 2018). Both *Alcaligenes denitrificans* and *P. dispersa* help in detoxifying *Xanthomonas albilineans*' virulent compound albicidin (Basnayake and Birch 1995; Walker et al. 1988). Similarly, fusaric acid toxicity was neutralized by *Burkholderia cepacia* and *Ralstonia solanacearum* (Toyoda et al. 1988). *T. viride*, *T. hamatum*, and *T. virens* were produced compound gliotoxin against *Rhizoctonia solani* (Reino et al. 2008). The yeast showed antagonistic properties against pathogens by biosynthesis of toxins called mycotoxins (Hatoum et al. 2012).

10.5.2.2 Phytohormones

Phytohormones such as indole-3-acetic acid, ethylene, cytokinins, and gibberellins play an important role against biotic and abiotic stresses (Khan et al. 2011; Shabir et al. 2016; Xu et al. 2018; Li et al. 2016a, b, 2017). Contreras-Cornejo et al. (2009) reported that *Trichoderma* spp. play a key role for auxin phytohormones secretion to promote plant the growth and development. *Aspergillus fumigatus* was found to secrete gibberellins and regulate the production of other phytohormones like abscisic acid (ABA), salicylic acid (SA), and jasmonic acid (JA) (Khan et al. 2011). Cosme et al. (2016) observed that *Piriformospora indica* is a mycorrhizal fungus-induced gibberellin hormone against root herbivory and biotic stress. *T. asperellum* has increased the indole's three acetic acid (IAA) content in maize plants and promoted their growth by activating the plasma membrane H+-ATPase (Coria et al. 2016). de Zelicourt et al. (2018) recently reported that ethylene hormone induced by endophyte *Enterobacter* sp. SA187 helped to improve the yield of alfalfa crops and growth *Arabidopsis*. Wang et al. (2018) reported that an endophytic *Streptomyces chartreusis* WZS021 has enhanced the phytohormones such as indole acetic acid, abscisic acid, and ethylene accumulation against drought stress in sugarcane plants.

10.5.3 Hyperparasitism Found in the Presence of Antagonistic Microbiota

Hyperparasitism is an interaction between specific microbes and target pathogens to either kill its propagules or suppress their growth and reproduction; it is also called parasitism. *Trichoderma harzianum* produced anthraquinone pachybasin that recognizes the host and increases the number of coils around *Rhizoctonia solani* (Lin et al. 2012). Some biocontrol fungi show predatory behavior in the production of enzymes that causes cell wall lysis of pathogens and helps to intake nutrient contents of pathogens (Junaid et al. 2013). *Trichoderma* sp. feed on sclerotia of *R. solani* by producing chitinase (Abbas et al. 2017). Lettuce *Sclerotinia* diseases were effectively controlled by using *Coniothyrium minitans* and *Sporidesmium sclerotivorum* (Jones et al. 2004). *Verticillium dahliae* was mycoparasitized by *T. harzianum* in vitro (Ruano-Rosa et al. 2016). *Trichoderma harzianum* parasitized the fungal hyphal of *Rhizoctonia solani*, causing black scurf and canker potato under in vitro conditions (Brewer and Larkin 2005; Ibrahim 2017). Arbuscular mycorrhizal fungi (AMF) has showed a symbiotic association with *Solanaceous* crops and helped to improve the plant health and defense system against *Potato virus* Y (PVY) (Sikora et al. 2019).

10.5.4 Competitions

Competition is a basic process that exhibited interconnections between nonpathogenic microorganisms and pathogens in rhizospheres (Brien 2017). Host plant hinders root surface from being exposing to pathogens and drawing nutrients from rhizosphere quickly, thus they reproduce faster than soilborne pathogens. Yeast has special antagonistic ability to colonize root rapidly before root exposed to the pathogens, e.g., 6 days preinoculation with *Candida valida* and *Trichosporon asahii* showed higher roots colonization after radicle emergence, while *Rhodotorula glutinis* pre-interaction colonization was initiated after 8 days. Although both pathogens and bioagents present in same rhizosphere but their association lead to competition (Duffy 2001). Abdallah et al. (2015) reported that nine isolates of *Aspergillus* sp. had showed antagonistic activity against *Fusarium sambucinum* (dry rot) and *Phytophthora erythroseptica* pink rot of potato tubers. *Bacillus* and *Pseudomonas* species are most commonly used as biocontrol agents against soilborne phytopathogens due to their strong compatibility with moist soil and high organic manure, e.g., *Fusarium* wilt of chickpea (Abed et al. 2016; Singh et al. 2014). Bubici et al. (2019) recently reported that the fusarium wilt of banana (FWB) caused by *Fusarium oxysporum* f. sp. *cubense* found to be managed by using *Pseudomonas* spp. up to 79% and *Trichoderma* spp. up to 70%. Mycophagous nematodes, for example, *Aphelenchus*, *Filenchus*, *Tylenchus*, and *Iotonchium*, have been found to feed on fungi and to be used as biocontrol agents for controlling phytopathogenic fungi (Tarique et al. 2017). Lagerlof et al. (2011) reported the *Aphelenchoides* spp. and *A. avenae* reduced the infection of *R. solani* inciting the damping-off disease in cauliflower seedlings. *Aphelenchoides besseyi* caused fungal cell wall degradation by feeding on it (Wang et al. 2014). Cetintas et al. (2018) examined the plant growth-promoting rhizobacteria strains which significantly reduced the *Meloidogyne incognita* infection on tomatoes. *Pseudomonas fluorescens* and *Rhizobium leguminosarum* were found to suppress the *Meloidogyne javanica* infection and enhanced growth of lentil, bean, chickpea, lentil, pea, and tomato (Saeedizadeh 2016; Tabatabaei and Saeedizadeh 2017). Xiang et al. (2017) demonstrated the antagonistic behavior of plant growth-promoting rhizobacteria (PGPR) by forming spore on *Heterodera glycines* cuticle in soybean. *Rhizobium* showed inhibitory effect on causing black rot disease of fava bean (Tamiru and Muleta 2018). *Trichoderma citrinoviride* (T33) and *H. semiorbis* (T15) used to control *Fusarium* wilt severity up to 19.77% under greenhouse condition (Al-Mekhlafi et al. 2019).

10.5.5 Biocontrol Agents Induced Host Resistance

Systemic acquired resistance (SAR) is facilitated by chemical compounds salicylic acid (SA) that led to produced pathogenesis-related proteins (PR protein) (Agrios 2005). Induced systemic resistance called secondary mediated pathway, which induced by signal compounds like jasmonic acid (JA) and ethylene secreted in the presence of antagonistic and non-pathogenic rhizomicrobes (Nie et al. 2017). PR proteins usually synthesize a variety of enzymes in host directly against the infection caused by invading pathogens via (1) lysing of invading host cells, (2) reinforcing host boundaries to resist infections, and (3) host-induce localized cell death (Vallad Goodman 2004). Various *Trichoderma* spp. can produce several plant defense elicitors such as xylanases, swollenins, peptaibols, and cerato-platanins (Harman et al. 2004; Druzhinina et al. 2011; Saldajeno et al. 2014). Microbiota-containing fungi belonging to genus *Trichoderma* are found most effective to stimulate plant growth and eliminate plant pathogens by producing specific and nonspecific antibiotics (Bhattacharjee and Dey 2014; Gebarowska et al. 2019). Beside these microorganisms showing antagonistic activity, they are also able to stimulate plants to defend themselves against phytopathogens by inducing systemic resistance (ISR) (Van Loon 2007). A nonpathogenic *Fusarium solani* can be used as a biological control agent by inducing defense response against pathogens (Vallad and Goodman 2004). Chitosan is a bioactive polymer found in fungal cell wall. Chitosan acts as an elicitor to stimulate defense reactions in plants against various soilborne pathogenic fungi, e.g., *Penicillium digitatum, Macrophomina phaseolina, Fusarium solani, Fusarium fujikuroi, Fusarium. oxysporum* f. sp. *lycopersici, Fusarium oxysporum* f. sp. *cubense, Phomopsis asparagi, Colletotrichum gloeosporioides, Rhizopus stolonifer, Sclerotium rolfsii, Rhizoctonia solani, Candida* spp., and *Cryptococcus* spp. (Ali et al. 2010; Al-Hetar et al. 2011; Oliveira Junior et al. 2012; Bhattacharya 2013; Long et al. 2014; Kim et al. 2016; Camacho et al. 2017; Garcia et al. 2018).

10.6 Mode of Application of Biocontrol Agents Against the Soilborne Phytopathogens

Pure and additive forms of biocontrol agents were used against soilborne phytopathogens. Additive biocontrols and surfactants have increased the efficacy of biocontrol agents as much as with fungicides to control many pathogenic plant diseases, e.g., 10% PelGel (polyox-N-10) is a binder which is used to coat the seed with *Trichoderma* sp. and enhance their growth and bioefficacy against *Pythium* sp. (Lo et al. 1997). Biocontrol must be applied in an active form in plant or soil ecosystem present around plant roots at the time when conditions become favorable for pathogenic disease cycle. The plant root surface is colonized rapidly by antagonistic microbes and unable to come in contact with pathogens infection. Therefore, the plant is protected via biofungicidal applications. Babeva and Belyanin (1966) reported that the culture filtrates of isolated nine yeast strain of *Torulopsis* sp. from cabbage rhizosphere were used to increase presoaked cabbage seed germination. *Trichoderma* species have some antimicrobial gene. Moreover, these genes can be isolated and cloned for commercial application on a large scale, as shown in Table 10.4.

Table 10.4 The beneficial gene of biocontrol agents associated with antagonism of soilborne phytopathogens

Biocontrol agent	Gene code for the lytic enzyme	Gene function against pathogen	References
Pseudomonas spp. and *Streptomyces* spp.	NRPS (non-ribosomal peptide-synthetase)	NRPS induce antibiotic (vancomycin and gramicidin) and lipopeptide (surfactin, iturin A, and bacillomycin) production against *Fusarium* wilt of banana	Zhao et al. (2018)
Clonostachys rosea	prs6 (serine protease)	Protease parasitized the plant pathogenic fungi *F. graminearum*	Iqbal et al. (2018)
Trichoderma virens	Tv-cht1	33-kDa endochitinases mycoparasitized the *R. solani*	Abbas et al. (2017)
T. harzianum	ChiB1 (chitinase)	Chitinase catalyze the transglycosylase of *Aspergillus fumigatus*	Andres et al. (2017)
T. brevicompactum IBT40841	Tbtri5 (trichodermin)	Trichodermin showed antimicrobial activity against *Saccharomyces cerevisiae* and *Kluyveromyces marxianus*	Tijerino et al. (2010)
T. virens	Tv-ech1 (endochitinase)	Endochitinase reduce the infection caused by *A. alternata* and *R. solani* in transgenic cotton plant	Emani et al. (2003) and Kumar et al. (2009)
T. harzianum	chit36 (chitinase)	Chitinase repress the *Alternaria radicina* and *Botrytis cinerea* in carrot and increase tolerance	Baranski et al. (2008)
T. atroviride	ech42, nag70 (chitinases) and gluc78 (β-1,3 glucanase)	Chitinase and β-1,3 glucanase inhibit the growth of *Rhizoctonia solani* and *Magnaporthe grisea* in rice	Liu et al. (2004)

10.7 Plant Growth-Promoting Rhizobacteria

Gray and Smith (2005) classified plant growth-promoting rhizobacteria into two categories based on their presence outside and inside the contact region. These included (1) extracellular plant growth-promoting rhizobacteria (ePGPR) such as *Agrobacterium*, *Azotobacter*, *Azospirillum*, *Azotobacter*, *Bacillus*, *Pseudomonas*, *Burkholderia*, *Caulobacter*, *Chromobacterium*, *Erwinia*, *Flavobacterium*, *Micrococcus*, and *Serratia* and (2) intracellular plant growth-promoting rhizobacteria (iPGPR) generally found inside cells of root nodule and belongs to family of Rhizobiaceae which includes *Allorhizobium*, *Bradyrhizobium*, *Mesorhizobium*, and *Rhizobium*, endophytes, and *Frankia* species. *Rhizobium* and *Frankia* species associated symbiotically with higher plants and can fix atmospheric nitrogen (Bhattacharyya and Jha 2012). Deshmukh and Shinde (2016) reported that PGPRs had been proven as vital biocontrol agents available in the form of natural enemies to protect plants against various soilborne phytopathogens. Rhizobium formed a symbiotic mutualistic relationship with leguminous plants. Rhizobium can be used considerably to improve the growth of legume plants through nitrogen fixation in nodules (Mus et al. 2016). PGRP is also used as biocontrol agents to induce host defense mechanism and suppress the disease produced by pathogen (Ahemad and Kibret 2014). Many antagonistic bacterial species, e.g., *Pseudomonas fluorescens*, *P. putida*, *P. aerofaciens*, *Burkholderia cepacia*, *Bacillus subtilis*, *B. Polymyxa*, *B. carries*, *Azotobacter chroococcum*, *Azospirillum lipoferum*, have been found to suppress or control cotton rhizospheres diseases (Kloepper et al. 2004; Heydari 2007; Ahemad and Kibret 2014).

10.8 Plant Growth-Promoting Rhizobacteria

PGPR help to promote and enhance the plant vigor under diverse environmental condition by various mechanisms. PGPR can increase nutrient uptake (nitrogen, phosphorus, potassium, and essential mineral elements) (Bargaz et al. 2018). PGPR community showed the direct or indirect antimicrobial mechanism in rhizosphere niche and included (1) direct promotion of plant growth by increasing biofertilizer activity and (2) indirect growth stimulation by biopesticide activity.

10.9 Microbivorous Nematodes

Nematodes are elongated, vermiform threadlike organisms abundantly found in soil and have a great impact in the maintenance of soil biodiversity (Yeates and Bongers 1999). Nematode appears as biocontrol agents showing a wide range of feeding types on the protoplasm of plants, fungal hyphae, bacterial cell contents, and protozoans (Yeates et al. 1993). Nematodes establish by antagonistic mechanism, viz., parasitoids, predators, pathogens, competition for space, and nutrients, by sharing same space habit. Nematodes belong to order Aphelenchida (*Aphelenchus*, *Aphelenchoides*, *Rhadinaphelenchus*, *and Bursaphelenchus*), and some belong to order Tylenchida (*Ditylenchus* spp.) which are called fungivorous nematodes (Yeates et al. 1993). They play a critical role in controlling soilborne phytopathogenic microbial communities (Yeates 2003). Further studies on nematode *Aphelenchus avenae* significantly reduces plant pathogens like *Fusarium moniliforme*, *Pythium butler*, and *Fusarium oxysporum* in soil (Gupta 1986; Okada 2006). Nematode feed on fungi, bacteria, diatoms, etc., *Aphelenchoides hamatum*, *A. composticola*, and *Ditylenchus*, feed on the spores of pathogenic fungi *Agaricus*, *Verticillium*, *Botrytis*, mushroom fungi, etc., while *Panagrolaimus* and *Poikilolaimus* species feed on numerous bacteria (Muschiol and Traunspurger 2007). Preinoculation of *Meloidogyne incognita* was protective against the bean root rot fungus caused by *Rhizoctonia solani* due to the competition of nutrient and space. Wolfarth et al. (2013) demonstrated that mutualistic interaction of fungivorous collembolans *Folsomia candida* and nematodes *Aphelenchoides saprophilus* was responsible for reducing mycotoxin deoxynivalenol released by *Fusarium culmorum* on infected wheat straw remaining on the soil surface compared to treatment with *Collembolans* and nematodes being separated. *Pochonia chlamydosporia* mycoparasitized the eggs of *Meloidogyne javanica* by producing chitosan and proteolytic enzyme (Martinez et al. 2016; Escudero et al. 2015). Fungivorous or bacterivorous nematodes were associated with microbiomes due to the presence of soilborne pathogen by feeding on their body (Morris et al. 2016; Elhady et al. 2017). Nematodes associated with microbes reduced the severity of apple replant disease (ARD) (Kanfra et al. 2018). Strom et al. (2019) reported that soybean cyst nematode *Heterodera glycines* increased population caused change in soil microbial community and correlated with yield loss. Soybean cyst nematode density attracted the nematode-trapping fungi and reduced the crop loss.

10.10 Conclusions

Beneficial microbiotic factors improve soil life. Generally, soilborne phytopathogens are the major constraint of the agriculture ecology and ecosystem by degrading soil fertility, causing human health hazard and groundwater contamination, and consequently disrupting the environment. Beneficial microorganisms present in microbiome system is promising aid and help to preserve soil fertility in a sustainable and eco-friendly manner. Moreover, various impacts of microbiotic factors have been discussed in this chapter. One of the most important mechanisms is competition in which beneficial organisms feed on soilborne pathogen propagules and thus reduce disease-causing potential. Bioremediation potential of microbiota detoxify pollutants like heavy metals and agrochemicals as biopesticides. These all have potential to control soilborne phytopathogens, stimulating defense response to resist pathogen infection, increase plant growth promoting factors, and allow to withstand under adverse environmental conditions.

10.11 Future Approaches

Rhizosphere shows great importance for the plant growth as it is interrelated with biotic and abiotic factors of the environment. The rhizosphere is full of viable microbes; some show noxious behavior toward the plant, and some show the beneficial effect that will cause to suppress many soilborne pathogen populations and help to stimulate defense response. Thus, beneficial phytomicrobiome overcomes the noxiousness of harmful microbes and promotes plant growth. Hence, natural microorganisms will offer the best safe pest management module and are the cheap source that we can easily get from the surrounding rhizosphere. Today farmers are more reliable on pesticides for disease management. Therefore, there is a need to manufacture large-scale important phytobeneficial biocontrol at minimum cost, is safe, has long shelf-life, and can easily be handled by farmers. Biocontrol agents applied as to coat the seed or too deep the seedling root in the biocontrol agent formulations before sowing. Biocontrol-treated planting material will help to improve the growth and provide strong protections against both abiotic and biotic stress. The microbiome is full of natural constituents that help to overcome the phytopathogenic potential by various mechanisms involved like mycoparasitism, competition, antibiosis, and self-defense system activation.

Acknowledgments We appreciate the support rendered by Department of Plant Protection, Aligarh Muslim University, Aligarh, India.

References

Abbas A, Jiang D, Fu YP (2017) *Trichoderma* Spp. as Antagonist of *Rhizoctonia solani*. J Plant Pathol Microbiol 8(3). https://doi.org/10.4172/2157-7471.100040

Abdallah AB, Jabnoun KRH, Mejdoub TB, Daami RM (2015) Soil-borne and compost-borne *Aspergillus* species for biologically controlling post-harvest diseases of potatoes incited by *Fusarium sambucinum* and *Phytophthora erythroseptica*. J Plant Pathol Microbiol 6:313. https://doi.org/10.4172/2157-7471.1000313

Abdel-Aziz Shadia M (2013) Extracellular metabolites produced by a novel strain, bacillus alvei NRC-14: 5. Multiple plant-growth promoting properties. J Basic Appl Sci Res 3(1):670–682

Abed H, Rouag N, Mouatassem D, Rouabhi A (2016) Screening for Pseudomonas and Bacillus antagonistic rhizobacteria strains for the biocontrol of Fusarium wilt of chickpea. Eur J Soil Sci 5(3):182–191. https://doi.org/10.18393/ejss.2016.3.182-191

Agrios NA (2005) Plant Pathology, 3rd edn. Academic Press, USA, pp 220–222

Ahemad M, Kibret M (2014) Mechanisms and applications of plant growth promoting rhizobacteria: current perspective. J King Saud Univ Sci 26:1–20

Ahemad M, Khan MS, Zaidi A, Wani PA (2009) Remediation of herbicides contaminated soil using microbes. In: Khan MS, Zaidi A, Musarrat J (eds) Microbes in sustainable agriculture. Nova Science Publishers, New York

Ahmed E, Holmstrom SJ (2014) Siderophores in environmental research: roles and applications. Microb Biotechnol 7(3):196–208

Al-Hetar MY, Abidin Z, Sariah M, Wong MY (2011) Antifungal activity of chitosan against *Fusarium oxysporum* f. sp. *cubense*. J Appl Polym Sci 120(4):2434–2439. https://doi.org/10.1002/app.33455

Ali A, Muhammad MTM, Sijam K, Siddiqui Y (2010) Potential of chitosan coating in delaying the postharvest anthracnose (*Colletotrichum gloeosporioides* Penz.) of Eksotika II papaya. Int J Food Sci Technol 45(10):2134–2140. https://doi.org/10.1111/j.1365-2621.2010.02389

Ali MA, Shahzadi M, Zahoor A, Dababat AA, Toktay H, Bakhsh A, Nawaz AM, Hongjie (2019) Resistance to cereal cyst nematodes in wheat and barley: an emphasis on classical and modern approaches. Int J Mol Sci 20(2):432. https://doi.org/10.3390/ijms20020432

Aliashkevich A, Alvarez L, Cava F (2018) New insights into the mechanisms and biological roles of D-amino acids in complex eco-systems. Front Microbiol 9. https://doi.org/10.3389/fmicb.2018.00683

Al-Mekhlafi N, Karem A, Abdullah YQ, Farhan AM, Saeed MA (2019) Efficacy of native *Trichoderma* spp. in controlling *Fusarium* wilt of tomato plants in green house, Yemen. Clin Biotechnol Microbiol 4(1)

Al-Naemi FA, Ahmed TA, Nishad R, Radwan O (2016) Antagonistic effects of *Trichoderma harzianum* isolates against *Ceratocystis radicicola*: pioneering a biocontrol strategy against black scorch disease in date palm trees. J Phytopathol 164(7–8):433–570

Alonso LM, Kleiner D, Ortega E (2008) Spores of the mycorrhizal fungus Glomus mosseae host yeasts that solubilize phosphate and accumulate polyphosphates. Mycorrhiza 18(4):197–204

Andres E, Boer H, Koivula A, Samain E, Driguez H, Armand S, Cottaz S (2017) Engineering chitinases for the synthesis of chitin oligosaccharides: catalytic amino acid mutations convert the GHfamily 18 glycoside hydrolases into transglycosylases. J Mol Catal B Enzym 74:89–96

Andrey YM (2018) Yeasts of the soil – obscure but precious. Yeast, published by John Wiley & Sons, Ltd, 35:369–378. https://doi.org/10.1002/yea.3310

Anoop K, Suseela BR, Shiva KN (2017) A survey on the in cadence of rhizome rot disease in major turmeric growing tracts of Sount India and isolation of associated organisms. Indian J Adv Plant Res (IJAPR) 1(16):17–23

Arora NK, Khare E, Ji Hoon O, Kang SC, Maheshwari DK (2008) Diverse mechanisms adopted by fluorescent Pseudomonas PGC2 during the inhibition of Rhizoctonia solani and Phytophthora capsici. World J Microbiol Biotechnol 24(4):581–585

Awad HM, El-Enshasy HA, Hanapi SZ, Hamed ER, Rosidi B (2014) A new chitinase-producer strain *Streptomyces glauciniger* WICC-A03: isolation and identification as a biocontrol agent for plants phytopathogenic fungi. Nat Prod Res 28(24):2273–2277

Babeva I, Belyanin AI (1966) Yeasts of the rhizosphere. Microbiologiya. Bacteria isolated from the root nodules of pea (*Pisum sativum* L.). World J Microbiol Biotechnol 30:719–725

Baranski R, Klocke E, Nothnagel T (2008) Chitinase CHIT36 from Trichoderma harzianum enhances resistance of transgenic carrot to fungal pathogens. J Phytopathol 156:513–521

Bargaz A, Lyamlouli K, Chtouki M, Zeroual Y, Dhiba D (2018) Soil microbial resources for improving fertilizers efficiency in an integrated plant nutrient management system. Front Microbiol 9:1606. https://doi.org/10.3389/fmicb.2018.01606

Basnayake S, Birch RG (1995) A gene from Alcaligenes denitrificans that confers albicidin resistance by reversible antibiotic binding. Article Microbiol 141(3):551–560. https://doi.org/10.1099/13500872-141-3-551

Bauermeister A, Amador IR, Pretti CP, Giese EC, Oliveira AL, Alves da CMA et al (2015) β-(1→3)-Glucanolytic yeasts from Brazilian grape microbiota: production and characterization of β-Glucanolytic enzymes by *Aureobasidium pullulans* 1WA1 cultivated on fungal mycelium. J Agric Food Chem 63(1):269–278

Beneduzi A, Ambrosini A, Passaglia MPL (2012) Plant growth-promoting rhizobacteria (PGPR): their potential as antagonists and biocontrol agents. Genet Mol Biol 35(4 (suppl)):1044–1051

Benyagoub M, Rhlid RB, Bèlanger RR (1996) Purification and characterization of new fatty acids with antibiotic activity produced by *Sporothrix flocculosa*. J Chem Ecol 22:405–413

Berg G (2009) Plant–microbe interactions promoting plant growth and health: perspectives for controlled use of microorganisms in agriculture. Appl Microbiol Biotechnol 84:11–18. https://doi.org/10.1007/s00253-009-2092-7

Berg G, Opelt K, Zachow C, Lottmann J, Gotz M, Costa R, Smalla K (2006) The rhizosphere effect on bacteria antagonistic towards the pathogenic fungus Verticillium differs depending on plant species and site. FEMS Microbiol Ecol 56:250–261

Bhattacharjee R, Dey U (2014) An overview of fungal and bacterial biopesticides to control plant pathogens/diseases. Afr J Microbiol Res 8(17):1749–1762. https://doi.org/10.5897/AJMR2013.6356

Bhattacharya A (2013) Fungicidal potential of chitosan against phytopathogenic *Fusarium solani*. J Exp Biol Agric Sci 1(4):259–263

Bhattacharyya PN, Jha DK (2012) Plant growth-promoting rhizobacteria (PGPR): emergence in agriculture. World J Microbiol Biotechnol 28:1327–1350

Brewer MT, Larkin RP (2005) Efficacy of several potential biocontrol organisms against *Rhizoctonia solani* on potato. Crop Prot 24(11):939–950

Brien OPA (2017) Biological control of plant diseases. Australas Plant Pathol 46:293–304

Bubici G, Kaushal M, Prigigallo IM, Carmen GLC, Jesus MB (2019) Biological control agents against Fusarium wilt of Banana. Front Microbiol. https://doi.org/10.3389/fmicb.2019.00616

Camacho E, Chrissian C, Cordero RJB, Liporagi LL, Stark RE, Casadevall A (2017) *N*-acetylglucosamine affects *Cryptococcus neoformans* cell-wall composition and melanin architecture. Microbiology 163:1540–1556. https://doi.org/10.1099/mic.0.000552

Cao Y, Pi H, Chandrangsu P, Li Y, Wang Y, Zhou H, Xiong H, Helmann DJ, Cai Y (2018) Antagonism of two plant-growth promoting *Bacillus velezensis* isolates against *Ralstonia solanacearum* and *Fusarium oxysporum*. Sci Rep 8:4360. https://doi.org/10.1038/s41598-018-22782-z

Cetintas R, Kusek M, Fateh SA (2018) Effect of some plant growth-promoting rhizobacteria strains on root-knot nematode, *Meloidogyne incognita*, on tomatoes. Egypt J Biol Pest Control 28:7

Chang PK, Hua SS, Sarreal SB et al (2015) Suppression of aflatoxin biosynthesis in *Aspergillus flavus* by 2-phenylethanol is associated with stimulated growth and decreased degradation of branched-chain amino acids. Toxins (Basel) 7:3887–3902

Chang KF, Hwang SF, Ahmed HU, Strelkov SE, Harding MW, Conner RL, McLaren DL, Gossen BD, Turnbull GD (2018) Disease reaction to Rhizoctonia solani and yield losses in soybean. Can J Plant Sci 98(1):115–124

Chatterton S, Punja KZ (2009) Chitinase and β-1,3-glucanase enzyme production by the mycoparasite *Clonostachys rosea* f. *catenulate* against fungal plant pathogens. Can J Microbiol 55(4):356–367. https://doi.org/10.1139/W08-156

Choudhury SR, Traquair JA, Jarvis WR (1994) 4-Methyl-7,11-heptadecadienal and 4-methyl-7,11-heptadecadienoic acids: new antibiotics from *Sporothrix flocculosa* and *Sporothrix rugulosa*. J Nat Prod 57:700–704

Cook RJ, Baker KF (1983) The nature and practice of biological control of plant pathogens. Am Phytopathol Soc:539

Coria ML, Mendoza JL, Hernandez S, Nieto S (2016) *Trichoderma asperellum* induces maize seedling growth by activating the plasma membrane H+-ATPase. Mol Plant-Microbe Interact 29:797–806

Cornejo CHA, Rodriguez ML, Penagos CC, Bucio LJ (2009) *Trichoderma virens*, a plant beneficial fungus, enhances biomass production and promotes lateral root growth through an auxin-dependent mechanism in Arabidopsis. Plant Physiol 149:1579–1592

Cosme M, Lu J, Erb M, Stout MJ, Franken P, Wurst S (2016) A fungal endophyte helps plants to tolerate root herbivory through changes in gibberellin and jasmonate signaling. New Phytol 211:1065–1076. https://doi.org/10.1111/nph.13957

Couillerot O, Combes-Meynet E, Pothier JF, Bellvert F, Challita E, Poirier M-A, Rohr R, Comte G, Moënne-Loccoz Y, Prigent-Combaret C (2011) The role of the antimicrobial compound 2,4-diacetylphloroglucinol in the impact of biocontrol pseudomonas fluorescens F113 on azospirillum brasilense phytostimulators. Microbiology 157(6):1694–1705

de Zelicourt A, Synek L, Saad MM, Alzubaidy H, Jalal R, Xie Y et al (2018) Ethylene induced plant stress tolerance by *Enterobacter* sp. SA187 is mediated by 2,keto-4-methylthiobutyric acid production. PLoS Genet 14(3):e1007273

Deshmukh P, Shinde S (2016) Beneficial role of rhizosphere Mycoflora in the field of agriculture: an overview. Int J Sci Res (IJSR) 5(8):529–533

Druzhinina IS, Seidl SV, Herrera EA, Horwitz BA, Kenerley CM, Monte E, Mukherjee PK, Zeilinger S, Grigoriev IV, Kubicek CP (2011) *Trichoderma:* the genomics of opportunistic success. Nat Rev Microbiol 9:749–759. https://doi.org/10.1038/nrmicro2637

Duffy BK (2001) Competition. In: Maloy OC, Murray TD (eds) Encyclopedia of plant pathology. John Wiley & Sons, Inc., New York, pp 243–244

Elhady A, Gine A, Topalovic O, Jacquiod S, Sorensen SJ, Sorribas FJ, Heuer H (2017) Microbiomes associated with infective stages of root-knot and lesion nematodes in soil. PLoS One 12(5):e0177145

El-Mehalawy AA, Hassanein NM, Khater HM, Karam E-DEA, Youssef YA (2004) Influence of maize root colonization by the rhizosphere actinomycetes and yeast fungi on plant growth and on the biological control of late wilt disease. Int J Agric Biol 6:599–605

El-Tarabily KA (2004) Suppression of *Rhizoctonia solani* diseases of sugar beet by antagonistic and plant growth-promoting yeasts. J Appl Microbiol 96:69–75

Emani C, Garcıa JM, Lopata FE, Pozo MJ, Uribe P, Kim DJ, Sunilkumar G, Cook DR, Kenerley CM, Rathore KS (2003) Enhanced fungal resistance in transgenic cotton expressing an endochitinase gene from *Trichoderma virens*. Plant Biotechnol J 1:321–336

Escudero N, Ferreira SR, Lopez MF, Naranjo OMA, Marin OA, Thornton CR, Lopez LLV (2015) Chitosan enhances parasitism of *Meloidogyne javanica* eggs by the nematophagous fungus *Pochonia chlamydosporia*. Fungal Biol 120:572–585. https://doi.org/10.1016/j.funbio.2015.12.005

Evidente A, Cabras A, Maddau L (2003) Viridepyronone, a new antifungal 6-Substituted 2H-Pyran-2-one produced by trichoderma viride. J Agric Food Chem 51(24):6957–6960

Ferreira SM, Formey D, Torres M, Wendy A, Padilla AE, Tromas A, Sohlenkamp C, Kátia RF, Schwan E, Mario SF (2018) Compounds released by the biocontrol yeast *Hanseniaspora opuntiae* protect plants against *Corynespora cassiicola* and *Botrytis cinerea*. Microbiolology. https://doi.org/10.3389/fmicb.2018.01596

Fu SF, Sun PF, Lu HY, Wei JY, Xiao HS, Fang WT et al (2016) Plant growth-promoting traits of yeasts isolated from the phyllosphere and rhizosphere of Drosera spatulata lab. Fungal Biol 120(3):433–448

Garcia LGS, Guedes GM, da Silva MLQ, Castelo BDSCM, Sidrim JJC, de Cordeiro RA, Rocha MFG, Vieira RS, Brilhante RSN (2018) Effect of the molecular weight of chitosan on its antifungal activity against *Candida* spp. in planktonic cells and biofilm. Carbohydr Polym 195:662–669. https://doi.org/10.1016/j.carbpol.2018.04.091

Gebarowska E, Pytlarz KM, Nofer J, Łyczko J, Adamski M, Szumny A (2019) The effect of *Trichoderma* spp. on the composition of volatile secondary metabolites and biometric parameters of coriander (*Coriandrum sativum* L.). J Food Qual. https://doi.org/10.1155/2019/5687032

Geisen S, Mitchell AD, Edward SA, Bonkowski M, Dunthorn M, Ekelund F, Fernandez DL, Jousset A, Krashevska V, Singer D, Spiegel WF, Walochnik J, Lara E (2018) Soil protists: a fertile frontier in soil biology research. FEMS Microbiol Rev 42:293–323. https://doi.org/10.1093/femsre/fuy006

Ghazalibiglar H et al (2016) Biological control of *Fusarium* wilt of tomato by *Trichoderma* isolates. New Zealand Plant Prot 69:57–63

Goncalves FAG, Colen G, Takahashi JA (2014) *Yarrowia lipolytica* and its multiple applications in the biotechnological. Sci World J 14:476207. https://doi.org/10.1155/2014/476207

Gray EJ, Smith DL (2005) Intracellular and extracellular PGPR: commonalities and distinctions in the plant-bacterium signaling processes. Soil Biol Biochem 37:395–412

Gu Q, Yang Y, Yuan Q, Shi G, Wu L, Lou Z, Huo R, Wu H, Borriss R, Gao X (2017) Bacillomycin D produced by *Bacillus amyloliquefaciens* is involved in the antagonistic interaction with the plant-pathogenic fungus *Fusarium graminearum*. Appl Environ Microbiol 83(19):2–17. https://doi.org/10.1128/AEM.01075-17

Guji J, Merga Y, Terefe H, Kidanu DE (2019) Yield loss of ginger (*Zingiber officinale*) due to bacterial wilt (*Ralstonia solanacearum*) in different wilt management systems in Ethiopia. Agric Food Secur 8(5):1–11. https://doi.org/10.1186/s40066-018-0245-6

Gupta MC (1986) Biological control of *Fusarium moniliforme* Sheldon and *Pythium butleri* Subramaniam by *Aphelenchus avenae* Bastian in chitin and cellulose-amended soils. Soil Biol Biochem 18(3):327–329. https://doi.org/10.1016/0038-0717(86)90069-6

Hajlaoui MR, Traquair JA, Jarvis WR, Belanger RR (1994) Antifungal activity of extracellular metabolites produced by *Sporothrixflocculosa*. Biocontrol Sci Tech 4:229–237

Hammerschmidt R, Kuc J (1995) Induced resistance to disease in plants. Kluwer Academic. https://doi.org/10.1007/978-94-015-8420-3

Harman GE, Howell CR, Viterbo A, Chet I, Lorito M (2004) *Trichoderma* species opportunistic avirulent plant symbionts. Nat Rev Microbiol 2(1):43–56

Hartmann T, Dumig M, Jaber BM, Szewczyk E, Olbermann P, Morschhauser J et al (2010) Validation of a self-excising marker in the human pathogen *Aspergillus fumigatus* by employing the β-rec/six site specific recombination system. Appl Environ Microbiol 76:6313–6317. https://doi.org/10.1128/AEM.00882-10

Hashem M, Ali E (2004) *Epicoccum nigrum* as biocontrol agent of Pythium damping-off and root-rot of cotton seedlings, Archives of Phytopathology and Plant Protection ISSN: 0323-5408 (Print) 1477-2906 (Online) Journal homepage: https://www.tandfonline.com/loi/gapp20. Arch Phytopathol Plant Protect 37:283–297

Hassani AM, Duran P, Hacquard S (2018) Microbial interactions within the plant holobiont. Microbiome 6:58. https://doi.org/10.1186/s40168-018-0445-0

Hatoum R, Labrie S, Fliss I (2012) Antimicrobial and probiotic properties of yeasts: from fundamental to novel applications. Front Microbiol 3:421. https://doi.org/10.3389/fmicb.2012.00421

Heydari A (2007) Biological control of Turfgrass fungal diseases. In: Pessarakli M (ed) Turfgrass management and physiology. CRC Press, Florida

Hiltner L (1904) Uber neue Erfahrungen und Probleme auf dem Gebiet der Bodenbakteriolgie und unter besonderes Berucksichtigung der Grundugungen und Brauche. Arb Dtsch Landwirt Ges Berl 98:59–78

Homma Y, Kato Z, Hirayama F, Konno K, Shirahama H, Suzui T (1989) Production of antibiotics by *Pseudomonas cepacia* as an agent for biological control of soilborne plant pathogens. Soil Biol Biochem 21:723–728

Horst RK (2001) Plant diseases and their pathogens. In: Westcott's plant disease handbook. Springer, Boston. https://doi.org/10.1007/978-1-4757-3376-1_3

Ibrahim ME (2017) *In-vitro* antagonistic activity of *Trichoderma harzianum* against *Rhizoctonia solani* the causative agent of potato black scurf and stem canker. International Conference of "*Plant & Microbial Biotech & their Role in the Development of the Society*", pp 173–185

Igiehon NO, Bbaloal OO (2018) Rhizosphere microbiome modulators: contributions of nitrogen fixing bacteri towards sustainable agriculture. International journal of environmental research and public health. Int J Environ Res Public Health 15:574. https://doi.org/10.3390/ijerph15040574. www.mdpi.com/journal/ijerph

Iqbal M, Dubey M, Gudmundsson M, Viketoft M, Jensen FD, Karlsson M (2018) Comparative evolutionary histories of fungal proteases reveal gene gains in the mycoparasitic and nematode-parasitic fungus *Clonostachys rosea*. BMC Evol Biol 18:171

Janisiewicz WJ, Peterson DJ (2004) Susceptibility of the stem pull area of mechanically harvest apples to blue mold decay and its control with a biocontrol agent. Plant Dis 88:662–664

Jones EE, Mead A, Whipps JM (2004) Effect of inoculum type and timing of application of *Coniothyrium minitans* on *Sclerotinia sclerotiorum*: control of sclerotinia disease in glasshouse lettuce. Plant Pathol 53:611–620

Junaid MJ, Nisar AD, Traiq AB, Arif HB, Mudasir AB (2013) Commercial biocontrol agents and their mechanism of action in the management of plant pathogens. Int J Mod Plant Anim Sci 1(2):39–57. Copyright © 2013 by Modern Scientific Press company, Florida, USA

Kanfra X, Liu B, Beerhues L, Sørensen JS, Heuer H (2018) Free-living nematodes together with associated microbes play an essential role in apple replant disease. Front Plant Sci 9:1666. https://doi.org/10.3389/fpls.2018.01666

Kashyap BK, Solanki MK, Pandey AK, Prabha S, Kumar P, Kumari B (2019) Bacillus as plant growth promoting rhizobacteria (PGPR): a promising green agriculture technology. In: Plant health under biotic stress, vol II. Springer, Singapore. https://doi.org/10.1007/978-981-13-6040-4_11

Kelly L (2017) Fusarium species associated with grain Sorghum and Mungbean in Queensland. MS thesis, The University of Queensland

Khan AL, Hamayun M, Kim YH, Kang SM, Lee JH, Lee IJ (2011) Gibberellins producing endophytic *Aspergillus fumigatus* sp. LH02influencedendogenous phytohormonal levels, isoflavonoids production and plant growth in salinity stress. Process Biochem 46:440–447. https://doi.org/10.1016/j.procbio.2010.09.013

Kim SW, Park JK, Lee CH, Hahn B, Koo JC (2016) Comparison of the antimicrobial properties of chitosan oligosaccharides (COS) and EDTA against *Fusarium fujikuroi* causing rice bakanae disease. Curr Microbiol 72(4):496–502. pmid: 26729353

Kloepper JW, Ryu CM, Zhang S (2004) Induce systemic resistance and promotion of plant growth by *Bacillus* spp. Phytopathology 94:1259–1266

Kumar G, Sarma BK (2016) Eco-friendly management of soil-borne plant pathogens through plant growth-promoting Rhizobacteria. Annu Tech Issue 20:167–171

Kumar V, Parkhi V, Kenerley CM, Rathore KS (2009) Defenserelated gene expression and enzyme activities in transgenic cotton plants expressing an endochitinase gene from Trichoderma virens in response to interaction with *Rhizoctonia solani*. Planta 230:277–291

Kumari B, Mallick MA, Solanki MK, Solanki AC, Hora A, Guo W (2019) Plant Growth-Promoting Rhizobacteria (PGPR): modern prospects for sustainable agriculture. In: Ansari RA, Mahmood I (eds) Plant health under biotic stress, vol II. Springer, Singapore. https://doi.org/10.1007/978-981-13-6040-4_6

Lagerlof J, Insunza V, Lundegårdh B, Ramert B (2011) Interaction between a fungal plant disease, fungivorous nematodes and compost suppressiveness. Acta Agric Scand Sect B Soil Plant Sci 61:372–377

Lanteigne C, Gadkar VJ, Wallon T, Novinscak A, Filion M (2012) Production of DAPG and HCN by *Pseudomonas* sp. LBUM300 contributes to the biological control of bacterial canker of tomato. Phytopathology 102:967–973

Leach JE, Triplett LR, Argueso CT, Trivedi P (2017) Communication in the Phytobiome. Cell 169:587–596. https://doi.org/10.1016/j.cell.2017.04.025

Lehner SM, Atanasova L, Neumann NK, Krska R, Lemmens M, Druzhinina IS, Schuhmacher R (2013) Isotope-assisted screening for iron-containing metabolites reveals a high degree of diversity among known and unknown siderophores produced by *Trichoderma* spp. Appl Environ Microbiol 79:18–31

Li C, Nong Q, Solanki MK, Liang Q, Xie J, Liu X, Li Y, Wang W, Yang L, Li Y (2016a) Differential expression profiles and pathways of genes in sugarcane leaf at elongation stage in response to drought stress. Sci Rep 6:25698

Li Y, Sun R, Yu J, Saravanakumar K, Chen J (2016b) Antagonistic and biocontrol potential of *Trichoderma asperellum*ZJSX5003 against the maize stalk rot pathogen *Fusarium graminearum*. Indian J Microbiol 56(3):318–327

Li C, Nong Q, Xie J, Wang Z, Liang Q, Solanki MK, Malviya MK, Liu X, Li Y, Htun R, Wei J, Li Y (2017) Molecular characterization and co-expression analysis of the SnRK2 gene family in sugarcane (*Saccharum officinarum* L.). Sci Rep 7:17659

Lin HF, Chen TH, Liu SD (2010) Bioactivity of antifungal substance iturin a produced by bacillus subtilis strain BS-99-H against pestalotiopsis eugeniae, a causal pathogen of wax apple fruit rot. Plant Pathol Bull 19:225–233

Lin YR, Lo CT, Li SY, Peng KC (2012) Involvement of pachybasin and emodin in selfregulation of *Trichoderma harzianum* mycoparasitic coiling. J Agric Food Chem 60:2123–2128

Liu M, Sun ZX, Zhu J, Xu T, Harman GE, Lorito M (2004) Enhancing rice resistance to fungal pathogens by transformation with cell wall degrading enzyme genes from *Trichoderma atroviride*. J Zhejiang Univ (Sci) 5:133–136

Lo CT, Nelson EB, Harman GE (1997) Biological control of Pythium, Rhizoctonia and Sclerotinia infected diseases of turfgrass with *Trichoderma harzianum*. Phytopathology 84:1372–1379

Lombardi N, Vitale S, Turra D, Reverberi M (2018) Root exudates of stressed plants stimulate and attract *Trichoderma* soil Fungi. Mol Plant-Microbe Interact 31(10):982–994

Long LT, Tien NTT, Trang NH, Ha TTT, Hieu NM (2014) Study on antifungal ability of water soluble chitosan against green mould infection in harvested oranges. J Agric Sci 6(8):205. https://doi.org/10.5539/jas.v6n8p205

Macias RL, Guzman GA, Garcia JP, Contreras CHA (2018) *Trichoderma atroviride* promotes tomato development and alters the root exudation of carbohydrates, which stimulates fungal growth and the biocontrol of the phytopathogen Phytophthora cinnamomi in a tripartite interaction system. FEMS Microbiol Ecol. https://doi.org/10.1093/femsec/fiy137

Malviya MK, Solanki MK, Li C-N, Htun R, Singh RK, Singh P, Yang L-T, Li Y-R (2019) Beneficial linkages of Endophytic *Burkholderia anthina* MYSP113 towards sugarcane growth promotion. Sugar Tech. https://doi.org/10.1007/s12355-019-00703-2

Manganiello G, Sacco A, Ercolano RM, Vinale F, Lanzuise S, Pascale A, Napolitano M, Lombardi N, Matteo L, Woo LLS (2018) Modulation of tomato response to *Rhizoctonia solani* by *Trichoderma harzianum* and its secondary metabolite Harzianic acid. Front Microbiol. https://doi.org/10.3389/fmicb.2018.01966

Manjunatha TG, Rai AB, Singh B (2017) Root Knot nematade: a threat to vegetables production and its management: Technical Bulletin No.76

Marik T, Tyagi C, Raci G, Rakk D, Szekeres A, Vagvolgyi C, Kredics L (2018) IDNew 19-residue Peptaibols from *Trichoderma Clade viride*. Microorganisms 6:85
Martinez AA, Escudero LN, Gonzalez N, Zavala EA, Llorca HB, Lopez LV (2016) CAZyme content of *Pochonia chlamydosporia* reflects that chitin and chitosan modification are involved in nematode parasitism: CAZome of *Pochonia chlamydosporia*. Environ Microbiol 18:4200–4215. https://doi.org/10.1111/1462-2920.13544
Mazzola M, Freilich S (2016) Prospects for biological soil-borne disease control: application of indigenous versus synthetic microbiomes. Phytopathology 107:256–263. https://doi.org/10.1094/PHYTO-09-16-0330-RVW
Meena KR, Kanwar SS (2015) Review article Lipopeptides as the antifungal and antibacterial agents: applications in food safety and therapeutics. BioMed Res Int Article ID 473050, 9 pages https://doi.org/10.1155/2015/473050
Mihajlovic M, Rekanovic E, Hrustic J, Grahovac M, Tanovic B (2017) Methods for management of soilborne plant pathogens. Pestic Phytomed (Belgrade) 32(1):9–24
Mohammadi P, Tozlu E, Kotan R, Kotan MS (2017) Potential of some bacteria for biological control of postharvest citrus green mold caused by *Penicillum digitatum*. Plant Prot Sci 53:1–10
Morris KA, Langston DB, Dutta B, Davis RF, Timper P, Noe JP et al (2016) Evidence for a disease complex between *Pythium aphanidermatum* and root-knot nematodes in cucumber. Plant Health Prog 17:200–201. https://doi.org/10.1094/PHP-BR-16-0036
Murali SP, Vanitha S, Kamalakannan A, Anantha RP, Jeyakumar P (2018) Prevalence of *Fusarium oxysporum* f. sp. *ciceris* causing wilt in chickpea and its pathogenic, cultural and morphological characterization. Int J Curr Microbiol App Sci 7(2):1301–1313
Mus F, Crook BM, Garcia K, Costas GA, Geddes AB, Kouri DE, Paramasivan P, Ryu HM, Oldroyd EDG, Poole SP, Udvardi KM, Voigt AC, Ane Michel J, Petersa WJ (2016) Symbiotic nitrogen fixation and the challenges to its extension to non-legumes. Appl Environ Microbiol 82(12):3698–3710
Muschiol D, Traunspurger W (2007) Life cycle and cal- culation of the intrinsic rate of natural increase of two bac- terivorous nematodes, *Panagrolaimus*sp. and *Poikilolaimus* sp. from chemoautotrophic Movile Cave, Romania. Nematology 9:271–284
Musoni M, Destain J, Thonart P, Bahama JB, Delvigne F (2015) Bioreactor design and implementation strategies for the cultivation of filamentous fungi and the production of fungal metabolites: from traditional methods to engineered systems. Biotechnol Agron Soc Environ 19(4):430–442
Nelson EB (2017) The seed microbiome: origins, interactions, and impacts. Plant Soil 422:7–34. https://doi.org/10.1007/s11104-017-3289-7
Neshev G (2008) Major soil-borne phytopathogens on tomato and cucumber in Bulgaria, and methods for their management. In: Labrada R (ed) Alternatives to replace methyl bromide for soil-borne pest control in East and Central Europe. FAO, UNEP, pp 1–14
Nie P, Li X, Wang S, Guo J, Zhao H, Niu D (2017) Induced systemic resistance against *Botrytis cinerea* by *Bacillus cereus*AR156 through a JA/ET- and NPR1-dependent signaling pathway and activates PAMP-triggered immunity in *Arabidopsis*. Front Plant Sci 8:238. https://doi.org/10.3389/fpls.2017.00238
Oerke EC (2006) Crop losses to pests. J Agric Sci 144:31–43
Okada H (2006) Ecology of fungivorous nematodes and their use for suppression of plant diseases. Bull Natl Agric Res Cent Tohoku 105:155–197. (published in Japanese with English summary)
Oliveira Junior EN, de Melo IS, de Franco TT (2012) Changes in hyphal morphology due to chitosan treatment in some fungal species. Braz Arch Biol Technol 55(5):637–646
Palma-Guerrero J, Jansson H-B, Salinas J, Lopez-Llorca L (2008) Effect of chitosan on hyphal growth and spore germination of plant pathogenic and biocontrol fungi. J Appl Microbiol 104:541–553. https://doi.org/10.1111/j.1365-2672.2007.03567.x
Patil HJ, Solanki MK (2016a) Microbial inoculant: modern era of fertilizers and pesticides. In: Singh D, Singh H, Prabha R (eds) Microbial inoculants in sustainable agricultural productivity: vol. 1: research perspectives. Springer, New Delhi, pp 319–343

Patil HJ, Solanki MK (2016b) Molecular prospecting: advancement in diagnosis and control of Rhizoctonia solani diseases in plants. Springer International Publishing, pp 165–185.

Perez MF, Contreras L, Garnica NM, Fernandez ZMV, Farias ME, Sepulveda M et al (2016) Native killer yeasts as biocontrol agents of postharvest fungal diseases in lemons. Public Libr Sci 11(10):e0165590

Peteira B, Martinez B, Muniz Y, Miranda I (2001) Molecular characterization of the genetic diversity of some promising isolates of *Trichoderma* spp. by RAPD. Revista-de-Proteccion-Vegetal 16:157–158. 23

Praveen Kumar D, Singh RK, Anupama PD, Solanki MK, Kumar S, Srivastava AK, Singhal PK, Arora DK (2012) Studies on exo-chitinase production from Trichoderma asperellum UTP-16 and its characterization. Indian J Microbiol 52:388–395

Raaijmakers JM, Mazzola M (2012) Diversity and natural functions of antibiotics produced by beneficial and plant pathogenic bacteria. Annu Rev Phytopathol 50:403–424. https://doi.org/10.1146/annurev-phyto-081211-172908

Rabbee FM, Ali SM, Choi J, Hwang HB, Jeong CS, Baek HK (2019) *Bacillus velezensis:* a valuable member of bioactive molecules within plant microbiomes review. Molecules 24(6):1046

Rai S, Solanki M, Solanki AC, Surapathrudu K (2019) Trichoderma as biocontrol agent: molecular prospectus and application. In: Ansari RA, Mahmood I (eds) Plant health under biotic stress, vol II. Springer, Singapore. https://doi.org/10.1007/978-981-13-6040-4_7

Ramanauskiene J, Dabkevicius Z, Tamosiunas K, Petraitiene E (2019) The incidence and severity of take-all in winter wheat and *Gaeumannomyces graminis* soil inoculum levels in Lithuania. Zemdirbyste-Agric 106:37–44. https://doi.org/10.13080/z-a.2019.106.005

Raza W, Ling N, Yang L, Huang Q, Shen Q (2016) Response of tomato wilt pathogen *Ralstonia solanacearum*to the volatile organic compounds produced by biocontrol strain *Bacillus amyloliquefaciens* SQR-9. Sci Report 6:24856

Recinosa AD, Sekedata VM, Hernandeza A, Cohen ST, Sakhtaha H, Prince SA, Price-Whelanc A, Dietricha EPL (2012) Redundant phenazine operons in Pseudomonas aeruginosa exhibit environment-dependent expression and differential roles in pathogenicity. 109(47):19420–19425. www.pnas.org/lookup/suppl/doi:10. 1073/pnas.1213901109/-/DCSupplemental

Reino JL, Guerrero RF, Hernandez GR, Collado IG (2008) Secondary metabolites from species of the biocontrol agent Trichoderma. Phytochem Rev 7:89–123

Rosane SC, Helder LSL, Gustavo ASP, Carlos ATG, Sueli R (2008) Effect of moisture on Trichoderma conidia production on corn and wheat bran by solid state fermentation. Food Bioprocess Technol 1:100–104

Ruano RD, Prieto P, Rincon AM, Gomez-Rodriguez MV, Valderrama R, Barroso JB et al (2016) Fate of *Trichoderma harzianum* in the olive rhizosphere: time course of the root colonization process and interaction with the fungal pathogen *Verticillium dahliae*. BioControl 61:269–282. https://doi.org/10.1007/s10526-015-9706-z

Saeedizadeh A (2016) *Trichoderma viride* and *Pseudomonas fluorescens* CHA0 against *Meloidogyne javanica* in the rhizosphere of tomato plants. Hellenic Plant Prot J 9:28–34

Sah S, Singh R (2015) Siderophore: structural and functional characterization –a comprehensive review. Agric (Polnohospodarstvo) 61:97–114. https://doi.org/10.1515/agri-2015-0015

Saldajeno MGB, Naznin HA, Elsharkawy MM, Shimizu M, Yakumachi MH (2014) Enhanced resistance of plants to disease using *Trichoderma* spp. Biotechnol Biol Trichoderma:477–493

Sansone G, Rezza I, Calvente V, Benuzzi D, Sanz de Tosetti MI (2005) Control of Botrytis cinereastrains resistant to iprodione in apple with rhodotorulic acid and yeasts. Postharvest Biol Technol 35:245–251

Saremi H, Amiri ME, Ashrafi J (2011) Epidemiological aspects of bean decline disease caused by *Fusarium* species and evaluation of the bean resistant cultivars to disease in Northwest Iran. Afr J Biotechnol 10(66):14954–14961. https://doi.org/10.5897/AJB11.2580

Saunders G, Cooke B, McColl K, Richard S (2010) Modern approaches for the biological control of vertebrate pests: an Australian perspective. Biol Control 52(3):288–295. https://doi.org/10.1016/j.biocontrol.2009.06.014

Schulz BK, Martin SL, Garbeva P (2017) Microbial volatiles: small molecules with an important role in intra- and inter-kingdom interactions. Front Microbiol 8:2484. https://doi.org/10.3389/fmicb.2017.02484

Shabir HW, Kumar V, Varsha S, Saroj KS (2016) Phytohormones and their metabolic engineering for abiotic stress tolerance in crop plants. Crop J 4:162–176

Shukla N, Negi AE, Singh A, Kabadwa BC, Sharma R, Kumar DJ (2019) Present status and future prospects of bio-agents in agriculture. Int J Curr Microbiol App Sci 8(4):2138–2153

Siddique MNA, Ahmmed ANF, Mazumder MGH, Khaiyam MO, Islam MR (2016) Evaluation of some fungicides and bio-agents against *Sclerotium rolfsii* and foot and root rot disease of eggplant (*Solanum melongena* L.). Agriculturists 14(1):92–97

Siddiqui Y, Meon S, Ismail R, Rahmani M (2009) Bio-potential of compost tea from agro-waste to suppress Choanephora cucurbitarum L. the causal pathogen of wet rot of okra. Biol Control 49(1):38–44

Sikora DE, Mercy L, Baum C, Hrynkiewicz K (2019) The contribution of Endomycorrhiza to the performance of *Potato Virus Y*-infected Solanaceous plants: disease alleviation or exacerbation review article. Front Microbiol. https://doi.org/10.3389/fmicb.2019.00516

Singh RK, Kumar DP, Solanki MK, Singh P, Srivastva AK, Kumar S, Kashayp PL, Saxena AK, Singhal PK, Arora DK (2012) Optimization of media components for chitinase production by chickpea rhizosphere associated *Lysinibacillus fusiformis* B-CM18. J. Basic Microbiol 52:1–10

Singh RK, Kumar DP, Singh P, Solanki MK, Srivastava S, Kashyap PL, Kumar S, Srivastava AK, Singhal PK, Arora DK (2014) Multifarious plant growth promoting characteristics of chickpea rhizosphere associated Bacilli help to suppress soil-borne pathogens. Plant Growth Regul 73:91–101

Singh V, Upadhyay RS, Sarma BK, Singh HB (2016) Seed biopriming with Trichoderma asperellum effectively modulate plant growth promotion in pea. Int J Agric Environ Biotechnol 9(3):361–365

Singh P, Song Q-Q, Singh R, Li H-B, Solanki M, Malviya M, Verma K, Yang L-T, Li Y-R, Singh P, Song Q-Q, Singh RK, Li H-B, Solanki MK, Malviya MK, Verma KK, Yang L-T, Li Y-R (2019) Proteomic analysis of the resistance mechanisms in sugarcane during Sporisorium scitamineum infection. Int J Mol Sci 20:569

Smith RS, Osburn RM (2016) Combined used of Lipo-Chitooligosaccharides and Chitinous compounds for enhanced plant growth and yield. U.S. Patent 9,253,989, 9

Smith KP, Handelsman J, Godman RM (1999) Genetic basis in plants for interactions with disease-suppressive bacteria. Proc Natl Acad Sci U S A 9:4786–4790

Smitha C, Finosh GT, Rajesh R, Abraham PK (2014) Induction of hydrolytic enzymes of phytopathogenic fungi in response to Trichoderma viride influence biocontrol activity. Int J Curr Microbiol App Sci 3(9):1207–1217

Solanki MK, Singh N, Singh RK, Singh P, Srivastava AK, Kumar S, Kashyap PL, Arora DK (2011) Plant defense activation and management of tomato root rot by a chitin-fortified Trichoderma/Hypocrea formulation. Phytoparasitica 39:471–481

Solanki MK, Kumar S, Pandey AK, Srivastava S, Singh RK, Kashyap PL, Srivastava AK, Arora DK (2012a) Diversity and antagonistic potential of Bacillus spp. associated to the rhizosphere of tomato for the management of *Rhizoctonia solani*. Biocontrol Sci Tech 22:203–217

Solanki MK, Robert AS, Singh RK, Kumar S, Pandey AK, Srivastava AK, Arora DK (2012b) Characterization of Mycolytic enzymes of Bacillus strains and their bio-protection role against Rhizoctonia solani in tomato. Curr Microbiol 65:330–336

Solanki MK, Singh RK, Srivastava S, Kumar S, Kashyap PL, Srivastava AK, Arora DK (2014) Isolation and characterization of siderophore producing antagonistic rhizobacteria against Rhizoctonia solani. J Basic Microbiol 54:585–597

Solanki M, Singh RK, Srivastava S, Kumar S, Kashyap PL, Srivastava AK (2015) Characterization of antagonistic-potential of two Bacillus strains and their biocontrol activity against Rhizoctonia solani in tomato. J Basic Microbiol 55:82–90

Solanki MK, Malviya MK, Wang Z (2016) Actinomycetes bio-inoculants: a modern prospectus for plant disease management. In: Subramaniam G, Arumugam S, Rajendran V (eds) Plant growth promoting Actinobacteria. Springer Singapore, Singapore, pp 63–81

Solanki MK, Wang Z, Wang FY, Li CN, Lan TJ, Singh RK, Singh P, Yang LT, Li YR (2017) Intercropping in sugarcane cultivation influenced the soil properties and enhanced the diversity of vital Diazotrophic Bacteria. Sugar Tech 19:136–147

Solanki MK, Yandigeri MS, Kumar S, Singh RK, Srivastava AK (2018) Nutrient competition mediated antagonism of microbes against *Rhizoctonia solani*. Notulae Scientia Biologicae 10:392–399

Solanki MK, Kashyap BK, Solanki AC, Malviya MK, Surapathrudu K (2019a) Helpful linkages of *Trichoderma* in the Mycoremediation and Mycorestoration. In: Ansari RA, Mahmood I (eds) Plant health under biotic stress, vol II. Springer, Singapore. https://doi.org/10.1007/978-981-13-6040-4_2

Solanki MK, Wang F-Y, Wang Z, Li C-N, Lan T-J, Singh RK, Singh P, Yang L-T, Li Y-R (2019b) Rhizospheric and endospheric diazotrophs mediated soil fertility intensification in sugarcane-legume intercropping systems. J Soils Sediments 19:1911–1927

Srivastava M, Shahid M, Pandey S, Singh A, Kumar V, Gupta S, Maurya M (2014) *Trichoderma* genome to genomics: a review. J Data Min Genomics Proteomics 5(3):1000172

Strom N, Hu W, Haarith D, Chen S, Bushley K (2019) Fungal communities in soybean cyst nematode-infested soil under long term corn and soybean monoculture and crop rotation. https://doi.org/10.1101/516575

Tabatabaei FS, Saeedizadeh A (2017) Rhizobacteria cooperative effect against *Meloidogyne javanica* in rhizosphere of legume seedlings. Hellenic Plant Prot J 10:25–34

Tamiru G, Muleta D (2018) The effect of rhizobia isolates against black root rot disease of Faba bean (*Vicia faba* L) caused by *Fusarium solani*. Open Agric J 12:131–147. https://doi.org/10.2174/1874331501812010131

Tarique HA, Mahfouz MM, Abd E (2017) Beneficial nematodes in agroecosystems: a global perspective. Entomopathogenic and Slug Parasitic Nematodes, Biocontrol Agents

Thakore Y, Ehlers R-U (2006) Biopesticide market for global agricultural use. Ind Biotechnol 2:194–208

Thangavelu R, Mustaffa M (2012) Current advances in the Fusarium wilt disease management in banana with emphasis on biological control. INTECH, Shanghai, pp 274–287

Thaware DS, Kohire OD, Gholve VM (2015) Survey of chickpea wilt (Fusarium oxysporum f. sp. ciceri) disease in Marathwada region of Maharashtra state. Adv Res J Crop Improv 6(2):134–138

Thaware DS, Gholve VM, Ghante PH (2017) Screening of chickpea varieties, cultivars and genotypes against *Fusarium oxysporium* f. sp. *ciceri*. Int J Curr Microbiol App Sci 6(1):896–904. https://doi.org/10.20546/ijcmas.2017.601.105

Thomashow LS, Weller DM, Bonsall RF et al (1990) Appl Environ Microbiol 56:908–912

Thrall PH, Hochberg ME, Burdon JJ, Bever JD (2007) Coevolution of symbiotic mutualists and parasites in a community context. Trends Ecol Evol 22:120–126

Tijerino A, Cardoza RE, Moraga J, Malmierca GM, Vicente F, Aleu J, Collado GI, Gutierrez S, Monte E, Hermosa R (2010) Overexpression of the trichodiene synthase gene Tbtri5 increases trichodermin production and antimicrobial activity in *Trichoderma brevicompactum*. Fungal Genet Biol 48(3):285–296

Tiwari N, Ahmed S, Kumar S, Sarker A (2018) Fusarium wilt: Akiller disease of lentil review. Books, Fusarium – plant diseases, pathogen diversity, genetic diversity, resistance and molecular markers. Agric Microbiol:119–138. https://doi.org/10.5772/intechopen.72508

Toufiq N, Tabassum B, Bhatti MU, Khan A, Tariq M, Shahid N, Nasir IA, Husnain T (2017) Improved antifungal activity of barley derived chitinase i gene that overexpress a 32 kDa recombinant chitinase in *Escherichia coli* host. Braz J Microbiol 42:414–421

Toyoda H, Hashimoto H, Utsumi R, Kobayashi H, Ouchi S (1988) Detoxification of Fusaric acid by a Fusaric acid- resistant mutant of *Pseudomonas solanacearum* and its application to biological control of Fusarium wilt of tomato. Phytopathology 78:1307–1311

Tsegaye Z, Assefa F, Tefera G, Alemu T, Gizaw B, Abatenh E (2018) Concept, principle and application of biological control and their role in sustainable plant diseases management strategies. Int J Res Stud Biosci 6(4):18–34
Vallad GE, Goodman RM (2004) Goodman systemic acquired resistance and induced systemic resistance in conventional agriculture. Crop Sci 44:1920–1934
van Agtmaal M, Straathof AL, Termorshuizen A, Lievens B, Hoffland E, de Boer W (2018) Volatile-mediated suppression of plant pathogens is related to soil properties and microbial community composition. Soil Biol Biochem 117:164–174. https://doi.org/10.1016/j.soilbio.2017.11.015
Van Loon LC (2007) Plant responses to plant growth promoting bacteria. Eur J Plant Pathol 119:243–254
Velázquez-Becerra C, Macías-Rodríguez LI, López-Bucio J, Flores-Cortez I, Santoyo G, Hernández-Soberano C, Valencia-Cantero E (2013) The rhizobacterium arthrobacter agilis produces dimethylhexadecylamine, a compound that inhibits growth of phytopathogenic fungi in vitro. Protoplasma 250(6):1251–1262
Verma VC, Singh SK, Prakash S (2011) Bio-control and plant growth promotion potential of siderophore producing endophytic *Streptomyces* from *Azadirachta indica A. Juss*. J Basic Microbiol 51:550–556. https://doi.org/10.1002/jobm.201000155
Vessey JK (2003) Plant growth promoting rhizobacteria as biofertilizers. Plant Soil 255:571–586
Vinale F, Sivasithamparam K, Ghisalberti EL, Woo SL, Nigro M, Marra R, Lombardi N, Pascale A, Ruocco M, Lanzuise S, Manganiello G, Lorito M (2014) Trichoderma secondary metabolites active on plants and fungal pathogens. Open Mycol J 8:127–139
Walker MJ, Birch RG, Pemberton JM (1988) Mol Microbiol 2:443–454
Wang F, Li D, Wang Z, Dong A, Liu L, Wang B, Chen Q, Liu X (2014) Transcriptomic analysis of the Rice white tip nematode, *Aphelenchoides besseyi* (Nematoda: Aphelenchoididae). PLoS One 9(3):e91591. https://doi.org/10.1371/journal.pone.009159.1
Wang Z, Solanki MK, Pang F, Singh RK, Yang L-T, Li Y-R, Li H-B, Zhu K, Xing Y-X (2017) Identification and efficiency of a nitrogen-fixing endophytic actinobacterial strain from sugarcane. Sugar Tech 19:492–500
Wang Z, Solanki MK, Yu Z-X, Yang L-T, An Q-L, Dong D-F, Li Y-R (2018) Draft genome analysis offers insights into the mechanism by which streptomyces chartreusis WZS021 increases drought tolerance in sugarcane. Front Microbiol 9:3262
Whipps J (2001) Microbial interactions and biocontrol in the rhizosphere. J Exp Bot 52:487–511
Wild CP, Gong YY (2010) Mycotoxins and human disease: a largely ignored global health issue. Carcinogenesis 31:71–82
Wolfarth FS, Schrader E, Oldenburg E, Weinert J (2013) Nematode-collembolan- interaction promotes the degradation of Fusarium biomass and deoxynivalenol according to soil texture. Soil Biol Biochem 57:903–910
Xiang N, Lawrence KS, Kloepper JW, Donald PA, McInroy JA (2017) Biological control of *Heterodera glycines* by spore-forming plant growth-promoting rhizobacteria (PGPR) on soybean. PLoS One 12:e0181201
Xiao L, Xie C-c, Cai J, Lin Z-J, Chen Y-H (2009) Identification and characterization of a chitinase-produced bacillus showing significant antifungal activity. Curr Microbiol 58(5):528–533
Xu L, Wu C, Oelmuller R, Zhang W (2018) Review article: role of phytohormone in *Piriformospora indica* induced growth promotion and stress tolerance in plants: more questions than answers. Front Microbiol. https://doi.org/10.3389/fmicb.2018.01646
Yandigeri MS, Malviya N, Solanki MK, Shrivastava P, Sivakumar G (2015) Chitinolytic Streptomyces vinaceusdrappus S5MW2 isolated from Chilika lake, India enhances plant growth and biocontrol efficacy through chitin supplementation against Rhizoctonia solani. World J Microbiol Biotechnol 31:1217–1225
Yeates GW (2003) Nematodes as soil indicators: functional and biodiversity aspects
Yeates GW, Bongers T (1999) Nematode diversity in agro ecosystems. Agriculture
Yeates GW, Bongers T, Goede RGM, de Freckman DW, Georgieva SS (1993) Feeding habits in soil nematode families and genera – an outline for soil ecologists. J Nematol 25:315–331

Yu T (2008) Effect of chitin on the antagonistic activity of Cryptococcus laurentii against Penicillium expansum in pear fruit. Int J Food Microbiol 122(1–2):44–48
Yurkov AM (2018) Yeasts of the soil – obscure but precious. Yeast 35(5):369–378
Zaidi NW, Singh US (2013) 14 Trichodermain plant health management. *Trichoderma*: biology and applications. CAB International, London, p 230
Zakaria AZ (2018) Applied environmental science and engineering for a sustainable future. © Springer Nature Singapore Pte Ltd. http://www.springer.com/series/13085
Zhao M, Yuan J, Zhang R, Dong M, Deng X, Zhu C et al (2018) Microflora that harbor the NRPS gene are responsible for Fusarium wilt disease-suppressive soil. Appl Soil Ecol 132:83–90

Website

https://www.koppert.com/trianum-p/

Zinc-Solubilizing Microbes for Sustainable Crop Production: Current Understanding, Opportunities, and Challenges

11

Prity Kushwaha, Prem Lal Kashyap, K. Pandiyan, and Ajay Kumar Bhardwaj

Abstract

Zinc (Zn) is an essential and prime micronutrient needed in diminutive amount by agricultural crops for complete growth and development. It acts as an essential and key constituent of a variety of enzymatic reactions, carbohydrate metabolism, synthesis of proteins and auxin, and maintenance of cellular membrane veracity in plants. Zn is also an essential element in human diet as its deficiency affects normal development and functioning of nervous, immune, and skeletal systems. Crop plants and their consumable parts serve as major sources of Zn in human diet. Plants can uptake Zn as divalent cation, but a major portion of it exists in insoluble form in the soil and very little Zn becomes available to the plants. There are wide varieties of microbes which employ myriads of biological processes to make Zn available to plants from unavailable sources. These zinc-solubilizing microbes (ZSM) can be utilized as prospective alternatives to conventional less-efficient fertilizer application for enhancing Zn availability in soils. Owing to the naturally available source of Zn in soil and high cost of synthetic Zn fertilizers, the demand of ZSM is escalating with time. The injudicious application of chemical fertilizers can be minimized by using ZSM in crop production that can lead to environmental and agricultural sustainability. At the global level, several researchers have recognized the importance of ZSM for crop growth, health, and development. The current article illustrates the role of ZSM in improving plant production in an economical, environment-friendly, and sustainable manner. The mechanisms used by ZSM for Zn solubilization have been

P. Kushwaha · K. Pandiyan
ICAR-National Bureau of Agriculturally Important Microorganisms (NBAIM), Mau, Uttar Pradesh, India

P. L. Kashyap (✉)
ICAR- Indian Institute of Wheat and Barley Research (IIWBR), Karnal, Haryana, India

A. K. Bhardwaj
ICAR-Central Soil Salinity Research Institute (CSSRI), Karnal, Haryana, India

© Springer Nature Singapore Pte Ltd. 2020

M. K. Solanki et al. (eds.), *Phytobiomes: Current Insights and Future Vistas*,
https://doi.org/10.1007/978-981-15-3151-4_11

explained. An attempt has been made to provide a comprehensive global overview of research initiatives made in the field of sustainable crop production through ZSM, and further opportunities and challenges for use of ZSM-based technology in agriculture have been discussed.

Keywords
Agriculture · Bacteria · Fungi · Microbes · Nutrients · Sustainability · Zinc

11.1 Introduction

Zinc (Zn) is an essential and prime micronutrient needed in small amounts (5–100 mg Kg^{-1}) by agricultural crops for complete growth, development, and nutrition (Singh et al. 2019; Khan et al. 2019; Yuvaraj and Subramanian 2015; Tripathi et al. 2015; Broadley et al. 2007). In general, Zn deficiency in plants slows down the biological processes of photosynthesis and nitrogen metabolism, reduces complete growth and development of blossoms and fruits, diminishes the production process of carbohydrates and phytohormones, hinders crop ripening duration, and ultimately reduces final plant yield and nutritional grain quality (Mandal and Das 2013; Sadeghzadeh 2013; Nielsen 2012; Alloway 2008; Alloway 2004). Several reports have indicated that at global scale approximately half of the land under agriculture cultivation is deficient in available Zn content and as a consequence resulted in a significant reduction in both nutritional quality as well as the potential yield of agriculturally important crops (Mumtaz et al. 2017; Cakmak 2008; Welch and Graham 2004). Due to the inadequate quantity of Zn contents in the arable soil, Zn deficiency has been reported as one of the most prominent and insidious micronutrient deficiencies in several parts of the globe including India, Bangladesh, and Nepal (Jasrotia et al. 2018). It is estimated that in India, approximately 48% area of arable land is Zn deficient (Singh 2009). Further, Venkatakrishnan et al. (2003) have reported that within a week exogenously applied water-soluble Zn fertilizers are rapidly converted to 96–99% of insoluble Zn forms, although Zn conversion rate to insoluble forms varied with soil type and soil physicochemical properties. To overcome Zn deficiency, several approaches have been developed and evaluated. Zn-EDTA and zinc sulfate ($ZnSO_4$) are the most commonly employed fertilizers in the agricultural fields (Doolette et al. 2018; Karak et al. 2005; White and Broadley 2005), but their rapid transformation (within 7 days) into insoluble complex forms after application makes them uneconomical and environmentally unsafe (Rattan and Shukla 1991). Besides this, several other agronomic interventions (e.g., crop rotations and intercropping strategies), traditional crop breeding, and transgenic and genetic engineering have been exploited for boosting plant Zn uptake (Gunes et al. 2007; Zuo and Zhang 2009; Gustin et al. 2009; Cakmak et al. 2010; Mhatre et al. 2011; Tan et al. 2015). Unfortunately, these technologies are costly, laborious, and slower (Hussain et al. 2018). Therefore, it becomes utmost important to devise a strategy for enhancing availability of native soil Zn which is abundant but unavailable to plants. In this context, biofortification of agricultural crops by employing zinc-solubilizing microbes (ZSM) is advocated as a novel strategy not only to fulfill the optimum requirement of micronutrient in food crops but also to enhance crop production on less fertile soils.

Recently, zinc solubilization by microbes (Fig. 11.1) has drawn significant attention in the agriculture sector. Microbial strains with such capabilities are generally termed as zinc-solubilizing microbes (ZSM). These microbes, either of fungal or bacterial nature, are reasonably supportive in augmentation of Zn availability in the soil and also play a crucial role in mobilizing Zn to the edible parts of the plants by enhancing Zn uptake capabilities of the plant. More importantly, these microbial strains have the capability to facilitate rapid solubilization of insoluble Zn compounds [zinc sulfate (ZnS), zinc carbonate ($ZnCO_3$), and zinc oxide (ZnO)] in soil by producing organic acids that confiscate the Zn^{2+} cations and alter the soil pH in close proximity of plant rhizosphere (Alexander 1997). Besides this, the anions also have the capability to augment Zn solubility by chelation phenomenon (Jones and Darrah 1994). Other mechanisms associated with Zn solubilization include effective production of siderophores, chelated ligands, proton, and oxidation-reduction systems on plant cell membranes (Saravanan et al. 2011; Chang et al. 2005; Wakatsuki 1995). Several microbes associated with different kinds of agricultural crops have been observed to boost growth and Zn contents of plants when applied exogenously (Table 11.1). Among them, prominent bacterial genera documented as potential Zn solubilizers include *Agrobacterium*, *Azospirillum*, *Burkholderia*, *Bacillus*, *Gluconacetobacter*, *Enterobacter*, *Microbacterium*, *Rhizobium*, *Serratia*, *Pseudomonas*, and *Thiobacillus* (Vidyashree et al. 2018a, b; Jamali et al. 2018; Khanghahi et al. 2018; Khande et al. 2017; Naz et al. 2016; Pawar et al. 2015; Hussain et al. 2015; Abaid-Ullah et al. 2015; Vaid et al. 2014; Ramesh et al. 2014; Deepak et al. 2013; Subramanian et al. 2009; Saravanan et al. 2007; Whiting et al.

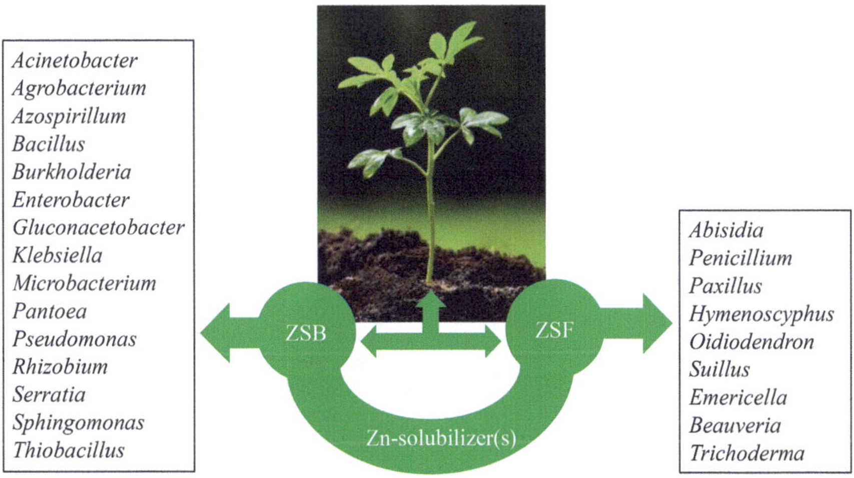

Fig. 11.1 Principal microbial genus associated with Zn-solubilization and uptake by plants

Table 11.1 Effect of ZSM on crop growth, yield, and uptake

Crop(s)	Microbes	Experimental condition(s)	Effect on crops	References
Maize	*Azotobacter*, *Azospirillum*	Greenhouse	Increase in Zn contents of grain	Biari et al. (2008)
Maize	*B. aryabhattai*, *B. subtilis*, and *Bacillus* sp.	Growth chamber	Enhancement in the root length, root fresh biomass, and root dry	Mumtaz et al. (2017)
Maize	*Gluconacetobacter diazotrophicus*	Pot	Increase the available zinc content in the soil	Sarathambal et al. (2010)
Maize	*Enterobacter cloacae*	Pot assay	Increase in dry weight and Zn uptake	Omara et al. (2016)
Maize	*Bacillus* sp.	Growth room	Enhancement in growth and physiology of plants	Hussain et al. (2015)
Maize	*Pseudomonas* sp. P29, *Pseudomonas* sp. P33, and *Bacillus* sp. B40	Pot assay	Increase in total dry mass and uptake of Zn	Goteti et al. (2013)
Rice	*Sphingomonas* sp., *Enterobacter* sp.	Greenhouse	Increase in Zn bioavailability in rhizosphere soils and grain yields and Zn densities in grains	Wang et al. (2014)
Rice	*Acinetobacter* sp. and *Serratia* sp.	Pot assay	Increase in plant growth and root development	Othman et al. (2017)
Rice	*Burkholderia* and *Acinetobacter*	Greenhouse	Enhancement in the total Zn uptake and reduction in phytate	Vaid et al. (2014)
Rice	*A. lipoferum*, *Pseudomonas* sp., *Agrobacterium* sp. (consortium)	Field	Increase in growth, physiology, and yield	Tariq et al. (2007)
Rice	*A. lipoferum*, *Pseudomonas* sp., *Agrobacterium* sp. (consortium)	Field	Increase in growth, physiology, and yield	Hafeez et al. (2002)
Rice	*Pseudomonas aeruginosa*, *Ralstonia pickettii*, *Burkholderia cepacia*, *Klebsiella pneumonia*	Pot	Increment in shoot and root lengths as well as higher dry weights of root and shoot	Gontia-Mishra et al. (2017)
Rice	*Bacillus* sp. AZ6 (consortium with ZnO)	Pot	Improved growth, physiology, and yield parameters of rice; improved the quality of rice grains and bioaccumulation of Zn in various parts of rice	Zeb et al. (2018)

(continued)

Table 11.1 (continued)

Crop(s)	Microbes	Experimental condition(s)	Effect on crops	References
Soybean	*Bacillus firmus* KHBD-6 and *Bacillus amyloliquefaciens KHBAR-1*	Microcosm	Increase in the zinc concentration in soybean seeds	Sharma et al. (2012)
Soybean and wheat	*Bacillus aryabhattai*	Microcosm	Improved mobilization of zinc and its concentration in edible portion, yield of soybean and wheat	Ramesh et al. (2014)
Tomato	*Bacillus aryabhattai*, *Pseudomonas taiwanensis*, *Bacillus* sp., *Enterobacter oryzae*, and *Bacillus aerophilus*	Pot assay	Increase in growth, yield, and better quality parameters	Vidyashree et al. (2018a, b)
Wheat	*Pseudomonas fragi*, *Pantoea dispersa*, *Pantoea agglomerans*, *E. cloacae*, and *Rhizobium* sp.	Pot assay	Enhancements in total Zn uptake by plant	Kamran et al. (2017)
Wheat	*Providencia* sp., *Calothrix* sp., and *Anabaena* sp.	Field	Improvement in the nutritional quality of wheat grains, in terms of protein content	Rana et al. (2012)

2001; Fasim et al. 2002). Similarly, fungal strains belonging to genera *Absidia, Penicillium*, *Paxillus*, *Hymenoscyphus*, *Oidiodendron*, *Suillus*, *Emericella*, *Beauveria*, and *Trichoderma* have been identified as promising zinc-solubilizing fungi (ZSF) (Pawar et al. 2015; Gadd 2007; Fomina et al. 2005a, b; Schöll et al. 2006; Martino et al. 2003). Therefore, exploitation of these microbial inoculants for boosting available Zn content in the soil as well as its enhanced plant uptake could be one of the prospective options. Hence, this chapter provides a comprehensive global overview of research initiatives made in the field of sustainable crop cultivation through ZSM and brings forth the opportunities along with the challenges that exist for the use of ZSM-based technology in agriculture.

11.2 Role and Function of Zn in Plant

Zinc (Zn) is an essential element for plant growth and sustainable crop production, as the plants need all the essential nutrients in appropriate proportion for proper growth functions and optimum yield (Gupta et al. 2016; Hafeez et al. 2013; Sadeghzadeh 2013). Most importantly, in more than 1200 proteins, such as zinc-finger proteins, RNA polymerases, and DNA polymerases, existing inside the plant system, Zn works as a cofactor (Lucini and Bernardo 2015; Figueiredo et al. 2012;

Lopez-Millan et al. 2005). Further, it is required for the efficient functioning of carbohydrate and auxin metabolism in plants (Alloway 2008). Besides this, Zn-finger transcription factors also regulate the growth and development of leaf, flower, fruit, and fertilization (Epstein and Bloom 2005). Zn is also involved in the regulation of various plant physiological processes like repair processes of photosystem (PS)-II complex during photo-inhibition, maintenance of carbon-dioxide concentration in the mesophyll, hormone secretion and mitogen-activated protein kinases (MAPK)-based signal transduction, etc. (Hansch and Mendel 2009; Lin et al. 2005; Bailey et al. 2002). It has been observed that in the majority of crop plants, Zn deficiency results in stunted stem growth, reduced leaf size, and chlorosis and also negatively influences the root development, water uptake and transport inside the plant, pollen formation, and grain yield (Hefferon 2019; Tavallali et al. 2010; Alloway 2004). More specifically, in barley and wheat, a noteworthy decline in crop growth and yield has been observed in Zn-deficient soils (McDonald et al. 2001). Moreover, Zn deficiency also diminishes the amount of Zn contents in consumable parts of crops and resulted in poor nutritional quality of food (Costerousse et al. 2018; Welch and Graham 2004). In the case of rice, Zn deficiency causes multiple deformations in rice seedlings. For instance, rice leaves develop brown blotches and streaks over leaves and result in stunted plant growth. Moreover, under severe conditions, death of plants may occur, while those plants that recuperate will show considerable hindrance in maturity and will lead to substantial grain yield loss (Hafeez et al. 2013).

11.3 Zinc-Solubilizing Microorganisms (ZSM)

11.3.1 Zinc-Solubilizing Fungi (ZSF)

Zinc solubilization is performed by a wide range of fungi. The significance of Zn in the fungal growth and nutrition was noticed for the first time in *Aspergillus niger*, which was unable to grow without Zn content (Raulin 1869). Strains belonging to *Absidia cylindrospora*, *A. spinosa*, *A. glauca*, *Penicillium aurantiogriseum*, *P. brevicompactum*, and *P. simplicissimum* efficiently solubilized ZnO and $Zn_3(PO_4)_2$ forms of Zn present in the soil (Coles et al. 2001). Among the mycorrhiza, *Beauveria caledonica*, *Hymenoscyphus ericae*, *Oidiodendron maius*, *Paxillus involutus*, *Suillus bovinus*, and *S. luteus* showed prospective role in the solubilization of insoluble Zn compounds (Fomina et al. 2005a, b; Schöll et al. 2006; Gadd 2007). Solubilization of Zn compounds of water-insoluble nature by *O. maius* isolated from heavy-metal-polluted locations has been reported by Martino et al. (2003). Anitha et al. (2015) identified *Emericella rugulosa* ZSF-2, *Penicillium citrinum* ZSF-5, *Aspergillus candidus* ZSF-7, *A. terreus* ZSF-9, and *A. niger* ZSF-16 as potent candidates for efficient solubilization of three insoluble sources of zinc [ZnO, $ZnCO_3$, and $Zn(PO_4)_3$]. In another similar study, Pawar et al. (2015) studied solubilization of insoluble zinc sources [ZnO, $ZnCO_3$, and $Zn_3(PO_4)_2$] by *Trichoderma viride* and *T. harzianum* and further confirmed their prospective utility as ZSF.

11.3.2 Zinc-Solubilizing Bacteria (ZSB)

Bacteria is one of the most prospective alternatives that could assist plants in fulfilling its optimum requirement of Zn by solubilizing complex Zn forms occurring in the soil. A wide range of bacterial species belonging to genera *Acinetobacter*, *Bacillus*, *Burkholderia*, *Gluconacetobacter*, *Klebsiella*, *Microbacterium*, *Pantoea*, *Pseudomonas*, *Ralstonia*, *Serratia*, and *Thiobacillus* have been reported to solubilize Zn (Fig. 11.1). Hutchins and associates (1986) noticed that facultative thermophilic iron oxidizers, *Thiobacillus thiooxidans* and *T. ferrooxidans*, have the potential to solubilize Zn from complex sulfide ore (sphalerite) forms. Similarly, Simine et al. (1998) reported zinc phosphate solubilization potential of *P. fluorescens* isolated from a forest soil. They noticed production of high concentration of gluconic acid when *P. fluorescens* 3a was cultivated in a medium containing zinc phosphate. Similarly, ZnO solubilizing strains of *Microbacterium saperdae*, *Enterobacter cancerogenus*, and *Pseudomonas monteilii* were screened and identified in the Zn hyperaccumulating rhizosphere of *Thlaspi caerulescens* plant (Whiting et al. 2001). Fasim et al. (2002) reported solubilization of Zn salts by *Pseudomonas aeruginosa* CMG 823 derived from the ambient environment of a tannery. In their study, they noticed that the solubilization of zinc oxide by bacterium occurred due to a rapid gain in the H^+ concentration by brisk synthesis of 2-ketogluconic acid. A similar conclusion was reached by Saravanan et al. (2007) in their studies where *Gluconacetobacter diazotrophicus* was identified as a potential agent for effective solubilization of insoluble Zn compounds. Hussain et al. (2015) revealed the ZnO solubilizing capabilities of *Bacillus sp.* AZ6 and *Bacillus sp.* AZ6 strains. Othman et al. (2017) identified and established *Acinetobacter* sp. TM56 and *Serratia* sp. TM9 as prospective zinc solubilizer inoculants for rice growth promotion. Recently, *E. cloacae* PBS 2, *Pseudomonas fragi* EPS 1, *Pantoea dispersa* EPS 6, *P. agglomerans* EPS 13, and *Rhizobium* sp. LHRW1 were identified and classified as promising zinc solubilizers (Kamran et al. 2017). Similar studies of Khande et al. (2017) also identified *B. anthracis*, *B. cereus*, *B. tequilensis*, *B. thuringiensis*, and *B. subtilis* subsp. *inaquasorum* as promising ZSB for enhancing seed yield and Zn fortification in the seeds of soybean and wheat crops. Gontia-Mishra and associates (2017) identified Zn solubilizing capabilities of *Burkholderia cepacia*, *Klebsiella pneumonia*, *P. aeruginosa*, and *Ralstonia pickettii*. Similarly, *B. aryabhattai* S10, *Bacillus* sp. ZM20, *Bacillus aryabhattai* ZM31, and *Bacillus subtilis* ZM63 and these strains were ascertained as potential zinc-solubilizing candidates for biofortification in maize (Mumtaz et al. 2017).

11.3.3 Mechanism of Zn Solubilization by Microbes

The principal mechanisms of Zn solubilization by microbes include (i) acidification, (ii) chelation, and (iii) chemical transformation. Majority of research articles documented that different kind of microbes including bacteria excrete diverse kinds of organic acids (Table 11.2) and extrude protons to reduce the rhizosphere pH (Wu

Table 11.2 Organic acids produced by ZSM

ZSM(s)	Organic acid	References
Aspergillus terreus	Gluconic acid	Anitha et al. (2015)
Bacillus aryabhattai, *Pseudomonas taiwanensis*, and *Bacillus* sp.	Lactic acid, malonic acid, and citric acid	Vidyashree et al. (2018a, b)
Bacillus sp.	Cinnamic acid, ferulic acid, caffeic acid, chlorogenic acid, syringic acid, and gallic acid	Hussain et al. (2004)
Gluconacetobacter diazotrophicus	5-ketogluconic acid	Saravanan et al. (2007)
Pseudomonas aeruginosa	2-ketogluconic acid	Fasim et al. (2002)
Pseudomonas fluorescens	Gluconic acid and 2-ketogluconic acid	Simine et al. (1998)
Pseudomonas sp.	Acetic acid and gluconic acid	Jaivel et al. (2017)
Streptomyces cinnamonensis, *Streptomyces scabiei*, and *Streptomyces netropsis*	Formic acid, malonic acid, lactic acid, citric acid, and propionic acid	Poovarasan et al. (2015)
Pseudomonas chlororaphis strain 44, *Pseudomonas moraviensis* strain 106, *Pseudomonas syringae* strain 24, *Stenotrophomonas rhizophila* NZSB, *Curtobacterium oceanosedimentum* strain 81, *Streptomyces narbonensis* strain 68, *Plantibacter flavus* strain 5, and *Plantibacter flavus* strain 42	Gluconic acid, citric acid, fumaric acid, malic acid, 2-oxoglutaric acid, and succinic acid	Costerousse et al. (2018)

et al. 2006). Particularly, *Bacillus* and *Pseudomonas* have the capacity to generate organic acids which decrease the soil pH, thereby ensuring sufficient Zn availability to the plant (Saravanan et al. 2004). It has been observed that rhizobacteria produce gluconate or the derivatives of gluconic acids (e.g., 2-ketogluconic acid and 5-ketogluconic acid) for Zn solubilization (Saravanan et al. 2011; Tariq et al. 2007). Hussain et al. (2004) reported that *Bacillus* sp. AZ6 has the capability to solubilize insoluble source of Zn by synthesizing various types of organic acids (e.g., gallic acid, caffeic acid, syringic acid, cinnamic acid, chlorogenic acid, and ferulic acid). All of the above-described mechanisms facilitate enhanced Zn availability in soil and its uptake by plants. Other indirect effects on plant growth have also been reported. For example, 5-ketogluconic acid synthesized by *G. diazotrophicus* stimulated the Zn solubilization route, and the available Zn^{2+}cations have been noticed to enhance the nematicidal activity of *G. diazotrophicus* against *Meloidogyne incognita* in tomato (Saravanan et al. 2007).

In case of fungi, Martino et al. (2003) reported that mycorrhizal fungi produced organic acids to convert insoluble zinc phosphate [$Zn_3(PO_4)_2$] and zinc oxide [ZnO] into water-soluble Zn. Similarly, Subramanian et al. (2009) reported that inoculation of arbuscular mycorrhizae reduced the pH of rhizospheric soil and facilitated quick release of Zn from mineral fractions. However, the level of rhizosphere pH sinking differed among microbes (Giri et al. 2005). Wu et al. (2006) recorded a significant decline of 0.47 units in pH, owing to the fact that bacterial inoculation helped in the release of organic acids and H^+, which eventually led to the rapid Zn solubilization and in turn improved plant Zn uptake efficiency. Entomopathogenic strains of ectomycorrhizal fungi like *Beauveria caledonica* displayed excellent potential to solubilize chemical forms of Zn (e.g., zinc phosphate) in contrast to mineral forms (e.g., pyromorphite) by employing acidolysis and complexolyis mechanisms. Besides this, oxalic acid production has also been found related to Zn solubilization (Fomina et al. 2004). The role of the siderophores in Zn solubilization has been reported in the case of *Fusarium solani* (Hong et al. 2010).

Zinc chelation is another mechanism employed by *E. cancerogenus*, *Microbacterium saperdae*, and *P. monteilii* for enhancing Zn bioavailability and its uptake efficiency by plant roots (Whiting et al. 2001). Generally, bacterial metabolites form complexes with Zn^{2+} and diminish their response within the soil system (Tarkalson et al. 1998). Subsequently, these newly formed Zn chelates travel toward the roots and liberate chelating ligands (Zn^{2+}) at the root surface, making them independent to chelate with other Zn^{2+} ions from the soil solution. In another independent study, Tariq et al. (2007) observed that a biofertilizer comprised of *Agrobacterium* sp. Ca-18, *Azospirillum lipoferum* JCM-1270, *A. lipoferum* ER-20, and *Pseudomonas* sp.96-51 helped to discharge fixed Zn quickly and also made it bioavailable to rice seedlings for a longer period by discharging Zn-EDTA as a chelating molecule. Plant inoculation with *Penicillium bilaji* has also been reported to improve Zn bioavailability to plants via chelating mechanisms (Kucey 1987). The results of a metal mobilization experiment performed by Li et al. (2010) demonstrated that despite the presence of glucose in the growth medium, *Burkholderia cepacia* released oxalic, tartaric, formic, and acetic acids which facilitated Zn solubilization. Recently, Costerousse et al. (2018) explained a range of Zn solubilizing processes adopted by bacteria presented in the wheat rhizosphere. They reported proton extrusion and organic acid synthesis by bacterial strains as the major mechanisms linked with Zn solubilization. Further, they observed that in ZnO liquid solubilization assays, *Curtobacterium*, *Plantibacter*, *Pseudomonas*, *Stenotrophomonas*, and *Streptomyces* triggered the rapid synthesis of organic acids, and as a consequence medium acidification took place, which further assisted in effective ZnO solubilization in the presence of glucose. Interestingly, they observed that Zn solubilization by *Streptomyces* and *Curtobacterium* occurred due to a rapid production of six and seven different organic acids, while other strains involve only gluconic, malonic, and oxalic acids for Zn solubilization. Similarly, *Plantibacter* strains performed ZnO dissolution by proton extrusion via ammonia consumption in the absence of glucose. On the other hand, *Curtobacterium* strains employed complexation processes by involving glutamic acid (Costerousse et al. 2018). Therefore, it is clearly evident that solubilization mechanisms vary among ZSM.

11.4 Techniques for Identification and Determining Zinc-Solubilizing Capabilities of Microbes

Zinc-solubilizing microbes can be screened for their ability to solubilize Zn from mineral salts in agar medium [Glucose-10 g, $(NH_4)_2SO_4$-1.0 g, KCl-0.2 g, K_2HPO_4-0.1 g, $MgSO_4$-0.2 g, and H_2O-1000 ml with pH 7.0]. Petri plates supplemented with either insoluble zinc oxide (ZnO) or zinc carbonate ($ZnCO_3$) or zinc phosphate in the concentration of 0.1% have been commonly used. Spot inoculation with actively growing culture (5 μl) onto the medium and incubation at 25–28 °C for 3–5 days has been found suitable. The appearance of transparent and clear zone reveals the zinc-solubilizing capability of spotted strain. The zone diameter is measured and zinc-solubilizing index (ZSI) is calculated using the following formula:

ZSI = Colony diameter + Diameter of clear and transparent halozone / colony diameter

Generally, atomic absorption spectrophotometer (AAS) is used to estimate the available Zn in the supernatant. In general, the microbial solubilization of Zn is strongly influenced by pH, oxygen, microbial strains used, and kind of Zn-bearing minerals, and therefore, optimal conditions for Zn solubilization by microbes need to be standardized for appropriate results.

11.5 Application of ZSM in Sustainable Crop Production

Several research reports highlighted that ZSM in soil health restoration and sustainable crop production is gaining significant importance day by day. Whiting et al. (2001) studied the efficiency of ZSB strains derived from the rhizosphere of Zn hyperaccumulating plant (*Thlaspi caerulescens*). In their study, they observed that ZSB enhanced the water-soluble Zn fraction in the soil as well as there was 22–67% increment in Zn content in shoots and roots when inoculations were performed in the rhizosphere of the germinating seeds of *Thlaspi* plants.

Sarathambal et al. (2010) demonstrated that the application of *G. diazotrophicus* with ZnO in maize showed better uptake of the nutrient, irrespective of soil types. Further, they noticed that *G. diazotrophicus* solubilized Zn better in unsterile Zn-deficient soil. Similarly, a pot experiment conducted by Goteti et al. (2013) revealed the impact of seed treatment with zinc solubilizing, plant growth promoting bacteria (P29, P33, and B40) on maize. They observed significant improvement in plant biomass as well as in mineral nutrient (N, K, Mn, and Zn) uptake when seed bacterization was performed with P29 strain @ 10 g kg^{-1} maize seed.

Vaid et al. (2014) reported that effects of individual or combined inoculation of rice plants with ZSB (*Acinetobacter* sp. SG2, *Acinetobacter* sp. SG3, and *Burkholderia* sp. SG1) resulted in significant increase in the number of panicles (13.3%), productive tillers per plant (15.1%), number of grains per panicle (12.8%), straw yield (12.4%), grain yield (17.0%), and mean dry matter yield per pot (12.9%), over the control. Further, they also noticed that bacterial inoculation enhanced the grain methionine concentration (38.8%) as well as total Zn uptake efficiency per pot

(52.5%). In a similar study, Shakeel and colleagues (2015) documented that inoculation with *Bacillus* sp. provided effective Zn movement to rice grains and resulted in rice grain yield gain of 22–49% and 18–47% in the case of basmati-385 and super basmati rice varieties, respectively. Similarly, a significant gain in Zn accumulation in rice grains as well as an increase in seedling growth of rice as a result of inoculation with endophytic strains of *Burkholderia* sp. SaZR4, *Burkholderia* sp. SaMR10, *Enterobacter* sp. SaCS20, *Sphingomonas* sp. SaMR12, and *Variovorax* sp. SaNR1 was obtained for *Sedum alfredii* plant by Wang et al. (2014). Further, they observed that inoculation with SaMR12 and SaCS20 under hydroponic conditions resulted in the elevation of Zn concentration by 73.6% and 83.4% in roots, respectively, and by 44.4% and 51.1% in shoots, respectively. Similar results with endophytic inoculation using SaMR12 and SaCS20 strains were observed in the case of polished rice. In the same study, a significant gain in grain yields and Zn concentrations have been observed in the case of brown rice (20.3% and 21.9%, respectively) and polished rice (13.7% and 11.2%), respectively. Moreover, inoculation with SaMR12 and SaCS20 resulted in the accumulation of 10.4% and 20.6% more DTPA-Zn in the rhizosphere soil of rice seedlings, respectively (Wang et al. 2014).

Ramesh et al. (2014) conducted a pot experiment with soil containing low levels of plant available Zn to reveal the effectiveness of seed inoculation with different *Bacillus aryabhattai* strains on Zn solubilization and plant growth promotion. They observed that *B. aryabhattai* inoculations resulted in the elevation of Zn concentration in wheat grains from 42 to 61 mg kg^{-1}, compared to control. They further explained that this elevation was due to the capability of the inoculated ZSB to convert insoluble Zn in the soil by producing organic acids and decrease in the pH (Ramesh et al. 2014). On parallel lines, Hussain and associates (2015) also noticed the significant increment in plant growth attributes, mainly fresh and dry biomass of roots, fresh and dry biomass of shoots, shoot length and root length, etc. as a result of the inoculation of ZSB (*Bacillus* sp. AZ6). Recent research findings of Mumtaz et al. (2017) also revealed that inoculation with *Bacillus* sp. ZM20, *B. aryabhattai* S10, *B. aryabhattai* ZM31, and *B. subtilis* ZM63 enhanced maize growth significantly. They observed that all these isolates have excellent multifarious plant growth promoting attributes and able to colonize plant roots effectively and, therefore, could be utilized as a potential microbe-based eco-friendly alternative for enhancing the productivity as well as nutritional status of maize grains.

The improvement in crop growth and soil fertility of maize plants was observed when seedlings were inoculated with *Enterobacter cloacae* with ZnO (Omara et al. 2016). However, the available Zn content was greater in non-sterile than in the sterile soil conditions, which pointed to the fact that microorganisms other than *E. cloacae* may also be involved in the solubilization of insoluble zinc sources. The research findings documented by Othman et al. (2017) revealed the potential of ZSB (*Acinetobacter* sp. TM56) with zinc sulfate @ 0.2 mg L^{-1} for improving the various plant growth parameters as well as root development of rice. This bacterium had the capacity to solubilize a considerable fraction of insoluble form of Zn under field conditions.

The enhanced content of Zn in the seeds of wheat and soybean crops with effective inoculations of *B. cereus* has been reported by Khande et al. (2017). Under microcosm conditions, they observed that seed priming with the alone application of *B. cereus* BBU-5, *B. cereus* KMR-5, and *B. cereus* DBU-1 significantly enhanced Zn contents in both seed and straw of soybean. Similarly, wheat seed inoculation with *B. cereus* KMR-5, *B. cereus* KBY-5, *B. cereus* BHKD-6, and *B. cereus* DBU-1 provided a high amount of Zn in both seed and straw in contrast to uninoculated control plants. They further explained that the inoculation of ZSB significantly declined phytic-P content in the seeds of soybean and, as a consequence, enhanced bioavailability of Zn in seeds. Similarly, in another experiment, Kamran et al. (2017) studied the effect of ZSB (*E. cloacae* PBS 2, *P. dispersa* EPS 6, *P. fragi* EPS1, *P. agglomerans* EPS 13, and *Rhizobium* sp. LHRW1) on the seedling growth, development, and Zn uptake. They observed that one month after ZSB inoculation, highest shoot and root dry weight and shoot length were observed in seedlings inoculated with *Rhizobium* sp. LHRW1, whereas increased shoot and root length was noticed in *E. cloacae* PBS 2-inoculated plants. Maximum Zn concentration was recorded in shoots of *E. cloacae* PBS 2-inoculated plants and in roots of *P. agglomerans* EPS 13-inoculated plants followed by sole zinc carbonate ($ZnCO_3$) amended control. Interestingly, even after 3 months post ZSB inoculation, a significant elevation in shoot dry weights was detected in plants inoculated with *P. dispersa* EPS6, *P. agglomerans* EPS13, and *E. cloacae* PBS2. Significant increment in terms of root dry weight and maximum Zn content was obtained in the case of *P. fragi* EPS1-inoculated plants derived from the wheat rhizosphere. Interestingly, highest Zn content for roots was obtained in the control plants demonstrating the plant's incapability to mobilize Zn from roots to grains (Kamran et al. 2017).

Gontia-Mishra and associates (2017) demonstrated that the inoculation of rice seedlings with ZSB (*Klebsiella pneumonia* Zn8, *P. aeruginosa* Zn2, and *R. pickettii* Zn3) resulted in a significant increase in plant biomass in contrast to only water-inoculated control seedlings. Additionally, they also reported that treatment of rice plants with individual strain or with bacterial consortium and grown in soils containing an abundant source of an insoluble form of Zn or in Zn-deprived soils with added insoluble Zn compounds enhanced the plant Zn uptake efficiency.

Zeb et al. (2018) demonstrated that compost enriched with *Bacillus* sp. AZ6 inoculum and ZnO enhanced Zn availability and showed a positive effect on growth, physiology, and yield of rice, in comparison to $ZnSO_4$. Further, they observed that Zn-enriched compost with *Bacillus* sp. AZ6 (Zn-EC60:40), in comparison to $ZnSO_4$, significantly enhanced growth, physiology, and yield parameters of rice through a slow and steady release of Zn from ZnO. Additionally, the microbial inoculation also resulted in the bioaccumulation of Zn in various parts of rice as well as rice grain quality.

Based on the basic and applied research knowledge generated till date, it emerges that the application of ZSM can be a promising technique to solubilize the unavailable Zn reserves in the soil and facilitate easy Zn accessibility to the plants, resulting in enhanced plant growth with the minimum application of Zn fertilizers.

11.6 Opportunities and Challenges

Zinc-solubilizing microbes (ZSM) could be an excellent option and practical technology to solubilize insoluble Zn into a soluble form; however, their application in the agricultural sector is still uncommon due to the following reservations:

(i) Based on the information generated so far, ZSM have been less studied under field conditions with rare multilocation testing. Besides this, very limited information is available on the application methodology and delivery methods of ZSM under field conditions. As the results obtained under in vitro environments or under greenhouse conditions may divert from actual field conditions, therefore, special attention and emphasis should be made on testing under natural field conditions to evaluate the potential of ZSM-based technologies for sustainable crop production. Further research is needed to evaluate the influence of other plant growth promoting microbes (PGPM) such as IAA producers, ACC deaminase producers, phosphate solubilizers, and N_2 fixers on the availability of Zn in the soil. Additionally, further understanding of synergistic and antagonistic interactions between ZSM and other PGPM and determination of the optimal conditions for ZSM activity is also needed. Interaction between efficient plant species and ZSM would also be useful yet still unexplored. More specifically, understanding the plant-ZSM interaction specific mechanisms under environmental variables and controls needs to be outlined for potential crops. Additionally, the impact assessment of ZSM on different crops grown under the diverse agroecological conditions and geographical locations needs to be assessed. The role of ZSM in increased availability of other nutrients under the influence of soil pH variation and content of other nutrients (e.g., P, N, Fe, K, etc.) also needs to be studied in detail.

(ii) Limited knowledge and awareness among the farmers regarding the application and benefits of ZSM-based biofertilizers in crop production is one of the biggest hurdles for its practical field application. A major section of the farming community is unaware of the Zn biofertilizers and their merits and demerits in enhancing crop yields. Perhaps, they are also unaware of the negative impacts of continuous application of inorganic zinc fertilizers on the ecosystem functions.

(iii) At present, limited attention is given by the scientific community toward development of Zn-based biofertilizer technologies. Specifically, deficiency in technology with respect to carrier suitability and product formulations to enhance the shelf life of the Zn-based biofertilizer is one of the biggest challenges.

(iv) Unavailability of suitable ZSM strains due to lack of their availability in culture collections or microbial banks is yet another hurdle.

11.7 Conclusions

Zinc mobilization in soils and into plants using ZSM has emerged as an important tool to enhance Zn bioavailability. Judicious application of ZSM is remarkably an efficient, environmentally sound, and low-cost strategy to enhance Zn concentrations in the edible parts of agricultural crops. Microbial inoculation has shown immense potential to enrich plants with Zn micronutrient, with and without fertilizer sources. Due to lots of ambiguities and dilemma in getting success in enriching seed grains with Zn by breeding or agronomic strategies, biofortification using ZSM should be considered as one of the best approaches with quick effects for short-term adoption. In the future, there is a need to strengthen microbial research programs on the development and field evaluations of ZSM-mediated biofortification methods at critical plant growth stages for enhancing Zn uptake efficiency and elevating Zn accrual in food grains. Studying the bioavailability of Zn inside grain, resulting from ZSM applications in single or consortium mode, will be an interesting aspect to investigate. Significant progress has been made in the ZSM-based Zn fortification of crops, and it has convincingly proven that micronutrition malnutrition can be effectively tackled with ZSM-based management of food crop systems.

References

Abaid-Ullah M, Hassan MN, Jamil M, Brade G, Shah MKN, Sessitsch A (2015) Plant growth promoting rhizobacteria: an alternate way to improve yield and quality of wheat (*Triticum aestivum*). Int J Agric Biol 17:51–60

Alexander M (1997) Introduction to soil microbiology. Wiley, New York

Alloway BJ (2004) Zinc in soils and crop nutrition. Publication of the International Zinc Association. http://www.iza.com/Documents/Communications/Publications/ALLOWAY_PRINT.pdf. Accessed on 8 June 2019

Alloway BJ (2008) Zinc in soils and crop nutrition, 2nd edn. IZA and IFA, Brussels/Paris

Anitha S, Padma DSN, Sunitha KK (2015) Isolation and identification of zinc solubilizing fungal isolates from agriculture fields. Indian J Agric Sci 85(12):1638–1642

Bailey SE, Thompson PJ, Horton NP, Mullineaux CW, Robinson C, Mann NH (2002) A critical role for the Var2 FtsH homologue of Arabidopsis thaliana in the photosystem II repair cycle *in vivo*. J Biol Chem 277:2006–2011

Biari A, Gholami A, Rahmani HA (2008) Growth promotion and enhanced nutrient uptake of maize (*Zea mays* L.) by application of plant growth promoting rhizobacteria in arid region of Iran. J Biol Sci 8:1015–1020

Broadley MR, White PJ, Hammond JP, Zelko I, Lux A (2007) Zinc in plants. New Phytol 173:677–702

Cakmak I (2008) Enrichment of cereal grains with zinc: agronomic or genetic biofortification? Plant Soil 302:1–17

Cakmak I, Pfeiffer WH, McClafferty B (2010) Biofortification of durum wheat with zinc and iron. Cereal Chem 87:10–20. https://doi.org/10.1094/CCHEM-87-1-0010

Chang HB, Lin CW, Huang HJ (2005) Zinc induced cell death in rice (*Oryza sativa* L.) roots. Plant Growth Regul 46:261–266. https://doi.org/10.1007/s10725-005-0162-0

Coles KE, David JC, Fisher PJ, Lappin-Scott HM, Macnair MR (2001) Solubilization of zinc compounds by fungi associated with the hyperaccumulator *Thlaspi caerulescens*. Bot J Scotl 51:237–247

Costerousse B, Schonholzer-Mauclaire L, Frossard E, Thonar C (2018) Identification of heterotrophic zinc mobilization processes among bacterial strains isolated from wheat rhizosphere (*Triticum aestivum* L.). Appl Environ Microbiol 84:e01715–e01717
Deepak J, Geeta N, Sachin V, Anita S (2013) Enhancement of wheat growth and Zn content in grains by zinc solubilizing bacteria. Int J Agric Environ Biotechnol 6:363–370. https://doi.org/10.5958/j.2230-732X.6.3.004
Doolette CL, Read TL, Li C, Scheckel KG, Donner E, Kopittke PM, Schjoerring JK, Lombi E (2018) Foliar application of zinc sulphate and zinc EDTA to wheat leaves: differences in mobility, distribution, and speciation. J Exp Bot 69(18):4469–4481
Epstein E, Bloom AJ (2005) Mineral nutrition of plants: principles and perspectives, 2nd edn. Sinauer Associates, Sunderland, pp 17–39
Fasim F, Ahmed N, Parsons R, Gadd GM (2002) Solubilization of zinc salts by a bacterium isolated from the air environment of a tannery. FEMS Microbiol Lett 213:1–6. https://doi.org/10.1111/j.1574-6968.2002.tb11277.x
Figueiredo DD, Barros PM, Cordeiro AM, Serra TS, Lourenço T, Chander S, Oliveira MM, Saibo NJ (2012) Seven zinc-finger transcription factors are novel regulators of the stress responsive gene OsDREB1B. J Exp Bot 63:3643–3656
Fomina M, Alexander IJ, Hillier S, Gadd GM (2004) Zinc phosphate and pyromorphite solubilization by soil plant- symbiotic fungi. Geomicrobiol J 21:351–366
Fomina MA, Alexander IJ, Colpaert JV, Gadd GM (2005a) Solubilization of toxic metal minerals and metal tolerance of mycorrhizal fungi. Soil Biol Biochem 37:851–866
Fomina M, Hillier S, Charnock JM, Melville KI, Alexander J, Gadd GM (2005b) Role of oxalic acid overexcretion in transformations of toxic metal minerals by *Beauveria caledonica*. Appl Environ Microbiol 71:371–381
Gadd GM (2007) Geomycology: biogeochemical transformations of rocks, minerals, metals and radionuclides by fungi, bioweathering and bioremediation. Mycol Res 111:3–49
Giri B, Giang PH, Kumari R, Prasad R, Varma A (2005) Microbial diversity in soils. In: Buscot F, Varma S (eds) Micro-organisms in soils: roles in genesis and functions. Springer, Heidelberg, pp 195–212
Gontia-Mishra I, Sapre S, Tiwari S (2017) Zinc solubilizing bacteria from rhizosphere of rice as a prospective modulator of zinc biofortification in rice. Rhizosphere 3:185–190
Goteti PK, Emmanuel LDl A, Desai S, Shaik MHA (2013) Prospective zinc solubilising bacteria for enhanced nutrient uptake and growth promotion in maize (*Zea mays* L.). Int J Microbiol, Article ID 869697(7). https://doi.org/10.1155/2013/869697
Gunes A, Inal A, Adak MS, Alpaslan M, Bagci EG, Erol T (2007) Mineral nutrition of wheat, chickpea and lentil as affected by mixed cropping and soil moisture. Nutr Cycl Agroecosyst 78:83–96. https://doi.org/10.1007/s10705-006-9075-1
Gupta N, Ram H, Kumar B (2016) Mechanism of zinc absorption in plants: uptake, transport, translocation and accumulation. Rev Environ Sci Biotechnol 15:89–109
Gustin JL, Loureiro ME, Kim D, Na G, Tikhonova M, Salt DE (2009) MTP1-dependent Zn sequestration into shoot vacuoles suggests dual roles in Zn tolerance and accumulation in Zn-hyperaccumulating plants. Plant J 57:1116–1127. https://doi.org/10.1111/j.1365-313X.2008.03754.x
Hafeez FY, Hameed S, Zaidi AH, Malik KA (2002) Biofertilizer for sustainable agriculture. In: Azam F, Iqbal MM, Inayatullah C, Malik KA (eds) Techniques for sustainable agriculture. ISBN/NIAB, Faisalabad, pp 67–73
Hafeez B, Khanif YM, Saleem M (2013) Role of zinc in plant nutrition- a review. Am J Agric Econ 3(2):374–391
Hansch R, Mendel RR (2009) Physiological functions of mineral micronutrients (Cu, Zn, Mn, Fe, Ni, Mo, B, Cl). Curr Opin Plant Biol 12:259–266
Hefferon K (2019) Biotechnological approaches for generating zinc-enriched crops to combat malnutrition. Nutrients 11(2):253. https://doi.org/10.3390/nu11020253
Hong JW, Park JY, Gadd GM (2010) Pyrene degradation and copper and zinc uptake by *Fusarium solani* and *Hypocrea lixii* isolated from petrol station soil. J Appl Microbiol 108(6):2030–2040

Hussain D, Haydon MJ, Wang Y, Wong E, Sherson SM, Young J, Camakaris J, Harper JF, Cobbett CS (2004) P-type ATPase heavy metal transporters with roles in essential zinc homeostasis in Arabidopsis. Plant Cell 16:1327–1339

Hussain A, Arshad M, Zahir ZA, Asghar M (2015) Prospects of zinc solubilizing bacteria for enhancing growth of maize. Pak J Agric Sci 52:915–922

Hussain A, Zahir ZA, Asghar HN, Ahmad M, Jamil M, Naveed M, Akhtar MFUZ (2018) Zinc solubilizing Bacteria for zinc biofortification in cereals: a step toward sustainable nutritional security. In: Meena VS (ed) Role of rhizospheric microbes in soil. Springer, Singapore, pp 203–227

Hutchins SR, Davidson MS, Brierey JA, Brierley CL (1986) Microorganisms in reclamation of metals. Annu Rev Microbiol 40:311–336

Jaivel N, Siva kumar U, Marimuthu P (2017) Characterization of zinc solubilization and organic acid detection in *Pseudomonas sp.* RZ1 from rice phyllosphere. Int J Chem Stud 5(6):272–277

Jamali H, Sharma A, Kushwaha P, Kashyap PL, Roohi SAK (2018) Exploitation of multifarious abiotic stresses, antagonistic activity and plant growth promoting attributes of *Bacillus amyloliquefaciens* AH53 for sustainable agriculture production. Int J Curr Microbiol App Sci 7(10):751–763

Jasrotia P, Kashyap PL, Bhardwaj AK, Kumar S, Singh GP (2018) Scope and applications of nanotechnology for wheat production: a review of recent advances. Wheat Barley Res 10(1):1–14

Jones DL, Darrah PR (1994) Role of root-derived organic acids in the mobilization of nutrients from the rhizosphere. Plant Soil 166:247–257. https://doi.org/10.1007/BF00008338

Kamran S, Shahid I, Baig DN, Rizwan M, Malik KA, Mehnaz S (2017) Contribution of zinc solubilizing bacteria in growth promotion and zinc content of wheat. Front Microbiol 8:2593. https://doi.org/10.3389/fmicb.2017.02593

Karak T, Singh UK, Das S, Das DK, Kuzyakov Y (2005) Comparative efficacy of ZnSO4 and Zn-EDTA application for fertilization of rice (*Oryza sativa* L.). Arch Agron Soil Sci 51:253–264. https://doi.org/10.1080/03650340400026701

Khan A, Singh J, Upadhayay VK, Singh AV, Shah S (2019) Microbial biofortification: a green technology through plant growth promoting microorganisms. In: Shah S et al (eds) Sustainable green technologies for environmental management. Springer, Singapore, pp 255–268

Khande R, Sushil KS, Ramesh A, Mahaveer PS (2017) Zinc solubilizing *Bacillus* strains that modulate growth, yield and zinc biofortification of soybean and wheat. Rhizosphere 4:126–138. https://doi.org/10.1016/j.rhisph.2017.09.002

Khanghahi MY, Ricciuti P, Allegretta I, Terzano R, Crecchio C (2018) Solubilization of insoluble zinc compounds by zinc solubilizing bacteria (ZSB) and optimization of their growth conditions. Environ Sci Pollut R 25(26):25862–25868

Kucey RMN (1987) Increased phosphorus uptake by wheat and field beans inoculated with a phosphorus-solubilizing *Penicillium bilaji* strain and with vesicular-arbuscular mycorrhizal fungi. Appl Environ Microbiol 53:2699–2703

Li WC, Ye ZH, Wong MH (2010) Metal mobilization and production of short-chain organic acids by rhizosphere bacteria associated with a Cd/Zn hyperaccumulating plant, *Sedum alfredii*. Plant Soil 326:453–467

Lin CW, Chang HB, Huang HJ (2005) Zinc induces mitogen-activated protein kinase activation mediated by reactive oxygen species in rice roots. Plant Physiol Biochem 43:963–968

Lopez-Millan AF, Ellis DR, Grusak MA (2005) Effect of zinc and manganese supply on the activities of superoxide dismutase and carbonic anhydrase in Medicago truncatula wild type and raz mutant plants. Plant Sci 168:1015–1022

Lucini L, Bernardo L (2015) Comparison of proteome response to saline and zinc stress in lettuce. Front Plant Sci 6:240

Mandal M, Das DK (2013) Zinc in rice-wheat irrigated ecosystem. J Rice Res 1:111. https://doi.org/10.4172/2375-4338.1000111

Martino E, Perotto S, Parsons R, Gadd GM (2003) Solubilization of insoluble inorganic zinc compounds by ericoid mycorrhizal fungi derived from heavy metal polluted sites. Soil Biol Biochem 35:133–141

McDonald GK, Graham RD, Lloyd J, Lewis J, Lonergan P, Khabas-Saberi H (2001) Breeding for improved zinc and manganese efficiency in wheat and barley. In: Rowe B, Donaghy D, Mendham N (eds) Science and technology: delivering results for agriculture? Proceedings of the 10th Australian Agronomy Conference, Hobart

Mhatre M, Srinivas L, Ganapathi TR (2011) Enhanced iron and zinc accumulation in genetically engineered pineapple plants using soybean Ferritin gene. Biol Trace Elem Res 144:1219–1228. https://doi.org/10.1007/s12011-011-9092-z

Mumtaz MZ, Ahmad M, Jamil M, Hussain T (2017) Zinc solubilizing *Bacillus spp.* potential candidates for biofortification in maize. Microbiol Res 202:51–60

Naz I, Ahmad H, Khokhar SN, Khan K, Shah AH (2016) Impact of zinc solubilizing bacteria on zinc contents of wheat. Am Eurasian J Agric Environ Sci 16:449–454. https://doi.org/10.5829/idosi.aejaes.2016.16.3.12886

Nielsen FH (2012) History of zinc in agriculture. Adv Nutr 3(6):783–789

Omara AA, GhazI AA, El-Akhdar IA (2016) Isolation and identification of zinc dissolving bacteria and their potential on growth of *Zea mays*. Egypt J Microbiol 51:29–43

Othman NMI, Othman R, MohdSaud H, Wahab PEM (2017) Effects of root colonization by zinc-solubilizing bacteria on rice plant (*Oryza sativa* MR219) growth. Agric Nat Res 51(6):532–537

Pawar A, Ismail S, Mundhe S, Patil VD (2015) Solubilization of insoluble zinc compounds by different microbial isolates in vitro condition. Int J Trop Agric 33:865–869

Poovarasan S, Mohandas S, Sita T (2015) Functional characterization of actinomycetes isolated from the AM fungal (*Glomus mosseae*) spores. Int J Curr Microbiol App Sci 4(9):598:612

Ramesh A, Sharma SK, Sharma MP, Yadav N, Joshi O (2014) Inoculation of zinc solubilizing *Bacillus aryabhattai* strains for improved growth, mobilization and biofortification of zinc in soybean and wheat cultivated in vertisols of central India. Agric Ecosyst Environ Appl Soil Ecol 73:87–96

Rana A, Joshi M, Prasanna R, Shivay YS, Nain L (2012) Biofortification of wheat through inoculation of plant growth promoting rhizobacteria and cyanobacteria. Eur J Soil Biol 50:118–126

Rattan RK, Shukla LM (1991) Influence of different zinc carrier on the utilization of 285 micro nutrients by rice. J Indian Soc Soil Sci 39:808–810

Raulin J (1869) Etudes chimique sur la vegetation (Chemical studies on plants). Annales des Sciences Naturrelles Botanique at Biologie Vegetale 11:293–299

Sadeghzadeh B (2013) A review of zinc nutrition and plant breeding. J Soil Sci Plant Nutr 13:905–927

Sarathambal C, Thangaraju M, Paulraj C, Gomathy M (2010) Assessing the zinc solubilization ability of *Gluconacetobacter diazotrophicus* in maize rhizosphere using labelled 65 Zn compounds. Indian J Microbiol 50:103–109

Saravanan VS, Subramoniam SR, Raj SA (2004) Assessing *in vitro* solubilization potential of different zinc solubilizing bacterial (ZSB) isolates. Brazil J Microbiol 35:121–125

Saravanan VS, Madhaiyan M, Thangaraju M (2007) Solubilization of zinc compounds by the diazotrophic, plant growth promoting bacterium *Gluconacetobacter diazotrophicus*. Chemosphere 66:1794–1798

Saravanan VS, Kumar MR, Sa TM (2011) Microbial zinc solubilization and their role on plants. In: Maheshwari DK (ed) Bacteria in agrobiology: plant nutrient management. Springer, Berlin, pp 47–63

Schöll LV, Smits MM, Hoffland E (2006) Ectomycorrhizal weathering of the soil minerals muscovite and hornblende. New Phytol 171:805–814

Sharma SK, Sharma MP, Ramesh A, Joshi OP (2012) Characterization of zinc-solubilizing *Bacillus* isolates and their potential to influence zinc assimilation in soybean seeds. J Microbiol Biotechnol 22(3):352–359

Shakeel M, Rais A, Hassan MN, Hafeez FY (2015) Root associated Bacillus sp. improves growth, yield and zinc translocation for basmati rice (Oryza sativa) varieties. Front Microbiol 6:1286. https://doi.org/10.3389/fmicb.2015.01286

Simine CDD, Sayer JA, Gadd GM (1998) Solubilization of zinc phosphate by a strain of *Pseudomonas fluorescens* isolated from a forest soil. Biol Fertil Soils 28:87–94

Singh MV (2009) Micronutrient nutritional problems in soils of India and improvement for human and animal health. Indian J Fertil 5(4):11–16

Singh P, Shukla AK, Behera SK, Tiwari PK (2019) Zinc application enhances superoxide dismutase and carbonic anhydrase activities in zinc-efficient and zinc-inefficient wheat genotypes. J Soil Sci Plant Nutr 19:477–487. https://doi.org/10.1007/s42729-019-00038-7

Subramanian KS, Tenshia V, Jayalakshmi K, Ramachandran V (2009) Role of arbuscular mycorrhizal fungus (*Glomus intraradices*) (fungus aided). Agric Biotechnol Sustain Dev 1:29–38

Tan S, Han R, Li P, Yang G, Li S, Zhang P (2015) Over-expression of the MxIRT1 gene increases iron and zinc content in rice seeds. Transgenic Res 2:109–122. https://doi.org/10.1007/s11248-014-9822-z

Tariq M, Hameed S, Malik KA, Hafeez FY (2007) Plant root associated bacteria for zinc mobilization in rice. Pak J Bot 39:245–253

Tarkalson DD, Jolley VD, Robbins CW, Terry RE (1998) Mycorrhizal colonization and nutrient uptake of dry bean in manure and composted manure treated subsoil and untreated topsoil and subsoil. J Plant Nutr 21:1867–1878

Tavallali V, Rahemi M, Eshghi S, Kholdebarin B, Ramezanian A (2010) Zinc alleviates salt stress and increases antioxidant enzyme activity in the leaves of pistachio (Pistacia vera L. 'Badami') seedlings. Turk J Agric For 34(4):349–359

Tripathi DK, Singh S, Singh S, Mishra S, Chauhan DK, Dubey NK (2015) Micronutrients and their diverse role in agricultural crops: advances and future prospective. Acta Physiol Plant 37:139

Vaid SK, Kumar B, Sharma A, Shukla AK, Srivastava PC (2014) Effect of zinc solubilizing bacteria on growth promotion and zinc nutrition of rice. J Soil Sci Plant Nutr 14(4):889–910. https://doi.org/10.4067/S0718-95162014005000071

Venkatakrishnan SS, Sudalayandy RS, Savariappan AR (2003) Assessing in vitro solubilization potential of different zinc solubilizing bacterial (ZSB) strains. Braz J Microbiol 34:121–125

Vidyashree DN, Muthuraju R, Panneerselvam P (2018a) Evaluation of zinc solubilizing bacterial (ZSB) strains on growth, yield and quality of tomato (*Lycopersicon esculentum*). Int J Curr Microbiol App Sci 7(4):1493–1502

Vidyashree DN, Muthuraju R, Panneerselvam P, Mitra D (2018b) Organic acids production by zinc solubilizing bacterial isolates. Int J Curr Microbiol App Sci 7(10):626–633

Wakatsuki T (1995) Metal oxidoreduction by microbial cells. J Ind Microbiol 14:169–177. https://doi.org/10.1007/BF01569900

Wang Y, Yang X, Zhang X, Dong L, Zhang J, Wei Y, Feng Y, Lu L (2014) Improved plant growth and Zn accumulation in grains of rice (*Oryza sativa* L.) by inoculation of endophytic microbes isolated from a Zn Hyperaccumulator, *Sedum alfredii* H. J Agric Food Chem 62:1783–1791

Welch RM, Graham RD (2004) Breeding for micronutrients in staple food crops from a human nutrition perspective. J Exp Bot 55:353–364

White PJ, Broadley MR (2005) Biofortifying crops with essential mineral elements. Trends Plant Sci 10:586–593. https://doi.org/10.1016/j.tplants.2005.10.001

Whiting SN, Souza MD, Terry N (2001) Rhizosphere bacteria mobilize Zn for hyper accumulator by Thlaspi caerulescens. Environ Sci Technol 35:3144–3150

Wu SC, Cheung KC, Luo YM (2006) Wong effects of inoculation of plant growth promoting rhizobacteria on metal uptake by *Brassica juncea*. Environ Pollut 140:124–135

Yuvaraj M, Subramanian KS (2015) Controlled-release fertilizer of zinc encapsulated by a manganese hollow core shell. Soil Sci Plant Nutr 61(2):319–326. https://doi.org/10.1080/00380768.2014.979327

Zeb H, Hussain A, Naveed M, Ditta A, Ahmad S, Jamshaid MU, Ahmad HT, Hussain MB, Aziz R, Haider MS (2018) Compost enriched with ZnO and Zn-solubilising bacteria improves yield and Zn-fortification in flooded rice. Ital J Agron 13(4):310–316

Zuo Y, Zhang F (2009) Iron and zinc biofortification strategies in dicot plants by intercropping with gramineous species. A review. Agron Sustain Dev 29:63–71. https://doi.org/10.1051/agro:2008055

Endophytic Phytobiomes as Defense Elicitors: Current Insights and Future Prospects

12

Satyendra Pratap Singh, Arpita Bhattacharya, Rupali Gupta, Aradhana Mishra, F. A. Zaidi, and Sharad Srivastava

Abstract

The endophytes are microbial organisms inhabiting within the plant body either intercellularly or intracellularly. Earlier these endophytes were considered either as pathogenic or having no significant role in the plant's physiology and metabolism, but with the advancement of research and technology, the critical roles played by them are emerging. These endophytes form a symbiotic relationship with their host where in exchange of nutrient and habitat, they provide the plant protection against various stresses both biotic and abiotic. They provide the protection or resistance through either direct mechanisms or indirectly by eliciting various pathways within their host plants against the stresses. The endophytes help the host plant to acclimatize under harsh conditions by eliciting the defense-related genes, which in response triggers the concerned pathways for the synthesis of secondary metabolites for plant defense. A better understanding of the mechanisms and the role of endophytes in stress management will help in designing defense strategies to cope with the stresses and to improve integrated strategies for stress management in agriculture. The chapter thus explores the various mechanisms with the endophytes eliciting both biotic and abiotic defense responses in their host plants under stress conditions and their future application in agriculture and crop sciences.

S. P. Singh (✉) · S. Srivastava
Pharmacognosy Division, CSIR-National Botanical Research Institute, Lucknow, Uttar Pradesh, India

A. Bhattacharya · A. Mishra
Division of Plant-Microbe Interaction, CSIR-National Botanical Research Institute, Lucknow, Uttar Pradesh, India

R. Gupta
Department of Plant Pathology and Weed Research, Agricultural Research Organization, The Volcani Center, Rishon LeZion, Israel

F. A. Zaidi
N.D. University of Agriculture and Technology, Faizabad, Uttar Pradesh, India

© Springer Nature Singapore Pte Ltd. 2020

M. K. Solanki et al. (eds.), *Phytobiomes: Current Insights and Future Vistas*,
https://doi.org/10.1007/978-981-15-3151-4_12

Keywords
Endophytes · Abiotic · Biotic · Elicitors · Resistance · Stress

12.1 Introduction

The plants in their natural habitat face various biotic and abiotic stresses, which they successfully cope with through their internal defense mechanisms (Gimenez et al. 2018; Mousavi et al. 2019). The cultivated or domesticated crops are more prone to damage compared to their wild relatives. With the ever-increasing food crisis and global target of achieving Sustainable Development Goals 2030 of zero hunger and for the conservation of nature, researchers are now looking for biological, safe alternatives for the management of various environmental stresses along with increasing the productivity of the crop plants without contaminating the environment with chemicals (Bengtsson et al. 2018). The microbial use has emerged as a key player in achieving these goals. Since the past few decades, our understanding of soil microflora has increased tremendously, and now the soil is considered as an equivalent to a living entity (Barman et al. 2019; Harman and Uphoff 2019). These understandings and relevant research have paved a new pathway of sustainable organic agriculture practices. In recent times, the role of rhizospheric bacteria has been largely explored, and they have been widely used for acclimatization of both biotic and abiotic stresses in the crop plants (Timmusk et al. 2017). Studies carried out during the last decade have highlighted that the plants are ubiquitously inhabited by microbes, which live within the host plant without causing any disease symptoms (Bacon 2018; Marsberg et al. 2017). These microbes are beneficial to their host plants and help them to cope with both biotic and abiotic stresses (Mishra et al. 2018a; Rho et al. 2018).

12.1.1 Plant Endophytes

The microorganisms living intercellularly or intracellularly within the host plants are termed as plant endophytes. The endophytes are microorganisms that inhabit and colonize within the plant tissues and share a symbiotic relationship with its host plant and do not manifest any disease symptoms in them (Bacon 2018). They are ubiquitous in distribution belonging to a varied group of microorganisms ranging from bacteria and fungi, including actinomycetes (Rho et al. 2018). Some commonly reported genera of bacterial and fungal plant endophytes include *Bacillus* sp., *Colletotrichum* sp., *Phomopsis* sp., etc. These microorganisms have been isolated from all parts of plants including leaves, stems, and roots and even from the floral tissue (Strobel 2018; Frank et al. 2017; Nissinen et al. 2012). The plant endophytes share a complex and multifaceted association with their host plants, where they have been reported to have a positive effect on their host. To adjust and to survive within the host tissues, the microbes develop a mutualistic relationship where both the host and inhabitant are benefited (Jia et al. 2016; Wani et al. 2015).

These microbes not only help the plant to respond and acclimatize during the biotic and abiotic stresses, but they also help to ameliorate the stress through triggering various pathways in the plant and regulating the responses (Mishra et al. 2018a; Pandey et al. 2018). They protect and prepare the host plants for pathogen attacks, environmental stress, and against herbivores as well. They have been reported to trigger secondary metabolite pathways in the plants, which leads to the production of various defense-related compounds in the host (Singh and Gaur 2017). Several endophytes are themselves involved in the production of commercially important secondary metabolites known to be produced by their host plants. For instance, the consortium-based application of *Pseudomonas fluorescens* and *Bacillus amyloliquefaciens* enhances the withanolide content in *Withania somnifera* under biotic stress condition (Mishra et al. 2018b), an endophyte, namely, *Phoma medicaginis* isolated from *Taxus wallichiana* var. *mairei* individually producing paclitaxel (Zaiyou et al. 2017). These metabolites are also valuable for humans commercially. These features make the endophytes a desirable choice as a source of commercially or medicinally valuable compounds that are used as drugs.

Some endophytes act as a biocontrol agent owing to their antimicrobial or mycoparasitic activities. The biocontrol activity is showing endophytes that either attack the pathogens directly through their antagonistic activity or indirectly through antibiosis by the production of various antimicrobial compounds like fengycin, iturin, etc. (Zouari et al. 2016; Brader et al. 2014). The endophytic microbes are known to produce various cellulosic and chitinolytic enzymes that degrade and disintegrate the cell wall of the pathogen (Naik 2019; Abdel-Rahim and Abo-Elyousr 2018) (Fig. 12.1).

12.1.2 Endophytic Microbes Interaction with Host Plants

The substantive evidence on plant-associated endophytes was found in the fossilized plant tissues which revealed that endophyte–host interactions may have evolved from the time of very existence of higher plants on the planet (Strobel 2003), signifying an essential part of plant evolution (Mendes et al. 2013; Philippot et al. 2013). In the past two decades, the researchers have made great efforts to assess the role of endophytes and their interaction with host plants. The ability of endophytes to colonize in the internal plant tissues has made them very crucial for the commercial as well as societal agricultural practices. The close mutualistic association of endophytes with the plant is an essential part of their existence or survival, and several advanced studies demonstrate that the host plants are dependent on the endophytes for several fundamental activities (Hardoim et al. 2015; Potshangbam et al. 2017). For instance, symbiotic relationship of rhizobia and the leguminous crop is the well-studied endophytic association; the bacterial endosymbiont fulfills the scarcity of nitrogen, whereas the host ensures the favorable condition for endophyte (Santoyo et al. 2016).

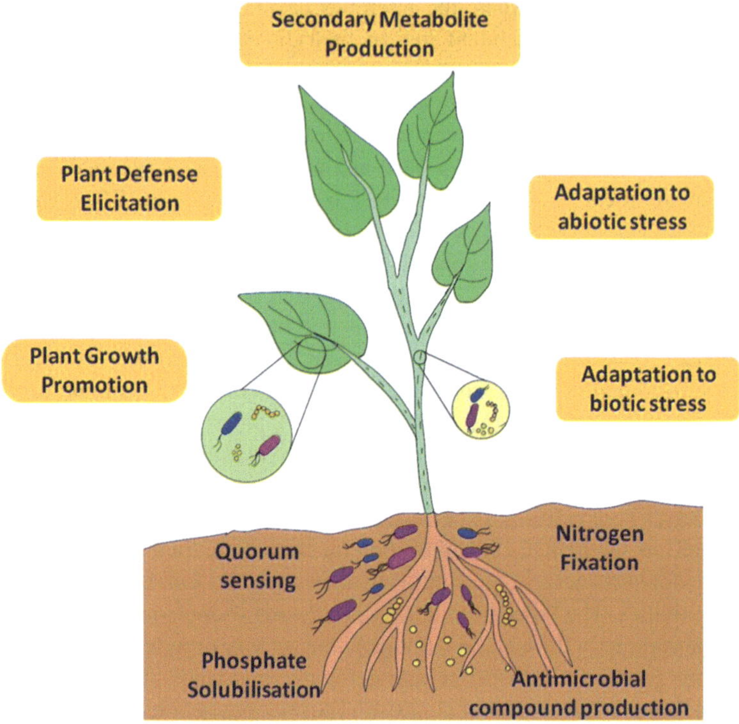

Fig. 12.1 Role of endophytes on plant growth promotion and stress management

Furthermore, several endophytes can convert the complex form of minerals and macro- and micronutrients into simplex form and avail to the host plant for the growth and development (Ma et al. 2011; Gaiero et al. 2013). The observations of two decades have projected that the endophytes migrated from the rhizospheric or seed-born microbial communities, and molecular tools based on advanced findings have revealed that these microbes are highly versatile and possess several functional genes that govern the several novel beneficial applications to the host plant (Ali et al. 2014). For example, the highly anti-cancerous compound Taxol was synthesized by endophytes of *Taxus baccata* (yew plant) which was not possible without endosymbiotic association (Somjaipeng et al. 2015). However, recognition of these specialized designated genes for the plant-microbe association is still the subject of investigation. To maintain the healthy symbiosis, endophytes maintain the optimum conditions for the host plant by protecting the host plants against several invaders, pre-immunizing against various biotic and abiotic stresses, and producing or inducing the host plant for the synthesis of metabolites to promote the growth and yield (Singh and Gaur 2016, 2017; Mishra et al. 2018a, b).

Moreover, an overwhelming number of studies highlight the considerably ponderability of microbial association with plants in seed germination, seedlings vigor, cellular development, and nutrient availability, ameliorate the biotic and abiotic stress resistance ability of the plant, and improve the synthesis of therapeutically important secondary metabolites and their productivity (Naveed et al. 2014; Forchetti et al. 2010; Khan et al. 2012). The perception of "plant symbionts" has re-modulated the conception of "plant microbiome," and therefore the current era emphasizes to investigate the coevolution of plants and their symbionts at the different environment and key factors and conditions, which govern the symbiosis between both partners (Turner et al. 2013). The upregulation of plant gene expression after the treatment of endophytes is one of the strongest evidence about the impact of endophytes on host plant (Mishra et al. 2018a, b; Berendsen et al. 2015). Furthermore, the modern technology-based studies, i.e., next-generation sequencing (NGS), metagenomics, metatranscriptomics, genome sequencing, and comparative genomics, may facilitate the much deeper knowledge about the multidimensional interactions between endophytes and their plant hosts.

12.2 Different Modes of Plant-Microbe Interaction

Endosymbionts are considered as a "second genome of plants" because they are very closely associated with host plants and independently play a very crucial role to accomplish the several definite tasks assigned by the host. Instead of this, host plants ensure the favorable condition of associated microbes by providing the required carbon and nutrients to proliferate competitively in a complex interactive environment. These microbial interactions encompasses the complete host plants as well as specific organ or region, i.e., roots, shoots, flowers, seeds, and leaves or rhizospheric, epiphytic, and endophytic (Turner et al. 2013; Rout and Southworth 2013; Wang et al. 2017; Malviya et al. 2019; Solanki et al. 2019). Other than this, the host plant also possesses an array of biological information that describes the association of endosymbionts including functional genes, transcripts, proteins, and several biologically active metabolites (Busby et al. 2017). Presently, several studies elaborated the significance of plant-associated microbiota on plant growth and development (Brussaard 2012). However, the highly focused investigation is still needed for the enumeration of potent plant-associated microflora.

Plants are closely associated with numerous range of microbiota both below and above the ground for their mutual benefits. This microbiome is characterized on the basis of their colonization and can be classified as rhizospheric (closely associated with the root), phyllospheric (associated with aerial part of the plant), and endophytic (inside the plant tissues) at their natural habitats (Truyens et al. 2015; Bakker et al. 2014). Out of them, the plant possesses microflora which are organ-specific. These endosymbiotic microbial loads are highly inconsistent because of the variation in the genome, environment, climate, and plant development. Several conditions, such as the activation of the innate immune system, can also be intervening in the modulation of the associated microbial load of the host plant.

To diagnose the impact of several factors on plant microbiomes, this chapter also summarizes the diversity of plant microbiome based on different compartments individually. These compartments are the major reservoir of plant microbiome and primarily known as rhizosphere, endosphere, and phyllosphere.

12.2.1 Rhizosphere

Out of the several natural habitats, soil system is the great reservoir of microbial load and exhibited about 10^8–10^9 CFU per gram of the soil (Chaparro et al. 2014). The rhizosphere is the most dynamic region which is extremely affected by the plant secretomes such as root exudates, mucilages, sloughed cells, and rhizodeposition (Spence et al. 2014). Plants regularly secrete several bioactive secondary metabolites as a rich source of carbohydrates, proteins, lipids, phenolic compounds, and organic acids, which can efficiently modulate the microbial dynamics (Andreote et al. 2014). Furthermore, the plant litters are also a good source of several organic molecules (i.e., amino acids, nucleic acids, and saccharides), cellulose, hemicelluloses, lignin, and polyphenolic compounds (Osono 2007, McGuire and Treseder 2010, Talbot and Treseder 2012). The microbial load of the rhizospheric region is completely different from the normal soil communities because of the availability of root secretomes and plant secondary metabolites that refurbished the conditions and nutrient availability. Molecules represent the root secretomes that play an essential role in the quorum sensing and may be also able to instigate the metabolic activity of soil-inhabiting microbiomes (Vandenkoornhuyse et al. 2015), even though few of them are secreted under special circumstances to magnetize the specific microbial community such as malic acid secreted by the plant to attract the *Bacillus* sp. (Rudrappa et al. 2008); rhizodeposition of plant litter especially lignin and other polyphenols mainly conserved the subkingdom Basidiomycota (Baldrian 2006). Furthermore, some plant secretomes also have growth-limiting abilities (Martiny 2016). On the one hand, the higher abundance of microbiome significantly induces the microbial load in a per gram of soil and can reach up to the 10^{11}, which can alter the environment of the rhizosphere. For example, deficiency of oxygen, as it is continuously utilized by the microbial biomass, can also lead to the alteration in the pH, the concentration of microbial enzymatic activity (Classen et al. 2015), and the organic contents present in the rhizosphere. On the other hand, the higher abundance of microbial diversity at the rhizosphere represents the higher genetic information which might be useful to diagnose several unrevealed metabolic pathways. The modulation in microbial diversity also explores the unresolved consequence of lifestyle, genetic adaptability, interaction with other communities, coevolution, and existence with organisms in the similar climatic conditions (Philippot et al. 2010; Berendsen et al. 2012). The mutualistic association between plants and associated microorganisms can also lead to several genetic re-modulation, horizontal gene transformation (HGT), etc. It follows the selection of highly appropriate microbial community under the influence of native environmental factors, and consequently the host plant communicates through the sharing of genetic information and possesses novel metabolic adaptability. The mutualistic association with the plant could be accomplished because of the availability of huge microbial diversity at rhizospheric as well as phyllospheric regions. Furthermore, the richness of highly transmitted genetic materials such as viruses, plasmids, and genes at the rhizospheric region can also facilitate an opportunity to exchange several unrevealed information for the plant-microbe interaction.

12.2.2 Phyllosphere

The phyllosphere can be considered as the second most primitive zone after the rhizosphere for plant-microbe interaction as it is the aerial part of the plant including leaves where the intensity of microbial diversity was found to be higher (Vorholt 2012; Copeland et al. 2015). Microbes associated at the phyllospheric region of the plant are also able to confer the health and wealth of the plant through the induction of mineralization of nutrients, biosynthesis of several phytohormones, and activation of the innate immune response of the host plants against several pathogenic invaders (Cappelletti et al. 2016). Carbon sequestration done by them is an important task which maintains the balance of the environment (Bulgarelli et al. 2013; Bringel and Couée 2015). Furthermore, they also perform the considerable role in the survival of plants under extreme conditions, i.e., nutrient-limiting conditions, high and low pH, unfavorable temperature, less humidity, UV, etc. (Whipps et al. 2008). Out of various microbes, bacterial communities are the most prominent colonizers at the phyllospheric region, although they are drifted severally. Because of the close attachment with several environmental factors, the microbial load at the phyllosphere is fluctuating very drastically even in same species of plants, under similar environmental condition, as well as at same developmental stage (Rastogi et al. 2013; Knief et al. 2012). Furthermore, the diversity of phyllospheric microflora also depends on the several key factors, i.e., leaf area, leaf secretome, the intensity of light and UV, airflow rate, moisture content, etc. For example, most of the pigments producing bacterial communities efficiently colonize at phyllospheric region because of the ability of higher tolerance toward the UV radiations. The origin of microbes is the next most considerable factor that imprinted the great versatility in microbial diversity at the phyllosphere. Other factors, especially air, wind, and water, are the important sources of microbial cells that are also able to alter the microbial dynamics at the phyllosphere (Bullgarelli et al. 2013). As due to the high airflow, several spore-forming microbes could colonize on the aerial part of the plant that is found far from the plant origin (Bullgarelli et al. 2013). Furthermore, the fluctuations in phytomicrobiome mostly rely on the available carbon substrates and nutrients on the aerial parts especially on the leaf surface including amino acids, glucose, xylose, volatiles, etc. (Lindow and Brandl 2003). Bacterial communities colonizing at the phyllosphere are also well studied for their biofilm formation to maintain the heterogeneity of the microbial population and for adherence under unfavorable circumstances. Among several bacterial communities, Proteobacteria, Actinobacteria, Bacteroidetes, and Firmicutes are the most common colonizers and contribute a decisive role which leads to the novel competency of the host plant (Redford et al. 2010; Vorholt 2012). Whereas, the consequence of the close associations and beneficiary assignments, the other microbes also acquire the several evolutionary pieces of evidence which describe the different modes of plant-microbe association in a time-dependent manner. Due to several constitutive and constructive applications, the knowledge about the longtime evolutionary relationship and association between the plant and microbiota will possibly open the new sights for the development of novel applications in the field of sustainable agricultural practices.

12.2.3 Endosphere

Every plant presents on earth at different habitats having their specific endophytes, and it assumed that the presence of endophytes in the inner tissues of plants is necessary for their existence (Rosenblueth and Martínez-Romero 2006). In summary, the host plants carry several closely associated endophytes at the same time under in vitro condition. However, the role of endophytes in plant growth promotion and molecular, biochemical, and physiological cross talk between the plant and endophytes to perceive the mechanisms during symbiotic association is still a matter of concern. Some findings reveal that endophytes can promote growth as well as reduce the disease incidence in the host plant. For example, the endophytic actinomycetes, namely, *Streptomyces diastaticus* SP2, *S. fradiae* SP4, *S. olivochromogenes* SP5, *S. collinus* SP8, *S. ossamyceticus* SP10, and *S. griseus* SP12 of different medicinal plants, were efficiently able to enhance the productivity and reduce the disease incidence by triggering the systemic resistance and mitigation of oxidative stress in *Cicer arietinum* against *S. rolfsii* (Singh and Gaur 2016, 2017). Endophytes isolated from sugarcane roots significantly enhance the level of amino acids (Ferrara et al. 2012), while mutant endophytic strain suppresses the pathogenicity of *Diatraea saccharalis* by the production of CRY1 Ac7protein (Quecine et al. 2014). However, out of several attempts on endophytes individually, few reports are also advocated about the aptitude of different endophytic communities to govern more than one function simultaneously. For example, the consortium mode application of endophytic bacterial isolates (*Bacillus amyloliquefaciens* and *Pseudomonas fluorescens*) is not only able to detoxify the excessive generation of ROS and RNS by the activation of defense-responsive genes and systemic resistance-related genes in *Withania somnifera* (L.) Dunal under *Alternaria alternata* stress (Mishra et al. 2018a) but also modulates the expression level of intermediate genes of withanolide biosynthetic pathway (MEV and MEP pathway) under biotic stress condition (Mishra et al. 2018b).

Still, limited information subjected to endophytic diversity creates discrepancies to understand the ecology of plant-associated microbes. In search of the origin of endophytes, several efforts are performed by researchers to track the path of GFP-tagged endophytes (Rouws et al. 2010; Compant et al. 2005; Germaine et al. 2004) or by GUS staining (James et al. 2002; Compant et al. 2005), and they advocated that endophytes are migrated through the rhizospheric or rhizoplane regions mainly, whereas few of them are transmitted *via* seeds (Hallmann et al. 1997; Saikkonen et al. 1998; Mitter et al. 2013) and entered through the opening during the formation of root hairs or the emergence of lateral root. However, very few of them access in the plant *via* phyllosphere region and entered by stomatal openings, wounds, and hydathodes. Moreover, various microbes possess the ability to secrete several plant cell wall lytic enzymes, i.e., cellulases, xylanases, pectinases, endoglucanases, etc., which enable the bacterial entry in plant internal tissues. The origin of endophytes is useful to understand the direct association with the plant and its proliferation and enumeration inside the inert plant tissues. Recently, it was also studied that specific endophytic communities can cross talk with plants, and strategies followed for their survival are evidence that their genome organization facilitates the survival and transmission.

Recently, few researchers have recognized that the endophytic microbes articulate the molecular signals through the horizontal and vertical transmissions. The majority of recognized plant endophytes is horizontally transmitted and produces a potential conflict of fitness where antagonistic coevolution of functional trait expression possibly arises between the plant and microbial symbionts (Wani et al. 2015). Vertically transmitted endophytes confer the increased host benefits over those horizontally transmitted that are significantly dependent on plant density (van Overbeek and Saikkonen 2016). Furthermore, several studies on the effect of habitat and environmental factors on genetic variation of microbes are noticed that microbes living freely in the water, soil, and air will require an array of genes to develop resistance against adverse environmental stresses, having larger genetic information, whereas microbes residing in irrevocable conditions, i.e., endosymbiotic condition, severally required less molecular adaptation, having comparatively small genome size (Andreote et al. 2012). Similarly, Mitter et al. (2013) revealed a great genetic drift in the diversity of endophytes, and they concluded that the endophytes are a group of different microbial communities which migrated from different habitats. Those which originate from variable ecosystems have a larger genome size, whereas the rest of them which originate from stable environments contained smaller genome size.

Out of the plant-microbe interaction, for the interpretation of endophytes as a phytoendobiome is still needed much deeper knowledge. However, many researchers noticed that in spite of host fitness-based genetic information, endosymbionts also possesses several clusters of functional genes that significantly evolved in the establishment, persistence, and proliferation of the host plant (Bulgarelli et al. 2012; Lundberg et al. 2012). Furthermore, it should also be noted that although the richness of endophytes mostly relies on the population dynamics of rhizospheric microbiota, few groups of microbial communities, i.e., Actinobacteria and Proteobacteria, immanently maintained their presence as an endophyte which annunciates about the specific selection-based criteria of the plant. The role of plant innate immunity, including recognition of microbe-associated molecular patterns (MAMPs), is the possible reason behind the microbial selection.

12.3 Cross Talk Between the Plant and Endophytes During Interaction

There are several crucial steps involved for a successful colonization of endosymbiont during plant-microbe interaction including the entry of endophyte in the inner tissue of the plant, the recognition of the plant, and finally the cross talk through the signaling molecules between the host plant and endophytes plants for colonization (Rosenblueth and Martínez-Romero 2006; Compant et al. 2010; Brader et al. 2014). Plant secretomes (exudates are very rich with water and nutrients and work as biomolecules) are facilitating the movement of the entire variant microflora toward the rhizospheric or phyllospheric region. And after that, host plant designates a few of the beneficial microbes as endophytes and allows to colonize and proliferate inside the plant tissue for growth and development. Flavonoids are the fascinating chemoattractants that perform a decisive role in the plant-endophyte interaction, including various plant secretomes as biomolecules. Flavonoids are well known for the accomplishment of meaningful association with legumes by rhizobia (Arora and Mishra 2016). Furthermore, flavonoids are also involved in the compatible association of non-rhizobial endophytes with the host plant, and these metabolites also affect the colonization of *Serratia* sp. EDA2 and *Azorhizobium caulinodans* ORS571 in the root of rice and wheat (Webster et al. 1998; Balachandar et al. 2006). Lipo-chitooligosaccharides (LCO) are reported as Nod factors, inducing the common symbiotic pathway (CSP) in arbuscular mycorrhizal associations as well as rhizobia-legume associations (Gough and Cullimore, 2011). As per the recent study, a beneficial plant endophyte *Mucor* sp. is associated with *Arabidopsis thaliana* because of the secretion of strigolactone (Rozpądek et al. 2018). Moreover, strigolactone also triggers the biosynthesis and secretion of chitin oligomers that induce the signaling pathways for the symbiotic association between the plant and endosymbiont (López-Ráez et al. 2017). Moreover, a superfamily of plant cell wall proteins, i.e., arabinogalactan (ABO) glycosylated members of the hydroxyproline-rich glycoprotein (HRGP), works as a receptor and signaling molecules participants to accomplish the symbiotic association during plant-microbe interaction (Nguema-Ona et al. 2013). Furthermore, some other basic plant biostimulants such as sugars, amino acids, organic acids, phenolic compounds, and other secondary metabolites provide the signal to endophytes for the mutualistic association (Chagas et al. 2017). However, the strategies followed by the plant to distinguish the plant beneficially and plant pathogenic microbes are still a matter of investigation. However, several studies favor the role of plant innate immune system in the identification of plant growth-promoting microorganisms and their colonization in inert plant tissues (Fesel and Zuccaro 2016). The current evidence on plant gene expression suggests that endophytes perform an essential role in the expression level of plant genes (Mishra et al. 2018b; Singh and Gaur 2017). However, the stimulation of defense-related signaling molecules such as ethylene (ET)/jasmonic acid (JA)/salicylic acid (SA) against different stresses relies on the defense strategies of plant endophytes (Singh and Gaur 2017). In the recent investigation, Mishra et al. (2018a) revealed that the consortium of bacterial endophytes, namely, *B. amyloliquefaciens* and *P. fluorescens*, suppresses the pathogenicity of *Alternaria alternata* in *W. somnifera* by the induction of SA and JA signaling-mediated induced systemic resistance (ISR) (Fig. 12.2). Furthermore, it also modulates the expression level of intermediate genes of MVA and MEP pathways (Mishra et al. 2018b). In contrast to the above findings, several plant defense-related genes were downregulated during the colonization of rhizobia and arbuscular mycorrhizal fungi (AMF) (Fouad et al. 2014; Benhiba et al. 2015; Sarkar et al. 2016). However, the stimulation of plant signaling molecules (SA/JA/ET) after the colonization of plant mutualistic partners controls the overwhelming actions of microbes (Plett and Martin 2018). It was also observed that several hormone pathways related to miRNA transcripts were induced for the proper colonization of endophytes (Formey et al. 2014). For example, the host plant adjourns the GA signaling pathway and induces the expression level of

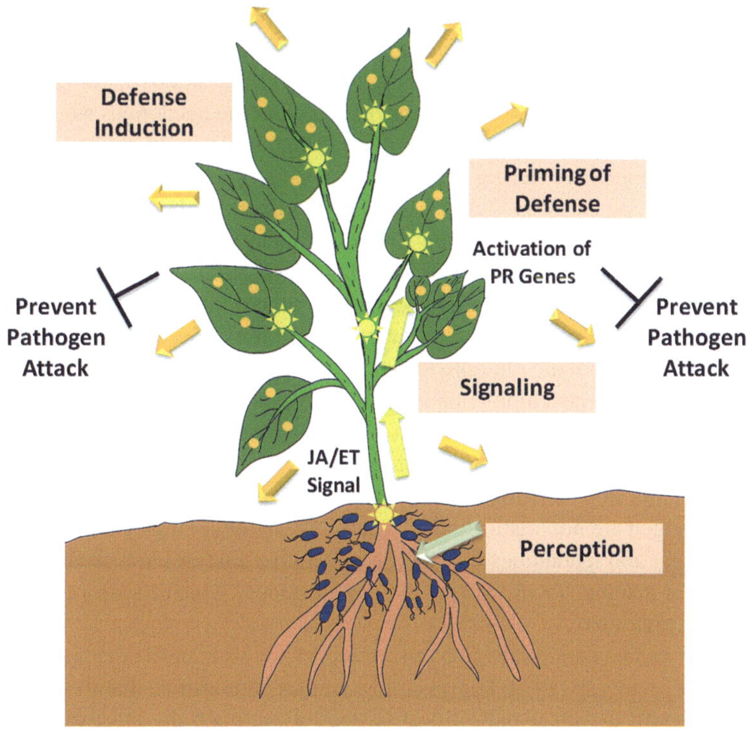

Fig. 12.2 Induced systemic resistance (ISR) in plants

miRNA – E4D3Z3Y01BW0TQ, to avoid the repressive action against mutualistic associations of AMF with the plant (Wu et al. 2016). The delay in the expression of plant defense-related genes might play a crucial role in the establishment of a mutualistic association with endophytes (Plett and Martin 2018). Several advanced techniques such as genomics- and metagenomics-based findings imprinted the novel insights subjected to the behavior of endophytes during the colonization in the endophytic regions of the plant. As per the findings of Hardoim et al. (2015), the genes involved in catabolic activities in endophytes are much more dominant during the invasion of phytopathogens as compared to the anabolic pathway-related genes (higher copy number in the genome) which are having multifunctional roles. The genes, namely, nitrogenase and ribulose bisphosphate carboxylase/oxygenase (RuBisCO) in combination, work as an indicator for endophytes for symbiotic nitrogen fixation-based association (Karpinets et al. 2014). The lateral gene transformation also performs a crucial role in acquiring several key features for the colonization of endophytes (Tisserant et al. 2013). For instance, several plant beneficiary bacterial endophytes possess the gene mannitol dehydrogenase through lateral gene transformation that not only significantly suppresses the proliferation of fungal phytopathogens but also provides an opportunity to colonize inside the plant (Wu et al. 2011). The existence and positions of several other genes designated for specific applications such as amino acid synthesis, iron transportation, hemolysin, and hemagglutinin also provide clues regarding the plant-microbe interaction as well as modulation in host and endophytes lifestyles (Taghavi et al. 2010; Xu et al. 2016; Shidore et al. 2010). Besides, the recent soil and plant metagenomics studies reveal that endophytes also depend on the host plants for several activities. Therefore, they reduce their genome size during evolution as a result of adaptation (Sessitsch et al. 2012; Brewer et al. 2016; Hottes et al. 2013). On the other hand, the species- and genotypic-based variations of the host also modulate the diversity of endophytes (Rodríguez-Blanco et al. 2015; Ding and Melcher 2016). Few endophytes modify themselves as per the response of the local environment (biotic and abiotic stresses), developmental stage, and genetic variation of the host (Bacon et al. 2008). For instance, an asymptomatic endophyte *Ramularia collo-cygni* can reside inside the host plant during the developmental stage; however, at the growing stage, it modulates itself as a necrotrophic pathogen (Walters et al. 2008). In the same way, *Fusarium verticillioides* is able to proliferate as an endophyte or pathogen in maize plant (Oren et al. 2003). However, key factors behind the transition of endophytes to phytopathogen are not much studied. Therefore, for better understanding, several advanced tools and techniques based on comparative analyses are needed that can provide novel insights about the specific circumstances which transform the endophyte into a pathogen.

The exact mechanism followed by the microbes to reside inside the plant tissue is still a case of an investigation. However, few studies demonstrate that the infiltration of endophytes in the inner plant tissue is possibly through the inactivation of the primary defense system of the host plant which leads the selection of endophytes by recognition of microbe-associated molecular patterns (MAMPs) (Newman et al. 2013; Cord-Landwehr et al. 2016). These MAMPs include several receptor molecules such as flagellin (Flg), elongation factor TU (EF-Tu), peptidoglycan (PGN), lipopolysaccharides (LPS), bacterial cold shock protein, several plant ROS scavenging molecules (SOD, GPx, APx, etc.), β-glycan (GE), β-glucans, oligopeptides (Pep-13), xylanase (EIX), and chitin (Newman et al. 2013; Shimizu et al. 2010), which are enucleated by the pattern recognition receptors (PRRs; present on the surface of plant cells). For example, the defensive reaction of the host plant is triggered by the recognition of chitin-specific receptors (PR-3; chitin oligomers) (Sanchez-Vallet et al. 2015). Furthermore, endophytes also synthesize several unknown MAMPs to defend themselves from the innate immune system of the plant and are non-detectable by PRRs (Vandenkoornhuyse et al. 2015; Cord-Landwehr et al. 2016). For example, *Burkholderia phytofirmans* (as an endophyte) introduces flagellin (FLS2) that is different from phytopathogens (i.e., *Pseudomonas aeruginosa or Xanthomonas campestris*) (Trda et al. 2015). Moreover, endophytes also produce several ROS/RNS scavenging molecules (such as superoxide dismutases, catalases, peroxidases, alkyl hydroperoxide reductases, and glutathione S-transferases) to protect themselves against nitro-oxidative bursts (Zeidler et al. 2004). As a part of these, endophytes are also able to modulate the host immune system by the formation of protein secretion systems (SSs). For instance, several pathogens deliver the effector proteins though the high stimulation of type III secretion system (T3SS) and type IV secretion system (T4SS) (Green and Mecsas 2016; Liu et al. 2017). However, endosymbionts communicate with host plants by least stimulation of T3SS and T4SS except few rhizobial and non-photosynthetic *Bradyrhizobium* strains where T3SS plays a decisive role in legumes for nodulation and rice for colonization, respectively (Okazaki et al. 2013; Ausmees et al. 2004; Piromyou et al. 2015). Similarly, type VI secretion systems (T6SSs) are present in several commensal plant symbionts for antimicrobial activity (Reinhold-Hurek and Hurek 2011; Bernal et al. 2018).

12.4 Role of Endophytes on ISR Elicitation

The new sustainable agricultural practices lead to the search for effective bioinoculants for the amelioration of productivity and health issues of plants. In this context, the effective bioinoculant is not only able to promote the growth and productivity of plants by the secretion of several phytohormones and other plant growth-promoting substances but is also involved in eliciting different defense responses in both abiotic (i.e., drought, salinity, heavy metals, extreme temperature, radiation, etc.) and biotic (such as insects, mites, aphids, nematodes, bacteria, fungi, viruses, etc.) stresses through different mode of actions. Such ability to suppress the infection of different plant invaders (through the secretion of antibiotics and antimicrobial substance) is termed as antagonism or biocontrol. It includes other specific mechanisms such as antibiosis, competition, etc. Alternatively, microbes can induce the innate immune system of the host plants through the synthesis of various secondary metabolites and in a way that the endophytes elicit the resistance of the plant toward several biotic and abiotic stresses. The microbe-mediated induction of systemic resistance of the plant is known as induced systemic resistance (ISR) (Fig. 12.2). Moreover, the resistance develops due to the chemical inducers or pathogenicity of necrotrophic pathogens is termed as systemic acquired resistance (SAR) (Fig. 12.3). Both pathways not only differ by the elicitors but also accomplished it by different signal transduction pathways. The ISR is activated through the signaling pathway of using jasmonic acid or ethylene (Matilla et al. 2010), whereas the stimulation of salicylic acid is responsible for the elicitation of SAR. In a few cases, jasmonic acid or ethylene also performs a decisive role in the activation of SAR. ISR cannot be distinguished based on SAR-based signal transduction pathway because some current findings have demonstrated that endophytes and other plant-associated beneficial microbes can induce the ISR which is completely dependent on the salicylic acid (van Loon et al. 2006).

Furthermore, bacterial endophytes are involved in the amelioration of various stress responses and are known to initiate responses like those of the rhizosphere-associated bacteria (Ryan et al. 2008). The mechanisms involved in eliciting abiotic defense responses have been extensively reviewed by Khan et al. (2015), and more recently by Lata et al. (2018). The endophytes have been reported to play a significant role in the amelioration of diverse abiotic stresses. The mechanisms involved in biotic stress defense responses have been thoroughly studied by Kloepper and Ryu (2006) and Busby et al. (2016), which involve the role of both bacterial and fungal plant endophytes against the invading plant pathogens.

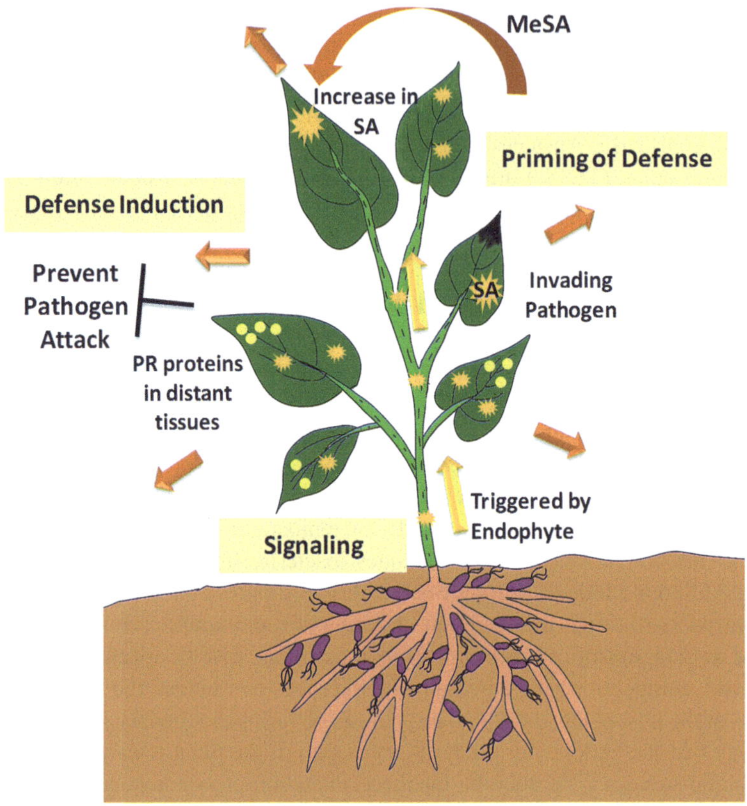

Fig. 12.3 Systemic acquired resistance (SAR) in plants

12.5 Role of Endophytes in Biotic Stresses

The plants in their natural habitat face several biotic stresses which include the attack of various macroscopic organisms including insects, mites, aphids, nematodes, etc. as well as various microscopic pathogens like bacteria, fungi, viruses, etc. It has been reported that the priming or treatment of plants with the endophytes has resulted in a decrease in damage caused by invading pathogens and pests (Romero et al. 2018; Pappas et al. 2018). The mechanisms related to the antimicrobial activity of potent endophytes have been studied severally (Mishra et al. 2018a; Singh and Gaur 2016, 2017). Many reports demonstrate that endophytes suppress the pathogenicity of the plant by direct antagonism (by the synthesis of several antimicrobial substances including antibiotics) or through the modulation in the severity of the host defense system (via ISR) against phytopathogens (Singh and Gaur 2017; Mishra et al. 2018a).

Furthermore, endosymbionts induce the stimulation of JA, SA, and ET (ISR and SAR pathway signaling molecules), which are severally known for the stress responses against phytopathogens (Khare et al. 2016).

Moreover, the application of endophytes also regulates the gene expression of host plants which leads to the enhanced physiological performance, higher expression in defense-related pathways, synthesis of antimicrobials, etc. (Mishra et al. 2018b). Phytohormones also play a decisive role in the plant defense system. For instance, gibberellin (GA_3) production by endophytes pre-immunizes the host plant toward the phytopathogens as well as insects (Waqas et al. 2015). Furthermore, seed coating with endophytic actinobacterial strains not only improves the strength and physiology of chickpea plants but also upregulates the expression level of defense-related genes against *S. rolfsii* (Singh and Gaur 2016). Endosymbionts are well known to synthesize several volatile organic compounds (VOCs) which have significant antimicrobial activity against fungal, bacterial, and viral phytopathogens and nematodes. Endophytes are also able to synthesize phytoalexins (low molecular weight antimicrobial molecules) that induce the production of phyto-stimulants in host plants. In a study, Gupta et al. (2017) reported that chitinolytic bacterial strains, namely, *Chitiniphilus* sp. and *Streptomyces* sp., inhibit the pathogenicity of *Meloidogyne incognita* (nematode) and induce the synthesis of secondary metabolites in *Bacopa monnieri*. Furthermore, findings of several workers reveal that the pretreatment of endophytes improves the cellulose content and lamina density, which reduces the attack by herbivores including leaf-cutting ants (Van Bael et al. 2012; Estrada et al. 2013). In this way, the endophytes not only develop resistance against plant invaders through ameliorating the host defense system but also stimulate several antimicrobial substances that have an antagonistic approach toward the phytopathogens.

12.5.1 Bacterial Endophytes

The niche provided by the host plants provides a safe harbor for the bacterial strains that can colonize *in planta* without causing diseases in their host plants (Bacon 2018). The bacterial endophytes are widely reported to inhabit within the intercellular and intracellular cells within the host tissue (Liu et al. 2017; Hardoim et al. 2015; Thomas and Sekhar 2014) (Table 12.1). Several endophytes have not only been isolated from the leaves, stems, and roots of the plants but also the seeds (White et al. 2018; Ambrose et al. 2018; Truyens et al. 2015). The endophyte isolation has been carried out from different plants like crop plants (Contreras-Cornejo et al. 2018; Abd-Allah et al. 2017; Ahmad et al. 2015), forest trees (Yao et al. 2017; Somjaipeng et al. 2015), and medicinally important plants (Mishra et al. 2018a; Singh and Gaur 2016) as well as from grasses (White et al. 2018).

The recent studies have shown that the endophytic bacteria play a critical role in eliciting defense responses in host plants against the invading pathogens through both direct and indirect mechanisms (Strobel 2018; Lata et al. 2018; Singh and Gaur 2017). The mechanisms of eliciting defense by endophytes are reported to be similar to that of the rhizospheric bacteria (Brader et al. 2014). The prior introduction or priming of plants through the treatment of roots or seed treatments prevents the host

Table 12.1 Endophytes defense mechanism against the phytopathogens

S.no.	Endophytes	Sources	Pathogens	Crops	Defence responses	References
Bacterial endophytes						
1.	*Bacillus amyloliquefaciens* (GB03) *and Microbacterium imperiale* (MAIIF2a)	Cassava *(Manihot esculenta)*	*Fusarium solani*	Cassava *(Manihot esculenta)*	ISR	Freitas et al. (2019)
2.	*Streptomyces* and *Pseudomonas*	Fruit trees	*Agrobacterium tumefaciens*	Honggengansutao and Okinawa	ISR and SAR	Li et al. (2019)
3.	*Piriformospora indica*	*Anthurium andraeanum*	*Ralstonia solanacearum*	*Anthurium andraeanum*	ISR	Lin et al. (2019)
4.	Endophytic bacteria strain REB01	Rice seed	*Rhizoctonia solani*	Rice	ISR	Mao et al. (2019)
5.	*Bacillus*, *Paenibacillus*, *Pediococcus*, *Lactococcus*, and *Pantoea*	Seeds of diverse cucurbits	Fungal and oomycete pathogens including powdery mildew	Cucurbits and vegetables	ISR	Khalaf and Raizada (2018)
6.	*B.amyloliquefaciens* and *P. florescence*	*W. somnifera* and *R. serpentina*	*Alternaria alternata*	*Withania somnifera*	SAR and ISR	Mishra et al. (2018a, b)
7.	*Pseudomonas protegens* Pf-5 and *Pseudomonas* sp. G22	*Triticum aestivum*	*Magnaporthe oryzae B157 and Rhizoctonia solani*	Rice, sorghum, and wheat	ISR	Patel and Archana (2018)
8.	*Actinobacteria*	Medicinal plants	*S. rolfsii*	*Chickpea*	ISR	Singh and Gaur (2017)
9.	*Pseudozyma churashimaensis*	*Pepper*	*Xanthomonas axonopodis, Cucumber mosaic virus, Pepper mottle virus, Pepper mild mottle virus*	*Pepper*	ISR	Lee et al. (2017)
10.	*Streptomyces* sp.	*Medicinal plants*	*S. rolfsii*	*Chickpea*	ISR	Singh and Gaur (2016)
11.	*Rhodococcus* sp.	Plant leaf	*Ceratocystis fimbriata*	*Sweet potato*	ISR	Hong et al. (2016)

12.	*Pantoea eucalyptii*	*Solanum lycopersici*	*Botrytis cinerea*	*Arabidopsis thaliana*	ISR	Romero et al. (2016)
13.	*Pseudomonas fluorescens* PICF7	Olive roots	*Verticillium dahliae*	*Arabidopsis thaliana*	ISR	Maldonado-González et al. (2015)
14.	*B. amyloliquefaciens*	*Ginkgo biloba*	*Phytophthora infestans*	*Pepper*	ISR	Yang et al. (2015)
15.	*Bacillus pumilus*	*Cucumis sativus*	*Ralstonia solanacearum*	*Piper nigrum*	ISR	Yi et al. (2013)
16.	*B. amyloliquefaciens, B. pumilus*	*Solanum lycopersicum*	*Xanthomonas vesicatoria*	*Solanum lycopersicum*	ISR	Lanna-Filho et al. (2013)
17.	*Azospirillum* sp.	*Oryza sativa*	*Magnaporthe oryzae*	*Oryza sativa*	SAR	Yasuda et al. (2009)
18.	*Streptomyces* sp.	*Triticum aestivum*	*Erwinia carotovora*	*Arabidopsis thaliana*	SAR	Conn et al. (2008)
19.	*Actinobacteria*	*Triticum aestivum*	*Fusarium oxysporum*	*Arabidopsis thaliana*	SAR	Conn et al. (2008)
20.	*Pseudomonas fluorescence*	*Triticum aestivum*	*Fusarium* wilt	*Dianthus caryophyllus*	ISR	Van et al. (1991)
Fungal endophytes						
1.	*Beauveria bassiana*	*Vitis vinifera* L.	*Planococcus ficus*	*Vitis vinifera* L.	ISR	Moloinyane and Nchu (2019)
2.	*Aspergillus fumigatus* SG-17	*M. laxiflora* root	Oxidative stress	*M. laxiflora*	ISR and antioxidant ROS scavenging	Qin et al. (2019)
3.	*Fusarium solani*	*Solanum lycopersicum*	*Tetranychus urticae*	*Solanum lycopersicum*	Indirect tomato defense by the production of volatiles	Pappas et al. (2018)
4.	*Trichoderma atroviride*	*Zea mays*	*Spodoptera frugiperda*	*Zea mays*	ISR	Contreras-Cornejo et al. (2018)

(continued)

Table 12.1 (continued)

S.no.	Endophytes	Sources	Pathogens	Crops	Defence responses	References
5.	*Rhexocercosporidium* sp. and *F. solani*	*Sophora tonkinensis*	*Pathogenic F. solani*	*Panax notoginseng*	Production of secondary metabolites	Yao et al. (2017)
6.	*Epichloë* sp.	*Achnatherum robustum*	*Aphids*	*Achnatherum robustum*	Alkaloid Production	Shymanovich et al. (2015)
7.	*Phomopsis* sp.	*Fallopia japonica*	*Puccinia polygoni-amphibii* var. *tovariae*	*Fallopia japonica*	*Suppressive interaction*	Kurose et al. (2012)
8.	*Trichoderma* spp.	*Arabidopsis thaliana*	*Botrytis cinerea*	*Arabidopsis thaliana*	ISR	Contreras-Cornejo et al. (2011)
9.	*Fusarium solani*	*Solanum lycopersicum*	*Septoria lycopersici*	*Solanum lycopersicum*	ISR	Kavroulakis et al. (2007)
10.	*Fusarium oxysporum* strain 162	*Solanum lycopersicum*	*Meloidogyne incognita*	*Solanum lycopersicum*	ISR	El-Fattah Adnan Dababat and Alexander Sikora (2007)
11.	*Fusarium oxysporum*	*Solanum lycopersicum*	*Radopholus similis*	*Musa acuminata*	ISR	Vu et al. (2006)
12.	*Piriformospora indica*	*Prosopis juliflora*	*Rhynchosporium secalis*	*Hordeum vulgare*	ISR	Waller et al. (2005)

plant from diseases caused by bacterial and fungal phytopathogens (Mishra et al. 2018a; Singh and Gaur 2017). The effects of priming of plants with endophytes have been reported to act as a defense against various insects and pest as well (Vidal and Jaber 2015). The bacterial endophytes are reported to activate induced systemic resistance (ISR) similar to that of the rhizospheric bacteria (Berendsen et al. 2018; Shoresh et al. 2010). ISR is a plant defense pathway that is triggered by jasmonic acid or ethylene-based pathways. It is similar in response to systemic acquired resistance (SAR), which is salicylic acid-dependent pathway for the defense against the invading phytopathogen in the host plants (Shoresh et al. 2010). The SAR is activated when a pathogen invades which in turn activates the first line of defense in the host plant and elicits hypersensitive reaction to restrict the spreading and propagation of the pathogen further into the plant cells. The hypersensitive reaction acts suicidal response causing the death of its cells where the pathogen invaded, resulting in restricting the spread of the pathogen through the formation of localized necrotic zones (Gao et al. 2015). The ISR, on the other hand, primes the plants' second line of defenses and thus does not cause visible symptoms like those formed during SAR (Pieterse et al. 2014).

The bacterial endophytes not only trigger defense responses in the host; they are also involved in direct antagonistic action against the invading pathogens through the secretion of various enzymes and antimicrobial compounds. The enzyme produced by the endophytes includes fungal cell wall degrading chitinase and cellulose or for bacterial pathogen β-glucanases (Khare et al. 2018). The endophytes also are a rich repository of various antimicrobial compounds which are used as antibiotics and are known to be antifungal. Some endophytes are belonging to genera *Bacillus*, *Pseudomonas*, and *Burkholderia* that are known to produce a wide range of antimicrobial compounds against the invading pathogens (Martinez-Klimova et al. 2017; Gouda et al. 2016).

Antibacterial compounds like munumbicins A–D (Hasegawa et al. 2006); kakadumycins (Castillo et al. 2003); coronamycins, a peptide antibiotic (Ezra et al. 2004); and dimethyl novobiocins (Kurosawa et al. 2006) have been reported from endophytic strain of *Streptomyces* spp. compounds like pumilacidin, an antifungal compound (Melo et al. 2009); and antifungal iturin A, fengycin, and bacillomycin (Gond et al. 2015) are reported to be produced by *Bacillus* sp. Among the *Pseudomonas*, antifungal volatile organic compounds (Hernández-León et al. 2015) are reported to be produced along with compounds like ecomycins B and C (Miller et al. 1998) and pseudomycins (Harrison et al. 1991).

12.5.2 Fungal Endophytes

Several fungal strains have been reported as endophytes (Table 12.1). These endophytes are known for their hyperparasitic or mycoparasitic activity against the invading fungal phytopathogens. Endophytic *Trichoderma* sp. is known to parasitize the phytopathogen *Rhizoctonia solani* and other phytopathogens by coiling around the pathogen and producing chitinase enzyme leading to the disintegration of the fungal cell wall (Grosch et al. 2006; Chen et al. 2016)

The endophytic fungal isolates have been isolated from several medicinal plants such as *Paraconiothyrium* from *Taxus baccata* (Somjaipeng et al. 2015); *Bacillus amyloliquefaciens* and *Pseudomonas fluorescens* from *W. somnifera* and *R. serpentina* (Mishra et al. 2018a); bacterial strains from *Ginkgo biloba* (Yang et al. 2015); *Trichoderma gamsii* from *Panax notoginseng* (Chen et al. 2016) as well as commercially important plants, i.e., *Trichoderma atroviride* from maize (Contreras-Cornejo et al. 2018); *Pestalotiopsis* sp. from rice (Cord-Landwehr et al. 2016); and *Pseudomonas* sp. from *Solanum lycopersicum* (Subramanian et al. 2015), and these endophytes have been extensively utilized owing to their ability of production of secondary metabolites. Studies have revealed that these fungal endophytes mimic the secondary metabolites produced by their host plant, thus providing an additional layer of defense to their host plant against the invading phytopathogens, pest, and herbivores. However, the exact mechanism involved in the mimicking the production of the secondary metabolites of host plants by the endophytes is still not clear. The fungal endophytes are known to produce a more diverse group of compounds as compared to the native soil fungi (Nisa et al. 2015). Endophytic *Trichoderma harzianum* isolated from *Ilex cornuta* has been reported to produce antimicrobial compound trichodermin against phytopathogens like *Alternaria solani* and *R. solani* (Shentu et al. 2010). Cycloepoxylactone and cycloepoxytriol B are produced endophytic fungus *Phomopsis* sp. isolated from *Laurus azorica* (Hussain et al. 2009). Pestalachloride A isolated from the endophytic fungus *Pestalotiopsis adusta* inhabiting stem of an unknown Chinese tree is reported to be antagonistic against the phytopathogens like *Fusarium culmorum*, *Gibberella zeae* (anamorph *F. graminearum*), and *Verticillium albo-atrum* (Li et al. 2008).

The fungal endophytes like their bacterial counterparts are known to induce ISR mechanism in their host plants to elicit defense responses. The ISR elicited by the fungal endophytes may also be linked to the expression of pathogenesis-related genes. PR genes PR5 and PR7 in tomato plant roots are triggered by the endophytic fungal isolate *Fusarium solani* against the tomato foliar pathogen *Septoria lycopersici* (Kavroulakis et al. 2007). Nonpathogenic strains of *Fusarium oxysporum* are known to elicit ISR in banana plants against *Radopholus similis* (Vu et al. 2006). Endophytic *Trichoderma spp.* have been reported to protect *Arabidopsis thaliana* against *Botrytis cinerea* through ISR (Contreras-Cornejo et al. 2011).

Other mechanisms used by the endophytic fungal isolates include eliciting higher lignin deposition and enhanced activity of both antioxidant and defense enzymes. It has been reported in *Citrullus lanatus* and *Cucumis sativus* primed with the nonpathogenic mutant of *Colletotrichum magna* against the phytopathogens *Colletotrichum orbiculare* and *Fusarium oxysporum* (Redman et al. 1999). In *Cicer arietinum* colonization by endophytic fungi, *Piriformospora indica* has shown enhanced production of the antioxidant enzyme against the phytopathogen *Botrytis cinerea* (Narayan et al. 2017). *Trichoderma* T22 has shown to enhance the expression of the defense enzyme phenylalanine ammonia lyase (PAL) which is involved in the lignifications of the host plant's cell wall in maize plants (Shoresh et al. 2010).

Nematodes are also identified as one of the major agricultural pathogens that lead to severe crop loss annually. Stirling (2017) defined the biological control of nematodes as "reduction in nematode damage by organisms antagonistic to nematodes through the regulation of nematode populations and/or a reduction in the capacity of nematodes to cause damage, which occurs naturally or is accomplished through the manipulation of the environment or by the mass introduction of antagonists" (Stirling 2017). Although nematologists have been extensively researching biological control of nematodes since the 1920s till date, only a few biological control agents have been reported, but some fungal endophytes have been reported to be antagonistic to the invading nematodes. *Clonostachys rosea* has been reported to be a parasitic nematode fungus through the production of fungal proteases (Iqbal et al. 2018).

12.6 Role of Endophytes in Abiotic Stresses

Plants were grown under the surroundings of inanimate components of environments (i.e., light, water, carbon, nutrients, etc.) which are associated with climatic, edaphic, and physiographic factors. Categorically the extreme conditions of these factors (i.e., drought, salinity, temperature, heavy metals, poor soil nutrient, and oxidative burst) considerably affect the plant growth and cause abiotic stress. Endophytes successfully ameliorate the negative impact of these abiotic factors by the implementation of several mechanisms. The interaction of endosymbionts with the plant is very crucial not only for the induction of ISR against biotic stresses but also in terms of adaptation and tolerance of both participants against abiotic stresses. The endosymbiotic-mediated abiotic stress responses of the plant are termed as *induced systemic tolerance (IST)*.

In past decades, the emerging role of endophytes for the alleviation of plant abiotic stresses has attained a great response (Nadeem et al. 2014; Souza et al. 2014). Endophytes possess several fundamental metabolic and genetic abilities that significantly perform a vital role in the amelioration of abiotic stresses in the host plant (Hardoim et al. 2015). Several plant-associated beneficial microbes including *Pseudomonas* sp. (Otieno et al. 2015), *Bacillus* sp. (Vardharajula et al. 2011), *Rhizobium* sp. (Remans et al. 2008), *Bradyrhizobium* sp. (Aung et al. 2013), *Streptomycetes* sp. (Naylor et al. 2017), *Trichoderma* sp. (Mishra et al. 2019; Ahmad et al. 2015), *Methylobacterium* sp. (Meena et al. 2012), and *Cyanobacteria* sp. (Singh et al. 2011) are reported for the mitigation of abiotic stresses in the plant. Currently, Mishra et al. (2019) have reported the efficacy of impulsive micromanager *Trichoderma reesei* MTCC5659 on CO_2 stress amelioration in drought-resistant and sensitive rice cultivars because of the upregulation of dehydrin (*dhy*), glutathione S-transferases (*gst*), universal stress protein (*usp*), late embryogenesis protein (*lea*), and no epical meristematic (*nam*) genes along with several fundamental processes (i.e., photosynthesis, transpiration, respiration, stomatal conductance, and water-use efficiency). The association of endophytes leads to several modifications in the host plant, i.e., level of phytohormones, defense-related proteins, synthesis of antioxidants, exopolysaccharide, and enzymes which ensure the tolerance of the host plant against drought stress condition (Hubbard et al. 2014). For example, the application of *Trichoderma reesei* not only enhances the germination of *Cicer arietinum* in diesel fuel-spiked soil but also elevates the dehydrogenase activity and microbial dynamics of rhizospheric soil (Mishra and Nautiyal 2009).

Soil salinity is also a very effective abiotic factor and causes substantive harm to plant growth and productivity. The bacterial endophytes such as *Pseudomonas* sp. and *Bacillus subtilis* (BERA 17) are reported to produce the higher content of ACC deaminase and IAA which confirms the growth, osmo-regularity, photosynthetic rate, and productivity of tomato and chickpea plant under high salinity condition of soil, respectively (Win et al. 2018; Abd-Allah et al. 2017). Likewise, *Burkholderia phytofirmans* strain PsJN ameliorates the salinity stress in *Arabidopsis thaliana* (Pinedo et al. 2015). Similarly, the application of *Streptomyces* sp. confers the NaCl tolerance and growth in tomato plant (Palaniyandi et al. 2014). Besides, these several toxic metalloids (arsenic, cadmium, lead, etc.) and other hazardous compounds are regularly deposited in the soil because of several anthropogenic activities and industrial effluents. They not only affect the growth and productivity of plants but also limit the nutritional value of the soil.

Furthermore, several findings demonstrate that the traces of heavy metals are accruing in the crops and vegetable from the contaminated irrigating water or growing soil (Sharma et al. 2006; Dwivedi et al. 2010; Awasthi et al. 2017). These toxic metalloids are nondegradable; therefore, they can never be eliminated from the environment completely. However, studies revealed that few beneficial microbes efficiently reduce the toxicity through the biotransformation of toxic metalloids in less or nontoxic forms (Lakshmanan et al. 2016; Singh and Gaur 2016). For instance, the endophytic bacterial strains isolated from the *Betula celtiberica* have the great potential to improve the phytoremediation of As (Mesa et al. 2017).

Similarly, *Salix caprea* plant is well known regarding their potential to support heavy metal phytoextraction. Different actinobacterial strains of *Salix caprea* can synthesize the Zn and Cd mobilizing metabolites and therefore enhance the allocation of metals in the plant (Kuffner et al. 2010). On the other hand, the treatment of *Trichoderma* not only reduces the As deposition in chickpea plant but also enhances the nutritional value of the plant yield (Tripathi et al. 2013). Likewise, arsenic (As)-resistant strain *Staphylococcus arlettae* reduces the As uptake in Indian mustard (*Brassica juncea*) along with plant growth-promoting abilities (Srivastava et al. 2015).

Apart of that, endosymbionts also able to trigger the innate immunity of plants through the secretion of extracellular secondary metabolites, and consequently several defense-related activities have been taking place in each and every plant cell to reduce the negative impact of stress conditions. Similar to other microscopic endosymbionts, mycorrhiza also played a decisive role in the plant growth regulation and abiotic stress amelioration. The fine threads like networking of mycorrhizal hyphae are beneficial in the nutrient uptake management by plant roots. An endophytic mycorrhiza, namely, *Piriformospora indica*, ameliorates the level of antioxidants which subsequently improves the NaCl and drought tolerance ability of barley

(Baltruschat et al. 2008) and Chinese cabbage (Sun et al. 2010). On the one hand, endophytes triggered the systemic resistance for the survival of the plant under unfavorable environmental conditions, while on the other they improve the health and yielding attributes of the host plant by phytohormones stimulation, synthesis of phyto-inducers and antimicrobial substances, production of chelating ions, and fixation of N and other factors. All of these traits of endosymbionts make their presence necessary for the amelioration of abiotic stresses and thus prove themselves as an excellent option for crop management strategies.

12.7 Future Outlook

Endophytes remain untapped, not widely explored the community of microbes residing within the host plants. A few decades ago, they were thought not to affect their host plants. But in the recent years, the extensive research done on the role of endophytes in regulating various physiological aspects in its host plants have brought to light their critical role and have opened up new avenues for their utilization of various sectors including in pharmaceutical industries, mediating various plant protection strategies and abiotic stress amelioration. The unique ability of the endophytes to mimic the secondary metabolite synthesis pathways of their host plants makes them ideal candidates for the source of several medicinally and pharmaceutically important drugs which earlier could only be obtained through harvesting the host plants. This technique is turning in the conservation of several important plant species that are being threatened to extinction due to overexploitation by human interventions. Thus, these endophytes serve as an economically attractive solution for obtaining various valuable drugs and medicines without harming the environment. However, how and why the endophytes develop these abilities to mimic their host remains the question of further research.

In the past decade, the development of science and technologies through the facilitation of various "omics" studies has opened doors of new possibilities to understand these microbes better and to utilize them for the betterment of human society. The omics studies like transcriptomics, proteomics, and metabolomics help to elucidate the various pathways involved and to understand better the mechanisms that they employ during the tripartite interaction of the endophyte and host and the invading pathogen (Laur et al. 2018). The studies have revealed that these endophytes that are involved in inducing disease resistance in their host plants have a wide variety of tools at their disposal (Mishra et al. 2018a, b; Singh and Gaur 2017). The myriad of metabolites like bacillomycin, phenazines, iturin, fengycin, etc., produced by these endophytes are found to have antimicrobial properties, which they use against the invading pathogens. These compounds in the future could be identified and isolated for the development of pathogen-specific drugs for agricultural purposes (Zouari et al. 2016; Brader et al. 2014). The endophytes are also known to produce fungal cell wall degrading enzymes that contribute to their antagonistic ability against various invading phytopathogens which can be employed in designing targeted, integrated disease management strategies (Mishra et al. 2018a; Singh and Gaur 2017).

These strategies will not only be effective in controlling the diseases but also play a critical role in promoting and achieving the goal of sustainable agriculture by replacing the use of synthetic chemicals. However, how different endophytic communities coordinate among one another in the microenvironment and coexist remains largely unknown. The better understanding of these aspects will further help in designing more precise disease management strategies (Card et al. 2016).

The endophytes are also known to regulate the physiology of the host plants to adapt better and to acclimatize with the surrounding environment. It has been found that the root endophytes help in drought resistance in their plants enhance the expression and activity of a vacuolar $H+$ -pumping pyrophosphatase (Vigani et al. 2018). Such endophytes can be widely used in abiotic stress mitigation in a cost-effective manner. Extensive research on their potential can unveil untapped aspects of their ability to combat the various constraints faced in the agriculture sector in cost-effective, eco-friendly, and environmentally sustainable manner.

12.8 Conclusion

As discussed in the different contents of the chapter, endophytes serve as great alternatives for the growth and development of host plants, including the enhancement of production of the novel active phyto-compounds. Moreover, several strains of endosymbionts (i.e., bacteria, actinomycetes, fungi, etc.) are performing a decisive role in the induction of systemic resistance in plants, thus leading the suppression of disease severity in the host plant. On the one hand, these findings on endophytic research describe the antagonistic mechanisms of microbes, whereas on the other hand it also illustrates the host responses during the signal transduction pathways that govern the disease protection. The induction of systemic resistance of plants is interconnected with plant growth and development. However, the determination of specific microbial traits and gene pools that ignite the systemic resistance-related signaling in the host plant is still the case of an investigation.

Furthermore, the understandings about the selection criteria followed by the host plant for the elicitation of systemic resistance and plant beneficial strategies in stress conditions also remain to be answered. Several researchers are encouraged to explore the diversity and applications for the endophytes for the upliftment of plant growth and yield under different stress conditions. However, the multifarious performance of endophytes and the way of communication with the host plant still need higher attention. Moreover, the endophytes also represent an incredible source of novel bioactive natural products. Several genes obtained from endophytes have great potentiality, which might be helpful in the development of biotic and abiotic resistant varieties and can also help to improve the understanding of the behavior and mechanisms of microbes during the interaction. Furthermore, currently, just 1–2% of plant species are studied for endophytic association (Strobel and Daisy 2003), and the rest of plant diversity remains unexplored (Strobel 2018). It is now needed to explore the remaining plant-associated microbial diversity at metabolomics as well as the genomic level to detect the biochemical and physiological aspects of endophytes. Moreover, the information regarding the secondary metabolites and novel bioactive compounds obtained from endophytes that can be facilitated as a cure of several problems are not available in any database.

By resolving these issues, we can exploit the endophytes as an emerging tool in agriculture, medical, and phytopharmaceuticals. In the upcoming scenario, the endophytes may revolutionize the agricultural practices in terms of plant growth and stress management. Furthermore, plant-based drugs can also be modified through the intervention of endophytes-derived bioactive molecules in the near future which can be useful for the amelioration of not only yield and growth of the plants for societal benefits but also the pre-immunization of host plants against several invaders. Similarly, the endophytes-treated plant varieties can also be introduced shortly that develop the plant resistance against phytopathogens. The application of endophytes can also be a better option for the preservation and conservation of indigenous plant species at their natural habitats. Moreover, the development of next-generation bio-formulations based on endophytes still needed more attention and much deeper knowledge on the following threatening issues: (i) genetic and molecular cross talk between both partners, (ii) important factors for the stable and healthy association between the plant and endophytes, and (iii) mode of the interaction.

Conclusively, endophytes are a very valuable biological tool, which can be explored in upcoming days to achieve the objectives of disease management, environmental sustainability, and novel drug development. There is a need to explore the molecular and genetic information of the plant and endophytes during the establishment of a plant-microbe association to obtain the higher benefits from this incredible event.

References

Abd-Allah EF, Alqarawi AA, Hashem A, Radhakrishnan R, Al-Huqail AA, Al-Otibi FA, et al. (2017). Endophytic bacterium *Bacillus subtilis* (BERA 71) improves salt tolerance in chickpea plants by regulating the plant defense mechanisms. J Plant Interact. 3:37–44. https://doi.org/10.1080/17429145.2017.1414321

Abdel-Rahim IR, Abo-Elyousr KA (2018) *Talaromyces pinophilus* strain AUN-1 as a novel mycoparasite of *Botrytis cinerea*, the pathogen of onion scape and umbel blights. Microbiol Res 212:1–9

Ahmad P, Hashem A, Abd-Allah EF, Alqarawi AA, John R, Egamberdieva D, Gucel S (2015) Role of *Trichoderma harzianum* in mitigating NaCl stress in Indian mustard (*Brassica juncea* L) through antioxidative defense system. Front Plant Sci 6:868

Ali S, Duan J, Charles TC, Glick BR (2014) A bioinformatics approach to the determination of genes involved in endophytic behavior in *Burkholderia* spp. J Theor Biol 343:193–198

Ambrose KV, Billings NA, Djonovic S, Flavell RB, Gulick TA, Johnston DM, Leff JW, Liva SM, Lyford J, Von Maltzahn G, Marquez LM (2018) Seed endophytes across cultivars and species, associated compositions, and methods of use thereof. U.S. Patent Appl 15(/540):961

Andreote FD, Jiménez DJ, Chaves D, Dias ACF, Luvizotto DM, Dini-Andreote F, Fasanella CC, Lopez MV, Baena S, Taketani RG, de Melo IS (2012) The microbiome of Brazilian mangrove sediments as revealed by metagenomics. PLoS One 7(6):e38600

Andreote FD, Gumiere T, Durrer A (2014) Exploring interactions of plant microbiomes. Scientia agrícola 71(6):528–539
Arora NK, Mishra J (2016) Prospecting the roles of metabolites and additives in future bioformulations for sustainable agriculture. Appl Soil Ecol 107:405–407
Aung TT, Tittabutr P, Boonkerd N, Herridge D, Teaumroong N (2013) Co-inoculation effects of bradyrhizobium japonicum and azospirillum sp. on competitive nodulation and rhizosphere eubacterial community structures of soybean under rhizobia-established soil conditions. Afr J Biotechnol 12(20):2850–2862
Ausmees N, Kobayashi H, Deakin WJ, Marie C, Krishnan HB, Broughton WJ et al (2004) Characterization of NopP, a Type III secreted effector of *Rhizobium* sp. strain NGR234. J Bacteriol 186:4774–4780
Awasthi S, Chauhan R, Srivastava S, Tripathi RD (2017) The journey of arsenic from soil to grain in rice. Front Plant Sci 8:1007
Bacon CW (2018) The potential role of endophytes in ecosystems. In: Biotechnology of endophytic fungi of grasses. CRC Press, Boca Raton, pp 85–98
Bacon CW, Glenn AE, Yates IE (2008) *Fusarium verticillioides*: managing the endophytic association with maize for reduced fumonisins accumulation. Toxin Rev 27:411–446
Bakker MG, Schlatter DC, Otto-Hanson L, Kinkel LL (2014) Diffuse symbioses: roles of plant–plant, plant–microbe and microbe–microbe interactions in structuring the soil microbiome. Mol Ecol 23(6):1571–1583
Balachandar D, Sandhiya GS, Sugitha TCK, Kumar K (2006) Flavonoids and growth hormones influence endophytic colonization and in planta nitrogen fixation by a diazotrophic *Serratia* sp. in rice. World J Microbiol Biotechnol 22(7):707–712
Baldrian P (2006) Fungal laccases–occurrence and properties. FEMS Microbiol Rev 30(2):215–242
Baltruschat H, Fodor J, Harrach BD, Niemczyk E, Barna B, Gullner G et al (2008) Salt tolerance of barley induced by the root endophyte *Piriformospora indica* is associated with a strong increase in anti-oxidants. New Phytol 180:501–510
Barman S, Das S, Bhattacharya SS (2019) The prospects of bio-fertilizer technology for productive and sustainable agricultural growth. In: New and future developments in microbial biotechnology and bioengineering. Elsevier, Amsterdam, pp 233–253
Bengtsson M, Alfredsson E, Cohen M, Lorek S, Schroeder P (2018) Transforming systems of consumption and production for achieving the sustainable development goals: moving beyond efficiency. Sustain Sci 13(6):1533–1547
Benhiba L, Fouad MO, Essahibi A, Ghoulam C, Qaddoury A (2015) Arbuscular mycorrhizal symbiosis enhanced growth and antioxidant metabolism in date palm subjected to long-term drought. Trees 29:1725–1733
Berendsen RL, Pieterse CM, Bakker PA (2012) The rhizosphere microbiome and plant health. Trends Plant Sci 17(8):478–486
Berendsen RL, van Verk MC, Stringlis IA, Zamioudis C, Tommassen J, Pieterse CM et al (2015) Unearthing the genomes of plant-beneficial *Pseudomonas* model strains WCS358, WCS374 and WCS417. BMC Genomics 16:539
Berendsen RL, Vismans G, Yu K, Song Y, Jonge R, Burgman WP, Burmølle M, Herschend J, Bakker PA, Pieterse CM (2018) Disease-induced assemblage of a plant-beneficial bacterial consortium. ISME J 12(6):1496
Bernal P, Llamas MA, Filloux A (2018) Type VI secretion systems in plant-associated bacteria. Environ Microbiol 20:1–15
Brader G, Compant S, Mitter B, Trognitz F, Sessitsch A (2014) Metabolic potential of endophytic bacteria. Curr Opin Biotechnol 27:30–37
Brewer T, Handley K, Carini P, Gibert J, Fierer N (2016) Genome reduction in an abundant and ubiquitous soil bacterial lineage. Nat Microbiol 2:16198
Bringel F, Couée I (2015) Pivotal roles of phyllosphere microorganisms at the interface between plant functioning and atmospheric trace gas dynamics. Front Microbiol 6:486
Brussaard L (2012) Ecosystem services provided by the soil biota. In: Soil ecology and ecosystem services. Oxford University Press, Oxford, pp 45–58

Bulgarelli D, Rott M, Schlaeppi K, van Themaat EVL, Ahmadinejad N, Assenza F, Rauf P, Huettel B, Reinhardt R, Schmelzer E, Peplies J (2012) Revealing structure and assembly cues for *Arabidopsis* root-inhabiting bacterial microbiota. Nature 488(7409):91

Bulgarelli D, Schlaeppi K, Spaepen S, Van Themaat EVL, Schulze-Lefert P (2013) Structure and functions of the bacterial microbiota of plants. Annu Rev Plant Biol 64:807–838

Busby PE, Ridout M, Newcombe G (2016) Fungal endophytes: modifiers of plant disease. Plant Mol Biol 90(6):645–655

Busby PE, Soman C, Wagner MR, Friesen ML, Kremer J, Bennett A, Morsy M, Eisen JA, Leach JE, Dangl JL (2017) Research priorities for harnessing plant microbiomes in sustainable agriculture. PLoS Biol 15:e2001793

Cappelletti M, Perazzolli M, Antonielli L, Nesler A, Torboli E, Bianchedi PL, Pindo M, Puopolo G, Pertot I (2016) Leaf treatments with a protein-based resistance inducer partially modify phyllosphere microbial communities of grapevine. Front Plant Sci 7:1053

Card S, Johnson L, Teasdale S, Caradus J (2016) Deciphering endophyte behaviour: the link between endophyte biology and efficacious biological control agents. FEMS Microbiol Ecol 92(8):21–35

Castillo U, Harper JK, Strobel GA, Sears J, Alesi K, Ford E, Lin J, Hunter M, Maranta M, Ge H, Yaver D (2003) Kakadumycins, novel antibiotics from *Streptomyces* sp. NRRL 30566, an endophyte of *Grevillea pteridifolia*. FEMS Microbiol Lett 224(2):183–190

Chagas MBO, Prazeres Dos Santos I, Nascimento da Silva LC, Correia MTDS, Magali de Araújo J (2017) Antimicrobial activity of cultivable endophytic fungi associated with *Hancornia speciosa* gomes bark. Open Microbiol J 11:179–188

Chaparro JM, Badri DV, Vivanco JM (2014) Rhizosphere microbiome assemblage is affected by plant development. ISME J 8(4):790

Chen JL, Sun SZ, Miao CP, Wu K, Chen YW, Xu LH, Guan HL, Zhao LX (2016) Endophytic *Trichoderma gamsii* YIM PH30019: a promising biocontrol agent with hyperosmolar, mycoparasitism, and antagonistic activities of induced volatile organic compounds on root-rot pathogenic fungi of *Panax notoginseng*. J Ginseng Res 40(4):315–324

Classen AT, Sundqvist M, Henning JA, Newman GS, Moore JAM, Cregger M, Moorhead LC, Patterson CM (2015) Direct and indirect effects of climate change on soil microbial and soil microbial-plant interactions: what lies ahead? Ecosphere 6:130

Compant S, Reiter B, Sessitsch A, Nowak J, Clément C, Barka EA (2005) Endophytic colonization of *Vitis vinifera* L. by plant growth-promoting bacterium *Burkholderia* sp. strain PsJN. Appl Environ Microbiol 71(4):1685–1693

Compant S, Clément C, Sessitsch A (2010) Plant growth-promoting bacteria in the rhizo-and endosphere of plants: their role, colonization, mechanisms involved and prospects for utilization. Soil Biol Biochem 42(5):669–678

Conn VM, Walker AR, Franco CMM (2008) Endophytic actinobacteria induce defense pathways in *Arabidopsis thaliana*. Mol Plant-Microbe Interact 21(2):208–218

Contreras-Cornejo HA, Macıas-Rodrıguez L, Beltran-Pena E, Herrera-Estrella A, Lopez-Bucio J (2011) *Trichoderma*-induced plant immunity likely involves both hormonal- and camalexin-dependent mechanisms in *Arabidopsis thaliana* and confers resistance against necrotrophic fungi *Botrytis cinerea*. Plant Signal Behav 6:1554–1563

Contreras-Cornejo HA, Macías-Rodríguez L, del-Val E, Larsen J (2018) The root endophytic fungus *Trichoderma atroviride* induces foliar herbivory resistance in maize plants. Appl Soil Ecol 124:45–53

Copeland JK, Yuan L, Layeghifard M, Wang PW, Guttman DS (2015) Seasonal community succession of the phyllosphere microbiome. Mol Plant-Microbe Interact 28(3):274–285

Cord-Landwehr S, Melcher RLJ, Kolkenbrock S, Moerschbacher BM (2016) A chitin deacetylase from the endophytic fungus *Pestalotiopsis* sp. efficiently inactivates the elicitor activity of chitin oligomers in rice cells. Sci Rep 6:38018

Ding T, Melcher U (2016) Influences of plant species, season and location on leaf endophytic bacterial communities of non-cultivated plants. PLoS One 11:e0150895

Dwivedi S, Tripathi RD, Srivastava S, Singh R, Kumar A, Tripathi P, Dave R, Rai UN, Chakrabarty D, Trivedi PK, Tuli R (2010) Arsenic affects mineral nutrients in grains of various Indian rice

(*Oryza sativa* L.) genotypes grown on arsenic-contaminated soils of West Bengal. Protoplasma 245(1-4):113–124

El-Fattah Adnan Dababat A, Alexander Sikora R (2007) Induced resistance by the mutualistic endophyte, *Fusarium oxysporum* strain 162, toward *Meloidogyne incognita* on tomato. Biocontrol Sci Tech 17(9):969–975

Estrada C, Wcislo WT, Van Bael SA (2013) Symbiotic fungi alter plant chemistry that discourages leaf-cutting ants. New Phytol 198:241–251

Ezra D, Castillo UF, Strobel GA, Hess WM, Porter H, Jensen JB, Condron MA, Teplow DB, Sears J, Maranta M, Hunter M (2004) Coronamycins, peptide antibiotics produced by a verticillate *Streptomyces* sp. (MSU-2110) endophytic on *Monstera* sp. Microbiology 150(4):785–793

Ferrara FI, Oliveira ZM, Gonzales HHS, Floh EIS, Barbosa HR (2012) Endophytic and rhizospheric enterobacteria isolated from sugar cane have different potentials for producing plant growth-promoting substances. Plant Soil 353(1-2):409–417

Fesel PH, Zuccaro A (2016) Dissecting endophytic lifestyle along the parasitism/mutualism continuum in *Arabidopsis*. Curr Opin Microbiol 32:103–112

Forchetti G, Masciarelli O, Izaguirre MJ, Alemano S, Alvarez D, Abdala G (2010) Endophytic bacteria improve seedling growth of sunflower under water stress, produce salicylic acid, and inhibit growth of pathogenic fungi. Curr Microbiol 61(6):485–493

Formey D, Sallet E, Lelandais-Brière C, Ben C, Bustos-Sanmamed P, Niebel A, Frugier F, Combier JP, Debellé F, Hartmann C, Poulain J (2014) The small RNA diversity from *Medicago truncatula* roots under biotic interactions evidences the environmental plasticity of the miRNAome. Genome Biol 15(9):457

Fouad OM, Essahibi A, Benhiba L, Qaddoury A (2014) Effectiveness of arbuscular mycorrhizal fungi in the protection of olive plants against oxidative stress induced by drought. Span J Agric Res 12:763–771

Frank A, Saldierna Guzmán J, Shay J (2017) Transmission of bacterial endophytes. Microorganisms 5(4):70

Freitas MA, Medeiros FH, Melo IS, Pereira PF, Peñaflor MFG, Bento JM, Paré PW (2019) Stem inoculation with bacterial strains *Bacillus amyloliquefaciens* (GB03) and *Microbacterium imperiale* (MAIIF2a) mitigates *Fusarium* root rot in cassava. Phytoparasitica 47(1):135–142

Gaiero JR, McCall CA, Thompson KA, Day NJ, Best AS, Dunfield KE (2013) Inside the root microbiome: bacterial root endophytes and plant growth promotion. Am J Bot 100:1738–1750. https://doi.org/10.3732/ajb.1200572

Gao QM, Zhu S, Kachroo P, Kachroo A (2015) Signal regulators of systemic acquired resistance. Front Plant Sci 6:228

Germaine K, Keogh E, Garcia-Cabellos G, Borremans B, Van Der Lelie D, Barac T, Oeyen L, Vangronsveld J, Moore FP, Moore ER, Campbell CD (2004) Colonization of poplar trees by gfp expressing bacterial endophytes. FEMS Microbiol Ecol 48(1):109–118

Gimenez E, Salinas M, Manzano-Agugliaro F (2018) Worldwide research on plant defense against biotic stresses as improvement for sustainable agriculture. Sustainability 10(2):391–399

Gond SK, Bergen MS, Torres MS, White JF Jr (2015) Endophytic *Bacillus* spp. produce antifungal lipopeptides and induce host defence gene expression in maize. Microbiol Res 172:79–87

Gouda S, Das G, Sen SK, Shin HS, Patra JK (2016) Endophytes: a treasure house of bioactive compounds of medicinal importance. Front Microbiol 7:1538

Gough C, Cullimore J (2011) Lipo-chitooligosaccharide signaling in endosymbiotic plant-microbe interactions. Mol Plant-Microbe Interact 24(8):867–878

Green ER, Mecsas J (2016) Bacterial secretion systems: an overview. Microbiol Spectr 4:10

Grosch R, Scherwinski K, Lottmann J, Berg G (2006) Fungal antagonists of the plant pathogen *Rhizoctonia solani*: selection, control efficacy and influence on the indigenous microbial community. Mycol Res 110(12):1464–1474

Gupta R, Singh A, Srivastava M, Singh V, Gupta MM, Pandey R (2017) Microbial modulation of bacoside A biosynthetic pathway and systemic defense mechanism in *Bacopa monnieri* under *Meloidogyne incognita* stress. Sci Rep 7:41867

Hallmann J, Quadt-Hallmann A, Mahaffee WF, Kloepper JW (1997) Bacterial endophytes in agricultural crops. Can J Microbiol 43:895–914

Hardoim PR, van Overbeek LS, Berg G, Pirttilä AM, Compant S, Campisano A et al (2015) The hidden world within plants: ecological and evolutionary considerations for defining functioning of microbial endophytes. Microbiol Mol Biol Rev 79:293–320
Harman GE, Uphoff N (2019) Symbiotic root-endophytic soil microbes improve crop productivity and provide environmental benefits. Scientifica 2019:1–25
Harrison L, Teplow DB, Rinaldi M, Strobel G (1991) Pseudomycins, a family of novel peptides from *Pseudomonas syringae* possessing broad-spectrum antifungal activity. Microbiology 137(12):2857–2865
Hasegawa S, Meguro A, Shimizu M, Nishimura T, Kunoh H (2006) Endophytic actinomycetes and their interactions with host plants. Actinomycetologica 20(2):72–81
Hernández-León R, Rojas-Solís D, Contreras-Pérez M, del Carmen Orozco-Mosqueda M, Macías-Rodríguez LI, Reyes-de la Cruz H, Valencia-Cantero E, Santoyo G (2015) Characterization of the antifungal and plant growth-promoting effects of diffusible and volatile organic compounds produced by *Pseudomonas fluorescens* strains. Biol Control 81:83–92
Hong CE, Jeong H, Jo SH, Jeong JC, Kwon SY, An D, Park JM (2016) A leaf-inhabiting endophytic bacterium, *Rhodococcus* sp. KB6, enhances sweet potato resistance to black rot disease caused by *Ceratocystis fimbriata*. J Microbiol Biotechnol 26(3):488–492
Hottes AK, Freddolino PL, Khare A, Donnell ZN, Liu JC, Tavazoie S (2013) Bacterial adaptation through loss of function. PLoS Genet 9:e1003617
Hubbard M, Germida JJ, Vujanovic V (2014) Fungal endophytes enhance wheat heat and drought tolerance in terms of grain yield and second-generation seed viability. J Appl Microbiol 116(1):109–122
Hussain H, Akhtar N, Draeger S, Schulz B, Pescitelli G, Salvadori P, Antus S, Kurtán T, Krohn K (2009) New bioactive 2, 3-epoxycyclohexenes and isocoumarins from the endophytic fungus *Phomopsis* sp. from *Laurus azorica*. Eur J Org Chem 2009(5):749–756
Iqbal M, Dubey M, Gudmundsson M, Viketoft M, Jensen DF, Karlsson M (2018) Comparative evolutionary histories of fungal proteases reveal gene gains in the mycoparasitic and nematode-parasitic fungus *Clonostachys rosea*. BMC Evol Biol 18(1):171
James EK, Gyaneshwar P, Mathan N, Barraquio WL, Reddy PM, Iannetta PP, Olivares FL, Ladha JK (2002) Infection and colonization of rice seedlings by the plant growth-promoting bacterium *Herbaspirillum seropedicae* Z67. Mol Plant-Microbe Interact 15(9):894–906
Jia M, Chen L, Xin HL, Zheng CJ, Rahman K, Han T, Qin LP (2016) A friendly relationship between endophytic fungi and medicinal plants: a systematic review. Front Microbiol 7:906
Karpinets TV, Park BH, Syed MH, Klotz MG, Uberbacher EC (2014) Metabolic environments and genomic features associated with pathogenic and mutualistic interactions between bacteria and plants. Mol Plant-Microbe Interact 27:664–677
Kavroulakis NS, Zervakis GI, Ehaliotis C, Haralampidis K, Papadopoulou KK (2007) Role of ethylene in the protection of tomato plants against soil-borne fungal pathogens conferred by an endophytic *Fusarium solani* strain. J Exp Bot 58:3853–3864
Khalaf EM, Raizada MN (2018) Bacterial seed endophytes of domesticated cucurbits antagonize fungal and oomycete pathogens including powdery mildew. Front Microbiol 9:42–52
Khan AL, Hamayun M, Kang SM, Kim YH, Jung HY, Lee JH, Lee IJ (2012) Endophytic fungal association *via* gibberellins and indole acetic acid can improve plant growth under abiotic stress: an example of *Paecilomyces formosus* LHL10. BMC Microbiol 12(1):3
Khan AL, Hussain J, Al-Harrasi A, Al-Rawahi A, Lee IJ (2015) Endophytic fungi: resource for gibberellins and crop abiotic stress resistance. Crit Rev Biotechnol 35(1):62–74
Khare E, Kim KM, Lee K-J (2016) Rice OsPBL1 (*Oryza sativa-Arabidopsis* PBS1-LIKE 1) enhanced defense of *Arabidopsis* against *Pseudomonas syringae* DC3000. Eur J Plant Pathol 146:901–910
Khare E, Mishra J, Arora NK (2018) Multifaceted interactions between endophytes and plant: developments and prospects. Front Microbiol 9
Kloepper JW, Ryu CM (2006) Bacterial endophytes as elicitors of induced systemic resistance. In: Microbial root endophytes. Springer, Berlin/Heidelberg, pp 33–52

Knief C, Delmotte N, Chaffron S, Stark M, Innerebner G, Wassmann R, Von Mering C, Vorholt JA (2012) Metaproteogenomic analysis of microbial communities in the phyllosphere and rhizosphere of rice. ISME J 6(7):1378
Kuffner M, De Maria S, Puschenreiter M, Fallmann K, Wieshammer G, Gorfer M, Strauss J, Rivelli AR, Sessitsch A (2010) Culturable bacteria from Zn-and Cd-accumulating *Salix caprea* with differential effects on plant growth and heavy metal availability. J Appl Microbiol 108(4):1471–1484
Kurosawa K, Bui VP, VanEssendelft JL, Willis LB, Lessard PA, Ghiviriga I, Sambandan TG, Rha CK, Sinskey AJ (2006) Characterization of *Streptomyces* MITKK-103, a newly isolated actinomycin X 2-producer. Appl Microbiol Biotechnol 72(1):145–154
Kurose D, Furuya N, Tsuchiya K, Tsushima S, Evans HC (2012) Endophytic fungi associated with *Fallopia japonica* (Polygonaceae) in Japan and their interactions with *Puccinia polygoni-amphibii* var. tovariae, a candidate for classical biological control. Fungal Biol 116(7):785–791
Lakshmanan V, Cottone J, Bais HP (2016) Killing two birds with one stone: natural rice rhizospheric microbes reduce arsenic uptake and blast infections in rice. Front Plant Sci 7:1514
Lanna-Filho R, Souza RM, Magalhães MM, Villela L, Zanotto E, Ribeiro-Júnior PM, Resende ML (2013) Induced defense responses in tomato against bacterial spot by proteins synthesized by endophytic bacteria. Trop Plant Pathol 38(4):295–302
Lata R, Chowdhury S, Gond SK, White JF Jr (2018) Induction of abiotic stress tolerance in plants by endophytic microbes. Lett Appl Microbiol 66(4):268–276
Laur J, Ramakrishnan GB, Labbé C, Lefebvre F, Spanu PD, Bélanger RR (2018) Effectors involved in fungal–fungal interaction lead to a rare phenomenon of hyperbiotrophy in the tritrophic system biocontrol agent–powdery mildew–plant. New Phytol 217(2):713–725
Lee G, Lee SH, Kim KM, Ryu CM (2017) Foliar application of the leaf-colonizing yeast *Pseudozyma churashimaensis* elicits systemic defense of pepper against bacterial and viral pathogens. Sci Rep 7:39432
Li E, Jiang L, Guo L, Zhang H, Che Y (2008) Pestalachlorides A–C, antifungal metabolites from the plant endophytic fungus *Pestalotiopsis adusta*. Bioorg Med Chem 16(17):7894–7899
Li Q, Guo R, Li Y, Hartman WH, Li S, Zhang Z, Tringe SG, Wang H (2019) Insight into the bacterial endophytic communities of peach cultivars related to crown gall disease resistance. Appl Environ Microbiol 85(9):e02931–e02918
Lin HF, Xiong J, Zhou HM, Chen CM, Lin FZ, Xu XM, Oelmüller R, Xu WF, Yeh KW (2019) Growth promotion and disease resistance induced in Anthurium colonized by the beneficial root endophyte *Piriformospora indica*. BMC Plant Biol 19(1):40–52
Lindow SE, Brandl MT (2003) Microbiology of the phyllosphere. Appl Environ Microbiol 69(4):1875–1883
Liu H, Carvalhais LC, Crawford M, Singh E, Dennis PG, Pieterse CMJ et al (2017) Inner plant values: diversity, colonization and benefits from endophytic bacteria. Front Microbiol 8:2552
López-Ráez JA, Shirasu K, Foo E (2017) Strigolactones in plant interactions with beneficial and detrimental organisms: the Yin and Yang. Trends Plant Sci 22(6):527–537
Lundberg DS, Lebeis SL, Paredes SH, Yourstone S, Gehring J, Malfatti S, Tremblay J, Engelbrektson A, Kunin V, Del Rio TG, Edgar RC (2012) Defining the core *Arabidopsis thaliana* root microbiome. Nature 488(7409):86
Ma Y, Rajkumar M, Luo Y, Freitas H (2011) Inoculation of endophytic bacteria on host and non-host plants—Effects on plant growth and Ni uptake. J Hazard Mater 195:230–237
Maldonado-González MM, Bakker PA, Prieto P, Mercado-Blanco J (2015) *Arabidopsis thaliana* as a tool to identify traits involved in *Verticillium dahliae* biocontrol by the olive root endophyte *Pseudomonas fluorescens* PICF7. Front Microbiol 6:266
Malviya MK, Solanki MK, Li C-N, Htun R, Singh RK, Singh P, Yang L-T, Li Y-R (2019) Beneficial linkages of endophytic Burkholderia anthina MYSP113 towards sugarcane growth promotion. Sugar Tech 21(5):737–748. https://doi.org/10.1007/s12355-019-00703-2
Mao J, Gong M, Guan Q (2019) Induced disease resistance of endophytic bacteria REB01 to bacterial blight of rice. In: AIP conference proceedings, vol 2079, no 1. AIP Publishing, Leiden, p 020017

Marsberg A, Kemler M, Jami F, Nagel JH, Postma-Smidt A, Naidoo S, Wingfield MJ, Crous PW, Spatafora JW, Hesse CN, Robbertse B (2017) *Botryosphaeria dothidea*: a latent pathogen of global importance to woody plant health. Mol Plant Pathol 18(4):477–488

Martinez-Klimova E, Rodriguez-Pena K, Sánchez S (2017) Endophytes as sources of antibiotics. Biochem Pharmacol 134:1–17

Martiny JBH (2016) History leaves its mark on soil bacterial diversity. mBio 7:e00784–e00716. https://doi.org/10.1128/mBio.00784-16

Matilla MA, Ramos JL, Bakker PA, Doornbos R, Badri DV, Vivanco JM, Ramos-González MI (2010) *Pseudomonas putida* KT2440 causes induced systemic resistance and changes in Arabidopsis root exudation. *Environ Microbiol Rep 2*(3):381–388

McGuire KL, Treseder KK (2010) Microbial communities and their relevance for ecosystem models: decomposition as a case study. *Soil Biol Biochem 42*(4):529–535

Meena KK, Kumar M, Kalyuzhnaya MG, Yandigeri MS, Singh DP, Saxena AK, Arora DK (2012) Epiphytic pink-pigmented methylotrophic bacteria enhance germination and seedling growth of wheat (*Triticum aestivum*) by producing phytohormone. Antonie Van Leeuwenhoek 101(4):777–786

Melo FMPD, Fiore MF, Moraes LABD, Silva-Stenico ME, Scramin S, Teixeira MDA, Melo ISD (2009) Antifungal compound produced by the cassava endophyte *Bacillus pumilus* MAIIIM4A. Sci Agric 66(5):583–592

Mendes R, Garbeva P, Raaijmakers JM (2013) The rhizosphere microbiome: significance of plant beneficial, plant pathogenic, and human pathogenic microorganisms. FEMS Microbiol Rev 37:634–663

Mesa V, Navazas A, González-Gil R, González A, Weyens N, Lauga B, Gallego JLR, Sánchez J, Peláez AI (2017) Use of endophytic and rhizosphere bacteria to improve phytoremediation of arsenic-contaminated industrial soils by autochthonous *Betula celtiberica*. Appl Environ Microbiol 83(8):e03411–e03416

Miller CM, Miller RV, Garton-Kenny D, Redgrave B, Sears J, Condron MM, Teplow DB, Strobel GA (1998) Ecomycins, unique antimycotics from *Pseudomonas viridiflava*. J Appl Microbiol 84(6):937–944

Mishra A, Nautiyal CS (2009) Functional diversity of the microbial community in the rhizosphere of chickpea grown in diesel fuel-spiked soil amended with *Trichoderma reesei* using sole-carbon-source utilization profiles. World J Microbiol Biotechnol 25(7):1175–1180

Mishra A, Singh SP, Mahfooz S, Singh SP, Bhattacharya A, Mishra N, Nautiyal CS (2018a) Endophyte-mediated modulation of defense-related genes and systemic resistance in *Withania somnifera* (L.) Dunal under *Alternaria alternata* stress. Appl Environ Microbiol 84(8):1–13

Mishra A, Singh SP, Mahfooz S, Bhattacharya A, Mishra N, Shirke PA, Nautiyal CS (2018b) Bacterial endophytes modulates the withanolide biosynthetic pathway and physiological performance in *Withania somnifera* under biotic stress. Microbiol Res 212:17–28

Mishra A, Singh SP, Mahfooz S, Shukla R, Mishra N, Pandey S, Dwivedi S, Pandey V, Shirke PA, Nautiyal CS (2019) External supplement of impulsive micromanager *Trichoderma* helps in combating CO_2 stress in rice grown under FACE. Plant Mol Biol Report:1–13

Mitter B, Petric A, Shin MW, Chain PS, Hauberg-Lotte L, Reinhold-Hurek B, Nowak J, Sessitsch A (2013) Comparative genome analysis of *Burkholderia phytofirmans* PsJN reveals a wide spectrum of endophytic lifestyles based on interaction strategies with host plants. Front Plant Sci 4:120

Moloinyane S, Nchu F (2019) The effects of endophytic *Beauveria bassiana* inoculation on infestation level of *Planococcus ficus*, growth and volatile constituents of potted greenhouse grapevine (*Vitis vinifera* L.). Toxins 11(2):72

Mousavi S, Regni L, Bocchini M, Mariotti R, Cultrera NG, Mancuso S, Googlani J, Chakerolhosseini MR, Guerrero C, Albertini E, Baldoni L (2019) Physiological, epigenetic and genetic regulation in some olive cultivars under salt stress. Sci Rep 9(1):1093

Nadeem SM, Ahmad M, Zahir ZA, Javaid A, Ashraf M (2014) The role of mycorrhizae and plant growth promoting rhizobacteria (PGPR) in improving crop productivity under stressful environments. Biotechnol Adv 32:429–448

Naik BS (2019) Potential roles for endophytic fungi in biotechnological processes: a review. In: Plant and human health, vol 2. Springer, Cham, pp 327–344

Narayan OP, Verma N, Singh AK, Oelmüller R, Kumar M, Prasad D, Kapoor R, Dua M, Johri AK (2017) Antioxidant enzymes in chickpea colonized by *Piriformospora indica* participate in defense against the pathogen *Botrytis cinerea*. Sci Rep 7(1):13553

Naveed M, Mitter B, Reichenauer TG, Wieczorek K, Sessitsch A (2014) Increased drought stress resilience of maize through endophytic colonization by *Burkholderia phytofirmans* PsJN and Enterobacter sp. FD17. Environ Exp Bot 97:30–39

Naylor D, DeGraaf S, Purdom E, Coleman-Derr D (2017) Drought and host selection influence bacterial community dynamics in the grass root microbiome. ISME J 11(12):2691

Newman MA, Sundelin T, Nielsen JT, Erbs G (2013) MAMP (microbe associated molecular pattern) triggered immunity in plants. Front Plant Sci 4:139

Nguema-Ona E, Vicré-Gibouin M, Cannesan MA, Driouich A (2013) Arabinogalactan proteins in root–microbe interactions. Trends Plant Sci 18(8):440–449

Nisa H, Kamili AN, Nawchoo IA, Shafi S, Shameem N, Bandh SA (2015) Fungal endophytes as prolific source of phytochemicals and other bioactive natural products: a review. Microb Pathog 82:50–59

Nissinen RM, Mannisto MK, van Elsas JD (2012) Endophytic bacterial communities in three arctic plants from low arctic fell tundra are cold-adapted and host-plant specific. FEMS Microbiol Ecol 82:510–522

Okazaki S, Kaneko T, Sato S, Saeki K (2013) Hijacking of leguminous nodulation signaling by the rhizobial type III secretion system. Proc Natl Acad Sci U S A 110:17131–17136

Oren L, Ezrati S, Cohen D, Sharon A (2003) Early events in the *Fusarium verticillioides*-maize interaction characterized by using a green fluorescent protein-expressing transgenic isolate. Appl Environ Microbiol 69:1695–1701

Osono T (2007) Ecology of ligninolytic fungi associated with leaf litter decomposition. Ecol Res 22(6):955–974

Otieno N, Lally RD, Kiwanuka S, Lloyd A, Ryan D, Germaine KJ, Dowling DN (2015) Plant growth promotion induced by phosphate solubilizing endophytic *Pseudomonas* isolates. Front Microbiol 6:745

Palaniyandi SA, Damodharan K, Yang SH, Suh JW (2014) *Streptomyces* sp. strain PGPA39 alleviates salt stress and promotes growth of 'Micro Tom' tomato plants. J Appl Microbiol 117(3):766–773

Pandey SS, Singh S, Pandey H, Srivastava M, Ray T, Soni S, Pandey A, Shanker K, Babu CV, Banerjee S, Gupta MM (2018) Endophytes of *Withania somnifera* modulate in planta content and the site of withanolide biosynthesis. Sci Rep 8(1):5450

Pappas ML, Liapoura M, Papantoniou D, Avramidou M, Kavroulakis N, Weinhold A, Broufas GD, Papadopoulou KK (2018) The beneficial endophytic fungus *Fusarium solani* strain K alters tomato responses against spider mites to the benefit of the plant. Front Plant Sci 9

Patel JK, Archana G (2018) Engineered production of 2, 4-Diacetylphloroglucinol in the diazotrophic endophytic bacterium *Pseudomonas* sp. WS5 and its beneficial effect in multiple plant-pathogen systems. Appl Soil Ecol 124:34–44

Philippot L, Andersson SG, Battin TJ, Prosser JI, Schimel JP, Whitman WB, Hallin S (2010) The ecological coherence of high bacterial taxonomic ranks. Nat Rev Microbiol 8(7):523

Philippot L, Raaijmakers JM, Lemanceau P, van der Putten WH (2013) Going back to the roots: the microbial ecology of the rhizosphere. Nat Rev Microbiol 11:789–799

Pieterse CM, Zamioudis C, Berendsen RL, Weller DM, Van Wees SC, Bakker PA (2014) Induced systemic resistance by beneficial microbes. Annu Rev Phytopathol 52:347–375

Pinedo I, Ledger T, Greve M, Poupin MJ (2015) *Burkholderia phytofirmans* PsJN induces long-term metabolic and transcriptional changes involved in *Arabidopsis thaliana* salt tolerance. Front Plant Sci 6:466

Piromyou P, Songwattana P, Greetatorn T, Okubo T, Kakizaki KC, Prakamhang J et al (2015) The type III secretion system (T3SS) is a determinant for rice-endophyte colonization by non-photosynthetic *Bradyrhizobium*. Microbes Environ 30:291–300

Plett JM, Martin FM (2018) Know your enemy, embrace your friend: using omics to understand how plants respond differently to pathogenic and mutualistic microorganisms. Plant J 93:729–746

Potshangbam M, Devi SI, Sahoo D, Strobel GA (2017) Functional characterization of endophytic fungal community associated with *Oryza sativa* L. and *Zea mays* L. Front Microbiol 8:325

Qin W, Liu C, Jiang W, Xue Y, Wang G, Liu S (2019) A coumarin analogue NFA from endophytic *Aspergillus fumigatus* improves drought resistance in rice as an antioxidant. BMC Microbiol 19(1):50

Quecine MC, Araújo WL, Tsui S, Parra JRP, Azevedo JL, Pizzirani-Kleiner AA (2014) Control of *Diatraea saccharalis* by the endophytic *Pantoea agglomerans* 33.1 expressing cry1Ac7. Arch Microbiol 196(4):227–234

Rastogi G, Coaker GL, Leveau JH (2013) New insights into the structure and function of phyllosphere microbiota through high-throughput molecular approaches. FEMS Microbiol Lett 348(1):1–10

Redford AJ, Bowers RM, Knight R, Linhart Y, Fierer N (2010) The ecology of the phyllosphere: geographic and phylogenetic variability in the distribution of bacteria on tree leaves. Environ Microbiol 12(11):2885–2893

Redman RS, Freeman S, Clifton DR, Morrel J, Brown G, Rodriguez RJ (1999) Biochemical analysis of plant protection afforded by a nonpathogenic endophytic mutant of *Colletotrichum magna*. Plant Physiol 119(2):795–804

Reinhold-Hurek B, Hurek T (2011) Living inside plants: bacterial endophytes. Curr Opin Plant Biol 14:435–443

Remans R, Ramaekers L, Schelkens S, Hernandez G, Garcia A, Reyes JL, Mendez N, Toscano V, Mulling M, Galvez L, Vanderleyden J (2008) Effect of *Rhizobium–Azospirillum* coinoculation on nitrogen fixation and yield of two contrasting *Phaseolus vulgaris* L. genotypes cultivated across different environments in Cuba. Plant Soil 312(1–2):25–37

Rho H, Hsieh M, Kandel SL, Cantillo J, Doty SL, Kim SH (2018) Do endophytes promote growth of host plants under stress? A meta-analysis on plant stress mitigation by endophytes. Microb Ecol 75(2):407–418

Rodríguez-Blanco A, Sicardi M, Frioni L (2015) Plant genotype and nitrogen fertilization effects on abundance and diversity of diazotrophic bacteria associated with maize (Zea mays L.). Biol Fertil Soils 51(3):391–402

Romero FM, Marina M, Rossi FR, Viaud M, Pieckenstain FL (2016) *Pantoea eucalyptii* induces systemic resistance in *Arabidopsis thaliana* against *Botrytis cinerea* by priming defense responses. In: XVII international botrytis symposium, p 80

Romero FM, Rossi FR, Gárriz A, Carrasco P, Ruíz OA (2018) A bacterial endophyte from apoplast fluids protects canola plants from different phytopathogens *via* antibiosis and induction of host resistance. Phytopathology 109(3):375–383. https://doi.org/10.1094/phyto-07-18-0262-r

Rosenblueth M, Martínez-Romero E (2006) Bacterial endophytes and their interactions with hosts. Mol Plant-Microbe Interact 19(8):827–837

Rout ME, Southworth D (2013) The root microbiome influences scales from molecules to ecosystems: the unseen majority. Am J Bot 100(9):1689–1691

Rouws LFM, Meneses CHSG, Guedes HV, Vidal MS, Baldani JI, Schwab S (2010) Monitoring the colonization of sugarcane and rice plants by the endophytic diazotrophic bacterium *Gluconacetobacter diazotrophicus* marked with gfp and gusA reporter genes. Lett Appl Microbiol 51(3):325–330

Rudrappa T, Czymmek KJ, Paré PW, Bais HP (2008) Root-secreted malic acid recruits beneficial soil bacteria. Plant Physiol 148(3):1547–1556

Ryan RP, Germaine K, Franks A, Ryan DJ, Dowling DN (2008) Bacterial endophytes: recent developments and applications. FEMS Microbiol Lett 278(1):1–9

Saikkonen K, Faeth SH, Helander M, Sullivan TJ (1998) Fungal endophytes: a continuum of interactions with host plants. Annu Rev Ecol Syst 29(1):319–343

Sanchez-Vallet A, Mesters JR, Thomma BP (2015) The battle for chitin recognition in plant-microbe interactions. FEMS Microbiol Rev 39:171–183

Santoyo G, Moreno-Hagelsieb G, Orozco-Mosqueda M, Glick BR (2016) Plant growth-promoting bacterial endophytes. Microbiol Res 183:92–99
Sarkar J, Ray A, Chakraborty B, Chakraborty U (2016) Antioxidative changes in *Citrus reticulata* L. induced by drought stress and its effect on root colonization by arbuscular mycorrhizal fungi. Eur J Biol Res 6:1–13
Sessitsch A, Hardoim P, Döring J, Weilharter A, Krause A, Woyke T et al (2012) Functional characteristics of an endophyte community colonizing rice roots as revealed by metagenomic analysis. Mol Plant-Microbe Interact 25:28–36
Sharma RK, Agrawal M, Marshall F (2006) Heavy metal contamination in vegetables grown in wastewater irrigated areas of Varanasi, India. Bull Environ Contam Toxicol 77(2):312–318
Shentu XP, Shi YJ, Yu XP (2010) Identification of an endophytic isolate from Ilex cornuta and its antagonism against rhizoctonia solani. Chin J Pest Sci 12:173–177
Shimizu T, Nakano T, Takamizawa D, Desaki Y, Ishii-Minami N, Nishizawa Y, Minami E, Okada K, Yamane H, Kaku H, Shibuya N (2010) Two LysM receptor molecules, CEBiP and OsCERK1, cooperatively regulate chitin elicitor signaling in rice. Plant J 64(2):204–214
Shoresh M, Harman GE, Mastouri F (2010) Induced systemic resistance and plant responses to fungal biocontrol agents. Annu Rev Phytopathol 48:21–43
Shymanovich T, Saari S, Lovin ME, Jarmusch AK, Jarmusch SA, Musso AM, Charlton ND, Young CA, Cech NB, Faeth SH (2015) Alkaloid variation among epichloid endophytes of sleepygrass (*Achnatherum robustum*) and consequences for resistance to insect herbivores. J Chem Ecol 41(1):93–104
Singh SP, Gaur R (2016) Evaluation of antagonistic and plant growth promoting activities of chitinolytic endophytic actinomycetes associated with medicinal plants against *Sclerotium rolfsii* in chickpea. J Appl Microbiol 121:506–518
Singh SP, Gaur R (2017) Endophytic *Streptomyces* spp. underscore induction of defense regulatory genes and confers resistance against *Sclerotium rolfsii* in chickpea. Biol Control 104:44–56
Singh RK, Tiwari SP, Rai AK, Mohapatra TM (2011) Cyanobacteria: an emerging source for drug discovery. J Antibiot 64(6):401–412
Solanki MK, Wang F-Y, Wang Z, Li C-N, Lan T-J, Singh RK, Singh P, Yang L-T, Li Y-R (2019) Rhizospheric and endospheric diazotrophs mediated soil fertility intensification in sugarcane-legume intercropping systems. J Soils Sediments 19:1911–1927
Somjaipeng S, Medina A, Kwasna H, Ortiz OJ, Magan N (2015) Isolation, identification, and ecology of growth and taxol production by an endophytic strain of *Paraconiothyrium* variabile from English yew trees (*Taxus baccata*). Fungal Biol 119:1022–1031
Souza A, Cruz JC, Sousa NR, Procópio ARL, Silva GF (2014) Endophytic bacteria from banana cultivars and their antifungal activity. Embrapa Amazônia Ocidental-Artigo em periódico indexado (ALICE)
Spence C, Alff E, Johnson C, Ramos C, Donofrio N, Sundaresan V, Bais H (2014) Natural rice rhizospheric microbes suppress rice blast infections. BMC Plant Biol 14(1):130
Srivastava S, Srivastava AK, Sablok G, Deshpande TU, Suprasanna P (2015) Transcriptomics profiling of Indian mustard (*Brassica juncea*) under arsenate stress identifies key candidate genes and regulatory pathways. Front Plant Sci 6:646
Stirling GR (2017) Biological control of plant-parasitic nematodes. In: Diseases of nematodes. CRC Press, Boca Raton, pp 103–150
Strobel GA (2003) Endophytes as sources of bioactive products. Microbes Infect 5(6):535–544
Strobel G (2018) The emergence of endophytic microbes and their biological promise. J Fungi 4:E57. https://doi.org/10.3390/jof4020057
Strobel G, Daisy B (2003) Bioprospecting for microbial endophytes and their natural products. Microbiol. Microbiol Mol Biol Rev 67:491–502
Subramanian P, Mageswari A, Kim K, Lee Y, Sa T (2015) Psychrotolerant endophytic Pseudomonas sp. strains OB155 and OS261 induced chilling resistance in tomato plants (*Solanum lycopersicum* Mill.) by activation of their antioxidant capacity. Mol Plant-Microbe Interact 28(10):1073–1081

Sun C, Johnson J, Cai D, Sherameti I, Oelmüeller R, Lou B (2010) *Piriformospora indica* confers drought tolerance in Chinese cabbage leaves by stimulating antioxidant enzymes, the expression of drought-related genes and the plastid-localized CAS protein. J Plant Physiol 167:1009–1017

Taghavi S, van der Lelie D, Hoffman A, Zhang YB, Walla MD, Vangronsveld J et al (2010) Genome sequence of the plant growth promoting endophytic bacterium *Enterobacter* sp. 638. PLoS Genet 6:e1000943

Talbot JM, Treseder KK (2012) Interactions among lignin, cellulose, and nitrogen drive litter chemistry–decay relationships. Ecology 93(2):345–354

Thomas P, Sekhar AC (2014) Live cell imaging reveals extensive intracellular cytoplasmic colonization of banana by normally non-cultivable endophytic bacteria. AoB Plants 6

Timmusk S, Behers L, Muthoni J, Muraya A, Aronsson AC (2017) Perspectives and challenges of microbial application for crop improvement. Front Plant Sci 8:49

Tisserant E, Malbreil M, Kuo A, Kohler A, Symeonidi A, Balestrini R, Charron P, Duensing N, dit Frey NF, Gianinazzi-Pearson V, Gilbert LB (2013) Genome of an arbuscular mycorrhizal fungus provides insight into the oldest plant symbiosis. Proc Natl Acad Sci U S A 110(50):20117–20122

Trda L, Boutrot F, Claverie J, Brule D, Dorey S, Poinssot B (2015) Perception of pathogenic or beneficial bacteria and their evasion of host immunity: pattern recognition receptors in the frontline. Front Plant Sci 6:219

Tripathi P, Singh PC, Mishra A, Chaudhry V, Mishra S, Tripathi RD, Nautiyal CS (2013) *Trichoderma* inoculation ameliorates arsenic induced phytotoxic changes in gene expression and stem anatomy of chickpea (*Cicer arietinum*). Ecotoxicol Environ Saf 89:8–14

Truyens S, Weyens N, Cuypers A, Vangronsveld J (2015) Bacterial seed endophytes: genera, vertical transmission and interaction with plants. Environ Microbiol Rep 7(1):40–50

Turner TR, James EK, Poole PS (2013) The plant microbiome. Genome Biol 14:209

Van Bael SA, Seid MA, Wcislo WT (2012) Endophytic fungi increase the processing rate of leaves by leaf-cutting ants (Atta). Ecol Entomol 37:318–321

van Loon LC, Geraats BP, Linthorst HJ (2006) Systemic acquired resistance and induced systemic resistance. Trends Plant Sci 4(11):184–191

van Overbeek LS, Saikkonen K (2016) Impact of bacterial–fungal interactions on the colonization of the endosphere. Trends Plant Sci 21(3):230–242

Vandenkoornhuyse P, Quaiser A, Duhamel M, Le Van A, Dufresne A (2015) The importance of the microbiome of the plant holobiont. New Phytol 206(4):1196–1206

Vardharajula S, Zulfikar Ali S, Grover M, Reddy G, Bandi V (2011) Drought-tolerant plant growth promoting *Bacillus* spp.: effect on growth, osmolytes, and antioxidant status of maize under drought stress. J Plant Interact 6(1):1–14

Vidal S, Jaber LR (2015) Entomopathogenic fungi as endophytes: plant–endophyte–herbivore interactions and prospects for use in biological control. Curr Sci 14:46–54

Vigani G, Rolli E, Marasco R, Dell'Orto M, Michoud G, Soussi A, Raddadi N, Borin S, Sorlini C, Zocchi G, Daffonchio D (2018) Root bacterial endophytes confer drought resistance and enhance expression and activity of a vacuolar H^+-pumping pyrophosphatase in pepper plants. Environ Microbiol 23:44–59

Vorholt JA (2012) Microbial life in the phyllosphere. Nat Rev Microbiol 10(12):828

Vu T, Hauschild R, Sikora RA (2006) *Fusarium oxysporum* endophytes induced systemic resistance against *Radopholus similis* on banana. Nematology 8(6):847–852

Waller F, Achatz B, Baltruschat H, Fodor J, Becker K, Fischer M, Heier T, Hückelhoven R, Neumann C, von Wettstein D, Franken P (2005) The endophytic fungus *Piriformospora indica* reprograms barley to salt-stress tolerance, disease resistance, and higher yield. Proc Natl Acad Sci 102(38):13386–13391

Walters DR, Havis ND, Oxley SJP (2008) *Ramularia collo-cygni*: the biology of an emerging pathogen of barley. FEMS Microbiol Lett 279:1–7

Wang Z, Solanki MK, Pang F, Singh RK, Yang L-T, Li Y-R, Li H-B, Zhu K, Xing Y-X (2017) Identification and efficiency of a nitrogen-fixing endophytic actinobacterial strain from sugarcane. Sugar Tech 19:492–500

Wani ZA, Ashraf N, Mohiuddin T, Riyaz-Ul-Hassan S (2015) Plant-endophyte symbiosis, an ecological perspective. Appl Microbiol Biotechnol 99(7):2955–2965

Waqas M, Khan AL, Hamayun M, Shahzad R, Kang SM, Kim JG et al (2015) Endophytic fungi promote plant growth and mitigate the adverse effects of stem rot: an example of *Penicillium citrinum* and *Aspergillus terreus*. J Plant Interact 10:280–287

Webster PJ, Magana VO, Palmer TN, Shukla J, Tomas RA, Yanai MU, Yasunari T (1998) Monsoons: processes, predictability, and the prospects for prediction. J Geophys Res Oceans 103(C7):14451–14510

Whipps J, Hand P, Pink D, Bending GD (2008) Phyllosphere microbiology with special reference to diversity and plant genotype. J Appl Microbiol 105(6):1744–1755

White JF, Kingsley KI, Kowalski KP, Irizarry I, Micci A, Soares MA, Bergen MS (2018) Disease protection and allelopathic interactions of seed-transmitted endophytic pseudomonads of invasive reed grass (*Phragmites australis*). Plant Soil 422(1–2):195–208

Win KT, Tanaka F, Okazaki K, Ohwaki Y (2018) The ACC deaminase expressing endophyte *Pseudomonas* spp. Enhances NaCl stress tolerance by reducing stress-related ethylene production, resulting in improved growth, photosynthetic performance, and ionic balance in tomato plants. Plant Physiol Biochem 127:599–607

Wu X, Monchy S, Taghavi S, Zhu W, Ramos J, Van der Lelie D (2011) Comparative genomics and functional analysis of niche specific adaptation in *Pseudomonas putida*. FEMS Microbiol Rev 35:299–323

Wu P, Wu Y, Liu CC, Liu LW, Ma FF, Wu XY, Wu M, Hang YY, Chen JQ, Shao ZQ, Wang B (2016) Identification of arbuscular mycorrhiza (AM)-responsive microRNAs in tomato. Front Plant Sci 7:429

Xu X, He Q, Chen C, Zhang C (2016) Differential communications between fungi and host plants revealed by secretome analysis of phylogenetically related endophytic and pathogenic fungi. PLoS One 11:e0163368

Yang R, Fan X, Cai X, Hu F (2015) The inhibitory mechanisms by mixtures of two endophytic bacterial strains isolated from *Ginkgo biloba* against pepper *phytophthora* blight. Biol Control 85:59–67

Yao YQ, Lan F, Qiao YM, Wei JG, Huang RS, Li LB (2017) Endophytic fungi harbored in the root of *Sophora tonkinensis* Gapnep: diversity and biocontrol potential against phytopathogens. Microbiol Open 6(3):e00437

Yasuda M, Isawa T, Shinozaki S, Minamisawa K, Nakashita H (2009) Effects of colonization of a bacterial endophyte, *Azospirillum* sp. B510, on disease resistance in rice. Biosci Biotechnol Biochem 73(12):2595–2599

Yi HS, Yang JW, Ryu CM (2013) ISR meets SAR outside: additive action of the endophyte *Bacillus pumilus* INR7 and the chemical inducer, benzothiadiazole, on induced resistance against bacterial spot in field-grown pepper. Front Plant Sci 4:122

Zaiyou J, Li M, Hu X (2017) An endophytic fungus efficiently producing paclitaxel isolated from taxus wallichiana var. mairei. Medicine 96(27):e7406

Zeidler D, Zahringer U, Gerber I, Dubery I, Hartung T, Bors W et al (2004) Innate immunity in *Arabidopsis thaliana*: lipopolysaccharides activate nitric oxide synthase (NOS) and induce defense genes. Proc Natl Acad Sci U S A 101:15811–15816

Zouari I, Jlaiel L, Tounsi S, Trigui M (2016) Biocontrol activity of the endophytic *Bacillus amyloliquefaciens* strain CEIZ-11 against *Pythium aphanidermatum* and purification of its bioactive compounds. Biol Control 100:54–62

Role of Biotechnology in the Exploration of Soil and Plant Microbiomes

13

Akhilendra Pratap Bharati, Ashutosh Kumar, Sunil Kumar, Deepak K. Maurya, Sunita Kumari, Dinesh K. Agarwal, and S. P. Jeevan Kumar

Abstract

The agricultural production security is a major challenge under global climatic changes as the agricultural production does not solely depend upon the abiotic factors like temperature and moisture but is also equally affected by the biotic factor present in rhizospheric zones known as soil-plant microbiomes. The importance of soil-plant microbiomes in agricultural production has been revealed by reviews and literatures. Rhizospheric microbiomes hold great ability to bestow the crops sustainable and to encourage the application of bio-formulation carrying potential microbiomes, which introduces additional or optional approaches to develop agricultural practices and finally enhances crop productivity. In current chapter, a brief outline of functional activities of rhizospheric microbiomes within their habitat have been focused and majorly on available modern approaches of biotechnological techniques and allied sciences used in identifications, exploration, and taxonomy in soil-plant microbiomes. Further, to conclude the theme of microbiomes exploration and identification, an endeavor has been made to cover some of the highly used tools of bioinformatics.

Keywords

Agriculture · Biotechnology · Identification · Microbiomes · Sequencing

Authors Akhilendra Pratap Bharati and Ashutosh Kumar have been equally contributed to this chapter.

A. P. Bharati · S. Kumar
ICAR-National Bureau of Agriculturally Important Microorganisms, Mau, Uttar Pradesh, India

A. Kumar (✉) · S. Kumari · D. K. Agarwal · S. P. J. Kumar
ICAR-Indian Institute of Seed Science, Mau, Uttar Pradesh, India

D. K. Maurya
Agarkar Research Institute, Pune, Maharashtra, India

© Springer Nature Singapore Pte Ltd. 2020

M. K. Solanki et al. (eds.), *Phytobiomes: Current Insights and Future Vistas*,
https://doi.org/10.1007/978-981-15-3151-4_13

13.1 Introduction

Intense agricultural practices have consequences in the enhanced agriculture production of crops like wheat, rice, and maize (Zhang et al. 2011; Deb et al. 2015; Kumar et al. 2019). Higher production of rice and wheat ensured food security for continuously growing world population under changing climatic conditions (Tilman 1999; West et al. 2014). Besides food security, there has been noticed an increasing awareness of undesirable environmental threats arising from the extensive application of modern agriculture techniques. Some of the highlighted impacts of intense agriculture are greenhouse gas emission, nutrient leaching, increase soil degradation, and worst part of modern agriculture techniques which is based on the incorporation of synthetic fertilizers and pesticides that deteriorate the ecosystem gradually (Matson et al. 1997; Kumar et al. 2018; Sabarwal et al. 2018). However, along with the maintenance of food security, it is also essential to the standardization of the modern agriculture practicing protocols and techniques, which stabilized the standard of soil quality through the inclusion of microbes which provide valuable ecosystem services. The presence of all soil nutrients in farmland is not only a single parameter for the healthy soil, but abundance of viable entities like microbes which perform various functions in the rhizosphere in order to maintain its fertility and soil sustainability is also equally important (Bhattacharyya et al. 2016; Singh 2015; Mosttafiz et al. 2012).

Soil is a heterogeneous complex system on earth, which provides residence to great diverse organisms ranging from unicellular to multicellular organisms. It has been concluded in the literature that single gram of soil contains an abundance of viable entities (bacteria) close to 10^5–10^6 different taxonomic groups with huge microbial diversity (Griffiths et al. 2016). The soil microbiome is a collection of microorganism population inhabiting in the animals, plant, rhizosphere, or food which includes fungi, bacteria, archaea, and viruses and other genera (Kurokawa et al. 2007; Smith et al. 2013; Coleman-Derr et al. 2016). Mainly, bacteria and fungi are core regulators of various ecosystem processes and play crucial roles in nutrient cycling (Bardgett et al. 2008). Microorganisms have a diverse responsibility in the soil like the decomposition of various compounds and transformation of nonsoluble nutrients to a soluble form and fixation of different nutrients, resulting in the promotion of plant growth (Bender et al. 2016).

Soil microbiomes are highly abundant in rhizospheric zones and equally diverse in nature (Blazewicz et al. 2013). They perform various beneficial functions in the soil, and their worthy functions need to be assessed, which demand their accurate identification. From the beginning of microbiology, identification of microbial strains was a very tedious and time-taking process as identification of microbes was purely based on the physical and morphological appearance. Microbial characterization based on the morphology was not at all precise and lacked accuracy. In the beginning year of microbiomes identification, in 18th century, the microbiologist of that era has started microbial characterization based on metabolic pathways. Further, soil microbiomes especially unicellular entity like bacteria showed enormous diversity about their metabolisms and cell structures, and hence, identification based on morphological and biochemical means becomes worthless (Gevers et al. 2005; Achtman and Wagner 2008).

In spite of deficiency in a consistent species-level classification, the well-timed categorization and identification of microorganisms remain to be significant in various sciences related to environmental and soil monitoring or clinical diagnosis (Liu et al. 2010). Especially, the current advances of molecular biology covering modern techniques in the field of genomics and proteomics which are potential to put forward attractive and easy alternatives over conventional microbiological procedures, which are used for the exploration, characterization, and identification of microbiomes (Hugenholtz, and Tyson 2008; Martínez-Porchas and Vargas-Albores 2017). From the last three decades, modern sciences like genomics and proteomics with the inclusion of computational science and informatics science have greatly strengthened and provide quick multidimensional data for quick microbial identification. The sequencing technologies coupled with bioinformatics has not only revealed rapid identification of microbiomes (Martínez-Porchas and Vargas-Albores 2017; Marshall et al. 2018) but also made possible rapid gene annotation of microbiomes within a short period of times (Liu et al. 2007; Logue et al. 2008).

Soil microbiomes communities and their related functions principally conclude the productivity of agriculture crops (Van Der Heijden et al. 2008; Dias et al. 2015; Kumar and Verma 2019). The present compositions of the rhizospheric microbiomes are the most indispensable factors in determining the bacterial and fungal association linked to plant roots (Boe et al. 2000; Berendsen et al. 2018), and hence the identification of microbiomes irrespective of origin is essential for the better understanding of their functions. Molecular identification is highly accurate and hardly creates cumbersome and also makes possible manipulation of microbial treasure for the better uses for human society. In the current chapter, the tools and techniques related to the soil microbiomes exploration and identification in the biotechnology era have been covered comprehensively. The current chapter also covered the putative role of bioinformatics to decipher the molecular taxonomy and phylogenetic studies.

13.1.1 Soil Microbiomes Interaction with Plant

In the last two decades, several milestones have been covered with respect to microbiome identification of different origin and the phrase "plant microbiome" or "rhizospheric microbiomes" has gained considerable attention, as rhizospheric microbiome affects the soil fertility in multiple ways. Soil microbes provide the protection to the plant against plant pathogens and promote goodness to the plant and finally enhance the crop productivity. The plant microbiome holds the dissimilar functional gene pool, linked from viruses, prokaryotes, and eukaryotes and also coupled with various residents of a plant host (Rout and Southworth 2013). The hot spot location of microbiomes and plant interaction is the rhizosphere which is always under the inclination of plant roots via accumulation of exudates, mucilages, and sloughed cells (Uren 2001; Bais et al. 2006; Moe 2013). Hence, plant roots are potential to manipulate the soil and to inhabit microbiomes. Further, the rhizosphere microbiomes can promote plant growth through the secretion of regulatory compounds like siderophore, hormone production as indole acetic acid, gibberellic acid, and stress regulation through 1-aminocyclopropane-1-carboxylate (ACC) deaminase activity (Shi et al. 2011; Philippot et al. 2013; Spence et al. 2014; Sarkar et al. 2018).

The rhizospheric microbiomes have inverse or direct (beneficial) influence on plant growth promotion and wellness. Rhizospheric microbiomes directly affected the attack of pathogens and mutualistic benefits through the decomposition of complex compounds and nutrient solubilization like phosphates, potassium, and other nutrient cycling (Glick 1995); discharge of plant growth hormones like auxin and gibberellins (Narula et al. 2006; Ortíz-Castro et al. 2008; Ali et al. 2009; Mishra et al. 2009); anti-pathogens activities (Kloepper et al. 2004); and stimulation and induction of the plant immune system through induced systemic resistance (Ramamoorthy et al. 2001; Vessey 2003; Rudrappa et al. 2010). The recent literature substantiates that plant host and developmental stage have a noteworthy influence on determining of the rhizospheric microbiome (Peiffer et al. 2013; Chaparro et al. 2014; del Carmen Orozco-Mosqueda et al. 2018).

13.2 Microbial Identification

There are diverse microorganisms which exhibit plant growth-promoting (PGP) activities and help in the crop growth and productions (Meena et al. 2017; Sahoo et al. 2019). Their application has lead to the use of microorganisms in agriculture, which has further needed the identification of these organisms having PGP activities. This has given a separate branch of study called systematics or taxonomy. Systematics or taxonomy consists of classification, nomenclature, and identification, where identification is the practical application of the taxonomy. Classification can be considered as an art, but the identification is purely a science (Cowan 1965). There are several methods for identification of microbes ranging from morphological studies to molecular studies (Zhang et al. 2010a; Sahoo et al. 2019). First of all, it is very important to isolate organisms from the pool of the microbes, and this purified single colony should be identified for accurate nomenclature with molecular-based methods (Colombo et al. 2009; Armstrong 2007; Wagner and Haider 2012).

Initially, the organisms were identified based on pathogenicity, whether they are pathogenic or nonpathogenic. Nungester (1963) started microbe identification based on fewer objectives: (a) susceptibility to the antimicrobial drug, (b) based on prognostic value, (c) potential danger to the people in contact to the infected person, and (d) source of infection, etc. Initially, these objectives were used for clinical microbiologist later amended and further utilized for the normal identification irrespective of the field. Before this, the identification involves a comparison of an unknown organism to known organism and eventually giving the name to the former. After that, more progressive methods (Cowan 1965) were used which aim to determine a few fundamental characteristics, so that the isolates can be placed in the same genera (Drahos 1991; Armstrong 2007; Colombo et al. 2009; Wagner and Haider 2012). Some of the modern techniques used in microbiome identification have been discussed in a further section.

13.2.1 Methods of Microbiomes Identification

The methods of identification of microbe can be categorized in (a) classical or traditional and (b) modern methods based on instrument and technique used (Fig. 13.1). Traditional methods rely on the phenotypical characterization of the microbe, which can be observed by the naked eye, including morphological and biochemical analysis. *Psychrobacter* spp., a nonmotile bacteria associated with freshwater fish spoilage, were characterized phenotypically (García-López et al. 2004). Morphological characteristics involve the morphology, shape, size, opacity, color, and transparency of the colony. These characters can be observed after incubating the microbe for a certain period on the solid agar plate. The primary distinguishing characteristics are whether it can grow in an aerobic or anaerobic environment. However, these phenotypic characterizations are not sound very much scientific and sensitive at the level of strain differentiation (Sagan et al. 1994; Collazo et al. 2005; Vernière et al. 1998). Figure 13.1 has briefly described the most common methods generally used in the microbial characterization.

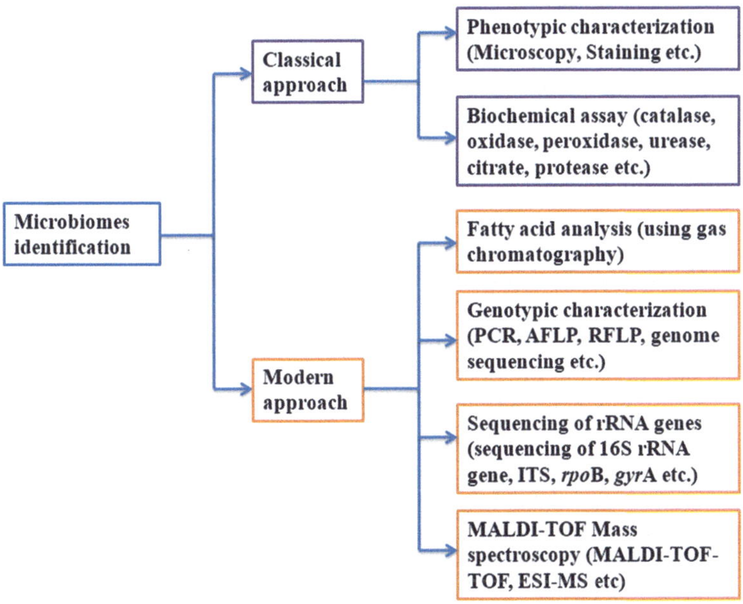

Fig. 13.1 Methods used to characterize genotypic and phenotypic characteristics and requirement of the characterization

Initially, microscopy was developed mainly for the phenotypical characterization. The phenotypical characterization includes the shape of the microbe whether it is a bacillus, coccus, spiral, or some other shapes and the study of the mobility of the microbe by identifying the flagella whether it is mono-, bi-, or multi-flagellated. Microbes can also be analyzed through the staining pattern of the bacteria, whether it is gram-negative or gram-positive. The presence of spore, capsule, inclusion body, and many more things can be detected. Some of the bacteria have a specific pigmentation and fluorescence, which can be utilized as a characteristic for identification (Giana et al. 2003). There is a variety of staining, i.e., gram staining, endospore staining, and Ziehl-Neelsen staining used for characterization (Hatamoto et al. 2007; Hugon et al. 2017). Besides the light and compound microscopy, there are several other microscopes, which have been proven advantageous in the characterization of the microbes, for example, confocal microscopy is being used to see the organelle of the microbe; besides this, for the greater resolution, SEM (scanning electron microscopy) and TEM (transmission electron microscopy) are being utilized (Hatamoto et al. 2007; Golding et al. 2016).

A variety of biochemical tests like catalase, oxidase, peroxidase, urease, citrate, protease, and others specify the organism at the genus and species level. In *Enterobacteriaceae* and *Pseudomonas* spp., identification of carbapenemase types using a biochemical test is one of the examples (Dortet et al. 2012). Similarly, the nitrate reduction test can be utilized in the identification of *Mycobacteria* (Virtanen 1960). Dichotomous identification is also one of the methods, where the whole kingdom is divided into two parts based on characteristics; as we choose multiple characters, it goes narrower (Noguerola and Blanch 2008). Traditional methods are widely used but have major drawbacks. It is only applicable to an organism which can be cultured in vitro, and there are various microbes which do not fit to a particular genus because of the unique biochemical properties. The modern methods mostly rely on the molecular biology, which can detect similarities and dissimilarities even at the molecular level, no need to be dependent on live culture, and can reveal minute differences between organisms which can escape in classical methods.

13.2.2 Molecular Techniques for Microbiomes Identification

Conventional methods of identification are labor-intensive and time-consuming and are often not enough to differentiate the species of the same genus (Verma et al. 2019). These limits are especially important for the medical diagnostic point of view. Rapid and proper identification of the microbe is essential in medical science. Therefore, there is the requirement of the alternating methods for quick identification. Nowadays, molecular taxonomy relies on the 16S ribosomal RNA gene sequencing and several PCR-based methods. The 16S ribosomal RNA gene sequencing has led to microbe identification at species and subspecies level (Adekambi and Drancourt 2004; Chen et al. 2000). These PCR-based sequencing methods are highly sensitive and reproducible. These sequencing methods also reduce the time and labor as it requires the genomic DNA of the microbe and PCR reaction.

Similarly, microarray-based methods use the probe specific to the organism. In the microarray, there are a number of probes (16S rRNA probes) from different organisms, so from a single microarray, it can identify several microorganisms in a microbiome as each probe corresponds to a specific microorganism (Peplies et al. 2003). The serology-based method also works on the same principle, but here we use the antigen-based enzyme-linked immunosorbent assay (ELISA) plate, and each antigen corresponds to the specific microbe. Besides the simple PCR, there are some recent advancements such as amplified fragment-length polymorphism (AFLP), repetitive element polymerase chain reaction, riboprinting, and multiplex PCR which have been proven useful in the identification (Emerson et al. 2008; Ricke et al. 2018).

13.2.3 Selection of Target Gene for Sequencing

Many gene targets have been identified in archaea, bacteria, fungi, and viruses. The gene sequence should have been conserved in the same species of bacteria or fungi. In bacterial identification, mostly 16S rRNA gene (16S rDNA) of ~1500 bp sequence is used for the identification of the species (Nocker et al. 2007). The 16S rDNA has conserved region as well as various regions (Fig. 13.2). In earlier studies, the scientist uses the whole 16S rDNA sequence for the identification. Nowadays, people are using the variable region only for the identification because they are more prone to the mutation and this region of DNA form only loops in the secondary structure of 16S ribosomal subunit (Fig. 13.2b). Some of the bacteria have multiple copies of the gene, while some microorganisms like *Bacillus cereus* and *Bacillus anthracis* have the same sequences. The multiple copies can make the interpretation difficult if every copy has a variation in base sequence. But, in clinical samples, the multiple copies are helpful in the amplification. Because of the difficulties in the interpretation in these cases, sometimes the sequence of the β subunit of the RNA polymerase II (*rpo B*) is used for the identification in place of 16S rDNA (Volokhov et al. 2012).

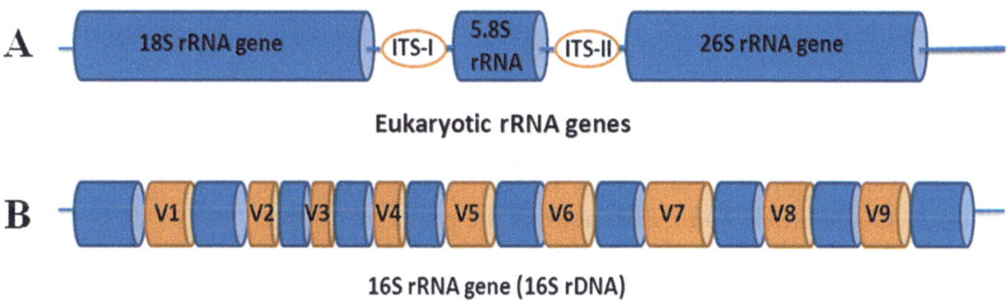

Fig. 13.2 (**a**) Schematic representation of ITS regions in fungi (eukaryotes) and (**b**) schematic representation of variable regions (orange) and conserved region (blue) present in 16S rRNA genes in bacteria

The *rpoB* gene has been proven useful in the identification of fast-growing bacteria like mycobacteria, i.e., *Mycobacterium chelonae* and *Mycobacterium abscessus*. These two species have indistinguishable 16S rRNA gene sequences but exhibit 13% sequence divergence with *rpoB* (Adékambi and Drancourt 2004; Poretsky et al. 2014; Nasiri et al. 2017). There are some other gene targets, which are the better option to distinguish the differences between species. These genes include tuf (elongation factor), *gyr* A (gyrase), gyr B, sod A (superoxide dismutase), HSP (heat-shock protein), etc. In the case of fungi, the scientist uses the internal transcribed region (ITS-I and ITS-II) sequence (Fig. 13.2a). These variable regions lie between 18S, 5.8S, and 26S rRNA genes. The variation between the ITS regions can be utilized for the identification of the fungi, such as *Saccharomyces*, *Candida*, *Trichosporon*, *Cryptococcus*, and *Aspergillus* (Chen et al. 2000; Ciardo et al. 2006). The ITS regions also have some limitations; because of that, the two variable domains D1 and D2 near the 5′ end of 28S ribosomal RNA genes (Dagar et al. 2011), elongation factor α (e.g., *Fusarium* sp.), and β-tabulin (e.g., *Phaeoacremonium* sp.) are being used as an alternative gene target (Stielow et al. 2015).

13.2.4 Gel-Based Identification and Characterization

Protein profiling is one of the recent methods for identification (Karlsson et al. 2018). For the protein profiling, first lyse the cell and then separate on the SDS-PAGE (polyacrylamide gel electrophoresis). Bacteria can be identify on the basis of the migration pattern of the protein on the gel, which can be compared with the reference. The 2DE (two-dimensional gel electrophoresis) is the fusion of the isoelectric focusing (IEF) and SDS-PAGE. In the isoelectric focusing, the proteins are first separated on the basis of the isoelectric point (charge). Then, the SDS-PAGE was performed to separate the protein on the basis of molecular weight. It means that the protein is first separated on the basis of charge and then on the basis of molecular weight. The 2DE map of different organisms can be stored, and a database can be prepared, which can be used as a reference and compared with test organisms (Malmström et al. 2002). The major drawback of this method is that it is labor-extensive and time-consuming and requires an ample amount of the protein. Proteome profile of organisms can be compared with the existing databases, and based on the comparison, the microorganism can be identified. Redmond and his colleague analyzed the exosporium of *Bacillus anthracis* spores by isolating the proteins from the outer casing of the spore using SDS-PAGE and analyzed the isolated protein. The team identified several proteins associated with the exosporium of *B. anthracis* (Redmond et al. 2004). The introduction of MALDI-TOF-MS greatly enhances this method. Using this method, the proteome of many microbes has been made available (Nouwens et al. 2000; Peng et al. 2005; Pieper et al. 2006).

13.2.5 Fingerprinting-Based Technology

Fingerprinting is the most widely used technology in the genomics for bacterial identification. The repetitive PCR, amplified fragment-length polymorphism (AFLP), random amplification of polymorphic DNA (RAPD), and multiplex PCR are widely used methods. These techniques utilize the PCR machine to amplify multiple copies of the DNA fragment using the defined set of primers, and then differences in these amplified fragments are utilized for the analysis and differentiation among the species of similar ecology (Lam et al. 2015). Similarly, restriction fragment-length polymorphism (RFLP) utilized a restriction enzyme. These methods are used to take advantage of the DNA polymorphism in the related organism, on the basis of which the organism can be differentiated in the mixed samples (Versalovic et al. 1994; Cocconcelli et al. 1995; Lin et al. 1996).

13.2.6 Microarray-Based Technology

Microarray is another technique to identify microbes at the species level and also impart the functional structure of a given microbial community (Bai et al. 2017; Thissen et al. 2019). The basic principle of this technique is to utilize specific probes which are spotted on a solid platform. This is called chip, which is further hybridized with fluorescently labeled DNA or RNA molecules from the microbial population. This genetic material (DNA & RNA) only hybridized with specific probe present on the specialized chip which can be detected. In the case of bacterial identification, several interactions of "polychip" have been used that utilize the small subunit of the ribosomal genes (Liu et al. 2001; Wilson et al. 2002). These chips are useful in the identification of a broad group of environmental bacteria (Loy et al. 2002).

13.2.7 Mass Spectrometry-Based Microbiomes Identification

Mass spectrometry is another latest next-generation tool for quick microbiome identification and classification. Matrix-assisted laser desorption ionization-time of flight mass spectroscopy (MALDI-TOF-MS) and electrospray ionization mass spectrometry (ESI-MS) have revolutionized the clinical identification of microorganism in a minute at the species level (Cherkaoui et al. 2010). Thomson (1899) first discovered this method to measure the mass to charge (m/z) ratio for the electron. Further, the mass spectrometry application was expended in physical and chemical characterization, including biological samples. In end of 19th century, the development of the soft ionization processes has made this useful for the large biological molecules (Fenn et al. 1989). MALDI-TOF is widely used in microbial identification. The bacterial cell can directly be utilized for the analysis. It can produce

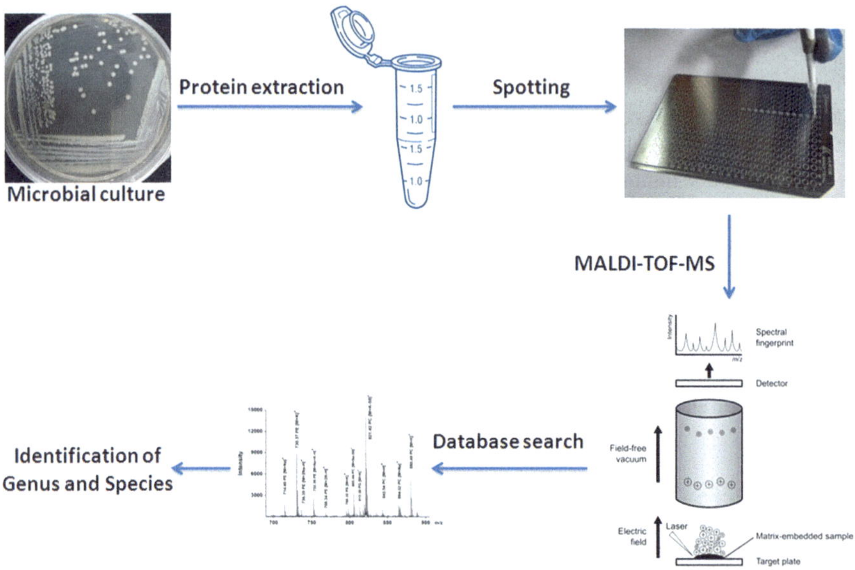

Fig. 13.3 Flowchart of microbial identification using MALDI-TOF-MS. The microbe was purified, and protein extraction was done from the whole cell lysate. The cell lysate was processed and mixed with the matrix, and the spot was done on the MS plate, which was further put inside the MALDI-TOF instrument. The spectra are analyzed on the database for the identification

the reproducible spectral pattern, which contains the information that can be used for the identification and characterization of the microbe. The basic procedures of the proteome analysis of the bacterial cell have been summarized in Fig. 13.3. For the ionization of the microbial sample, specific matrix has been used, which helps in the sample ionization. There are different types of matrix available for different types of sample. After the ionization of the microbe, particles are allowed to migrate in an electric field, and because of the field, each ionized particle acquires a path which gets detected by the analyzer. Finally, the analyzer gives spectra based on m/z ratio of the particle, which can be compared with the spectra of the known databases using an automated program (Dingle and Butler-Wu 2013). The most commonly used mass spectroscopy for the identification of microorganisms is MALDI-TOF (Krásný et al. 2013). Cell lysate can be utilized for protein profiling using MS. Using protein profiles to differentiate bacteria was first introduced by Cain et al. (1994).

13.2.8 Metagenomics

Metagenomics is derived from two major developments in biology. First is the development of next-generation DNA-sequencing technologies which has greatly enhanced capabilities for sequencing large meta-data sets, and second is the emerging appreciation for the importance of complex microbial communities in diverse environmental condition (Handelsman 2005; Petrosino et al. 2009). Metagenomics is the genome analysis of a population of microorganisms. It doesn't need the culturing of the organism. The genomic DNA can be isolated directly from the soil samples or the defined environment and cloned in the desired vector or plasmid. The idea of cloning DNA samples directly from the environment was first discovered by Pace et al. (1986), and then the next advance is the construction of the metagenomic DNA library of the isolated DNA from the mixture of the microbes and the library vector (Schmidt et al. 1991). Further, the clones can be analyzed or screened for the phylogenetic markers likes 16S ribosomal RNA genes, RecA, gyrA, rpoB, etc., or can be screened for specific traits like enzyme production and expression of specific genes (Stein et al. 1996).

Metagenomics is a tool that can provide an abundance of knowledge to understand the microbial population. It has further improved our understanding of many of the exotic and familiar habitats that are attracting the attention of microbial ecologists, including deep sea, hot springs, temperate, desert, and cold soils; frozen Antarctic lakes, plant rhizospheres, and phyllosphere; and fungi-lichen symbioses (Yin et al. 2018).

13.3 Bioinformatic Scope in Microbiome Identification

The modernistic approach of microbial identification is based on the application of modern technology, along with bioinformatics tools. Bioinformatics is an interdisciplinary field of science in which algorithms and software tools can be developed for a better understanding of biological data to accelerate and enhance biological research. In the recent past, the explosion in capabilities of high-throughput omics technologies, system biology, and deep-sequencing platforms is generating a huge amount of data which require computational approaches to manage and analyze (Jiang et al. 2013). There are several computational tools and techniques that are used to analyze the hunks of biological data more accurately and efficiently by automated processes. Bioinformatics played an important role in various fields, i.e., in sequencing and annotation of microbial genomes; it has also aided in the sequencing of observed mutations in the various organisms. These tools have proven useful in the comparative analysis of the microorganisms and understanding of the evolutionary aspect based on molecular biology (Zhang et al. 2010b). Hence, computational biology can be considered as a field of data science for solving problems in biology, medicine, and agriculture.

13.3.1 Next-Generation Sequencing of Microbes

Next-generation sequencing (NGS), a newly emerged technique, is being used a lot in various field of science. This technology has tremendous advantages over the Sanger sequencing method. The development in this technique has led to the high-throughput sequencing, which has further boost the several findings which were earlier seemed to be laborious and time-consuming by traditional methods. Millions of molecules can be sequenced simultaneously producing a huge amount of data. This technology is highly parallelized in such a way that it can analyze a huge amount of data and can give valuable information. Several algorithms have been written for data analysis. The NGS technology is high throughput, which is quick and economic enough to be considered as a tool for bacterial identification and characterization (Loman et al. 2012; Stahl and Lundeberg 2012; Srivastava et al. 2019a, b). Several algorithms have been written and assembled in a group, which further used as a tool for NGS data analysis. Short sequencing reads have limited power to resolve large repetitive regions, even within small microbial genomes (Chain et al. 2009; Nagarajan and Pop 2013). Generally, short-read technologies enable to resolve microbial genomes up to the high-quality draft standard (Treangen and Salzberg 2012), which sound applicable for understanding gene-coding potential, strain typing, or pan-genome analysis (Roberts et al. 2013). Draft genomes are the assemblies of fragmented short genomes which may contain incorrect gene calls, misassembled regions, and other artifacts. Fragmented assemblies are often attributed to repetitive DNA regions (such as rRNA operons) which are enormous in microbial genomes and present the greatest technical challenge to the assembly process, especially when the repetitive region is longer than the read lengths (Treangen and Salzberg 2012; Brown et al. 2014). Finished genome sequences are high quality by definition, represent more accurate genomic information, and can be adopted for model organisms and industrially important microbes (Fraser et al. 2002; Thomma et al. 2016; Boulund et al. 2018). The accurate analysis of NGS data revealed important clues in the quest for the treatment of various life-threatening diseases, improved crop varieties etc. There are several computational methods to analyze NGS data.

13.3.2 Data Cleaning and Preprocessing

The very first step is the cleaning or preprocessing of sequences of microbes retrieved from various sequencers. These sequences include the adapter and primers which can occur at both the ends of NGS reads. These adapter and primer sequences can hinder the correct alignment and mapping; because of that, it is important to remove these sequences. These ligated adapter sequences required the deletion from the sequence as the adapter can interfere with the original mapping of the reads and influence SNP calling and other downstream analyses. Various programs/applications are available for cleaning raw data by trimming low-quality reads and adapter contamination removal. Before analysis, these preliminary quality control checks can be applied. There are several open sources as well as commercial tools of data cleaning and processing. In Table 13.1, we have listed some of the online tools available.

Table 13.1 Tools of data cleaning and processing

Tools	Link
FastQC (Andrews 2010)	http://www.bioinformatics.bbsrc.ac.uk/projects/fastqc/
Fastx toolkit	http://hannonlab.cshl.edu/fastx_toolkit/download.html
Galaxy (Giardine et al. 2005)	https://usegalaxy.org/
Trimmomatic (Bolger et al. 2014)	http://www.usadellab.org/cms/?page=trimmomatic
CLC Bio (Workbench 2010)	Commercial

13.3.3 Genome Assembly

With the advent of many sequencing techniques, there is a massive overflow of genomic data in less time at low cost. Genome assembly is the process of making the original DNA sequence from a large number of short DNA sequences by putting all the sequences together. These sequences are aligned in such a way that the overlapping sequences read in contiguous sequences (contigs). This is called the de novo assembly, where there is no requirement of the reference genome. The most efficient assemblers for short-read sequences are typically those that employ de Bruijn graphs to produce an assembly (Compeau et al. 2011). Different algorithms are being used for the genome assembly. This algorithm takes all the pieces of DNA sequences, aligns them to one another, and searches for the overlapping sequences. These overlapping reads are merged with two paired reads, and assembly continues. Genome assembly is a difficult computational problem, as many genomes contain large numbers of repeat sequences. One of the first and most widely used de Bruijn graph assembler is the open-source program Velvet (Zerbino and Birney 2008). With further development to improve the resolution of repeats and scaffolding using paired-end and longer reads (Zerbino et al. 2009), Velvet remains one of the most used (and cited) assembler for bacterial genomes, being best suited to Illumina sequence reads (Velvet is included as the default assembler in the IlluminaMiSeq analysis suite).

There are two approaches to the assembly. First is de novo where no reference sequence is taken, and second is reference-guided where a reference genome is taken. Different types of software and algorithms are available for both the approaches of assembly. De novo approach relies on the fact that reads need to be assembled to generate a contiguous sequence either by overlap/layout/consensus graph (e.g., Celera Assembler, Arachne, CAP, and PCAP) or de Bruijn graph (e.g., Euler, Velvet, ABySS, AllPaths, SOAPdenovo, CLC Bio). Reference-guided assembly includes the use of reference genome to assemble reads into contigs. There are some assemblers which take reference genome as template to arrange reads helping in generating quick and accurate assembly (e.g., Velvet, DNASTAR's Lasergene Genomics Suite). This approach helps in identification of insertions and deletions.

13.3.4 Assembly Quality Assessment

After the genome assembly, it is very important to check the quality of the assembly. The maximum and minimum length of contigs/scaffolds, its length distribution, the total length of the assembly, etc., are the important criteria for the quality assessment. N50 and L50 are the important criteria as they represent the quality and size of the contigs or scaffolds in the assembly. There are several assembly tools available such as ABySS, Mimicking Intelligent Read Assembly (MIRA), CLC Bio Workbench, etc. The de novo assembly algorithm of CLC Genomics Workbench performs comprehensive support for a variety of data formats, including both short and long reads and mixing of paired reads (both insert size and orientation).

13.3.5 Gene Identification Tools

Gene prediction is one of the most important steps in understanding the sequenced genome of a species. At the computational point of view, gene prediction or gene findings are the process of identifying the regions of genomic DNA that encode genes. It includes protein-coding genes as well as RNA genes, but it may also include prediction of other functional elements, i.e., regulatory regions. At present, with comprehensive genome sequence and powerful computational resources available to the scientific communities, and gene finding has been redefined as a largely computational problem. The advances in bioinformatics research ensured the possibility to predict the function of a gene based on its sequence alone, and some of the popular tools have been listed in Table 13.2.

Table 13.2 Bioinformatics tools to predict the function of a gene based on its sequence

Name	Description	URL link
FRAMED	Find genes and frameshift in G+C rich prokaryotic sequences	https://omictools.com/framed-tool
GENIUS	Linking ORFs in complete genomes to protein 3D structures	http://genius.cbrc.jp/summary_linking.html
GENEID	Program to predict genes, exons, splice sites, and other signals along a DNA sequence	http://genome.crg.es/software/geneid/
GENEPARSER	Parse a DNA sequence into introns and exons	https://bio.tools/GeneParser
GeneMark.hmm	Gene prediction program for prokaryotes and eukaryotes	http://exon.gatech.edu/GeneMark/
GeneTack	Prediction of genes with frameshifts in prokaryotic genomes	http://83.149.211.146:23194/~ivan/cgi-bin/GeneTack/cgi/print_page.cgi?fn=home.html&title=Home
GLIMMER	Finding genes in microbial DNA	http://www.cbcb.umd.edu/software/glimmer/glimmer2.jun01.shtml

13.4 Conclusion

The classical identification of microbiomes was purely based on biochemical, which is insufficient to deal with the identification issue of diverse rhizospheric microbiomes. Hence, molecular level identification of rhizospheric microbiomes is essential and is covered only under the umbrella of biotechnology and its allied sciences. The advancement in the biotechnology tools has certainly played a crucial role in the accurate identification and phylogenic mapping of the diverse rhizospheric microbiomes as rhizospheric microbiomes have potential to increase crop productions. In last few decades, the application of biotechnology in a combination of with computational sciences like bioinformatics has refined the microbiomes phylogenic position. Advanced technologies like MALDI-TOF-MS, metagenomic studies of various environmental samples, nucleotides sequencing coupled with streams such as bioinformatics, molecular modeling, and docking have certainly led to several innovations and crucial progress in biological sciences. Further, biotechnology and bioinformatics helped in deciphering the structural and functional genes, which are involved in regulations of several physiological mechanisms of higher and lower organisms, including plant microbiomes.

References

Achtman M, Wagner M (2008) Microbial diversity and the genetic nature of microbial species. Nat Rev Microbiol 6(6):431–440

Adékambi T, Drancourt M (2004) Dissection of phylogenetic relationships among 19 rapidly growing *Mycobacterium* species by 16S rRNA, hsp65, sodA, recA and rpoB gene sequencing. Int J Syst Evol Microbiol 54(6):2095–2105

Ali B, Sabri AN, Ljung K, Hasnain S (2009) Auxin production by plant associated bacteria: impact on endogenous IAA content and growth of *Triticum aestivum* L. Lett Appl Microbiol 48(5):542–547

Andrews S (2010) FastQC: a quality control tool for high throughput sequence data

Armstrong D (2007) U.S. Patent No. 7,214,299. Washington, DC: U.S. Patent and Trademark Office

Bai R, Wang JT, Deng Y, He JZ, Feng K, Zhang LM (2017) Microbial community and functional structure significantly varied among distinct types of paddy soils but responded differently along gradients of soil depth layers. Front Microbiol 8:945. https://doi.org/10.3389/fmicb.2017.00945

Bais HP, Weir TL, Perry LG, Gilroy S, Vivanco JM (2006) The role of root exudates in rhizosphere interactions with plants and other organisms. Annu Rev Plant Biol 57:233–266

Bardgett RD, Freeman C, Ostle NJ (2008) Microbial contributions to climate change through carbon cycle feedbacks. ISME J 2(8):805

Bender SF, Wagg C, van der Heijden MG (2016) An underground revolution: biodiversity and soil ecological engineering for agricultural sustainability. Trends Ecol Evol 31(6):440–452

Berendsen RL, Vismans G, Yu K, Song Y, Jonge R, Burgman WP, Burmølle M, Herschend J, Bakker PA, Pieterse CM (2018) Disease-induced assemblage of a plant-beneficial bacterial consortium. ISME J 12(6):1496–1507

Bhattacharyya PN, Goswami MP, Bhattacharyya LH (2016) Perspective of beneficial microbes in agriculture under changing climatic scenario: a review. J Phytology 8:26–41

Blazewicz SJ, Barnard RL, Daly RA, Firestone MK (2013) Evaluating rRNA as an indicator of microbial activity in environmental communities: limitations and uses. ISME J 7(11):2061

Boe L, Danielsen M, Knudsen S, Petersen JB, Maymann J, Jensen PR (2000) The frequency of mutators in populations of *Escherichia coli*. Mutat Res Fundam Mol Mech Mutagen 448(1):47–55

Bolger AM, Lohse M, Usadel B (2014) Trimmomatic: a flexible trimmer for Illumina sequence data. Bioinformatics 30(15):2114–2120

Boulund F, Pereira MB, Jonsson V, Kristiansson E (2018) Computational and statistical considerations in the analysis of metagenomic data. In: Metagenomics. Academic, London, pp 81–102

Brown SD, Utturkar SM, Magnuson TS, Ray AE, Poole FL, Lancaster WA, Thorgersen MP, Adams MW, Elias DA (2014) Complete genome sequence of Pelosinus sp. strain UFO1 assembled using single-molecule real-time DNA sequencing technology. Genome Announc 2(5):e00881-14. https://doi.org/10.1128/genomeA.00881-14

Cain TC, Lubman DM, Weber WJ Jr, Vertes A (1994) Differentiation of bacteria using protein profiles from matrix-assisted laser desorption/ionization time-of-flight mass spectrometry. Rapid Commun Mass Spectrom 8(12):1026–1030

Chain PSG, Grafham DV, Fulton RS, Fitzgerald MG, Hostetler J, Muzny D, Ali J, Birren B, Bruce DC, Buhay C, Cole JR (2009) Genome project standards in a new era of sequencing. Science 326(5950):236–237

Chaparro JM, Badri DV, Vivanco JM (2014) Rhizosphere microbiome assemblage is affected by plant development. ISME J 8(4):790–803

Chen YC, Eisner JD, Kattar MM, Rassoulian-Barrett SL, LaFe K, Yarfitz SL, Limaye AP, Cookson BT (2000) Identification of medically important yeasts using PCR-based detection of DNA sequence polymorphisms in the internal transcribed spacer 2 region of the rRNA genes. J Clin Microbiol 38(6):2302–2310

Cherkaoui A, Hibbs J, Emonet S, Tangomo M, Girard M, Francois P, Schrenzel J (2010) Comparison of two matrix-assisted laser desorption ionization-time of flight mass spectrometry methods with conventional phenotypic identification for routine identification of bacteria to the species level. J Clin Microbiol 48(4):1169–1175

Ciardo DE, Schär G, Böttger EC, Altwegg M, Bosshard PP (2006) Internal transcribed spacer sequencing versus biochemical profiling for identification of medically important yeasts. J Clin Microbiol 44(1):77–84

Cocconcelli PS, Porro D, Galandini S, Senini L (1995) Development of RAPD protocol for typing of strains of lactic-acid bacteria and enterococci. Lett Appl Microbiol 21:376–379

Coleman-Derr D, Desgarennes D, Fonseca-Garcia C, Gross S, Clingenpeel S, Woyke T, North G, Visel A, Partida-Martinez LP, Tringe SG (2016) Plant compartment and biogeography affect microbiome composition in cultivated and native Agave species. New Phytol 209(2):798–811

Collazo C, Ramos PL, Chacón O, Borroto CJ, López Y, Pujol M, Thomma BP, Hein I, Borrás-Hidalgo O (2005) Phenotypical and molecular characterization of the tomato mottle Taino virus–*Nicotiana megalosiphon* interaction. Physiol Mol Plant Pathol 67(3–5):231–236

Colombo APV, Boches SK, Cotton SL, Goodson JM, Kent R, Haffajee AD, Socransky SS, Hasturk H, Van Dyke TE, Dewhirst F, Paster BJ (2009) Comparisons of subgingival microbial profiles of refractory periodontitis, severe periodontitis, and periodontal health using the human oral microbe identification microarray. J Periodontol 80(9):1421–1432

Compeau PE, Pevzner PA, Tesler G (2011) How to apply de Bruijn graphs to genome assembly. Nat Biotechnol 29(11):987–991

Cowan ST (1965) Principles and practice of bacterial taxonomy-a forward look. Microbiology 39(1):143–153

Dagar SS, Kumar S, Mudgil P, Singh R, Puniya AK (2011) D1/D2 domain of large-subunit ribosomal DNA for differentiation of *Orpinomyces* spp. Appl Environ Microbiol 77(18):6722–6725

Deb P, Kiem AS, Babel MS, Chu ST, Chakma B (2015) Evaluation of climate change impacts and adaptation strategies for maize cultivation in the Himalayan foothills of India. J Water Clim Change 6(3):596–614

del Carmen Orozco-Mosqueda M, del Carmen Rocha-Granados M, Glick BR, Santoyo G (2018) Microbiome engineering to improve biocontrol and plant growth-promoting mechanisms. Microbiol Res 208:25–31

Dias T, Dukes A, Antunes PM (2015) Accounting for soil biotic effects on soil health and crop productivity in the design of crop rotations. J Sci Food Agric 95(3):447–454

Dingle TC, Butler-Wu SM (2013) MALDI-TOF mass spectrometry for microorganism identification. Clin Lab Med 33(3):589–609

Dortet L, Poirel L, Nordmann P (2012) Rapid identification of carbapenemase types in *Enterobacteriaceae* and *Pseudomonas* spp. by using a biochemical test. Antimicrob Agents Chemother 56(12):6437–6440

Drahos DJ (1991) Methods for the detection, identification, and enumeration of microbes. In: Microbial ecology of leaves. Springer, New York, pp 135–157. https://doi.org/10.1007/978-1-4612-3168-4_7

Emerson D, Agulto L, Liu H, Liu L (2008) Identifying and characterizing bacteria in an era of genomics and proteomics. Bioscience 58(10):925–936

Fenn JB, Mann M, Meng CK, Wong SF, Whitehouse CM (1989) Electrospray ionization for mass spectrometry of large biomolecules. Science 246(4926):64–71

Fraser CM, Eisen JA, Nelson KE, Paulsen IT, Salzberg SL (2002) The value of complete microbial genome sequencing (you get what you pay for). J Bacteriol 184(23):6403–6405

García-López I, Otero A, García-López ML, Santos JA (2004) Molecular and phenotypic characterization of nonmotile gram-negative bacteria associated with spoilage of freshwater fish. J Appl Microbiol 96(4):878–886

Gevers D, Cohan FM, Lawrence JG, Spratt BG, Coenye T, Feil EJ, Stackebrandt E, Van de Peer Y, Vandamme P, Thompson FL, Swings J (2005) Re-evaluating prokaryotic species. Nat Rev Microbiol 3(9):733–739

Giana HE, Silveira L, Zângaro RA, Pacheco MT (2003) Rapid identification of bacterial species by fluorescence spectroscopy and classification through principal components analysis. J Fluoresc 13(6):489–493

Giardine B, Riemer C, Hardison RC, Burhans R, Elnitski L, Shah P, Zhang Y, Blankenberg D, Albert I, Taylor J, Miller W (2005) Galaxy: a platform for interactive large-scale genome analysis. Genome Res 15(10):1451–1455

Glick BR (1995) The enhancement of plant growth by free-living bacteria. Can J Microbiol 41(2):109–117

Golding CG, Lamboo LL, Beniac DR, Booth TF (2016) The scanning electron microscope in microbiology and diagnosis of infectious disease. Sci Rep 6:26516. https://doi.org/10.1038/srep26516

Griffiths RI, Thomson BC, Plassart P, Gweon HS, Stone D, Creamer RE, Lemanceau P, Bailey MJ (2016) Mapping and validating predictions of soil bacterial biodiversity using European and national scale datasets. Appl Soil Ecol 97:61–68

Handelsman J (2005) Metagenomics: application of genomics to uncultured microorganisms. Microbiol Mol Biol Rev 69(1):195–195

Hatamoto M, Imachi H, Ohashi A, Harada H (2007) Identification and cultivation of anaerobic, syntrophic long-chain fatty acid-degrading microbes from mesophilic and thermophilic methanogenic sludges. Appl Environ Microbiol 73(4):1332–1340

Hugenholtz P, Tyson GW (2008) Microbiology: metagenomics. Nature 455(7212):481

Hugon P, Lagier JC, Colson P, Bittar F, Raoult D (2017) Repertoire of human gut microbes. Microb Pathog 106:103–112

Jiang Q, Wang G, Jin S, Li Y, Wang Y (2013) Predicting human microRNA-disease associations based on support vector machine. Int J Data Min Bioinform 8(3):282–293

Karlsson R, Gonzales-Siles L, Gomila M, Busquets A, Salvà-Serra F, Jaén-Luchoro D, Jakobsson HE, Karlsson A, Boulund F, Kristiansson E, Moore ER (2018) Proteotyping bacteria: characterization, differentiation and identification of *Pneumococcus* and other species within the Mitis group of the genus *Streptococcus* by tandem mass spectrometry proteomics. PLoS One 13(12):e0208804

Kloepper JW, Ryu CM, Zhang S (2004) Induced systemic resistance and promotion of plant growth by *Bacillus* spp. Phytopathology 94(11):1259–1266

Krásný L, Hynek R, Hochel I (2013) Identification of bacteria using mass spectrometry techniques. Int J Mass Spectrom 353:67–79

Kumar A, Verma JP (2019) The role of microbes to improve crop productivity and soil health. In: Ecological wisdom inspired restoration engineering. Springer, Singapore, pp 249–265

Kumar A, Kumar SPJ, Kumar M, Singh AN, Srivastava AK (2018) Assessment of soil urease activity and its kinetic parameters in Chilli rhizosphere under the influence of pesticides. Biopesticides Int 14(1):47–57

Kumar A, Ramesh KV, Singh C, Sripathy KV, Agarwal DK, Pal G, Kuchlan MK, Singh RK, Prabha R, Kumar SPJ (2019) Bioprospecting nutraceuticals from soybean (Glycine max) seed coats and cotyledons. Indian J Agric Sci 89(12):2064–2068

Kurokawa K, Itoh T, Kuwahara T, Oshima K, Toh H, Toyoda A, Takami H, Morita H, Sharma VK, Srivastava TP, Taylor TD (2007) Comparative metagenomics revealed commonly enriched gene sets in human gut microbiomes. DNA Res 14(4):169–181

Lam K, Chan G, Xin GZ, Xu H, Ku CF, Chen JP, Yao P, Lin HQ, Dong T, Tsim K (2015) Authentication of *Cordyceps sinensis* by DNA analyses: comparison of ITS sequence analysis and RAPD-derived molecular markers. Molecules 20(12):22454–22462

Lin JJ, Kuo J, Ma J (1996) A PCR-based DNA fingerprinting technique: AFLP for molecular typing of bacteria. Nucleic Acids Res 24:3649–3650

Liu WT, Mirzabekov AD, Stahl DA (2001) Optimization of an oligo nucleotide microchip for microbial identification studies: a non-equilibrium dissociation approach. Environ Microbiol 3:619–629

Liu H, Du Z, Wang J, Yang R (2007) Universal sample preparation method for characterization of bacteria by matrix-assisted laser desorption ionization-time of flight mass spectrometry. Appl Environ Microbiol 73(6):1899–1907

Liu YQ, Moy B, Kong YH, Tay JH (2010) Formation, physical characteristics and microbial community structure of aerobic granules in a pilot-scale sequencing batch reactor for real wastewater treatment. Enzym Microb Technol 46(6):520–525

Logue JB, Bürgmann H, Robinson CT (2008) Progress in the ecological genetics and biodiversity of freshwater bacteria. AIBS Bull 58(2):103–113

Loman NJ, Constantinidou C, Chan JZ, Halachev M, Sergeant M, Penn CW, Robinson ER, Pallen MJ (2012) High-throughput bacterial genome sequencing: an embarrassment of choice, a world of opportunity. Nat Rev Microbiol 10(9):599–606

Loy A, Lehner A, Lee N, Adamczyk J, Meier H, Ernst J, Schleifer KH, Wagner M (2002) Oligonucleotide microarray for 16S rRNA gene-based detection of all recognized lineages of sulfate-reducing prokaryotes in the environment. Appl Environ Microbiol 68(10):5064–5081

Malmström L, Malmström J, Marko-Varga G, Westergren-Thorsson G (2002) Proteomic 2DE database for spot selection, automated annotation, and data analysis. J Proteome Res 1(2):135–138

Marshall IP, Karst SM, Nielsen PH, Jørgensen BB (2018) Metagenomes from deep Baltic Sea sediments reveal how past and present environmental conditions determine microbial community composition. Mar Genomics 37:58–68

Martínez-Porchas M, Vargas-Albores F (2017) Microbial metagenomics in aquaculture: a potential tool for a deeper insight into the activity. Rev Aquac 9(1):42–56

Matson PA, Parton WJ, Power AG, Swift MJ (1997) Agricultural intensification and ecosystem properties. Science 277(5325):504–509

Meena VS, Meena SK, Verma JP, Kumar A, Aeron A, Mishra PK, Bisht JK, Pattanayak A, Naveed M, Dotaniya ML (2017) Plant beneficial rhizospheric microorganism (PBRM) strategies to improve nutrients use efficiency: a review. Ecol Eng 107:8–32

Mishra PK, Mishra S, Selvakumar G, Kundu S, Shankar Gupta H (2009) Enhanced soybean (*Glycine max* L.) plant growth and nodulation by *Bradyrhizobium japonicum*-SB1 in presence of *Bacillus thuringiensis*-KR1. Acta Agric Scand Sect B Soil Plant Sci 59(2):189–196

Moe LA (2013) Amino acids in the rhizosphere: from plants to microbes. Am J Bot 100(9):1692–1705

Mosttafiz SB, Rahman M, Rahman M (2012) Biotechnology: role of microbes in sustainable agriculture and environmental health. Internet J Microbiol 10(1):1–6

Nagarajan N, Pop M (2013) Sequence assembly demystified. Nat Rev Genet 14(3):157–167
Narula N, Deubel A, Gans W, Behl RK, Merbach WPSE (2006) Paranodules and colonization of wheat roots by phytohormone producing bacteria in soil. Plant Soil Environ 52(3):119–129
Nasiri MJ, Shahraki AH, Fooladi AAI, Dabiri H, Feizabadi MM (2017) rpoB gene sequencing for identification of rapidly growing mycobacteria. Arch Pediatric Infect Dis 5(2):e40001. https://doi.org/10.5812/pedinfect.40001.
Nocker A, Burr M, Camper AK (2007) Genotypic microbial community profiling: a critical technical review. Microb Ecol 54(2):276–289
Noguerola I, Blanch AR (2008) Identification of Vibrio spp. with a set of dichotomous keys. J Appl Microbiol 105(1):175–185
Nouwens AS, Cordwell SJ, Larsen MR, Molloy MP, Gillings M, Willcox MD, Walsh BJ (2000) Complementing genomics with proteomics: the membrane subproteome of *Pseudomonas aeruginosa* PAO1. Electrophoresis 21:3797–3809
Nungester WJ (1963) Contributions of microbiology and immunology to medicine and some unfinished business. Tex Rep Biol Med 21:315–330
Ortíz-Castro RANDY, Martínez-Trujillo MIGUEL, López-Bucio JOSE (2008) N-acyl-L-homoserine lactones: a class of bacterial quorum-sensing signals alter post-embryonic root development in *Arabidopsis thaliana*. Plant Cell Environ 31(10):1497–1509
Pace NR, Stahl DA, Lane DJ, Olsen GJ (1986) The analysis of natural microbial populations by ribosomal RNA sequences. In: Advances in microbial ecology. Springer, Boston, MA, pp 1–55
Peiffer JA, Spor A, Koren O, Jin Z, Tringe SG, Dangl JL, Buckler ES, Ley RE (2013) Diversity and heritability of the maize rhizosphere microbiome under field conditions. Proc Natl Acad Sci 110(16):6548–6553
Peng X, Xu C, Ren H, Lin X, Wu L, Wang S (2005) Proteomic analysis of the sarcosine-insoluble outer membrane fraction of *Pseudomonas aeruginosa* responding to ampicilin, kanamycin, and tetracycline resistance. J Proteome Res 4:2257–2265
Peplies J, Glöckner FO, Amann R (2003) Optimization strategies for DNA microarray-based detection of bacteria with 16S rRNA-targeting oligonucleotide probes. Appl Environ Microbiol 69(3):1397–1407
Petrosino JF, Highlander S, Luna RA, Gibbs RA, Versalovic J (2009) Metagenomic pyrosequencing and microbial identification. Clin Chem 55(5):856–866
Philippot L, Raaijmakers JM, Lemanceau P, Van Der Putten WH (2013) Going back to the roots: the microbial ecology of the rhizosphere. Nat Rev Microbiol 11(11):789
Pieper R, Gatlin-Bunai CL, Mongodin EF, Parmar PP, Huang ST, Clark DJ, Fleischmann RD, Gill SR, Peterson SN (2006) Comparative proteomic analysis of *Staphylococcus aureus* strains with differences in resistance to the cell wall-targeting antibiotic vancomycin. Proteomics 6:4246–4258
Poretsky R, Rodriguez R, M L, Luo C, Tsementzi D, Konstantinidis KT (2014) Strengths and limitations of 16S rRNA gene amplicon sequencing in revealing temporal microbial community dynamics. PLoS One 9(4):e93827
Ramamoorthy V, Viswanathan R, Raguchander T, Prakasam V, Samiyappan R (2001) Induction of systemic resistance by plant growth promoting rhizobacteria in crop plants against pests and diseases. Crop Prot 20(1):1–11
Redmond C, Baillie LW, Hibbs S, Moir AJ, Moir A (2004) Identification of proteins in the exosporium of *Bacillus anthracis*. Microbiology 150(2):355–363
Ricke SC, Feye KM, Chaney E, Shi Z, Pavlidis HO, Yang Y (2018) Developments in rapid detection methods for the detection of foodborne *Campylobacter* in the United States. Front Microbiol 9:3280
Roberts RJ, Carneiro MO, Schatz MC (2013) The advantages of SMRT sequencing. Genome Biol 14(6):405
Rout ME, Southworth D (2013) The root microbiome influences scales from molecules to ecosystems: the unseen majority1. Am J Bot 100(9):1689–1691

Rudrappa T, Biedrzycki ML, Kunjeti SG, Donofrio NM, Czymmek KJ, Paul WP et al (2010) The rhizobacterial elecitor acetoin induces systemic resistance in *Arabidopsis thaliana*. Commun Integr Biol 3(2):130–138

Sabarwal A, Kumar K, Singh RP (2018) Hazardous effects of chemical pesticides on human health-cancer and other associated disorders. Environ Toxicol Pharmacol 63:103–114

Sagan M, Huguet T, Duc G (1994) Phenotypic characterization and classification of nodulation mutants of pea (*Pisum sativum* L.). Plant Sci 100(1):59–70

Sahoo B, Ningthoujam R, Chaudhuri S (2019) Isolation and characterization of a lindane degrading bacteria *Paracoccus* sp. NITDBR1 and evaluation of its plant growth promoting traits. Int Microbiol 22(1):155–167

Sarkar A, Ghosh PK, Pramanik K, Mitra S, Soren T, Pandey S, Mondal MH, Maiti TK (2018) A halotolerant *Enterobacter* sp. displaying ACC deaminase activity promotes rice seedling growth under salt stress. Res Microbiol 169(1):20–32

Schmidt TM, DeLong EF, Pace NR (1991) Analysis of a marine picoplankton community by 16S rRNA gene cloning and sequencing. J Bacteriol 173:4371–4378

Shi S, Richardson AE, O'Callaghan M, DeAngelis KM, Jones EE, Stewart A, Firestone MK, Condron LM (2011) Effects of selected root exudate components on soil bacterial communities. FEMS Microbiol Ecol 77(3):600–610

Singh JS (2015) Microbes: the chief ecological engineers in reinstating equilibrium in degraded ecosystems. Agric Ecosyst Environ 203:80–82

Smith MI, Yatsunenko T, Manary MJ, Trehan I, Mkakosya R, Cheng J, Kau AL, Rich SS, Concannon P, Mychaleckyj JC, Liu J (2013) Gut microbiomes of Malawian twin pairs discordant for kwashiorkor. Science 339(6119):548–554

Spence C, Alff E, Johnson C, Ramos C, Donofrio N, Sundaresan V, Bais H (2014) Natural rice rhizospheric microbes suppress rice blast infections. BMC Plant Biol 14(1):130. https://doi.org/10.1186/1471-2229-14-130

Srivastava AK, Sharma A, Srivastava R, Tiwari PK, Singh AK, Yadav J, Jamali H, Bharati AP, Srivastava AK, Kashyap PL, Chakdar H (2019a) Draft genome sequence of halotolerant bacterium chromohalobacter salexigens ANJ207, isolated from salt crystal deposits in pipelines. Microbiol Resour Announc 8(15):e00049–e00019

Srivastava AK, Saxena P, Sharma A, Srivastava R, Jamali H, Bharati AP, Yadav J, Srivastava AK, Kumar M, Chakdar H, Kashyap PL, Saxena AK (2019b) Draft genome sequence of a cold-adapted phosphorous-solubilizing pseudomonas koreensis P2 isolated from Sela Lake, India. 3 Biotech 9(7):256

Ståhl PL, Lundeberg J (2012) Toward the single-hour high-quality genome. Annu Rev Biochem 81:359–378

Stein JL, Marsh TL, Wu KY, Shizuya H, DeLong EF (1996) Characterization of uncultivated prokaryotes: isolation and analysis of a 40-kilobase-pair genome fragment front a planktonic marine archaeon. J Bacteriol 178:591–599

Stielow JB, Levesque CA, Seifert KA, Meyer W, Iriny L, Smits D, Renfurm R, Verkley GJ, Groenewald M, Chaduli D, Lomascolo A (2015) One fungus, which genes? Development and assessment of universal primers for potential secondary fungal DNA barcodes. Persoonia 35:242–263. https://doi.org/10.3767/003158515X689135

Thissen JB, Nicholas AB, McLoughlin K, Gardner S, Rack PG, Shapero MH, Rowland RR, Slezak T, Jaing CJ (2019) Axiom microbiome Array, the next generation microarray for high-throughput pathogen and microbiome analysis. PLoS One 14(2):e0212045

Thomma BP, Seidl MF, Shi-Kunne X, Cook DE, Bolton MD, van Kan JA, Faino L (2016) Mind the gap; seven reasons to close fragmented genome assemblies. Fungal Genet Biol 90:24–30

Thomson JJ (1899) LVIII. On the masses of the ions in gases at low pressures. Philos Mag 48(295):547–567

Tilman D (1999) Global environmental impacts of agricultural expansion: the need for sustainable and efficient practices. Proc Natl Acad Sci 96(11):5995–6000

Treangen TJ, Salzberg SL (2012) Repetitive DNA and next-generation sequencing: computational challenges and solutions. Nat Rev Genet 13(1):36–46

Uren NC (2001) Types, amounts and possible functions of compounds released into the rhizosphere by soil-grown plants. In: The rhizosphere. CRC Press, Boca Raton, pp 35–56
Van Der Heijden MG, Bardgett RD, Van Straalen NM (2008) The unseen majority: soil microbes as drivers of plant diversity and productivity in terrestrial ecosystems. Ecol Lett 11(3):296–310
Verma S, Dar JA, Malasiya D, Khare PK, Dayanandan S, Khan ML (2019) A MODIS-based spatiotemporal assessment of agricultural residue burning in Madhya Pradesh, India. Ecol Ind 105:496
Vernière C, Hartung JS, Pruvost OP, Civerolo EL, Alvarez AM, Maestri P, Luisetti J (1998) Characterization of phenotypically distinct strains of *Xanthomonas axonopodis* pv. citri from Southwest Asia. Eur J Plant Pathol 104(5):477–487
Versalovic J, Schneider M, de Bruijn FJ, Lupski JR (1994) Genomic fingerprinting of bacteria using repetitive sequence based PCR (rep-PCR). Methods Cell Mol Biol 5:25–40
Vessey JK (2003) Plant growth promoting rhizobacteria as biofertilizers. Plant Soil 255(2):571–586
Virtanen S (1960) A study of nitrate reduction by mycobacteria. The use of the nitrate reduction test in the identification of mycobacteria. Acta Tuber Scand 48(Suppl. 48):1–119
Volokhov DV, Simonyan V, Davidson MK, Chizhikov VE (2012) RNA polymerase beta subunit (rpoB) gene and the 16S–23S rRNA intergenic transcribed spacer region (ITS) as complementary molecular markers in addition to the 16S rRNA gene for phylogenetic analysis and identification of the species of the family Mycoplasmataceae. Mol Phylogenet Evol 62(1):515–528
Wagner M, Haider S (2012) New trends in fluorescence in situ hybridization for identification and functional analyses of microbes. Curr Opin Biotechnol 23(1):96–102
West PC, Gerber JS, Engstrom PM, Mueller ND, Brauman KA, Carlson KM, Cassidy ES, Johnston M, MacDonald GK, Ray DK, Siebert S, Siebert S (2014) Leverage points for improving global food security and the environment. Science 345(6194):325–328
Wilson KH, Wilson WJ, Radosevich JL, DeSantis TZ, Viswanathan VS, Kuczmarski TA, Andersen GL (2002) High-density microarray of small subunit ribosomal DNA probes. Appl Environ Microbiol 68:2535–2541
Workbench CG (2010) Now new v3. 6. version can be available at http://www.clcbio.com/products/clcgenomicsworkbench
Yin Q, Yang S, Wang Z, Xing L, Wu G (2018) Clarifying electron transfer and metagenomic analysis of microbial community in the methane production process with the addition of ferroferric oxide. Chem Eng J 333:216–225
Zerbino D, Birney E (2008) Velvet: algorithms for de novo short read assembly using de Bruijn graphs. Genome Res 18:821–829. https://doi.org/10.1101/gr.074492.107
Zerbino DR, McEwen GK, Margulies EH, Birney E (2009) Pebble and rock band: heuristic resolution of repeats and scaffolding in the velvet short-read de novo assembler. PLoS One 4(12):e8407. https://doi.org/10.1371/journal.pone.0008407
Zhang W, Li F, Nie L (2010a) Integrating multiple 'omics' analysis for microbial biology: application and methodologies. Microbiology 156(2):287–301
Zhang ZY, Pan LP, Li HH (2010b) Isolation, identification and characterization of soil microbes which degrade phenolic allelochemicals. J Appl Microbiol 108(5):1839–1849
Zhang F, Cui Z, Fan M, Zhang W, Chen X, Jiang R (2011) Integrated soil–crop system management: reducing environmental risk while increasing crop productivity and improving nutrient use efficiency in China. J Environ Qual 40(4):1051–1057

Plant-Parasitic Nematode Management by Phytobiomes and Application of Fly Ash

14

Gufran Ahmad, Mohammad Haris, Adnan Shakeel,
Abrar Ahmad Khan, and Asgar Ali

Abstract

Plant-parasitic nematode causes enormous damages to various crops. To manage this tiny pest through ecofriendly approaches is the need of the hour. Application of biocontrol agents in the management of phytopathogenic nematodes has been a promising tool. There are a wide range of biocontrol agents including fungi and rhizobacteria which can protect the crops from pathogens/pest attack and also promote the yield attributes. It also improves the germination of the seed, helps in the development of root, and much more important thing, it increases the water utilization rate in plants. Moreover, some of the particulate air pollutants are also recommended for the management of phytonematodes. Among the particulate matter, fly ash was found to be most resistant to nematodes. Application of fly ash improves the growth of many crop plants when amended in the soil at low levels due to the presence of almost all the macro- and micronutrients. Many of the researchers observed that fly ash suppress the nematode population or disease caused by nematode in roots of plants, and this is due to the presence of many heavy metals in it.

Keywords

Biocontrol · Fly ash · Fungi · PGPR · Phytonematode

G. Ahmad (✉) · M. Haris · A. Shakeel · A. A. Khan
Section of Environmental Botany, Department of Botany, Aligarh Muslim University, Aligarh, Uttar Pradesh, India

A. Ali
Department of Botany, D.S. College, Aligarh (Dr. Bhimrao Ambedkar University, Agra), Aligarh, Uttar Pradesh, India

© Springer Nature Singapore Pte Ltd. 2020

M. K. Solanki et al. (eds.), *Phytobiomes: Current Insights and Future Vistas*,
https://doi.org/10.1007/978-981-15-3151-4_14

14.1 Introduction

Plant-parasitic nematodes (PPN) cause enormous losses and therefore warranted to be managed with safe approaches (Bernard et al. 2017). Among the different stages of life cycle of root-knot nematodes, the second stage of juveniles (J2) penetrates and sucks the nutrients from host plants. (Gheysen and Mitchum 2011). Due to the high impact of PPN, many of the researches have been focused specially in the two groups of PPN like cyst nematodes (*Heterodera* and *Globodera* sp.) and root-knot nematodes (*Meloidogyne* sp.). The life cycle of cyst nematode completes in approximately 30 days of *Heterodera* sp. and 60 days of *Globodera* sp. in optimal situation and has six stages such as the egg, four stages of juvenile (J1, J2, J3, and J4), and lastly adult stage (Williamson and Gleason 2003). Out of four, the second stage causes the infection to the root system of the plants. After detecting the host root, it penetrates intracellularly with the help of cell wall-degrading enzymes secreted by the esophageal glands of the nematodes and migrates toward the vascular bundles. Here nematodes stimulate the formation of a metabolically active and sophisticated nematode feeding site (NFS), which is known as "syncytium" or "transfer cell" through secretion and injecting of enzymes and proteins inside the host cell via stylet (Baum et al. 2007). The main transcriptomic, metabolic, morphological, and physiological changes occur through the formation of NFS, like surrounding cell walls partially split-up, nuclei enlargement, organelles of cytoplasm increase in density, accumulation of endoplasmic reticulum, etc. (Hussey et al. 2002; Siddique and Grundler 2018). A female adult becomes swollen and sedentary and remains adjoining to the roots, and posterior extremity of the body comes outside from the ruptured cells of the root. The males maintain their vermiform form and leave the roots to fertilize. Mature female absorbs the nutrients from nurse cells (giant cell) to support the metabolic activities of the nematode and also helps in egg production. After completing the life cycle, wall of the female body becomes thick and makes a tough brown leather bag which is known as "cyst." The cysts serve as initial protection of around 200–500 eggs against adverse conditions until egg hatching takes place in favorable environmental conditions (Jasmer et al. 2003). In contrast, the duration of life cycle varies from several days to several months based on some factors like humidity, temperature, presence of the convenient host, and the species of RKN. The second stage of juvenile (J2) detects and penetrates inside the root of the host through signals from the host and the penetrating action of the stylet. The young one moves intercellularly within the tissues of the root and ultimately reaches to the vascular bundle. Thus, the nematode absorbs some secretions from the esophageal gland in the formation of 5–7 multinucleated and metabolically active nutritious or feeding cells known as "giant cells" that are the consequence of cytokinetic mitosis (Mhatre et al. 2015; Siddique and Grundler 2018). Male nematodes migrate towards the female for the copulation, on the other hand, apple-shaped swollen females remain sedentary within the roots, and eggs produce in the gelatinous matrix. Nematodes target the roots system of the host plant and cause the infection leading to intermittent supply of water and nutrient. Due to severe infestation, the plant growth and yield are significantly reduced.

Plant-parasitic nematodes recognized as hidden enemies are responsible for the heavy losses in various crops and horticultural. Moreover, PPN is responsible for approximately $ 157 billion yield loss in the world; a yield loss of $ 40.3 million is from India, which shows a 12.3% average loss of yield (Abad et al. 2008; Singh et al. 2015).

14.2 Management Strategies of Nematodes

Due to parasitic organisms, the management of plant disease has become a daunting task in the present scenario for researchers in sustainable agriculture. Approximately 4000 organisms which are parasitic have been identified and mostly present in the major biomes. Organisms as plant-parasitic absorb many nutrient content and water through the vascular tissues from their host plants (Press and Phoenix 2005). In the parasitic organisms, the management of nematodes is more difficult due to the habitat and their inhabitation and mode of parasitism (Gillet et al. 2017). In plants, nematodes attack mostly underground parts and cause a serious loss of yield. Many species of nematodes like *Meloidogyne* sp., *Heterodera* sp., *Globodera* sp., and *Pratylenchus* sp. are considered as the most important species from the economic point of view due to the damage and infection level to wide range of host.

Nematicides of chemical origin are responsible for the enhancement of agricultural capacity in terms of increased food and production of fiber. On the other hand, the application of chemically manufactured nematicides has resulted in environmental risk and also the impact on human health (Aktar et al. 2009). The use of synthetic nematicides is being banned because of their dangerous nature, which is harmful to untargeted organisms, a threat to environmental protection, and a cause of public health problems (Schneider et al. 2003). Therefore, it is the need of the hour to search an alternative that must be ecologically safe. Many ecological approaches that promote plant growth-promoting rhizobacteria (PGPR) can act as effective biological control of nematodes, as well as an agent that promotes plant growth and yield of various crops. Strains of rhizobacteria use organic compounds, sugars, and amino acids which are released from the roots of the plant for their growth and energy. In mutualism, strains of rhizobacteria fabricate different substances to improve plant growth and biological control activities in support of the host plant (Karthik et al. 2017). The above facts show that the chapter focused on the positive PGPR correlation for the biological control of nematodes and the enhancement of agricultural productivity. Among the different environment-friendly approaches, PGPR strain can act as effective biological control of the nematode, as well as play an important role in the plant growth-promoting agent for the growth of plants and yield improvement.

Several options are available here to limit the damage, such as eco-friendly use of some industrial waste, biological control, resistant cultivars, intercropping, deep plowing, crop rotation, and nematicides. However, the use of intercropping, crop rotation, and deep plowing is not beneficial in case of cyst formation, gelatinous matrix, various survival adaptations, and survival in the soil without presence of host. The chemical method has been very effective for controlling nematodes, but farmers avoid this method due to the expensive cost, environmental problems, and high risks of health, while the continued growth of resistant cultivars on the same plot of land leads to the reducing resistance capacity because of the continuous development of virulent pathogens. Removing the nematode stress through the most effective and efficient methods is biological control, which is also an aim to save the crops (Timper 2014). Biological control is defined as "a deficiency of the nematode population, which is accomplished by the activity of living organisms except for the nematode resistant host plant, which occurs through naturally or changes in the environment or the establishment of antagonists" (Tian et al. 2007). The main objective of biological control is reduction of nematode populations by enhancing the number of natural enemies which is present in the soil. The soil is the collection of microflora, which is majorly diverse in structure and activity. In the rhizosphere, microbial flora acts as a front line which performs defense mechanism against various pathogens and can be utilized as biocontrol agents (Mendes et al. 2013). Use of such kind of biological organisms is able to maintain ecological balance and is responsible for the pure environment. After the establishment, the biological agent is active in the soil for a long time and also leads to the concept of "oppressive soil," where microorganisms populate in the soil naturally suppressing the PPN population (Trudgill et al. 2000). Biological control of nematodes is also achieved from bacterial and fungal antagonists. Fungal antagonists to nematodes consist of a variety of microorganisms including endoparasitic fungi, toxin-producing fungi, nematode-trapping fungi, parasites, and vesicular-arbuscular mycorrhiza which feed on the sedentary nematodes, females, and eggs. (Tranier et al. 2014), while bacterial antagonists are made up of mainly three groups such as epiphytic, endophytic, and endoparasitic bacteria. Bacteria reach biological control by the mechanisms such as antibiotics, competition, and parasitism (Abd-Elgawad and Kabeil 2012).

14.3 Plant Growth-Promoting Rhizobacteria (PGPR)

The roots of the surrounding plant from the "rhizosphere" soil include many bacterial species, which promote plant growth regulators, improve the development of plants, and increase the availability of nutrients. Such bacteria are called PGPR (Fig. 14.1; Table 14.1). PGPR contain a large group of free-living bacteria that colonizes in rhizomes and contributes to the development of plant growth rather than improving the yield of agricultural crops (Kumar et al. 2016). In soil, diverse microorganisms such as *Rhizobium* sp., *Xanthomonas* sp., *Arthrobacter* sp., *Bacillus* sp., *Bradyrhizobium* sp., *Enterobacter* sp., *Frankia* sp., *Klebsiella* sp., *Proteus* sp., *Flavobacterium* sp., *Microbacterium* sp., *Pseudomonas* sp., *Serratia* sp., *Acinetobacter* sp., *Azospirillum* sp., *Alcaligenes* sp., *Agrobacterium* sp., *Azotobacter* sp., *Burkholderia* sp., *Cellulosimicrobium* sp., and *Erwinia* sp. are the common constituents of rhizosphere and make complete colony in the rhizosphere (Bhattacharyya and Jha 2012; Tailor and Joshi 2014; Karthik et al. 2016; Teymouri

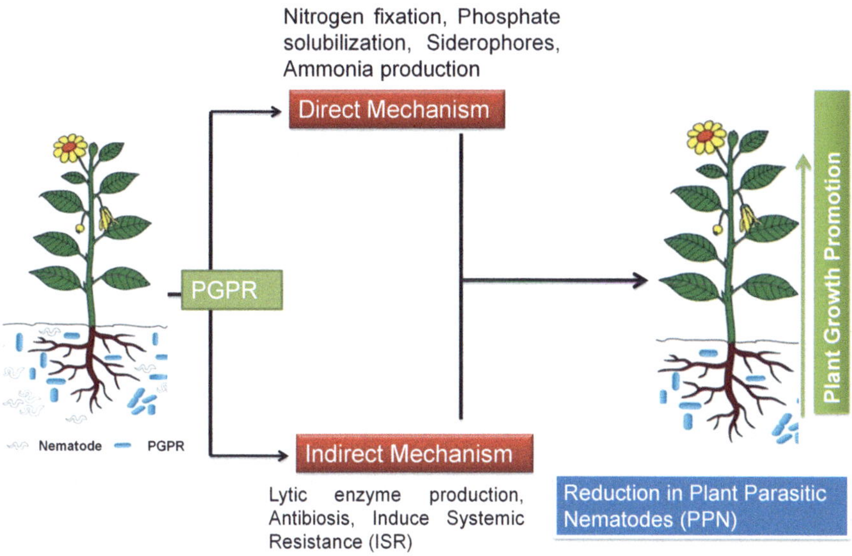

Fig. 14.1 PGPR help in plant growth and nematode control

et al. 2016). There are widely known commercial rhizosphere-colonized microorganisms among rhizobacterial genera like *Bacillus*, *Azospirillum, and Pseudomonas*. Use of rhizospheric bacteria play an important role in the biofertilizer, phytostimulation, biocontrol, and phytoremediation, but it depends on their formation of a colony in the rhizosphere. Microorganisms present in the rhizosphere have the efficiency to improve the plant growth through the production of multiple plant growth substances and also responsible for the inhibition of phytonematodes. To reduce the damage caused by plant-parasitic nematodes, PGPR also known as a potential agent (Tabatabaei and Saeedizadeh 2017; Rashad et al. 2015). Species of *Pseudomonas* and *Bacillus* belong to those bacterial groups which form endospore and are majorly antagonist to PPNs in the rhizosphere. Many *Bacillus* strains can inhibit nematodes and promote plant growth. Furthermore, it has been reported that *Bacillus* sp. was also directly antagonist toward the PPNs like *Meloidogyne*, *Heterodera*, and *Rotylenchulus* (Siddiqui and Mahmood 1999; Kokalis-Burelle et al. 2002; Li et al. 2005). Strains of *Pseudomonas* in the rhizosphere also show pathogenic mechanism against PPNs (Kerry 2000; Siddiqui et al. 2005). Some studies have been conducted to evaluate the mechanism involved in the decreasing populations of PPN during the interaction of *Pseudomonas* and PPNs by the production of induced systemic resistance (ISR) and antibiotics. (Siddiqui and Shaukat 2003c). These two major antagonists and many other rhizobacteria were also reported as PPN antagonists, including members of such kind of genera like *Agrobacterium*, *Arthrobacter*, *Alcaligenes*, *Azospirillum*, *Beijerinckia*, *Bradyrhizobium*, *Clostridium*, *Comamonas*, *Corynebacterium*, *Desulfovibrio*, *Gluconobacter*, *Flavobacterium*, *Phyllobacterium*, *Sphingobacterium*, *Serratia*, *Streptomyces*, *Variovorax*, *Actinomycetes*, *Aureobacterium*, *Azotobacter*, *Bacillus*,

Table 14.1 Effect of PGPR inoculation on plant-parasitic nematode biocontrol

Country	PGPR strains	Crops	Plant-parasitic nematodes	Mechanism of action	References
Egypt	*Bacillus subtilis*, *B. pumilus*, and *Pseudomonas fluorescens*	Cowpea	*Meloidogyne incognita*	Antibiosis, phytohormone and enzymes production	Abd-El-Khair et al. (2019)
Egypt	*Bacillus amyloliquefaciens*, *Lysinibacillus sphaericus*	–	*Meloidogyne incognita*	Phytohormone production	Abdel-Salam et al. (2018)
Turkey	*Paenibacillus castaneae*	Tomato	*Meloidogyne incognita*	Antibiosis	Cetintas et al. (2018)
USA	*Bacillus* sp.	Cotton and soybean	*Heterodera glycine*, *Meloidogyne incognita*	Siderophore formation	Xiang et al. (2017)
India	*Pseudomonas fluorescens*	Okra	*Meloidogyne incognita*	Antibiosis	Veronika and Khan (2015)
India	*Pseudomonas fluorescens* and *B. subtilis*	Rice	*Meloidogyne graminicola*	Antibiosis	Priya (2015)
China	*Bacillus subtilis*	Tomato	*Meloidogyne incognita*	Production of lytic enzymes	Wei et al. (2014)
China	*Bacillus cereus*, *Bacillus subtilis*, *Serratia* sp.	*Trichosanthes kirilowii*	*Meloidogyne incognita*	Production of lytic enzymes	Jiang et al. (2014)
Pakistan	*Bacillus spp.*, *Azotobacter sp.*	Tomato	*Meloidogyne incognita*	Nitrogen fixation	Anwarul Haq et al. (2011)
China	*Bacillus* sp.	In vitro	*Meloidogyne incognita*	Production of lytic enzymes	Wei et al. (2010)
India	*Methylobacterium fujisawaense*	–	*Meloidogyne incognita*	Antibiosis	Prabhu et al. (2009)
Pakistan	PGPR-H4, PGPR-K315, PGPR-Ax	Mash bean	*Meloidogyne incognita*	Antibiosis	Mahmood et al. (2009)
Pakistan	*Pseudomonas aeruginosa*	Mung bean	*Meloidogyne javanica*	Auxin, nematicidal activity, induce resistance	Ahmed et al. (2009)
Pakistan	*Bacillus cereus*	Cowpea, mash bean	*Meloidogyne javanica*	Phosphate solubilization	Dawar et al. (2008)
Pakistan	*P. variotii*	Banana	*Meloidogyne javanica*	Antibiosis-	Zarina (2007)
Pakistan	*B. circulane*	Guar	*Helicotylenchus indicus*	Phosphate solubilization	Khan et al. (2012)
Pakistan	*B. circulane*	Guar	*Helicotylenchus indicus*	Phosphate solubilization	Khan et al. (2012)

India	*Bacillus* sp.	Tomato	*Meloidogyne* spp.	Phosphate solubilization and nitrogen fixation	Siddiqui et al. (2005)
India	*B. thuringiensis*	Pigeon pea	*M. incognita*	Antibiosis	Dhawan et al. (2004)
Pakistan	*P. fluorescens*	In vitro	*Meloidogyne javanica*	Auxin, nematicidal activity, induce resistance	Hamid et al. (2003)
Pakistan	*P. fluorescens*	Soybean	*M. javanica*	Root colonization and acts as biocontrol	Siddiqui and Shaukat (2003a, b)
India	*Pseudomonas* spp. and *P. fluorescens*	Mustard, soybean, tomato	*Rotylenchulus reniformis*	Auxin, nematicidal activity, induce resistance	Asghar et al. (2002)
India	*Pseudomonas* spp. and *P. fluorescens*	Mustard, soybean, tomato	*Rotylenchulus reniformis*	Auxin, nematicidal activity, induce resistance	Niknam and Dhawan (2002)
USA	*Paenibacillus macerans* and *Bacillus amyloliquefaciens*	Pepper	*Meloidogyne incognita*	Phytohormone production, induced systemic resistance	Kokalis-Burelle et al. (2000)

Burkholderia, Chromobacterium, Clavibacter, Curtobacterium, Desulfovibrio, Enterobacter, Hydrogenophaga, Klebsiella, Methylobacterium, Pseudomonas, Rhizobium, and Stenotrotrophomonas (Tian et al. 2007; Siddiqui and Mahmood 1999; Wani 2015). Utilization of PGPR is beneficial in attaining substantial improvement in both plant growth and nematode suppression. Therefore, PGPR play an important role in the improvement of a sustainable agricultural system, being one of the key components in integrated nematode management (Mhatre et al. 2018; Abd-Elgawad and Kabeil 2012). Many of the PGPR strains like *Bacillus firmus* T11, *Bacillus aryabhattai* A08, *Paenibacillus barcinonensis* A10, *Paenibacillus alvei* T30, and *Bacillus cereus* N10w has to show the reduction in several galls. The greenhouse experiment conducted on carrots, treated with strain T30, caused a significant reduction in gall index and egg mass index as compared to non-treated control set. It was observed that the treatment with *B. aryabhattai* A08 gave significant results on tomato, reducing the gall index and egg mass index compared to the control set. It is concluded that the bacterial strains *P. alvei* T30 and *B. aryabhattai* A08 have potential as biological control agents of *M. incognita* on carrots and tomatoes, respectively (Viljoen et al. 2019).

Application of PGPR can improve the plant growth and yield related characterics with or without nematode infestations. PGPR can suspend the nematode activity and register significant enhancement in the plant health through nitrogen fixation, phosphate solubilization, siderophore production, lytic enzyme production, antibiosis, ISR, etc.

14.4 Mechanism of PGPR in Nematode Suppression

14.4.1 Direct Antagonism

Rhizobacteria plant-parasite exhibits a different mode of action in the rhizosphere to suppress nematodes. The mechanism for suppression of nematodes can be classified mainly into two major heads. In direct, enzymes are antagonistic to nematodes, toxic substances and other metabolic products regulate the nematode behavior, indirect effect, by changing root variation and motivating the production of repellents by the host which adversely affects. The recognition of the host, changes in the development of the nematode feeding site, and the sex ratio within the root system induce growth of plants, competing for essential nutrients and inducing systemic resistance (Siddiqui and Mahmood 1999; El-Nagdi and Youssef 2004; Singh et al. 2019).

14.4.1.1 Antibiosis

Microorganisms produce antibiotics, which are the organic compounds of low molecular weight. It plays a beneficial role in biocontrol of many pests through parasitism and competition (Raguchander et al. 2011). Most of the rhizobacteria work against PPN through the production of toxins, enzymes, and metabolic by-products. This helps in the survival, development, reproduction, and hatching of the nematodes (Siddiqui and Mahmood 1999). Furthermore, it is believed that *P. fluorescens* secretes secondary metabolites such as 2-4-diacetylphloroglucinol which reduces the cyst nematode population (Siddiqui and Shaukat 2003c), while some rhizobacteria were found to secrete compounds such as hydrogen cyanamide, which destroys detrimental organisms in the rhizosphere and is beneficial in creating a favorable environment for better development of the plants (Tian et al. 2007). Rose et al. (2012) also observed that the presence of *P. fluorescence* is able to decrease the growth rate of nematodes and is found to be most effective when used in combination with neem cake. In the study of PGPR, three genera, namely, *Azospirillium*, *Rhizobium*, and *Azotobacter* and mycorrhizal genus *Glomus* sp. have been reported to reduce galling in roots which is caused by the *M. javanica* in chickpea (Siddiqui and Mahmood 2001).

14.4.1.2 Production of Lytic Enzymes

The growth and development by the action of enzymes is another mechanism of PGPR through the production of some enzymes like chitinases, phenylalanine ammonia lyase, lipases, proteases, peroxidase, dehydrogenase, β-glucanase, phosphatases, etc. *Corynebacterium paurometabolous* produced chitinase and hydrogen sulfide, which are responsible for the inhibition of nematode egg hatching (Jamal et al. 2018; Patel et al. 2018; Mena and Pimentel 2002). Isolation of three bacteria *Pseudomonas* sp., *Stenotrophomonas maltophilia*, and *Bacillus mycoides* proven to be nematicidal reducing 56–74% population of Tricodorid infesting potato crop. In addition, these bacteria were a specialty for the oxidation of phenol and antifungal activity with the production of HCN and hydrolytic enzymes (Insunza et al. 2002). With the isolation of 16 potential PGPR from grapevine roots, 7 isolates, namely, *S. marcescens*, *C. acidovorans*, *A. piechaudii*, *S. plymuthica*, *Pantoea agglomerans*, *Sphingobacterium spiritivorum*, and *B. mycoides*, have been proved to suspend the reproduction of *Meloidogyne ethiopica*. Secondary metabolites of all the strains exhibit significant hatching prohibition of *M. ethiopica,* and from the isolates, *P. megatorium* and *P. putida* are found the most effective (Aballay et al. 2013).

14.4.1.3 Induced Systemic Resistance (ISR)

The induced resistance plant has increased defense capacity, which is achieved after proper stimulation against the parasite and broad-spectrum diseases. The increase in defense response due to the inducing agent causing the infection through pathogen is called induced systemic resistance (ISR) or systemic acquired resistance (SAR) (Van-Loon 2000). This induced resistance produces non-specific defense against various types of pathogens, such as fungus, bacteria, viruses, nematodes, and insects (Beneduzi et al. 2012). Several organic molecules are found to be associated with systemic resistance such as polyphenol oxidase (PPO), phenylalanine ammonia lyase (PAL), peroxidase (PO), lipoxygenase (LOX), superoxide dismutase (SOD), chitinase, catalase (CAT), ascorbate peroxidase (APX), proteinase, and β-1,3-glucanase (Pokhare et al. 2012; Mhatre et al. 2017). Such kinds of enzyme begin the stimulation of resistance through the production of phenolic and phytoalexin compounds (Viswanathan et al. 2003). Various studies observed that rhizobacteria inhibit nematode severity through inducing the plant systemic resistance (Ramamoorthy et al. 2001; Pieterse et al. 2002). This induced resistance is obtained through the mechanical and physical strength of the cell wall by the callose deposition, accumulation of phenolic compounds, and thickening of the cell wall. The synthesis of some biochemical compounds like phytoalexin, PR proteins, siderophores, salicylic acid (SA), lipopolysaccharides (LPSs), PO, jasmonic acid (JA), chitinase, and some other secondary metabolites that regulate the reaction provides the protections (Siddiqui and Mahmood 1999; Ramamoorthy et al. 2001). Another study showed that infections caused by the potato cyst nematodes and root-knot nematodes were reduced through induced systemic resistance by *Rhizobium etli* (Hallmann et al. 2001). *R. etli* have lipopolysaccharides that act as the inducing agent which is helpful in the systemic resistance and is found to play an important role in reducing the penetration and recognition of *Meloidogyne incognita* and *Globodera pallida* (Reitz et al. 2000; Mahdy et al. 2001). According to Meena et al. (2012), the maximum activity of enzymes is recorded in the low nematode population on tomato when treated by the consortium of PGPR formulation (*P. fluorescens*, Pf128+ *B. subtilis*, Bbv 57). Xiang et al. (2017) reported that 613 *Bacillus* strains caused mortality to *H. glycines* in vitro, and, specifically, *B. velezensis* strain Bve2 consistently reduced *H. glycines* cyst population density at 60 DAP in the greenhouse, microplot, and field trials. The *B. mojavensis* strain Bmo3 suppressed *H. glycines* cyst and total *H. glycines* population density under greenhouse conditions, and *B. safensis* strain Bsa27 and Mixture 1 (Bve2 + Bal13) reduced *H. glycines* cyst population density at 60 days after planting in the field trials. Moreover, Aljaafri et al. (2017) found that a harpin elicitor can function to turn on plant defenses in *H. glycines* in soybean and *M. incognita* and *R. reniformis* in cotton in susceptible cultivars.

14.4.2 Indirect Effects

PGPR promote the growth of plants by the phytohormone production like auxins, cytokines, ethylene, and abscisic acid. Moreover, siderophore improves the mineral uptake in the plants.

14.4.2.1 Phytohormone Production

Most rhizobacterial strains are capable of producing substances that promote growth and development of plant. PGPR produce some plant growth regulators like auxins, cytokines, gibberellic acid, abscisic acid, ethylene, polyamines, brassinosteroids, jasmonates, salicylic acid, strigolactones, and other compounds that help in the regulation of plant growth (Mhatre et al. 2018; Gopalakrishnan et al. 2015). It is believed that PGPR produce phytohormones and plays an important role in promoting plant growth and interaction between plant and bacteria. Microbial phytohormones responsible for the enhancement in the growth of the plant by the stimulation of cell division, cell elongation, and expansion of tissue indicate beneficial effects on the growth and yield of plants (Karthik et al. 2016; Khan et al. 2009). Indole acetic acid improves adventitious and lateral roots, which leads to increased nutrient and mineral uptake (Arora et al. 2013; Shaikh and Saraf 2016). It has been suggested that phytohormones formed by PGPR may inhibit the deadly effects of different environmental stresses. For example, *Streptomyces* strains produce phytohormones, which can improve the growth of eggplants through the suppression in the galls and egg masses of the nematode (Rashad et al. 2015). Similarly, Ruanpanun et al. (2010) reported that nematicidal activities of phytohormones generate *Streptomyces* sp. Therefore, any direct effects of bacteria on the production of phytohormones affect the efficiency of phytostimulation.

14.4.2.2 Nitrogen Fixation

Nitrogen is one of the important macronutrients, which is beneficial for the growth and development of the plant, also involved in photosynthesis and protein synthesis and acts as a nitrogenous base in nucleic acids. Due to the regular loss of nitrogen in the soil, agricultural land has a limited amount of nitrogen. However, the plants cannot directly use atmospheric nitrogen. In this situation, PGPR perform an important role in the fixation of nitrogen and nutrients supplementation. These microorganisms are categorized into two separate groups, such as symbiotic and free-living nitrogen-fixing microorganisms (Mhatre et al. 2018; Gopalakrishnan et al. 2015). PGPR play an important function in nitrogen fixation which shows a vital contribution in sustainable agriculture. Aggangan et al. (2013) found that inoculation of nitrogen-fixing bacteria significantly improves the growth of banana through inhibition of the population of *Radopholus similis* and *Meloidogyne incognita*. Similarly, El-Hadad et al. (2011) also suggest improvement in the growth of plants and nematicidal activity by the nitrogen-fixing microorganism, *Paenibacillus polymyxa*. El-Sayed et al. (2014) found the biocontrol activity of nematodes with the help of nitrogen-fixing PGPR strains. Some genera of PGPR such as *Bradyrhizobium*, *Mesorhizobium*, *Rhizobium*, *Frankia*, and *Sinorhizobium* are responsible for the fixation of atmospheric nitrogen which is utilized by the plants and beneficial for plant growth (Zahran 2001; Vessey 2003; Arora et al. 2012; Gaby and Buckley 2012; Dash et al. 2017).

14.4.2.3 Phosphate and Potassium Solubilization

After nitrogen, phosphate is another major macronutrient which can promote plant growth. Phosphate plays a pivotal role in plant growth, such as the composition of nucleic acid, protein synthesis, cell division, growth of new tissues, and association with energy transformation complex (Gopalakrishnan et al. 2015; Oves et al. 2013). PGPR also work in the conversion of unavailable forms of phosphorus from the soil in available form through the production of organic acids, chelation, and acidity (Mhatre et al. 2018; Gulati et al. 2010) and influence nutrient availability and growth of host plants. Several genera of PGPR such as *Bacillus*, *Burkholderia*, *Erwinia*, *Microbacterium*, *Rhizobium*, *Serratia*, *Arthrobacter*, *Beijerinckia*, *Enterobacter*, *Flavobacterium*, *Pseudomonas*, and *Rhodococcus* are already reported as solubilizers of phosphate (Zaidi et al. 2009; Sharma et al. 2013; Ahemad et al. 2009;

Oves et al. 2013). Guang-Can et al. (2008) reported that strain of bacteria such as *B. megaterium*, *B. caryophylli*, *B. cereus*, *P. cichorii*, and *P. syringae* perform such kind of activity like solubilizations of phosphate and efficiency of mineralization, thus enhancing phosphate bioavailability. Inoculation of solubilizing phosphate by the strain of *Mesorhizobium mediterraneum* improved the growth of barley and chickpea (Peix et al. 2001). El-Hadad et al. (2011) observed that inoculation of phosphate-solubilizing bacteria (*B. megaterium*) improved the length of shoot, dry weight of shoot, dry weight of root, and amount of N, P, K in the plant of tomato and decreased the population of *M. incognita* in the rhizosphere. Potassium is another important plant nutrient after phosphate and nitrogen which perform various functions in plants. In the growth of plants, potassium is responsible for the physiological and biochemical functions (Zhang and Kong 2014). However, the maximum amount of potassium nutrient in soil containing orthoclase, feldspar, mica, muscovite, biotite, and illite in the form of potassium, not easily utilized by the plants. Just as bacteria dissolve phosphate, few rhizospheric microorganisms dissolve insoluble form of potassium making it accessible to the plants for their growth and development. PGPR solubilize potassium through various mechanisms like chelation, organic acid, acidolysis, complex lysis, and reduction (Meena et al. 2016). Several genera of microbes like *Acidithiobacillus ferrooxidans*, *B. edaphicus*, *B. mucilaginosus*, *Burkholderia*, *Paenibacillus* sp., and *Pseudomonas* sp. are responsible for the solubilized potassium (Han and Lee 2006).

14.4.2.4 Production of Siderophores and Ammonia

Iron is another necessary plant nutrient element that is beneficial for the living organisms which performs many biological functions like photosynthesis, respiratory, electron transport, and cofactors for many enzymes, etc. (Aguado-Santacruz et al. 2012). Siderophores are iron-chelating agent which make the iron accessible to the plants. PGPR have the ability to develop some specific mechanisms like chelating the insoluble form of iron through the siderophores due to the production of low molecular weight (Dell'mour et al. 2012). Diverse groups of PGPR include *Aeromonas*, *Azadirachta*, *Azospirillum*, *Azotobacter*, *Bacillus*, *Burkholderia*, *Enterobacter*, *Pseudomonas*, *Rhizobium*, *Serratia*, and *Streptomyces* sp. beneficial for the production of the siderophore that transports iron in the cells of plants for growth (Cornelis 2010; Sujatha and Ammani 2013). Ruanpanun et al. (2010) reported that the production of siderophores inhibit the activity of nematodes by the *Streptomyces* sp. Similarly, El-Sayed et al. (2014) also reported that the *Bacillus* sp., *Enterobacter* sp., and *Pseudomonas* sp. improve multifarious growth of plants as well as nematicidal activity.

Plant growth promotes various substances by the strains of PGPR which directly affect the growth of host plant through the stimulation in cell division, development of tissue, and physio- and biochemical functions. However, these PGPR strains provide additional support to the host plant to cope with nematode infection. However, such strain of PGPR provides an additional endorsement to the host plant to cope up with the infection of nematode (Mhatre et al. 2018).

14.5 Fly Ash

Fly ash (FA) is one of the important types of particulate waste that is generated by the combustion of coal in thermal power plants for the generation of the increasing demand for electricity (Ahmaruzzaman 2010). With increasing demand of electricity in India, the US Energy Information Administration Department (EIAD) noted that the use of energy consumption growth would exceed and leave behind the countries like China, the USA, and Russia until the end of 2035 because 76.4% of electricity-generating power plants are coal-based (Singh et al. 2018). Production of FA between 2010 and 2011 increased to about 130 million tons, which was about 85.7% higher than 1996 to 1997. The FA utilization rate also increased from 9.63% in 1996 to 1997 to 54.53% in 2010 to 2011, an increase of 466% over 15 years (Singh and Gupta 2014). Atmospheric pollution due to FA has found to be harmful and affects the human health if the FA directly discharges from power plant chimneys into the atmosphere without treatment (Bartoňová 2015; Ding et al. 2015). In thermal power plants, the characteristics of FA depend on the type of coal and the burning conditions. The appearance of fly ash in the form of fine particle and the average diameter is less than 20 μm, bulk density is 0.54–0.86 g/cm^3, and specific surface area ranges between 300 and 500 m^2/kg (Lanzerstorfer 2018; Yao et al. 2015). There are some essential elements, including P, K, Mg, Zn, Fe, Mn, and others except N are found in FA. FA can improve the texture of the soil, making it more fertile and increasing the growth and yield of many crops (Parab et al. 2012; Kalra et al. 2000; Yeledhalli et al. 2010; Garg et al. 2005; Dhadse et al. 2008; Dermatas and Meng 2003). Besides this, the important oxide components, i.e., SiO_2, $Al2O_3$, CaO, MgO, Na_2O, and TiO_2, are also found in FA (Ahmaruzzaman 2010; Bartoňová 2015; Adriano et al. 1980). FA was found to reduce the gaseous air pollutant like SO_2 and VOCs after some modifications (Izquierdo and Rubio 2008; Zhou et al. 2015). Therefore, FA has a great concern to many researchers worldwide.

14.5.1 Management of Nematode by Fly Ash

Application of fly ash for the management of plant-parasitic nematodes improves the soil health. Khan (1989) reported that the soil application of fly ash inhibits the root penetration of juveniles and reduce the root-knot disease intensity caused by *M. incognita* on tomato. Similarly, according to Singh (1989), galling and the production of egg mass of *M. incognita* on lentil were found to be reduced. The higher concentration of fly ash suppressed the *M. javanica* in soybean and at 100% fly ash amendment exhibited no gallings and eggmasses of M. javanica (Singh 1993). Singh et al. (1994) studied that all the morphometric parameters of *M. javanica* decreased in pea plants grown in fly ash amended soil. According to Khan et al. (1997), the application of fly ash in soil (20–100%) adversely affected the root invasion by the larvae and decreased the disease intensity (gall and egg mass/root system) of root-knot nematodes, on tomato. Khan and Ghadirpour (1999) observed that fly ash reduced the disease severity caused by *M. incognita* on tomato, eggplant, and chili. Joshi et al. (2000) observed the effect of organic amendment and fly ash on the root-knot disease of tomato, and the galling index was reduced greatly. Tararmum et al. (2001) observed the effect of different levels of fly ash (0, 25, 50, 75, and 100%) on hatching, penetration, and development of *M. javanica* on chickpea. Hatching and penetration were greatly suppressed. At 50% onward, no J2 developed to the mature female stage. Similarly, Iram (2010) evaluated the different levels of fly ash amended soil in the management of root-knot nematode and observed all the levels of fly ash significantly inhibit the reproduction of nematodes. Iram (2006) studied the response of root-knot nematode, *M incognita*, to fly ash on pepper (*Capsicum annuum* L.). All the fly ash levels reduced the hatching and suppressed the development of juveniles but increased the mortality rate. Root penetration was inversely proportional to fly ash ratios. The highest increase in plant growth was observed at 20% level. Tanweer et al. (2007) observed the effect of fly ash amended soil (0, 10, 20, 30, 40 & 50%) on the development of root-knot nematode, *M. incognita*, on ivy gourd (*Coccinia cordifolia*). A number of galls and egg masses per plant were decreased in 10–50% fly ash levels as compared to control. According to Singh et al. (2011), root galling increased at 25 and 50%, but decreased at the 75 and 100% fly ash levels. Gradual suppression in egg mass production and fecundity were observed at all levels of fly ash. A pot experiment was conducted in a glasshouse to evaluate the effect of *Cassia tora* leaf extract on the growth and yield characteristics of tomato and on the reproduction of root-knot nematodes in fly ash amended soil. Data on plant growth, leaf area, yield characteristics, and root-knot and egg mass indices on tomato grown in different concentrations (0, 20, 30, and 50%) of fly ash amended soil with leaf extract of *Cassia tora* were recorded. The results indicated that the plant growth and yield were enhanced and the nematode population was reduced in the 20% fly ash treatment. Among all the treatments, 50% fly ash amended soil effectively reduced the galling and nematode population (Azam et al. 2013). Ahmad and Khan (2016) evaluated the effect of fly ash on hatching, mortality, and penetration of *M. incognita* in pumpkin roots. All the levels of fly ash were found toxic and reduced the hatching of juveniles, increased the mortality rate, and inhibited the penetration of juveniles. The maximum inhibition was observed at 50% level of fly ash. The development of juveniles of *Meloidogyne incognita* in the roots of pumpkin was significantly suppressed by all fly ash and soil mixtures. The J_2 developed to J_3/J_4 stages at all levels of fly ash amendment, but their number was less than in control and decreased with the increase of the fly ash up to 40% soil mixture. At the end of the first week, neither premature nor mature females were found. During the second week, J_2 developed to older stages. However, while premature females occurred in all roots, only a few mature females occurred in control and at the 5–10% levels of fly ash and none at larger proportions of the amendment. During the third week, the juveniles that had penetrated into the roots developed further. However, numbers of premature females were significantly suppressed by all proportions of fly ash, while mature females were significantly suppressed at 5–10% of this amendment and were still absent at larger proportions. After 4 weeks, all the J_3/J_4 had developed further, but premature females were still significantly less than in control at all proportions of the amendment, and a few mature females were observed only up to 20% of the amendment, with none at all at greater proportions (Ahmad et al. 2017).

14.5.2 Fly Ash and Phytobiome Consortia in the Management of Nematodes

Many researchers observed that fly ash in combination with PGPR and fungi caused a significant reduction in the reproduction of many plant-parasitic nematodes. Phytoparasitic nematodes usually cause physical damage to the roots, which may allow secondary infection by other pathogens (Sitaramaiah and Pathak 1993). Endoparasitic nematodes, *Meloidogyne* spp., create a wound in roots, allowing the other pathogens to become established (Siddiqui et al. 2012). There are many synergistic effects of fungi, bacteria, and nematode interactions are also investigated (Stansbury et al. 2001; Rubio-Cabetas et al. 2001; Partridge 2008; Mallesh et al. 2009). According to Khan and Siddiqui (2017), the galling of root and multiplication in the nematode population were decreased in the plant when inoculated with *R. solanacearum* or *P. vexans*. Moreover, inoculation of both *R. solanacearum* and *P. vexans* together caused a greater reduction in a number of galls and population of nematodes than inoculated singly with either *R. solanacearum* or *P. vexans*. Inoculation of *R. solanacearum* with *P. vexans* before *M. incognita* caused a maximum reduction in galling and nematode multiplication.

Glasshouse experiments were conducted twice to assess the ash amendments (0, 20, and 40% with soil), *Pseudomonas striata*, and a root-nodule bacterium *Rhizobium* sp. on the reproduction of root-knot nematode *Meloidogyne incognita* and the growth and transpiration of pea. 10%–50% levels of fly ash amended soil caused decreased egg masses and galls of *M. incognita* in carrot, while 50% fly ash amended soil completely inhibits the egg masses and galls formation (Haris et al. 2018). *Rhizobium* sp. was found to be much more effective and reduced the galling and nematode multiplication in comparison to *P. striata*. Moreover, the use of both organisms together also had a greater adverse effect on galling and nematode multiplication than caused by either of them alone. The 40% fly ash mixed soil with both *Rhizobium* sp. and *P. striata* showed the highest reduction in galling and nematode multiplication (Siddiqui and Singh 2005).

14.6 Conclusions and Future Prospects

Following conclusions from the present article can be drawn:

1. Various plant-parasitic nematodes such as *Meloidogyne*, *Heterodera*, *Trichodorus*, etc. are responsible for causing a greater loss in crop production.
2. Fly ash can be used in the management of plant-parasitic nematodes leading to enhanced plant growth and yield attributes.

3. An important rhizosphere component, PGPR also play an important role in the reduction of plant-parasitic nematodes infestations in various crops.
4. The most effective module has been found to be the consortium of root biome and fly ash which can be used in the management of plant-parasitic nematodes.
5. Application of these consortia not only improves the soil health but also enhances the crop productivity in a significant manner.
6. However, the presence of some heavy metals in the fly ash is the main problem which needs to be reckoned before its application in the field as they may be a source of soil pollution.

References

Abad P, Gouzy J, Aury JM, Castagnone-Sereno P, Danchin EGJ, Deleury E (2008) Genome sequence of the metazoan plant-parasitic nematode *Meloidogyne incognita*. Nat Biotechnol 26:909–915

Aballay E, Ordenes P, Martensson A, Persson P (2013) Effects of rhizobacteria on parasitism by *Meloidogyne ethiopica* on grapevines. Eur J Plant Pathol 135:137–145

Abd-Elgawad MM, Kabeil SA (2012) Biological control of *Meloidogyne incognita* by *Trichoderma harzianum* and *Serratia marcescens* and their related enzymatic changes in tomato roots. Afr J Biotechnol 11:16247–16252

Abd-El-Khair H, El-Nagdi WM, Youssef MM, Abd-Elgawad MM, Dawood MG (2019) Protective effect of Bacillus subtilis, B. pumilus, and Pseudomonas fluorescens isolates against root knot nematode Meloidogyne incognita on cowpea. Bull Nat Res Centre 43(1):64

Abdel-Salam MS, Ameen HH, Soliman GM, Elkelany US, Asar AM (2018) Improving the nematicidal potential of Bacillus amyloliquefaciens and Lysinibacillus sphaericus against the root-knot nematode Meloidogyne incognita using protoplast fusion technique. Egyp J Biol Pest Control 28:31

Adriano DC, Page AL, Elseewi AA, Chang AC, Straughan LA (1980) Utilization and disposal of fly ash and other coal residues in terrestrial ecosystem: A review. Environ Qual 9:333–344

Aggangan NS, Tamayao PJS, Aguilar EA, Anarna JA, Dizon TO (2013) *Arbuscular mycorrhizal* fungi and nitrogen fixing bacteria as growth promoters and as biological control agents against nematodes in tissue-cultured banana var. Lakatan Philipp Sci Lett 142:153–165

Aguado-Santacruz GAA, Moreno-Gómez BA, JiménezFrancisco BB, García-Moya EB, Preciado-Ortiz RE (2012) Impact of the microbial siderophores and phytosiderophores on the iron assimilation by plants: a synthesis. Rev Fitotec Mex 35:9–21

Ahemad M, Zaidi M, Khan MS, Oves M (2009) Biological importance of phosphorus and phosphate solubilizing microorganisms—an overview. In: Phosphate solubilizing microbes for crop improvement. Nova, New York, pp 1–4

Ahmad G, Khan AA (2016) Fly ash effect on hatching, mortality and penetration of root-knot nematode (*Meloidogyne incognita*) in pumpkin roots. Int J Environ 5(3):66–73

Ahmad G, Khan AA, Ansari S (2017) Interaction of a fly ash and root-knot nematode pathogens on Pumpkin (*Cucurbita moschata* Duch. ex Lam.). Trop Plant Res 4(3):449–455

Ahmaruzzaman M (2010) A review on the utilization of fly ash. Prog Energy Combust Sci 36:327–363

Ahmed N, Abbasi MW, Shaukat SS, Zaki MJ (2009) Induced systemic resistance in mung bean plant against root-knot nematode *Meloidogyne javanica* by DL-β-Amino butyric acid. Nematol Mediterr 37:67–72

Aktar MW, Sengupta D, Chowdhury A (2009) Impact of pesticides use in agriculture: their benefits and hazards. Interdiscip Toxicol 2:1–12

Aljaafri WAR, McNeece BT, Lawaju BR, Sharma K, Niruala PM, Pant SR, Long DH et al (2017) A harpin elicitor induces the expression of a coiled-coil nucleotide binding leucine rich repeat (CC-NB-LRR) defense signaling gene and others functioning during defense to parasitic nematodes. Plant Physiol Biochem 121:161–175
Anwarul Haq M, Anwar S, Shahid M, Javed N (2011) Management of root knot nematode Meloidogyne incognita by plant growth promoting rhizobacteria on tomato. Pak J Zool 43(6):1027–1031
Arora NK, Tewari S, Singh S, Lal N, Maheshwari DK (2012) PGPR for protection of plant health under saline conditions. In: Maheshwari DK (ed) Bacteria in agrobiology: stress management. Springer, Berlin, pp 239–258
Arora NK, Tewari S, Singh R (2013) Multifaceted plant-associated microbes and their mechanisms diminish the concept of direct and indirect PGPRs. In: Arora NK (ed) Plant microbe symbiosis: fundamentals and advances. Springer, New Delhi, pp 411–449
Asghar HN, Zahir ZA, Arshad M, Khaliq A (2002) Relation between *in-vitro* production auxins by rhizobacteria and their growth promoting activities in *Brassica juncea* L. Biol Biofertil Soils 35:231–237
Azam T, Hisamuddin, Akhtar MS, Akhtar MS (2013) Effect of Cassia tora leaf extract on the growth and yield characteristics of tomato and on the reproduction of root-knot nematode in fly ash amended soil. Adv Sci Eng Med 5:177–181
Bartoňová L (2015) Unburned carbon from coal combustion ash: an overview. Fuel Process Technol 134:136–158
Baum TJ, Hussey RS, Davis EL (2007) Root-knot nematode parasitism genes: the molecular basis of plant parasitism. Genet Eng 28:17–43
Beneduzi A, Ambrosini A, Passaglia LMP (2012) Plant growth-promoting rhizobacteria (PGPR): their potential as antagonists and biocontrol agents. Genet Mol Biol 35:1044–1051
Bernard GC, Egnin M, Bonsi C (2017) The impact of plant-parasitic nematodes on agriculture and methods of control. In: Shah MM, Mahamood M (eds) Nematology – concepts, diagnosis and control. IntechOpen, Rijeka (August 16). https://doi.org/10.5772/intechopen.68958. Available from https://www.intechopen.com/books/nematology-concepts-diagnosis-and-control/the-impact-ofplant-parasitic-nematodes-on-agriculture-and-methods-of-control
Bhattacharyya PN, Jha DK (2012) Plant growth-promoting rhizobacteria (PGPR): emergence in agriculture. World J Microbiol Biotechnol 28:1327–1350
Cetintas R, Kusek M, Fateh SA (2018) Effect of some plant growth-promoting rhizobacteria strains on root-knot nematode, Meloidogyne incognita, on tomatoes. Egypt J Biol Pest Control 28:7
Cornelis P (2010) Iron uptake and metabolism in pseudomonads. Appl Microbiol Biotechnol 86:1637–1645
Dash NP, Kumar A, Kaushik MS, Abraham G, Singh PK (2017) Agrochemicals influencing nitrogenase, biomass of N2-fixing cyanobacteria and yield of rice in wetland cultivation. Biocatal Agric Biotechnol 9:28–34
Dawar S, Tariq M, Zaki MJ (2008) Application of *Bacillus* species in control of *Meloidogyne javanica* (Treub) Chitwood on cowpea and mash bean. Pak J Bot 40:439–444
Dell'mour M, Schenkeveld W, Oburger E, Fischer L, Kraemer S, Puschenreiter M, Lammerhofer M, Koellensperger G, Hann S (2012) Analysis of iron-phytosiderophore complexes in soil related samples: LC-ESI-MS/MS versus CE-MS. Electrophoresis 33:726–733
Dermatas D, Meng X (2003) Utilization of fly ash for stabilization/solidification of heavy metal contaminated soils. Eng Geol 70:377–394
Dhadse S, Kumari P, Bhagia L (2008) Fly ash characterization, utilization and government initiatives in India—a review. J Sci Ind Res 67:11–18
Dhawan SC, Kaur S, Singh A (2004) Effect of *Bacillus thuringiensis* on the mortality of root-knot nematode, *Meloidogyne incognita*. Indian J Nematol 34:98–99
Ding J, Tang Z, Yang L, Wang Y, Zheng S, Zhang Y, Shen S, Xie Z (2015) A novel process for synthesis of tobermorite fiber from high-alumina fly ash. Cem Concr Compos 65. https://doi.org/10.1016/j.cemconcomp.2015.10.01

El-Hadad ME, Mustafa ME, Selim Sh M, El-Tayeb TS, Mahgoob AEA, Norhan H, Aziz A (2011) The nematicidal effect of some bacterial biofertilizers on *Meloidogyne incognita* in sandy soil. Braz J Microbiol 42:105–113

El-Nagdi WMA, Youssef MMA (2004) Soaking faba bean seed in some bio-agent as prophylactic treatment for controlling *Meloidogyne incognita* root-knot nematode infection. J Pest Sci 77:75–78

El-Sayed WS, Akhkha A, El-Naggar MY, Elbadry M (2014) *In vitro* antagonistic activity, plant growth promoting traits and phylogenetic affiliation of rhizobacteria associated with wild plants grown in arid soil. Front Microbiol Plant Microbe Interact 5:1–11

Gaby JC, Buckley DH (2012) A comprehensive evaluation of PCR primers to amplify the *nifH* gene of nitrogenase. PLoS One 7:42149

Garg RN, Pathak H, Das D, Tomar R (2005) Use of flyash and biogas slurry for improving wheat yield and physical properties of soil. Environ Monit Assess 107:1–9

Gheysen G, Mitchum MG (2011) How nematodes manipulate plant development pathways for infection. Curr Opin Plant Biol 14:415–421

Gillet FX, Bournaud C, Junior JDAS, Grossi-de-Sa MF (2017) Plant-parasitic nematodes: towards understanding molecular players in stress responses. Ann Bot 119:775–789

Gopalakrishnan S, Sathya A, Vijayabharathi R, Varshney RK, Gowda CLL, Krishnamurthy L (2015) Plant growth promoting rhizobia: challenges and opportunities. 3 Biotech 5:355–377

Guang-Can T, Shu-Jun T, Miao-Ying C, Guang-Hui X (2008) Phosphate-solubilizing and -mineralizing abilities of bacteria isolated from soils. Pedosphere 18:515–523

Gulati A, Sharma N, Vyas P, Sood S, Rahi P, Pathania V, Prasad R (2010) Organic acid production and plant growth promotion as a function of phosphate solubilization by *Acinetobacter rhizosphaerae* strain BIHB 723 isolated from the cold deserts of the trans-Himalayas. Arc Microbiol 192:975–983

Hallmann J, Quadt-Hallmann A, Miller WG, Sikora RA, Lindow SE (2001) Endophytic colonization of plants by the biocontrol agent *Rhizobium etli* G12 in relation to *Meloidogyne incognita* infection. Phytopathology 91:415–422

Hamid M, Siddiqui IA, Shaukat SS (2003) Improvement of *Pseudomonas fluorescens* CHAO biocontrol activity against root-knot nematode by the addition of ammonium molybdate. Lett Appl Microbiol 36:239–244

Han HS, Lee KD (2006) Effect of co-inoculation with phosphate and potassium solubilizing bacteria on mineral uptake and growth of pepper and cucumber. Plant Soil Environ 52:130–136

Haris M, Ahmad G, Shakeel A, Khan AA (2018) Utilization of fly ash to improve the growth and the management of root-knot nematode on carrot. Haya Saudi J Life Sci 4(7):221–226

Hussey RS, Davis EL, Baum TJ (2002) Secrets in secretions: genes that control nematode parasitism in plants. Braz J Plant Physiol 14:183–194

Insunza V, Alstrom S, Eriksson KB (2002) Root bacteria from nematicidal plants and their biocontrol potential against trichodorid nematodes in potato. Plant Soil 241:271–278

Iram (2006) Response of root-knot nematode, *M. incognita* to fly ash on pepper (*Capsicum annuum* L.). M. Phil. dissertation, A.M.U., Aligarh, India

Iram (2010) Utilization of fly ash for the management of root-knot nematodes on some vegetable crops, Ph.D. thesis. Aligarh Muslim University, Aligarh

Izquierdo MT, Rubio B (2008) Carbon-enriched coal fly ash as a precursor of activated carbons for SO2 removal. J Hazard Mater 155:199–205. https://doi.org/10.1016/j.jhazmat.2007.11.047

Jamal Q, Lee YS, Munir S, Malik MS, Kim KY (2018) The combined use of five bacteria including Bacillus amyloliquefaciens Y1 as biofertilizer in compost improved low bush blueberry growth, rhizosphere bacteria and enzyme at various pH. Pak J Agric Sci 55(2)

Jasmer DP, Goverse A, Smant G (2003) Parasitic nematode interactions with mammals and plants. Annu Rev Phytopathol 41:245–270

Jiang CH, Fang W, Yu ZY et al (2014) Study on screening and antagonistic mechanisms of *Bacillus amyloliquefaciens* 54 against bacterial fruit blotch (BFB) caused by *Acidovoraxavenae* subsp. *citrulli*. Microbiol Res 170:95–104

Joshi SD, Tadhav AS, Patil NB, Kurundkar BP (2000) Effect of organic amendment and fly ash on root-knot disease of tomato. J Mahar Agric Univ 25:84–85

Kalra N, Harit R, Sharma S (2000) Effect of flyash incorporation on soil properties of texturally variant soils. Bioresour Technol 75:91–93

Karthik C, Oves M, Thangabalu R, Sharma R, Santhosh SB, Arulselvi PI (2016) *Cellulosimicrobium funkei*-like enhances the growth of *Phaseolus vulgaris* by modulating oxidative damage under Chromium (VI) toxicity. J Adv Res 7:839–850

Karthik C, Elangovan N, Senthil Kumar T, Govindharaju S, Barathi S, Oves M, Indra Arulselvi P (2017) Characterization of multifarious plant growth promoting traits of rhizobacterial strain AR6 under Chromium (VI) stress. Microbiol Res 204:65–71

Kerry BR (2000) Rhizosphere interactions and exploitation of microbial agents for the biological control of plant-parasitic nematodes. Annu Rev Phytopathol 38:423–441

Khan MR (1989) Studies on root-knot nematodes in relation to environmental pollution on vegetable crops. Ph.D. thesis A.M.U., Aligarh, India

Khan MR, Ghadirpour MH (1999) Management of root-knot nematodes of vegetables by fly ash application in soil. J Plant Dis Prot 106:632–634

Khan M, Siddiqui ZA (2017) Effects of fly ash amendments, Ralstonia solanacearum, Meloidogyne incognita and Phomopsis vexans on the growth of Solanum melongena. Acta Phytopathol Entomol Hung. https://doi.org/10.1556/038.52.2017.017

Khan MR, Khan MW, Singh K (1997) Management of root-knot disease of tomato by the application of fly ash in soil. Plant Pathol 46:33–43

Khan MS, Zaidi A, Wani PA, Ahemad M, Oves M (2009) Functional diversity among plant growth-promoting rhizobacteria: current status. In: Microbial strategies for crop improvement. Springer, Heidelberg, pp 105–132

Khan MR, Khan MM, Anwer MA, Haque Z (2012) Laboratory and field performance of some soil bacteria used as seed treatments on *Meloidogyne incognita* in chickpea. Nematol Mediterr 40:143–151

Kokalis-Burelle N, Fuentes-Borquez P, Adams P (2000) Effects of reduced risk alternatives on nematode populations and crop yield. In: Proceedings of Annual International Research Conference on Methyl Bromide Alternatives and Emissions Reductions, pp 61.1–61.2

Kokalis-Burelle N, Vavrina CS, Rosskopf EN, Shelby RA (2002) Field evaluation of plant growth-promoting rhizobacteria amended transplant mixes and soil solarization for tomato and pepper production in Florida. Plant Soil 238:257–266

Kumar A, Singh M, Singh PP, Singh SK, Singh PK, Pandey KD (2016) Isolation of plant growth promoting rhizobacteria and their impact on growth and curcumin content in Curcuma longa L. Biocatal Agric Biotechnol 8:1–7

Lanzerstorfer C (2018) Fly ash from coal combustion: dependence of the concentration of various elements on the particle size. Fuel 228:263–271

Li B, Xie GL, Soad A, Coosemans J (2005) Suppression of *Meloidogyne javanica* by antagonistic and plant growth promoting rhizobacteria. J Zhejiang Univ Sci B 6:496–501

Mahdy M, Hallmann J, Sikora RA (2001) Influence of plant species on the biological control activity of the antagonistic rhizobacterium *Rhizobium etli* strain G12 toward the root knot nematode *Meloidogyne incognita*. Meded Rijksuniv Gent Fak Landbouwkd Toegep Biol Wet 66:655–662

Mahmood Y, Khan MA, Javed N (2009) Evaluation of PGPR to control root-knot of mash bean infected by *Meloidogyne incognita* (Kofoid & White) Chitwood. Pak J Phytopathol 21:144–147

Mallesh SB, Lingraju S, Byadgi AS, Hegde YR, Mokashi AN, Krishnaraj PU (2009) Bioefficacy of rhizobacteria on root-knot/wilt disease complex in coleus and ashwganda. Karnataka J Agric Sci 22:1116–1120

Meena KS, Jonathan EI, Devrajan K, Raguchander T (2012) *Pseudomonas fluorescens* induced systemic resistance in tomato against *Meloidogyne incognita*. Indian J Nematol 42:5–10

Meena VS, Bahadur I, Maurya BR, Kumar A, Meena RK, Meena SK, Verma JP (2016) Potassium-solubilizing microorganism in evergreen agriculture: an overview. In: Meena VS et al (eds) Potassium solubilizing microorganisms for sustainable agriculture. Springer, New Delhi. https://doi.org/10.1007/978-81-322-2776-21

Mena J, Pimentel E (2002) Mechanism of action of *Corynebacterium pauronetabolum* strain C-924 on nematodes. Nematology 4:287

Mendes R, Garbeva P, Raaijm JM (2013) The rhizosphere microbiome: significance of plant beneficial, plant pathogenic, and human pathogenic microorganisms. Fems Microbiol Rev 37:634–663

Mhatre PH, Pankaj MSK, Kaur S, Singh AK, Mohan S, Sirohi A (2015) Histopathological changes and evaluation of resistance in Asian rice (*Oryza sativa*) against rice root knot nematode, *Meloidogyne graminicola*. Indian J Genet Plant Breed 75:41–48

Mhatre PH, Pankaj, Kumar J, Shakil NA, Kumar R, Adak T (2017) New formulations of salicylic acid and their bioefficacy evaluation on wheat against cereal cyst nematode. Indian J Nematol 47:155–165

Mhatre PH, Karthik C, Kadirvelu K, Divya KL, Venkatasalam EP, Srinivasan S et al (2018) Plant growth promoting Rhizobacteria (PGPR): a potential alternative tool for nematodes bio-control. Biocatal Agric Biotechnol:119–128

Niknam GR, Dhawan SC (2002) *In vitro* study on the efficacy of *Bacillus subtilis* strain BSt cell concentrations and cell-free filtrates on hatching and mobility of *Rotylenchulus reniformis*. Indian J Nematol 32:9–15

Oves M, Khan MS, Zaidi A (2013) Chromium reducing and plant growth promoting novel strain *Pseudomonas aeruginosa* OSG41 enhance chickpea growth in chromium amended soils. Eur J Soil Biol 56:72–83

Parab N, Mishra S, Bhonde S (2012) Prospects of bulk utilization of fly ash in agriculture for integrated nutrient management. Bull Nat Inst Ecol 23:31–46

Partridge JE (2008) Bacterial wilt of Alfalfa. Department of plant pathology, University of Nebraska-Lincoln. http://nudistance.unl.edu/homer/disease/agron/alfalfa/

Patel V, Reddy MN, Patel K (2018) Isolation and characterization of lead resistant indole acetic acid producing Rhizobacteria from fly ash contaminated site near thermal power plant

Peix A, Rivas-Boyero AA, Mateos PF, Rodriguez-Barrueco C, Martínez-Molina E, Velazquez E (2001) Growth promotion of chickpea and barley by a phosphate solubilizing strain of *Mesorhizobium mediterraneum* under growth chamber conditions. Soil Biol Biochem 33:103–110

Pieterse CMJ, VanWees SCM, Ton J, VanPelt JA, VanLoon LC (2002) Signalling in rhizobacteria-induced systemic resistance in *Arabidopsis thaliana*. Plant Biol 4:535–544

Pokhare S, Pankaj, Shakil NA, Kumar J, Singh K (2012) Foliar application of chemical elicitors induces biochemical changes in wheat against the cereal cyst nematode, *Heterodera avenae*. Nematol Mediterr 40:181–187

Prabhu S, Kumar S, Subramanian S, Sundaram SP (2009) Suppressive effect of Methylobacterium fujisawaense against root-knot nematode, *Meloidogyne incognita*. Indian J Nematol 39:165–169

Press MC, Phoenix GK (2005) Impacts of parasitic plants on natural communities. New Phytol 166:737–751

Priya SM (2015) Biomanagement of rice root knot nematode, Meloidogyne graminicola golden and brichfield in aerobic rice. Int J Manag Soc Sci 3:591–598

Raguchander T, Saravanakumar D, Balasurramanian P (2011) Molecular approaches to improvement of biocontrol agents of plant diseases. J Biol Control 25:71–84

Ramamoorthy V, Viswanathan R, Raguchander T, Prakasam V, Samiyappan R (2001) Induction by systemic resistance by plant growth promoting rhizobacteria in crop plants against pests and diseases. Crop Prot 20:1–11

Rashad FM, Fathy HM, El-Zayat AS, Elghonaimy AM (2015) Isolation and characterization of multifunctional *Streptomyces* species with antimicrobial, nematicidal and phytohormone activities from marine environments in Egypt. Microbiol Res 175:34–47

Reitz M, Rudolph K, Schroder I, Hoffmann-Hergarten S, Hallmann J, Sikora RA (2000) Lipopolysaccharides of *Rhizobium etli* strain G12 act in potato roots as an inducing agent of systemic resistance to infection by the cyst nematode *Globodera pallid*. Appl Environ Microbiol 66:3515–3518

Rose R, Irshad M, Tiyagi SA, Zehra K (2012) Conjoint effect of oil-seed cakes and *Pseudomonas fluorescens* on the growth of chickpea in relation to the management of plant-parasitic nematodes. Braz Arch Biol Technol 55:801–808

Ruanpanun P, Tangchitsomkid N, Hyde KD, Lumyong S (2010) Actinomycetes and fungi isolated from plant-parasitic nematode infested soils: screening of the effective biocontrol potential, indole-3-acetic acid and siderophore production. World J Microbiol Biotechnol 26:1569–1578

Rubio-Cabetas M, Minot J, Voisin R, Esmenjaud D (2001) Interaction of root-knot nematodes (RKN) and the bacterium Agrobacterium tumefaciens in roots of Prunus cerasifera: evidence of the protective effect of the Ma RKN resistance genes against expression of crown gall symptoms. Eur J Plant Pathol 107:433–441

Schneider SM, Rosskopf EN, Leesch JG, Chellemi DO, Bull CT, Mazzola M (2003) Research on alternatives to methyl bromide: pre-plant and post-harvest. Pest Manag Sci 59:814–826

Shaikh S, Saraf M (2016) Biofortification of Triticum aestivum through the inoculation of zinc solubilizing plant growth promoting rhizobacteria in field experiment. Biocatal Agric Biotechnol 9:120–126

Sharma SB, Sayyed RZ, Trivedi MH, Gobi TA (2013) Phosphate solubilizing microbes: sustainable approach for managing phosphorus deficiency in agricultural soils. Springer Plus 2:1–14

Siddique S, Grundler FMW (2018) Parasitic nematodes manipulate plant development to establish feeding sites. Curr Opin Microbiol 46:102–108

Siddiqui ZA, Mahmood I (1999) Role of bacteria in the management of plant parasitic nematodes: A review. Bioresour Technol 69:167–179

Siddiqui ZA, Mahmood I (2001) Effects of rhizobacteria and root symbionts on the reproduction of *Meloidogyne javanica* and growth of chickpea. Bioresour Technol 79:41–46

Siddiqui IA, Shaukat SS (2003a) Role of iron in rhizobacteria-mediated suppression of root-infecting fungi and root-knot nematode in tomato. Nematol Mediterr 31:11–14

Siddiqui IA, Shaukat SS (2003b) Endophytic bacteria: prospects and opportunities for the biological control of plant-parasitic nematodes. Nematol Mediterr 31:111–120

Siddiqui IA, Shaukat SS (2003c) Suppression of root-knot disease by *Pseudomonas fluorescens* CHA0 in tomato: importance of bacterial secondary metabolite 2,4- diacetylphloroglucinol. Soil Biol Biochem 35:1615–1623

Siddiqui ZA, Singh LP (2005) Effect of fly ash, Pseudomonas striata and Rhizobium on the reproduction of nematode, Meloidogyne incognita and on the growth and transpiration of pea. J Environ Biol 26:117–122

Siddiqui IA, Haas D, Heeb S (2005) Extracellular protease of *Pseudomonas fluorescens* CHA0, a biocontrol factor with activity against the root-knot nematode *Meloidogyne incognita*. Appl Environ Microbiol 71:5646–5649

Siddiqui ZA, Nesha R, Singh N, Alam S (2012) Interactions of plant parasitic nematodes and plant pathogenic bacteria. In: Maheshwari DK (ed) Bacteria in agrobiology. Plant probiotics. Springer, Berlin/Heidelberg, pp 251–267. ISBN 978-3-642-27515-9

Singh SK (1989) Studies on interaction of air pollutants and root-knot nematodes on some pulse crops. Ph. D. thesis, A.M.U., Aligarh

Singh K (1993) Impact assessment of root-knot nematode on air pollution stressed plants. Ph.D. thesis, A.M.U., Aligarh

Singh RK, Gupta NC (2014) Value added utilization of fly ash- prospective and sustainable solutions. Int J Appl Sci Eng Res 3:1–16

Singh H, Singh SN, Yunus M, Ahmad KS (1994) Growth response and element accumulation in *Beta vulgaris* L. raised in fly ash amended soils. Ecotoxicology 3:287–229

Singh K, Khan AA, Safiuddin (2011) Effect of fly ash on growth, yield and root-knot disease of soybean. Nematol Mediterr 39:127–131

Singh S, Singh B, Singh AP (2015) Nematodes: a threat to sustainability of agriculture. Procedia Environ Sci 29:215–216

Singh J, Mantha SS, Phalle VM (2018) Characterizing domestic electricity consumption in the indian urban household sector. Energ Buildings 170:74–82

Singh PP, Kujur A, Yadav A, Kumar A, Singh SK, Prakash B (2019) Mechanisms of plant-microbe interactions and its significance for sustainable agriculture. In: PGPR amelioration in sustainable agriculture. Woodhead Publishing, Duxford, pp 17–39

Sitaramaiah K, Pathak KN (1993) Nematode bacterial disease interactions. In: Khan MW (ed) Nematode interactions. Chapman and Hall, New York, pp 232–250 377

Stansbury C, MCKirdy S, Mackie A, Power G (2001) Bacterial Wilt Ralstonia Solanacearum Race-3, Exotic Threat to Western Australia. Fact Sheet ISSN 1443- 7783 No. 7

Sujatha N, Ammani K (2013) Siderophore production by the isolates of fluorescent Pseudomonads. Int J Curr Res Rev 5:1–7

Tabatabaei FS, Saeedizadeh A (2017) Rhizobacteria cooperative effect against *Meloidogyne javanica* in rhizosphere of legume seedlings. Hellenic Plant Protect J 10:25–34

Tailor AJ, Joshi BH (2014) Harnessing plant growth promoting rhizobacteria beyond nature: A review. J Plant Nutr 37:9

Tanweer A, Hisamuddin, Niyaz T, Robab MI (2007) Effect of fly ash amended soil on the development of root-knot nematode, Meloidogyne incognita and growth of ivy gourd, Coccinia cordifolia. Indian J of Nematol 37:119–122

Tararmum A, Khan AA, Diva I, Khan B (2001) Impact of fly ash on hatching, penetration and development of root-knot nematode, *Meloidogyne javanica*. Nematol Mediterr 29:215–218

Teymouri M, Ebrahimipour G, Karkhane M, Marzban A (2016) Metal resistant and phosphate solubilizing bacterium improves maize (*Zea mays*) growth and mitigates metal accumulation in plant. Biocatal Agric Biotechnol 8:13–17

Tian BY, Yang JK, Lian LH, Wang CY, Zhang KQ (2007) Role of neutral protease from *Brevibacillus laterosporus* in pathogenesis of nematode. Appl Microbiol Biotechnol 74:372–380

Timper P (2014) Conserving and enhancing biological control of nematodes. J Nematol 46:75–89

Tranier MS, Pognant-Gros J, Quiroz RDC, Gonzalez CNA, Mateille T, Roussos S (2014) Commercial biological control agents targeted against plant-parasitic root-knot nematodes. Braz Arch Biol Technol 57:831–841

Trudgill DL, Bolla G, Blok VC, Daudi A, Davies KG, Gowen SR, Fargette M, Madulu JD, Mateille T, Mwagenui W, Netscher C, Phillips MS, Sawadogo A, Trivino CG, Voyoukallou E (2000) The importance of tropical root-knot nematodes (*Meloidogyne* spp.) and factors affecting the utility of *Pasteuria penetrans* as a biocontrol agent. Nematology 2:823–845

Van-Loon LC (2000) Systemic induced resistance. In: Slusarenko AJ, Fraser RSS, van Loon LC (eds) Mechanisms of resistance to plant diseases. Dordrechet, Kluwer, pp 521–574

Veronika K, Khan MR (2015) Biomanagement of root-knot nematode (Meloidogyne incognita) infecting okra in West Bengal. India Indian J Nematol 45:178–183

Vessey JK (2003) Plant growth promoting rhizobacteria as biofertilizers. Plant Soil 255:571–586

Viljoen JJF, Labuschagne N, Fourie H et al (2019) Biological control of the root-knot nematode *Meloidogyne incognita* on tomatoes and carrots by plant growth-promoting rhizobacteria. Trop Plant Pathol. https://doi.org/10.1007/s40858-019-00283-2

Viswanathan R, Rajitha R, Ramesh Sundar A, Ramamoorthy V (2003) Isolation and identification of endophytic bacterial strains from sugarcane stalks and their *In Vitro* antagonism against the red rot pathogen. Sugarcane 5:2529

Wani AM (2015) Plant growth-promoting rhizobacteria as biocontrol agents of phytonematodes. In: Askary TH, Martinelli PRP (eds) Biocontrol agents of phytonematodes. CAB International, Wallingford, pp 339–362

Wei LH, Xu QY, Wei BQ, Wang YM, Li SM, Guo JH (2010) Screening of antagonistic bacterial strains against *Meloidogyne incognita* using protease activity. Biocontrol Sci Tech 20:739–750

Wei L, Shao Y, Wan J et al (2014) Isolation and characterization of a rhizobacterial antagonist of root-knot nematodes. PLoS One 9(1):e85988

Williamson VM, Gleason C (2003) Plant-nematode interactions. Curr Opin Plant Biol 6:327–333

Xiang N, Lawrence KS, Donal PA (2017) Biological control potential of plant growth-promoting rhizobacteria suppression of Meloidogyne incognita on cotton and Heterodera glycines on soybean: a review. J Phytopathol 166:449–458

Yao ZT, Ji XS, Sarker PK, Tang JH, Ge LQ, Xia MS et al (2015) A comprehensive review on the applications of coal fly ash. Earth Sci Rev 141:105–121

Yeledhalli N, Prakash S, Gurumurthy S, Ravi M (2010) Coal fly ash as modifier of physico-chemical and biological properties of soil. Karnataka J Agric Sci 20:531–534

Zahran HH (2001) Rhizobia from wild legumes: diversity, taxonomy, ecology, nitrogen fixation and biotechnology. J Biotechnol 91:143–153

Zaidi A, Khan MS, Ahemad M, Oves M (2009) Plant growth promotion by phosphate solubilizing bacteria. Acta Microbiol Immunol Hung 56:263–284

Zarina B (2007) Management of root-knot nematode (*Meloidogyne javanica*) on banana plantation. Pak J Nematol 25:73–78

Zhang C, Kong F (2014) Isolation and identification of potassium-solubilizing bacteria from tobacco rhizospheric soil and their effect on tobacco plants. Appl Soil Ecol 82:18–25

Zhou C, Gao Q, Luo W et al (2015) Preparation, characterization and adsorption evaluation of spherical mesoporous Al-MCM-41 from coal fly ash. J Taiwan Inst Chem Eng 52:147–157

Phytobiome Engineering and Its Impact on Next-Generation Agriculture

15

Baby Kumari, Mahendrakumar Mani, Anjali Chandrol Solanki, Manoj Kumar Solanki, Amandeep Hora, and M. A. Mallick

Abstract

Phytobiomes exist in plant biome and consist of diverse microbial communities. In the recent past, a new field emphasizing on the characterization of the plant-associated microbiome, referred to as the phytobiome, is seen as a solution toward the green agriculture methods as these phytobiomes can be easily manipulated for the betterment of agricultural practices in a eco-friendly method of crop production. Microbiome engineering plays a fundamental role in plant's requirements of nutrients and disease management, and it helps in maintaining the soil conditions for agricultural production. This chapter discusses about the influential role of microbiome on the plant and related environment. This chapter focuses on the role of quorum sensing and signaling and its application in the eco-friendly method of farming that is sustainable agriculture.

Keywords

Phytobiomes · Eco-friendly · Quorum sensing · Sustainable agriculture

B. Kumari (✉)
University Department of Biotechnology, Vinoba Bhave University, Hazaribagh, Jharkhand, India

M. Mani · A. Hora
Department of Biotechnology, Guru Nanak College (Autonomous), Chennai, Tamil Nadu, India

A. C. Solanki
Soil Science and Agriculture Chemistry, Jawaharlal Nehru Agricultural University, Jabalpur, Madhya Pradesh, India

M. K. Solanki
Department of Food Quality & Safety, Institute for Post-harvest and Food Sciences, The Volcani Center, Agricultural Research Organization, Rishon LeZion, Israel

M. A. Mallick
Advance Science and Technology Research Centre, Vinoba Bhave University, Hazaribagh, Jharkhand, India

© Springer Nature Singapore Pte Ltd. 2020

M. K. Solanki et al. (eds.), *Phytobiomes: Current Insights and Future Vistas*,
https://doi.org/10.1007/978-981-15-3151-4_15

15.1 Introduction

In the recent past, a new field emphasizing on the characterization of the plant-associated microbiome is referred to as the phytobiome. These phytobiomes can impart its role in the production of enhanced yield and disease-resistant and ecologically adapted crops, along with preserving the natural ecosystem. Phytobiome consists of all living organisms in, on, or around plants (e.g., microorganisms, animals, other plants) and the environment (i.e., soil, air, water, and climate). The phytomicrobiomes are vibrant in composition when evaluated with the atmosphere in which they thrive (Hawkes and Connor 2017). The plant microflora is composed of a diversity of microbes concomitant within a typical habitat representing an ecological balance (Jasim et al. 2014).

Soil microorganisms play fundamental roles in achieving food demand as per increasing global demand, and it helps in maintaining the soil quality (Kumari et al. 2017; Majeed et al. 2015; Rosier et al. 2018). This method has further made us comprehend that microbes could momentously influence host improvement and evolutionary dynamics of plant populations. Overall, possibilities of a new green revolution rely on the ecological management and engineering of this microbiome and its practical applications in an eco-friendly manner (Gupta and Dikshit 2010; Patil and Solanki 2016).

The phytobiome complex communicates within its network (consisting microflora, environment, and the host plant) through various mechanisms like nutrient recycling, competition for nutrients, antagonism, and chemical signals. These chemical signal molecules are responsible for the maintenance of ecological balance in the soil community. There is a need for understanding these signal systems for its developing new strategies of breeding, to combat different stresses (biotic and abiotic) in the agricultural system. Recently the researchers are focusing on new horizon of metagenomics to get highly efficient molecular database of phytobiomes and other omics technologies to get a better understanding of plant microbiome and its communication within the plant system and with the surroundings as well. Therefore, there is an urgent need of interdisciplinary, system approach for the study of phytobiome's dynamics and interactions.

15.2 Interdependency of Plant and Microbiome: Novel Opportunity

The mutual interaction of microbiome and plant in the rhizosphere has led nutrient and disease management in plants. The plant directly or indirectly influences the microbes around its rhizosphere; this potential of plant provides us a novel opportunity toward understanding their interaction and microbiome engineering. The role of these rhizospheric microorganisms becomes important for ecologists and researchers as it can benefit plant health, leading us toward sustainable methods of farming. The microbiome helps plants in nutrient enhancement, disease resistance, and resistance toward different biotic and abiotic stresses, ozone depletion and increasing carbon-dioxide level is one of them (Bacon et al. 2015; Kasim et al. 2016). The application of microbiome on main agricultural crops like wheat, maize, rice, and barley is extensively studied during the last two decades (Chen et al. 2012; Moshynets et al. 2019; Mücke et al. 2019; Rankl et al. 2016; Sivakumar et al. 2018; Yuan et al. 2010) (Table 15.1). The microbes (bacteria, fungus, viruses, and nematodes) and rhizosphere have great impact on attaining sustainable agriculture goals as new techniques like high-throughput analysis, genome banking, metagenomics, proteomics, and micro-engineering assist crop production methods (Bais et al. 2006; Doornbos et al. 2012; Finkel et al. 2017).

15.2.1 Role of the Plant

Recently, researchers and scientists are focusing on two major aspects to promote potential and helpful microbiome utilization near rhizosphere so as to get improved agricultural products (Bakker et al. 2012). The first aspect relies on manipulating the specific microbial diversity as per the requirement of the crop plant; it helps in nutrient uptake and provides disease resistance for the plant. Another aspect utilizes the plant's ability to attract specific microbiome for maintaining the promotion of plant growth and susceptibility toward adverse ecological conditions. In the nearest future, direct manipulation of the soil microbiome would lead us to attain green agriculture goal (Kumari et al. 2019a; Pandin et al. 2017; Rognes et al. 2016).

15.2.2 Role of the Microbiome: Bacterial Microbiome

For decades, rhizobia were believed to be the only nitrogen-fixing population of legume nodules. However, other bacteria, which are not typical rhizobia, are habitually perceived within nodules obtained from soil, thus exposing the existence of a phytomicrobiome where the interaction among the individuals not only is complex but also likely affects the behavior and fitness of the host plant (Borriss 2015; Igual et al. 2001; Kashyap et al. 2018; Lemanceau et al. 2017). It is also noted that the incredibly diverse population of bacteria residing within nodules induce neither nodulation nor nitrogen fixation. This community exists within the nodule, albeit outnumbered by nitrogen-fixing rhizobia. This phytomicrobiome has the potential to enhance plant survival, particularly under environmental stress conditions. This knowledge has paved the way for research in bringing out strategies to formulate bio-inoculants utilizing these rhizobia (Albareda et al. 2006; Sang et al. 2018).

15.2.3 Fungal Microbiome

A variety of fungal microbiomes have been reported in plants and soil atmosphere. Several reports have demonstrated the role of fungal association with plants like *Alternaria, Fusarium* spp. *Acremonium*, *Cladosporium*, *Epicoccum* (Bailey et al.

Table 15.1 Phytobiomes and their interaction with majors crops triggering beneficial effects

Phytobiomes	Occurrence	Targeted gene/plant	Sequencing methods/platform	Significance	References
Maize (*Zea maize*)					
Frigoribacterium sp., *Microbacterium* sp., *Pantoea* sp., *Sphingomonas* sp., *Bacillus* sp., *Paenibacillus* sp.	Associated with kernel of maize rhizosphere	ACC deaminase, IAA pathway	Sanger sequencing	Antifungal	Rijavec et al. (2007)
Alcaligenes, Bacillus, Erwinia, Methylobacterium, Microbacterium, Tsukamurella, and Rhodococcus	Enhanced colonization in roots exposed to high-salinity conditions	ACC deaminase, IAA, *acdS* gene	Sanger sequencing		Rosenblueth et al. (2012), Bouffaud et al. (2018)
Pantoea agglomerans	Enhanced colonization in roots exposed to high-salinity conditions	Type 2 (*PIP2-1*) gene	16 s RNA	Salt tolerance	Gond et al. (2015)
Burkholderia phytofirmans	Enhanced colonization in roots exposed to high-salinity conditions	ACC deaminase, IAA	16 s RNA	Salt tolerance	Santoyo et al. (2016)
Azospirillum lipoferum CRT 1, *A. brasilense*	Positively affects metabolome (sugar metabolism)	PGP, phytohormone production, IAA pathway	16 s RNA	PGP	Rozier et al. (2017), Steenhoudt and Vanderleyden (2000)
P. thivervalensis, Serratia marcescens	Enhanced colonization in rhizosphere under the fertilized condition	PGP, phytohormone production, IAA pathway	16 s RNA	PGP, phytohormone production	Shahzad et al. (2013)

Pseudomonas putida KT2440		Aminotransferase	16S and ITS2, in vitro expression technology (IVET), rRNA gene amplicon analysis	Capacity to remove xenobiotics, improved soil fitness	Ramos-González et al. (2005)
Pseudomonas aeruginosa Pseudomonas monteilii, *Enterobacter asburiae*	Enhanced colonization in roots exposed to drought conditions	Indole acetic acid, siderophore, phosphate solubilization, hydrolytic enzymes	Biochemical and 16S rDNA gene sequencing	Antifungal, under drought stress	Sandhya et al. (2017)
Rice (*Oryza sativa*)					
Azospirillum spp., *Pseudomonas sp.*	Enhanced colonization in roots, increased nitrogen assimilation	Indole acetic acid, siderophore, phosphate solubilization, hydrolytic enzymes	16 s RNA	PGP, and biofertilizer	Gond et al. (2015)
Bacillus amyloliquefaciens	Improve metabolism in abiotic stress condition	Proline, sugar, glycine metabolism	3-D QT-PCR	Improved saline drought and stress tolerance	Tiwari et al. (2017)
Bacillus subtilis	Improve metabolism	Malic acid pathway	OSMS	PGP, phytohormone production	Rekha et al. (2017)

(continued)

Table 15.1 (continued)

Phytobiomes	Occurrence	Targeted gene/plant	Sequencing methods/platform	Significance	References
Acinetobacter sp., *Curtobacterium citreum*, *Curtobacterium* sp., *Microbacterium* sp., *Paenibacillus* sp., *Pantoea agglomerans*, *Pantoea ananatis*, *Pantoea* sp., *Pseudomonas* sp., *Rhizobium larrymoorei*, *Sphingomonas* sp., and *Staphylococcus cohnii*	Improve metabolism in abiotic stress condition	Indole acetic acid, siderophore, phosphate solubilization, hydrolytic enzymes	3-D QT-PCR, high-throughput methods	PGP, phytohormone production, PO_4 solubilization, and antifungal metabolites production	Ruiza et al. (2011)
Agromyces mediolanus, *Curtobacterium citreum*, *Curtobacterium herbarum*, *Curtobacterium* sp., *Flavobacterium johnsoniae*	Enhanced colonization in roots, increased nitrogen assimilation	Indole acetic acid, siderophore, phosphate solubilization, hydrolytic enzymes	Sanger sequencing	PGP, mitigating of biotic and abiotic stresses	Hardoim et al. (2012)
Wheat (*Triticum aestivum*)					
Bacillus sp., *Paenibacillus* sp., and *Pantoea* sp.	PO_4 solubilization, PGP, IAA antifungal compound and siderophore		Sanger sequencing		Díaz Herrera et al. (2016)
Indian mustard and pumpkin					
Enterobacter cloacae, *Pseudomonas sp.*, *Bacillus sp.*	Phytohormone production, nutrient enrichment	PGP, phytohormone production, IAA pathway		Enhanced plant growth and reduced metal contamination, Cr (VI) content	Ahemad and Kibret (2014)

Soybean *(Glycine max)*					
Pseudomonas fluorescens japonicum CB 1809 and USDA110, Azospirillum sp., Bacillus pumilus, Rhizobium japonicum, Azotobacter chroococcum, and Azospirillum brasilense				More nodulation, increased growth and seed protein, water stress tolerance, drought resistance	Aung et al. (2013), Masciarelli et al. (2014), Naseri et al. (2013), Zahedi and Abbasi (2015)
Bradyrhizobium japonicum, Bacillus thuringiensis	The highest and most consistent increase in nodulation, shoot dry weight and root fresh weight, and root volume				Mishra et al. (2009)
B. japonicum, A. brasilense	Indole acetic acid, siderophore, phosphate solubilization, hydrolytic enzymes			Seed germination, nodule formation, and early development of soybean seedlings	Cassan et al. (2009)
Barley *(Hordeum vulgare)*					
Bacillus amyloliquefaciens, Bacillus megaterium M3 and MIX (Bacillus subtilis OSU142, B. megaterium M3), Azospirillum brasilense Sp245	Improve metabolism	Biofilm formation	–	Increased yield, growth of seedlings, salinity tolerance	Kasim et al. (2016)

(continued)

Table 15.1 (continued)

Phytobiomes	Occurrence	Targeted gene/plant	Sequencing methods/platform	Significance	References
Gram (*Cicer arietinum* L.)					
Pseudomonas fluorescens, *Mesorhizobium* sp.	Enhanced colonization in roots	ACC deaminase, IAA	16 s RNA	The maximum increase in nodule number, dry matter, and nutrient content	Verma et al. (2012)
Bacillus sp., *Mesorhizobium* sp.,	–	–	–	Increase in plant P uptake	Wani et al. (2007)
Pseudomonas jessenii, *Mesorhizobium ciceri*	Rhizospheric nodulation	PGP, phytohormone production, IAA pathway	–	Increased nodule size, weight, N contents in shoot and seed yield	Valverde et al. (2006)
Pseudomonas sp., *Mesorhizobium sp.*	Rhizospheric nodulation	PGP, phytohormone production	–	Enhancement of nodulation and stimulation of plant growth	Malik and Sindhu (2011)

2006; Baker 1988; Chaparro et al. 2011; Mohammad et al. 2012; Mohsenzadeh and Shahrokhi 2014), *Penicillium* spp. Eurotiomycetes, Dothideomycetes, Leotiomycetes, Sordariomycetes, Tremellomycetes, *Aureobasidium, Cladosporium, Chaetomium* (Almeida et al. 2007; Cook 1993; Harman et al. 2004; Jetiyanon et al. 2003), *Fusarium* and its teleomorphs, *Microdochium, Stemphylium,* and *Xylaria* (Yadav et al. 2015).

Many fungi form vesicular-arbuscular mycorrhizal (VAM) relationship with plants. These fungi provide nutrient absorption in root and help as a carbohydrate source for the fungi in return (Harman et al. 2004). So it is possible to have two-thirds of all plants form a relationship with VAM fungi in their root systems (Tilman et al. 2002). Some VAM fungal relationships with plants depend on fungi to provide sufficient levels of phosphorus (Svenningsen et al. 2018). Phosphorus absorption is not the only benefit derived by the plant from the mycorrhizal relationship, but also VAM fungi protect plant roots from pathogens.

15.2.4 Other Insects

Along with symbiotic and asymbiotic PGPR and diversity of fungal microbiomes, there is a special class of organisms like insects which affects plant's aboveground parts like leaves and stems. These may act as causative agents of different plant diseases. Meanwhile some can be beneficial for the plants. Some insect's behavior helps plants in its defense mechanism, as they may feed on some harmful bacteria or fungal strains and, therefore, the plants are indirectly benefitted. Such insects include caterpillar, aphids, and grasshoppers. Recently Hannula et al. (2019) studied the behavior of insects in maintaining plant-soil interactions. They hypothesized that plant-mediated changes in soil microbiome would affect the microbiome of caterpillars feeding on plants that grow later in these soils, through modifications of the microbiome of their host plants. After analyzing the data, they found that insect microbiome relied on soil microbiome and those effects of plants on soil microbiome. This mutual interaction of plant and soil microbiome affects the aboveground insects which later on feeds the other plants (Hannula et al. 2019).

15.3 Role of the Environment

15.3.1 Mechanism of Communication

15.3.1.1 Plant-Microbe Signaling

Plant-microbe interacts through chemical signal system produced as exudates and the microorganisms in the soil. Plants root exudates attract the microbiome, which is often called rhizodeposition. The rhizodeposition mainly consists of carbon compounds, mucilage, soluble root exudates, and some organic carbon both volatile and nonvolatile (Bais et al. 2006; Doornbos et al. 2012; Finkel et al. 2017; Hartmann et al. 2009). These compounds help the host plant to attract a diversity of bacteria, including both prokaryotes and eukaryotes (Doornbos et al. 2012). Plant secrets amino acid and carbohydrate as exudates through roots which attract a variety of bacteria in the rhizosphere as compared to bulk soil (Kumari et al. 2019b); besides, a number of chemical signals triggered by bacteria and plants interact to communicate. Mark et al. (2005) characterized expression of some genes responsible for the mutualistic association of *Pseudomonas* strain in the soil with beetroot plant, and It induced more colonization of *Pseudomonas aeruginosa* strain PAO1in the host plant rhizosphere. This condition was caused due to some exudates secreted from the host plant (Mark et al. 2005). Similarly, Han et al. (2016) reported that two isolates from banana (*B. subtilis* N11) and cucumber rhizosphere (*B. amyloliquefaciens* SQR9) which were reciprocally inoculated in same plants indicated more enrichment of microbiome colonization among cucumber and banana plant-soil interaction zone. Chemotaxis and biofilm formation were reported as the reason of this increased interaction with reciprocal rhizospheric soil and plant environment. Plants root exudates trigger signals which led to communication among the microbes and neighboring plants in the rhizospheric zone. Some plants secrete a chemical (mucilage) which acts as biocontrol agent and prevents the growth of pathogenic microbes around the meristamic zone of root tip and elongating cells (Walker et al. 2003). A compound canavanine, which mostly resembles arginine, is exudates from certain leguminous roots. Canavanine has toxic effects on native bacteria around the host plants, but a different strain of bacteria in the same rhizosphere can detoxify this compound. In this way, a specific strain of bacteria is considered beneficial for the rhizospheric colonization among the legumes (Cai et al. 2009). The signals produced by phytobiomes are known as quorum sensing signals (QSS). Transcriptosome and proteome of host plant play active role responding to the signals produced by microbiomes. These responses are brought about by specific changes in the plant protein in response to structure and specific concentration of quorum sensing molecules. Quorum sensing and biofilm formation are considered key feature of microbiome engineering as it provides the agronomist a platform to manipulate the microbiome as per the requirement of modern agricultural method in a sustainable manner. Table 15.2 shows a recent advancement in the field of microbiome engineering and quorum sensing (QS) molecular studies in different plants and agricultural crops.

15.3.1.2 Quorum Sensing and Biofilm Formation

Cell-to-cell communication between bacteria is mediated via diffusible chemical signals, known as quorum sensing (Abisado et al. 2018). During the process of swarming, virulence, and biofilm formation, regulation of microbial genes is through the quorum sensing molecules. The complex extracellular matrix of exopolysaccharides and proteins is mainly involved in the formation of biofilm (Gond et al. 2015; Rosier et al. 2018). This embedded matrix of compounds helps the bacterial population in the adherence on or inside plant tissue and cells (Kashyap et al. 2019; Rankl et al. 2016). It is reported that in axenic conditions, *Bacillus amyloliquefaciens* sp. plantarum FZB42 transposon mutagenesis leads to gene expression related to bacterial swarming, biofilm formation, root colonization, and plant

Table 15.2 Quorum-sensing-mediated communication mechanism of phytobiome (special reference to staple crops)

Bacteria	Occurrence	Effect	Method of analysis	References
Pantoea stewartii	Maize rhizosphere	The causative agent of Stewart's wilt	Transcriptome EsaR regulon, whole genome sequencing	Ramachandran et al. (2014), Tan et al. (2015)
Burkholderia sp.	Maize rhizosphere	PGP and antifungal activity	Whole genome sequencing	Abisado et al. (2018)
Pseudomonas aylmerense sp.	Maize rhizosphere	Xenobiotic degrader and PGP	16S rRNA sequence analysis, DNA fingerprinting, BLAST	Tchagang et al. (2018)
Pseudomonas aeruginosa PGPR2	Maize rhizosphere	PGP and disease resistance activity	INSeq technology	Sivakumar et al. (2018)
Xanthomonas oryzae pv. *oryzae* (Xoo)	Rice rhizosphere	Causes bacterial blight disease in rice through its flagellin	Solexa/Illumina sequencing	Yu et al. (2014)
Xanthomonas oryzae pv. *oryzae* (Xoo)	Rice rhizosphere	XA13, COPT1, and COPT5 proteins are linked to causing copper redistribution in rice plant	qRT-PCR	Yuan et al. (2010)
Xanthomonas oryzae pv. *oryzae* *(Xoo*	*Rice rhizosphere*	*Set of proteins involved in causing virulence in rice*	*TALEs target genes were identified, causing virulence in rice through* in silico *methods*	Mücke et al. (2019)
Magnaporthe oryzae (fungus)	*Rice rhizosphere,*	*Causes rice blast,*	*Avirulence genes identified,*	Yoshida et al. (2009)
Bacillus, Pseudomonas spp. *(AHLs) in wheat*	*wheat rhizosphere*	*improves growth and resistance in wheat*	*qRT-PCR*	Moshynets et al. (2019)
Burkholderia glumae	*Rice rhizosphere*	*Causes blight of rice*	*luxI homolog, tofI, and luxR homolog, tofR by DNA sequencing*	Chen et al. (2012)
Pseudomonas ultra-performance liquid chromatography (UPLC)'s spp.	*Rice, wheat, maize*	*Responsible for PGP*	*Acylated homoserine lactone was identified*	Venturi (2006)

(continued)

Table 15.2 (continued)

Bacteria	Occurrence	Effect	Method of analysis	References
PGPRs of barley	*Barley*	*Responsible for lateral root formation, increased K uptake*	*AHLs were identified*	Rankl et al. (2016)
PGPRs of barley	*Barley*	*Responsible for lateral root formation, increased K uptake*	AHLs identified by UPLC	Götz et al. (2007)
Acidovorax radicis N35	*Primed to barley seeds*	*Responsible for PGP*	*AHL compound detection through pcr + biosensor strain*	Han et al. (2016)
Paenibacillus, Pantoea, and *Pseudomonas* spp.	*Primed to barley seeds*	*Phytohormones stimulation*	*Illumina Miseq*	Yang et al. (2017)
Paenibacillus, Pantoea, and *Pseudomonas* spp.	*Primed to barley seeds*	*PGP improved mineral nutrition and induced resistance against the fungal pathogen Blumeria graminis*	*Ion Torrent (gyrB gene)*	Rahman et al. (2018)

PGP plant growth promotion, AHL *N*-acyl homoserine lactone, *PCR* polymerase chain reaction, *UPLC* ultra-performance liquid chromatography, *TALE* transcription activator-like effector

growth promotion (Budiharjo et al. 2014). The bacterial strain which is involved in film formation is quite more tolerant to antimicrobial compounds; they are different from other nitrogen-fixing bacteria (free living) phenotypically and physiologically as well (Kavamura and de Melo 2014). Free-living diazotroph *Azospirillum brasilense* found in the wheat rhizosphere promotes growth through biofilm formation in the root zone (Aßmus et al. 1995). It is suggested that root exudates are initiator and manipulator molecules for the biological and physiological interactions between microorganisms and the host plant root system (Bais et al. 2004). Bais et al. reported about the role of biofilm formation on biocontrol in an experiment carried out on *Arabidopsis*. He observed that when *Bacillus subtilis* mutant strain was inoculated with *Arabidopsis*, there was no biofilm formation and *Arabidopsis* was infected with *Pseudomonas syringae*, whereas when another mutant *B. subtilis* subtilis strain 6051 was inoculated it was able to form biofilm and triggered the biocontrol on to *P. syringae*, *B. subtilis* strain 6051 produced some specific antimicrobial compound (a lipopeptide) and surfactin which were lethal to *P. syringae*. It could be concluded that biofilm formation is an essential factor in the biocontrol activity of phytobiomes (Bais et al. 2004a).

15.3.1.3 Phytobiome Engineering

According to the United Nation's recent report, the global population will overpass the data of more than nine billion by twenty-second century (Kumari et al. 2017). To meet the ever-increasing demand for food, there must be more agricultural yield per unit area of available agricultural land. The role of plant microbiome has been now seen as an effective way to enhance the global food production, in an eco-friendly manner. This practice would need the engineering of phytobiomes. Upon effective way of manipulation, these phytobiomes may encounter the problems arose due to excessive use of inorganic fertilizers. The phytobiome engineering is creating an opportunity to enhance the nutritional requirement of plants in a sustainable manner, beside that they are also enabling the host plant to combat the physiological stresses like heat, salinity, and other stresses (Abhinandan et al. 2018; Bouffaud et al. 2018). Thus, the significance of evolving approaches that involve engineering of phytobiomes for plant's growth promotion, enhanced nutrition availability, biocontrol, and disease resistance is currently acknowledged (Kashyap et al. 2018; Kumari et al. 2019a; Pandin et al. 2017; Rosier et al. 2018).

The richness of microbial diversity in the rhizospheric soil is greatly enhanced and altered with the crop rotation technique of agriculture, which allows the phytobiome manipulation. Altered microbial diversity in the root zone enables the plant to avail better nutrients and more disease resistance as well (Cook 1993; Doornbos et al. 2012; Finkel et al. 2017). Researchers observed that the altered microbial strains affected the cultivars especially corn (Foo et al. 2017; Pham et al. 2017; Rekha et al. 2017; Rozier et al. 2017). The other way of manipulating the microbial community is the co-inoculating with other strains; this strategy is found to be useful in plant disease management also (Khezri et al. 2011; Pandin et al. 2017; Sumi et al. 2015). Further, the growth management of inoculated microbiome is crucial as the reduction of time in its exploration would directly affect the colonizing capacity around the root zone; co-inoculation of beneficial strains could be a crucial factor in achieving the proper exploration of inoculated phytobiomes. The co-inoculation of specific strain is also vital as they may secret antibiotics and disease-resistant compounds (Larran et al. 2016; Rafique et al. 2015; Sudha et al. 2016; Tao et al. 2014). Some microorganisms like *Rhizobium* spp. have the capacity to enhance the nutritional availability of the soil, and ultimately, the plant gets its benefit. Rhizobium spp. are reported to have the potential of enhancing nitrogen contents in soil, especially in leguminous plants (Bach et al. 2016; Han et al. 2016; Manjunath et al. 2016; Timmusk et al. 2014). Some inoculants have the potential to improve the phosphate level in the soil; these inoculations could be incorporated with rock phosphates (Chakdar et al. 2018; Geetha and Joshi 2013). Although this manipulation has excellent possibilities for sustainable methods of attaining desired yield goals, few things may be considered as limiting factors. The maintenance of the desired density of inoculated strain is a point of consideration, as they start decreasing by their density with time. Another significant consent is as follows: are these engineered phytobiomes free from other chemicals like their metabolites and antibiotic compounds? So these methods of manipulating phytobiomes must assure their ethical adaptability from the human health point of view (Amundson et al. 2015; Morawicki and Díaz González 2018).

Table 15.1 describes the mutual association and beneficial activities of the microbiome and the target plant. Scientists are trying to establish the best delivery system for enabling the PGPR and other microbes to communicate in more important perspectives. Researchers are nowadays focusing on the development of better carrier molecules and multi-strain inoculation environment. Another focused area is the formulation of low-cost yet sustainable encapsulation systems and a better transition of it from laboratory to the farmer's field. These technologies have to be implemented soon for the development of transgenic crops (Bargaz et al. 2018; Kafle et al. 2019; Ribeiro et al. 2018; Rosier et al. 2018; de Souza et al. 2015).

Several PGPR inoculums products are currently in the market globally (Kumari et al. 2019b). The main problem in fungal microbiome engineering is actually due to difficulties in culturing them in the absence of their host/obligate symbionts. In this condition, inoculums production gets limited, though several global companies are launching AM inoculums products, useful in forestry, agriculture, and horticulture. New developments like in vitro monoxenic root organ cultures have enabled the scientists to establish AM fungal culture successfully. Molecular identification and different in silico lab techniques are enabling us to understand the communication among them (Hunter et al. 2017; Jacoby et al. 2017; de Souza et al. 2015).

15.4 Its Application on Sustainable Agriculture

15.4.1 Soil Health

Recent advances in the field of microbial identification and its metabolite interaction (signals) attracted the researchers toward its practical applications for sustainable and integrated approaches to agriculture. This microbiome has potential to increase plant nutrient uptake (Abhinandan et al. 2018; Santoyo et al. 2016) and enhance plant growth (Aung et al. 2013; Baltrus 2017; Díaz Herrera et al. 2016; Zahedi and Abbasi 2015). Benefits of these living microorganisms are vibrant and potentially self-sustaining, which ultimately reduce the need for repeated applications, pests attack, and pathogens that evolve resistance to the treatments (Lucas 2011). Phytobiomes help the host plant by increasing their growth (enhances the phytohormone biosynthesis, nutrient availability) directly or indirectly (Brown and Saa 2015; du Jardin 2015). They also enable the plant against different biotic and abiotic stress. Different microorganism-based bioformulations are available in the agricultural market. These microbiomes are engineered or co-inoculated PGPRs like *Serratia*, *Variovorax*, and *Azotobacter* species. Endophytic bacteria and fungal formulations are beneficial for the plant (Nakkeeran et al. 2006; Barea 2015; Bishnoi 2015; FAO 2016; Le Mire et al. 2016).

15.4.2 Disease Control

Consistent use of inorganic fertilizer and pesticides has led to dangerous conditions of soil due to its residual persistence, which raises food safety concerns among the consumers (Ahemad and Kibret 2014; Anand et al. 2016). Recently PGPRs and other microflora have been proven to be better biocontrol agents as they help the plant in fighting against pathogens and by electing antagonistism (Beattie 2006). These microorganisms are designed now with advanced systems like proteomics and high-throughput data analysis methods. These methods are economically acceptable, and as per the host's requirement, they are easily degradable. These formulations are a better solution to integrated pest management (IPM) programs. Being safe to use, ecologically adaptable, and easy to handle, it is an excellent opportunity for a new green agricultural method of farming. This microbiome increases the availability of nutrients and effects positively on the plant hormone biosynthesis. They produce siderophores enabling the plant to fight against the harmful pathogens (Kumari et al. 2016; Yang et al. 2009; Haghighi et al. 2011; Anand et al. 2016; Le Mire et al. 2016; Rubin et al. 2017). A variety of bacteria, specially *Bacillus*, are reported to possess the growth-promoting activities along with disease resistance in agricultural and medicinal plants (Kashyap et al. 2019; Kumari et al. 2017). Nanomaterial-based biocontrol strain engineering is another fascinating area on which researchers are working to synthesize nanoparticle or nanofiber-based sensor tool and formulation development with minimal impact on natural rhizosphere (Kashyap et al. 2019; Kumari et al. 2019b).

15.5 Challenges of Phytobiome Engineering

The modern technologies like high-throughput sequence analysis methods have influenced the characterization methods of plant microbiome. These technologies have an impact on promoting high-level research on plant health in a stressed condition.

Another technological advancement that took place recently is metagenomic analysis. This technology is enabling the identification of a functional group of microbes based on their metabolic potential. This technology helps us in the study of different types of secretions from plant root (exudates) and their interaction toward plant microbiome and surroundings.

Besides these developments, there are some challenges in the field of phytobiome engineering, and this may be due to a large number of genomic data and diversity of data as well. The influence of environment is also an important factor that controls the changes in the microbial diversity in plant biome. The biggest limitation of micro-engineering-based agricultural advancement is its ethical acceptance. The farmer is properly not well educated about the costiveness of this recent agricultural facility. Another drawback is this engineered product is not easily available for commercialization.

15.6 Future Prospects

Presently, diverse research tactics are being addressed to discover the aspect of microbiome engineering parameters while averting the presence of pathogens. Upon getting fruitful rhizosphere, new advantageous mechanistic approaches can lead to new advancement in its engineering. By exploiting the useful microbial services, we could enhance rhizobial performance or persistence. It would decrease the demand of agrochemical fertilizers and pesticides in our agricultural system. A combination of all of these approaches can increase our understanding of how to enrich the effectiveness and tenacity of bacteria in the rhizosphere to finally improve plant health and agroecosystem productivity.

15.7 Conclusion

It is evident that all plants can acquire a variety of microorganisms throughout all phases of their lifetime. Phytobiome characterizes plant growth and evolutionary change, and thus, a thorough understanding of the relationship among plant and their associated microbes could facilitate the engineering of these communities. Microbial inoculants could be a boon to the agricultural sector if implemented on following aspects: (i) to increase the scientific/technical aspects of inoculum production; (ii) to produce precise normative for the preparation of inoculum, either from seed or on soil, or the plant on which inoculum is to experiment; (iii) to reduce the unpredictability of the field results; and (iv) to increase knowledge and propagation by elaborating its usefullness and limitations to the society

References

Abhinandan K, Skori L, Stanic M, Hickerson NMN, Jamshed M, Samuel MA (2018) Abiotic stress signaling in Wheat – an inclusive overview of hormonal interactions during abiotic stress responses in wheat. Front Plant Sci 9:734

Abisado RG, Benomar S, Klaus JR, Dandekar AA, Chandler JR (2018) Bacterial quorum sensing and microbial community interactions. MBio 9:e02331–e02317. https://doi.org/10.1128/mBio.02331-17

Ahemad M, Kibret M (2014) Mechanisms and applications of plant growth promoting rhizobacteria: current perspective. J King Saud Univ – Sci 26:1–20. https://doi.org/10.1016/j.jksus.2013.05.001

Albareda M, Dardanelli MS, Sousa C, MegÃas M, Temprano F, RodrÃguez-Navarro DN (2006) Factors affecting the attachment of rhizospheric bacteria to bean and soybean roots. FEMS Microbiol Lett 259:67–73. https://doi.org/10.1111/j.1574-6968.2006.00244.x

Almeida FB d R, Cerqueira FM, Silva R d N, Ulhoa CJ, Lima AL (2007) Mycoparasitism studies of Trichoderma harzianum strains against Rhizoctonia solani: evaluation of coiling and hydrolytic enzyme production. Biotechnol Lett 29:1189–1193. https://doi.org/10.1007/s10529-007-9372-z

Aßmus B, Hutzler P, Kirchhof G, RI A, Lawrence J, Hartmann A (1995) In situ localization of Azospirillum brasilense in the rhizosphere of wheat with fluorescently labeled, rRNA-targeted oligonucleotide probes and scanning confocal laser microscopy, applied and environmental microbiology. Appl Environ Microbiol 61(3):1013–1019

Amundson R, Berhe AA, Hopmans JW, Olson C, Sztein AE, Sparks DL (2015) Soil and human security in the 21st century. Science 348:1261071–1261071. https://doi.org/10.1126/science.1261071

Anand K, Kumari B, Mallick M (2016) Phosphate solubilizing microbes: an effective and alternative approach as biofertilizers. Int J Pharm Pharm Sci 8:37–40

Aung TT, Buranabanyat B, Piromyou P, Longtonglang A (2013) Enhanced soybean biomass by co-inoculation of Bradyrhizobium japonicum and plant growth promoting rhizobacteria and its effects on microbial community structures. African J Microbiol Res 7:3858–3873. https://doi.org/10.5897/AJMR2013.5917

Bach E, dos Santos Seger GD, de Carvalho Fernandes G, Lisboa BB, Passaglia LMP (2016) Evaluation of biological control and rhizosphere competence of plant growth promoting bacteria. Appl Soil Ecol 99:141–149. https://doi.org/10.1016/j.apsoil.2015.11.002

Bacon CW, Palencia ER, Hinton DM (2015) Abiotic and biotic plant stress-tolerant and beneficial secondary metabolites produced by Endophytic Bacillus species BT. In: Arora NK (ed) Plant microbes symbiosis: applied facets. Springer India, New Delhi, pp 163–177. https://doi.org/10.1007/978-81-322-2068-8_8

Bailey BA, Bae H, Strem MD, Roberts DP, Thomas SE, Crozier J, Samuels GJ, Choi I-Y, Holmes KA (2006) Fungal and plant gene expression during the colonization of cacao seedlings by endophytic isolates of four Trichoderma species. Planta 224:1449–1464. https://doi.org/10.1007/s00425-006-0314-0

Bais HP, Fall R, Vivanco JM (2004) Biocontrol of bacillus subtilis against infection of arabidopsis roots by pseudomonas syringae is facilitated by biofilm formation and surfactin production. Plant Physiol 134(1):307–319. https://doi.org/10.1104/pp.103.028712

Bais HP, Park S-W, Weir TL, Callaway RM, Vivanco JM (2004) How plants communicate using the underground information superhighway. Trends Plant Sci 9:26–32. https://doi.org/10.1016/j.tplants.2003.11.008

Bais HP, Weir TL, Perry LG, Gilroy S, Vivanco JM (2006) The role of root exudates in rhizosphere interactions with plants and other organisms. Annu Rev Plant Biol 57:233–266. https://doi.org/10.1146/annurev.arplant.57.032905.105159

Baker R (1988) *Trzchoderma* SPP. as plant-growth stimulants. Crit Rev Biotechnol 7:97–106. https://doi.org/10.3109/07388558809150724

Bakker MG, Manter DK, Sheflin AM, Weir TL, Vivanco JM (2012) Harnessing the rhizosphere microbiome through plant breeding and agricultural management. Plant Soil 360(1):1–13. https://doi.org/10.1007/s11104-012-1361-x

Baltrus DA (2017) Adaptation, specialization, and coevolution within phytobiomes. Curr Opin Plant Biol 38:109–116. https://doi.org/10.1016/j.pbi.2017.04.023

Barea JM (2015) Future challenges and perspectives for applying microbial biotechnology in sustainable agriculture based on a better understanding of plant-microbiome interactions. J Soil Sci Plant Nutr 15:261–282. https://doi.org/10.4067/S0718-95162015005000021

Bargaz A, Lyamlouli K, Chtouki M, Zeroual Y, Dhiba D (2018) Soil microbial resources for improving fertilizers efficiency in an integrated plant nutrient management system. Front Microbiol 9:1606. https://doi.org/10.3389/fmicb.2018.01606

Beattie G (2006) Plant-associated bacteria: survey, molecular phylogeny, genomics and recent advances. In: Plant-Associated Bacteria, pp 1–56. https://doi.org/10.1007/978-1-4020-4538-7_1

Bishnoi U (2015) PGPR interaction: an ecofriendly approach promoting the sustainable agriculture system. Elsevier Ltd. https://doi.org/10.1016/bs.abr.2015.09.006

Borriss R (2015) Bacillus, a plant-beneficial bacterium. In: Principles of plant-microbe interactions. Springer International Publishing, Cham, pp 379–391. https://doi.org/10.1007/978-3-319-08575-3_40

Bouffaud M-L, Renoud S, Dubost A, Moënne-Loccoz Y, Muller D (2018) 1-Aminocyclopropane-1-carboxylate deaminase producers associated to maize and other Poaceae species. Microbiome 6:114. https://doi.org/10.1186/s40168-018-0503-7

Brown P, Saa S (2015) Biostimulants in agriculture. Front Plant Sci 6:671. https://doi.org/10.3389/fpls.2015.00671

Budiharjo A, Chowdhury SP, Dietel K, Beator B, Dolgova O, Fan B, Bleiss W, Ziegler J, Schmid M, Hartmann A, Borriss R (2014) Transposon mutagenesis of the plant-associated Bacillus amyloliquefaciens ssp. plantarum FZB42 revealed that the nfrA and RBAM17410 genes are involved in plant-microbe-interactions. PLoS One 9:1–13. https://doi.org/10.1371/journal.pone.0098267

Cai T, Cai W, Zhang J, Zheng H, Tsou AM, Xiao L, Zhong Z, Zhu J (2009) Host legume-exuded antimetabolites optimize the symbiotic rhizosphere. Mol Microbiol 73:507–517. https://doi.org/10.1111/j.1365-2958.2009.06790.x

Chakdar H, Dastager SG, Khire JM, Rane D, Dharne MS (2018) Characterization of mineral phosphate solubilizing and plant growth promoting bacteria from termite soil of arid region. 3 Biotech 8:463. https://doi.org/10.1007/s13205-018-1488-4

Chaparro AP, Carvajal LH, Orduz S (2011) Fungicide tolerance of Trichoderma asperelloides and T. harzianum strains Agric Sci 2:301–307

Chen R, Barphagha IK, Karki HS, Ham JH (2012) Dissection of quorum-sensing genes in Burkholderia glumae reveals non-canonical regulation and the new regulatory gene tofM for Toxoflavin production. PLoS One 7:e52150

Cook RJ (1993) Making greater use of introduced microorganisms for biological control of plant pathogens. Annu Rev Phytopathol 31:53–80. https://doi.org/10.1146/annurev.py.31.090193.000413

de Souza R, Ambrosini A, Passaglia LMP (2015) Plant growth-promoting bacteria as inoculants in agricultural soils. Genet Mol Biol 38:401–419. https://doi.org/10.1590/S1415-475738420150053

Díaz Herrera S, Grossi C, Zawoznik M, Groppa MD (2016) Wheat seeds harbour bacterial endophytes with potential as plant growth promoters and biocontrol agents of Fusarium graminearum. Microbiol Res 186:37–43. https://doi.org/10.1016/J.MICRES.2016.03.002

Doornbos RF, van Loon LC, Bakker PAHM (2012) Impact of root exudates and plant defense signaling on bacterial communities in the rhizosphere. A review. Agron Sustain Dev 32:227–243. https://doi.org/10.1007/s13593-011-0028-y

du Jardin P (2015) Plant biostimulants: definition, concept, main categories and regulation. Sci Hortic (Amsterdam) 196:3–14. https://doi.org/10.1016/j.scienta.2015.09.021

FAO (2016) The state of food and agriculture. Fixers N, solubilizers P fertecon biofertilizers. FAO, Rome. http://www.fao.org/publications/sofa/2016/en/

Finkel OM, Castrillo G, Herrera Paredes S, Salas González I, Dangl JL (2017) Understanding and exploiting plant beneficial microbes. Curr Opin Plant Biol 38:155–163. https://doi.org/10.1016/J.PBI.2017.04.018

Foo JL, Ling H, Lee YS, Chang MW (2017) Microbiome engineering: current applications and its future. Biotechnol J 12:1600099. https://doi.org/10.1002/biot.201600099

Geetha SJ, Joshi SJ (2013) Engineering rhizobial bioinoculants: a strategy to improve iron nutrition. ScientificWorldJournal 2013:315890. https://doi.org/10.1155/2013/315890

Gond SK, Torres MS, Bergen MS, Helsel Z, White JF Jr (2015) Induction of salt tolerance and up-regulation of aquaporin genes in tropical corn by rhizobacterium Pantoea agglomerans. Lett Appl Microbiol 60:392–399. https://doi.org/10.1111/lam.12385

Götz C, Fekete A, Gebefuegi I, Forczek ST, Fuksová K, Li X, Englmann M, Gryndler M, Hartmann A, Matucha M, Schmitt-Kopplin P, Schröder P (2007) Uptake, degradation and chiral discrimination of N-acyl-D/L-homoserine lactones by barley (Hordeum vulgare) and yam

bean (Pachyrhizus erosus) plants. Anal Bioanal Chem 389:1447–1457. https://doi.org/10.1007/s00216-007-1579-2
Gupta S, Dikshit AK (2010) Biopesticides: an ecofriendly approach for pest control. J Biopest 3:186–188. https://doi.org/10.4172/2168-9881.S1.012
Haghighi BJ, Alizadeh O, Firoozabadi AH (2011) The role of plant growth promoting rhizobacteria (PGPR) in sustainable agriculture. Adv Environ Biol 5:3079–3083
Han S, Li D, Trost E, Mayer KF, Vlot AC, Heller W, Schmid M, Hartmann A, Rothballer M (2016) Systemic responses of barley to the 3-hydroxy-decanoyl-homoserine lactone producing plant beneficial Endophyte Acidovorax radicis N35. Front Plant Sci 7:1868. https://doi.org/10.3389/fpls.2016.01868
Han T, You C, Zhang L, Feng C, Zhang C, Wang J, Kong F (2016) Biocontrol potential of antagonist Bacillus Subtilis Tpb55 against tobacco black shank. BioControl 61(2):195–205. https://doi.org/10.1007/s10526-015-9705-0
Hannula SE, Zhu F, Heinen R, Bezemer TM (2019) Foliar-feeding insects acquire microbiomes from the soil rather than the host plant. Nat Commun 10:1–9. https://doi.org/10.1038/s41467-019-09284-w
Hardoim PR, Hardoim CCP, van Overbeek LS, van Elsas JD (2012) Dynamics of seed-borne Rice Endophytes on early plant growth stages. PLoS One 7:e30438. https://doi.org/10.1371/journal.pone.0030438
Harman G, Howell CR, Viterbo A, Chet I, Lorito M (2004) Trichoderma species — opportunistic, avirulent plant symbionts. Nat Rev Microbiol 2:43–56. https://doi.org/10.1038/nrmicro797
Hartmann A, Schmid M, van Tuinen D, Berg G (2009) Plant-driven selection of microbes. Plant Soil 321:235–257. https://doi.org/10.1007/s11104-008-9814-y
Hawkes CV, Connor EW (2017) Translating Phytobiomes from theory to practice: ecological and evolutionary considerations. Phytobiomes J 1:57–69. https://doi.org/10.1094/PBIOMES-05-17-0019-RVW
Hunter MC, Smith RG, Schipanski ME, Atwood LW, Mortensen DA (2017) Agriculture in 2050: recalibrating targets for sustainable intensification. Bioscience 67:386–391. https://doi.org/10.1093/biosci/bix010
Igual J, Valverde A, Cervantes E, Velazquez E (2001) Phosphate-solubilizing bacteria as inoculants 979 for agriculture: use updated molecular techniques in their study. Agronomie 21:561–568
Jacoby R, Peukert M, Succurro A, Koprivova A, Kopriva S (2017) The role of soil microorganisms in plant mineral nutrition-current knowledge and future directions. Front Plant Sci 8:1617. https://doi.org/10.3389/fpls.2017.01617
Jasim B, Joseph AA, John CJ, Mathew J, Radhakrishnan EK (2014) Isolation and characterization of plant growth promoting endophytic bacteria from the rhizome of Zingiber officinale. 3 Biotech 4:197–204. https://doi.org/10.1007/s13205-013-0143-3
Jetiyanon K, Fowler WD, Kloepper JW (2003) Broad-Spectrum protection against several pathogens by PGPR mixtures under field conditions in Thailand. Plant Dis 87:1390–1394. https://doi.org/10.1094/PDIS.2003.87.11.1390
Kafle A, Cope K, Raths R, Krishna Yakha J, Subramanian S, Bücking H, Garcia K, Kafle A, Cope KR, Raths R, Krishna Yakha J, Subramanian S, Bücking H, Garcia K (2019) Harnessing soil microbes to improve plant phosphate efficiency in cropping systems. Agronomy 9:127. https://doi.org/10.3390/agronomy9030127
Kashyap PL, Srivastava AK, Tiwari SP, Kumar S (eds) (2018) Microbes for climate resilient agriculture. Wiley, Hoboken. https://doi.org/10.1002/9781119276050
Kashyap BK, Solanki MK, Pandey AK, Prabha S, Kumar P, Kumari B (2019) Bacillus as plant growth promoting Rhizobacteria (PGPR): a promising green agriculture technology BT. In: Ansari RA, Mahmood I (eds) Plant health under biotic stress: volume 2: microbial interactions. Springer Singapore, Singapore, pp 219–236. https://doi.org/10.1007/978-981-13-6040-4_11
Kasim WA, Gaafar RM, Abou-Ali RM, Omar MN, Hewait HM (2016) Effect of biofilm forming plant growth promoting rhizobacteria on salinity tolerance in barley. Ann Agric Sci 61:217–227. https://doi.org/10.1016/j.aoas.2016.07.003

Kavamura VN, de Melo IS (2014) Effects of different osmolarities on bacterial biofilm formation. Braz J Microbiol 45:627–631. https://doi.org/10.1590/s1517-83822014000200034

Khezri M, Ahmadzadeh M, Jouzani GS, Behboudi K, Ahangaran A, Mousivand M, Rahimian H (2011) Characterization of some biofilm-forming Bacillus subtilis strains and evaluation of their biocontrol potential against Fusarium culmorum. J Plant Pathol 93:373–382. https://doi.org/10.4454/JPP.V93I2.1192

Kumari B, Hora A, Mallick M (2017) Stimulatory effect of PGPR (plant growth promoting Rhizospheric Bacteria) on medicinal and growth properties of a potential medicinal herb Chlorophytum borivilianum: a review. J Plant Sci Res 33(2):151–156

Kumari B, Mallick MA, Hora A (2016) Plant growth-promoting rhizobacteria (PGPR): their potential for development of sustainable agriculture. In: Trivedi PC (ed) Bio-exploitation for sustainable agriculture, Ist edn. Avinskar publishing house, Jaipur, pp 1–19

Kumari B, Mallick MA, Solanki MK, Solanki AC, Hora A, Guo W (2019a) Plant Growth Promoting Rhizobacteria (PGPR): modern prospects for sustainable agriculture. In: Plant health under biotic stress. Springer Singapore, Singapore, pp 109–127. https://doi.org/10.1007/978-981-13-6040-4_6

Kumari B, Mallick MA, Solanki MK, Solanki AC, Hora A, Guo W (2019b) Plant growth promoting Rhizobacteria (PGPR): modern prospects for sustainable agriculture BT. In: Ansari RA, Mahmood I (eds) Plant health under biotic stress: volume 2: microbial interactions. Springer Singapore, Singapore, pp 109–127. https://doi.org/10.1007/978-981-13-6040-4_6

Larran S, Simón MR, Moreno MV, Santamarina Siurana MP, Perelló A (2016) Endophytes from wheat as biocontrol agents against tan spot disease. Biol Control 92:17–23. https://doi.org/10.1016/j.biocontrol.2015.09.002

Le Mire G, Nguyen ML, Fassotte B et al (2016) Review : implementing plant biostimulants and biocontrol strategies in the agroecological management of cultivated ecosystems review: implementing plant biostimulants and biocontrol strategies in the agroecological management of cultivated ecosystems. Biotechnol Agron Société Environ 20:299–313

Lemanceau P, Barret M, Mazurier S, Mondy S, Pivato B, Fort T, Vacher C (2017) Plant communication with associated microbiota in the spermosphere, rhizosphere and phyllosphere. Adv Bot Res 82:101–133. https://doi.org/10.1016/BS.ABR.2016.10.007

Lucas JA (2011) Advances in plant disease and pest management. J Agr Sci 149:91–114

Manjunath M, Kanchan A, Ranjan K, Venkatachalam S, Prasanna R, Ramakrishnan B, Hossain F, Nain L, Shivay YS, Rai AB, Singh B (2016) Beneficial cyanobacteria and eubacteria synergistically enhance bioavailability of soil nutrients and yield of okra. Heliyon 2:e00066. https://doi.org/10.1016/j.heliyon.2016.e00066

Majeed A, Abbasi MK, Hameed S, Imran A, Rahim N (2015) Isolation and characterization of plant growth-promoting rhizobacteria from wheat rhizosphere and their effect on plant growth promotion. Front Microbiol 6:198. https://doi.org/10.3389/fmicb.2015.00198

Malik DK, Sindhu SS (2011) Production of indole acetic acid by Pseudomonas Sp.: effect of Coinoculation with Mesorhizobium Sp. Cicer on nodulation and plant growth of chickpea (Cicer Arietinum). Physiol Mol Biol Plants 17(1):25–32. https://doi.org/10.1007/s12298-010-0041-7

Mark GL, Dow JM, Kiely PD, Higgins H, Haynes J, Baysse C, Abbas A, Foley T, Franks A, Morrissey J, O'Gara F (2005) Transcriptome profiling of bacterial responses to root exudates identifies genes involved in microbe-plant interactions. Proc Natl Acad Sci U S A 102:17454–17459. https://doi.org/10.1073/pnas.0506407102

Masciarelli O, Llanes A, Luna V (2014) A new PGPR co-inoculated with Bradyrhizobium japonicum enhances soybean nodulation. Microbiol Res 169:609–615. https://doi.org/10.1016/j.micres.2013.10.001

Mishra VK, Kumar A (2009) Impact of metal nanoparticles on the plant growth promoting rhizobacteria. Dig J Nanomater Biostruct 4(3):587–592

Mohammad N, Alam MZ, Kabbashi NA, Ahsan A (2012) Effective composting of oil palm industrial waste by filamentous fungi: a review. Resour Conserv Recycl 58:69–78. https://doi.org/10.1016/j.resconrec.2011.10.009

Mohsenzadeh F, Shahrokhi F (2014) Biological removing of cadmium from contaminated media by fungal biomass of Trichoderma species. J Environ Health Sci Eng 12:102. https://doi.org/10.1186/2052-336X-12-102

Morawicki RO, Díaz González DJ (2018) Food sustainability in the context of human behavior. Yale J Biol Med 91:191–196

Moshynets OV, Babenko LM, Rogalsky SP, Iungin OS, Foster J, Kosakivska IV, Potters G, Spiers AJ (2019) Priming winter wheat seeds with the bacterial quorum sensing signal N-hexanoyl-L-homoserine lactone (C6-HSL) shows potential to improve plant growth and seed yield. PLoS One 14:e0209460

Mücke S, Reschke M, Erkes A, Schwietzer C-A, Becker S, Streubel J, Morgan RD, Wilson GG, Grau J, Boch J (2019) Transcriptional reprogramming of Rice cells by Xanthomonas oryzae TALEs. Front Plant Sci 10:162. https://doi.org/10.3389/fpls.2019.00162

Nakkeeran S, Fernando WGD, Siddiqui ZA (2006) Plant growth promoting rhizobacteria formulations and its scope in commercialization for the management of pests and diseases. In: PGPR Biocontrol biofertilization, pp 257–296. https://doi.org/10.1007/1-4020-4152-7_10

Naseri R, Moghadam A, Darabi F, Hatami A, G.R.T (2013) The effect of deficit irrigation and Azotobacter chroococcum and Azospirillum brasilense on grain yield, yield components of maize (SC 704) as a second cropping in western Iran. Bull Env Pharmacol Life Sci 2:104–112

Pandin C, Le Coq D, Canette A, Ephane Aymerich S, Briandet R (2017) Minireview should the biofilm mode of life be taken into consideration for microbial biocontrol agents? Microb Biotechnol 10:719–734. https://doi.org/10.1111/1751-7915.12693

Patil HJ, Solanki MK (2016) Microbial inoculant: modern era of fertilizers and pesticides. In: Singh D, Singh H, Prabha R (eds) Microbial inoculants in sustainable agricultural productivity, Research perspectives, vol 1. Springer, New Delhi, pp 319–343. https://doi.org/10.1007/978-81-322-2647-5_19

Pham VTK, Rediers H, Ghequire MGK, Nguyen HH, De Mot R, Vanderleyden J, Spaepen S (2017) The plant growth-promoting effect of the nitrogen-fixing endophyte Pseudomonas stutzeri A15. Arch Microbiol 199:513–517. https://doi.org/10.1007/s00203-016-1332-3

Rafique M, Hayat K, Mukhtar T, Khan A, Afridi M, Hussain T, Sultan T, Munis M, Imran M, Chaudhary H (2015) Bacterial biofilm formation and its role against agricultural pathogens. In: Mendez-Vilas A (ed) The battle against microbial pathogens: basic science, technological advances and educational programs. Formatex Research Center, Spain, pp 373–382

Rahman MM, Flory E, Koyro H-W, Abideen Z, Schikora A, Suarez C, Schnell S, Cardinale M (2018) Consistent associations with beneficial bacteria in the seed endosphere of barley (Hordeum vulgare L.). Syst Appl Microbiol 41:386–398. https://doi.org/10.1016/j.syapm.2018.02.003

Ramachandran R, Burke AK, Cormier G, Jensen RV, Stevens AM (2014) Transcriptome-based analysis of the Pantoea stewartii quorum-sensing regulon and identification of EsaR direct targets. Appl Environ Microbiol 80:5790 LP–5795800. https://doi.org/10.1128/AEM.01489-14

Ramos-González MI, Campos MJ, Ramos JL (2005) Analysis of Pseudomonas putida KT2440 gene expression in the maize rhizosphere: in vitro expression technology capture and identification of root-activated promoters. J Bacteriol 187:4033 LP–4034041. https://doi.org/10.1128/JB.187.12.4033-4041.2005

Rankl S, Gunsé B, Sieper T, Schmid C, Poschenrieder C, Schröder P (2016) Microbial homoserine lactones (AHLs) are effectors of root morphological changes in barley. Plant Sci 253:130–140. https://doi.org/10.1016/j.plantsci.2016.09.014

Rekha K, Baskar B, Srinath S, Usha B (2017) Plant-growth-promoting rhizobacteria Bacillus subtilis RR4 isolated from rice rhizosphere induces malic acid biosynthesis in rice roots. Can J Microbiol 64:20–27. https://doi.org/10.1139/cjm-2017-0409

Ribeiro VP, Marriel IE, Morais De Sousa S, Gomes U, Lana P, Mattos BB, Abreu De Oliveira C, Gomes EA (2018) Environmental microbiology Endophytic Bacillus strains enhance pearl millet growth and nutrient uptake under low-P. https://doi.org/10.1016/j.bjm.2018.06.005

Rijavec T, Lapanje A, Dermastia M, Rupnik M (2007) Isolation of bacterial endophytes from germinated maize kernels. Can J Microbiol 53:802–808. https://doi.org/10.1139/W07-048

Rognes T, Flouri T, Nichols B, Quince C, Mahé F (2016) VSEARCH: a versatile open source tool for metagenomics. PeerJ 4:e2584. https://doi.org/10.7717/peerj.2584

Rosenblueth M, López-López A, Martínez J, Rogel MA, Toledo I, Martínez-Romero E (2012) Seed bacterial endophytes: common genera, seed-to-seed variability and their possible role in plants. Acta Hortic:39–48. https://doi.org/10.17660/ActaHortic.2012.938.4

Rosier A, Medeiros FHV, Bais HP (2018) Defining plant growth promoting rhizobacteria molecular and biochemical networks in beneficial plant-microbe interactions. Plant Soil 428:35–55. https://doi.org/10.1007/s11104-018-3679-5

Rozier C, Hamzaoui J, Lemoine D, Czarnes S, Legendre L (2017) Field-based assessment of the mechanism of maize yield enhancement by Azospirillum lipoferum CRT1. Sci Rep 7:7416. https://doi.org/10.1038/s41598-017-07929-8

Rubin RL, van Groenigen KJ, Hungate BA (2017) Plant growth promoting rhizobacteria are more effective under drought: a meta-analysis. Plant Soil 416:309–323. https://doi.org/10.1007/s11104-017-3199-8

Ruiza D, Agaras B, de Werrab P, Wall LG, Valverde C, Wall LG, Valverde C (2011) Characterization and screening of plant probiotic traits of bacteria isolated from rice seeds cultivated in Argentina. J Microbiol 49:902–912. https://doi.org/10.1007/s12275-011-1073-6

Sandhya V, Shrivastava M, Ali SZ, Sai Shiva Krishna Prasad V (2017) Endophytes from maize with plant growth promotion and biocontrol activity under drought stress. Russ Agric Sci 43:22–34. https://doi.org/10.3103/S1068367417010165

Sang MK, Jeong J-J, Kim J, Kim KD (2018) Growth promotion and root colonisation in pepper plants by phosphate-solubilising *Chryseobacterium* sp. strain ISE14 that suppresses Phytophthora blight. Ann Appl Biol 172:208–223. https://doi.org/10.1111/aab.12413

Santoyo G, Moreno-Hagelsieb G, del Carmen Orozco-Mosqueda M, Glick BR (2016) Plant growth-promoting bacterial endophytes. Microbiol Res 183:92–99. https://doi.org/10.1016/j.micres.2015.11.008

Shahzad S, Arif M, Riaz M, Iqbal Z, Ashraf M (2013) PGPR with varied ACC-deaminase activity induced different growth and yield response in maize (Zea mays L.) under fertilized conditions. Eur J Soil Biol. https://doi.org/10.1016/j.ejsobi.2013.04.002

Sivakumar R, Ranjani J, Vishnu US, Jayashree S, Lozano GL, Miles J, Broderick NA, Guan C, Gunasekaran P, Handelsman J, Rajendhran J (2018) Evaluation of InSeq to identify genes essential for Pseudomonas aeruginosa PGPR2 corn root colonization. bioRxiv. https://doi.org/10.1101/377168

Steenhoudt O, Vanderleyden J (2000) Azospirillum, a free-living nitrogen-fixing bacterium closely associated with grasses: genetic, biochemical and ecological aspects. FEMS Microbiol Rev 24:487–506. https://doi.org/10.1111/j.1574-6976.2000.tb00552.x

Sudha ML, Viswanath P, Siddappa V, Rajarathnam S, Shashirekha MN (2016) Control of rope spore forming bacteria using carambola (Averrhoa carambola) fruit pomace powder in wheat bread preparation. Qual Assur Saf Crop Food 8:555–564. https://doi.org/10.3920/QAS2014.0409

Sumi CD, Yang BW, Yeo I-C, Hahm YT (2015) Antimicrobial peptides of the genus Bacillus : a new era for antibiotics. Can J Microbiol 61:93–103. https://doi.org/10.1139/cjm-2014-0613

Svenningsen NB, Watts-Williams SJ, Joner EJ, Battini F, Efthymiou A, Cruz-Paredes C, Nybroe O, Jakobsen I (2018) Suppression of the activity of arbuscular mycorrhizal fungi by the soil microbiota. ISME J 12:1296–1307. https://doi.org/10.1038/s41396-018-0059-3

Tan W-S, Chang C-Y, Yin W-F, Chan K-G (2015) Understanding the quorum-sensing bacterium Pantoea stewartii strain M009 with whole-genome sequencing analysis. Genome Announc 3:e01509–e01514. https://doi.org/10.1128/genomeA.01509-14

Tao A, Pang F, Huang S, Yu G, Li B, Wang T (2014) Characterisation of endophytic Bacillus thuringiensis strains isolated from wheat plants as biocontrol agents against wheat flag smut. Biocontrol Sci Tech 24:901–924. https://doi.org/10.1080/09583157.2014.904502

Tchagang CF, Xu R, Overy D, Blackwell B, Chabot D, Hubbard K, Doumbou CL, Bromfield ESP, Tambong JT (2018) Diversity of bacteria associated with corn roots inoculated with Canadian woodland soils, and description of Pseudomonas aylmerense sp. nov. Heliyon 4:e00761. https://doi.org/10.1016/j.heliyon.2018.e00761

Tilman D, Cassman KG, Matson PA, Naylor R, Polasky S (2002) Agricultural sustainability and intensive production practices. Nature 418:671

Timmusk S, Abd El-Daim IA, Copolovici L, Tanilas T, Kännaste A, Behers L, Nevo E, Seisenbaeva G, Stenström E, Niinemets Ü (2014) Drought-tolerance of wheat improved by Rhizosphere Bacteria from harsh environments: enhanced biomass production and reduced emissions of stress volatiles. PLoS One 9:e96086. https://doi.org/10.1371/journal.pone.0096086

Tiwari S, Prasad V, Chauhan PS, Lata C (2017) Bacillus amyloliquefaciens confers tolerance to various abiotic stresses and modulates plant response to Phytohormones through Osmoprotection and gene expression regulation in Rice. Front Plant Sci 8:1510. https://doi.org/10.3389/fpls.2017.01510

Valverde A, Burgos A, Fiscella T, Rivas R, Velazquez E, Rodriguez-Barrueco C, Cervantes E, Chamber M, Igual JM (2006) Differential effects of coinoculations with Pseudomonas jessenii PS06 (a phosphate-solubilizing bacterium) and mesorhizobium ciceri C-2/2 strains on the growth and seed yield of chickpea under greenhouse and field conditions. Plant Soil 287:43–50

Verma JP, Yadav J, Tiwari KN (2012) Enhancement of nodulation and yield of chickpea by co-inoculation of indigenous mesorhizobium spp. and plant growth–promoting rhizobacteria in eastern Uttar Pradesh. Commun Soil Sci Plant Anal 43(3):605–621. https://doi.org/10.1080/00103624.2012.639110

Venturi V (2006) Regulation of quorum sensing in Pseudomonas. FEMS Microbiol Rev 30:274–291. https://doi.org/10.1111/j.1574-6976.2005.00012.x

Wani PA, Khan MS, Zaidi A (2007) Effect of metal tolerant plant growth promoting Bradyrhizobium Sp. (Vigna) on growth, symbiosis, seed yield and metal uptake by greengram plants. Chemosphere 70(1):36–45. https://doi.org/10.1016/j.chemosphere.2007.07.028

Walker TS, Bais HP, Grotewold E, Vivanco JM (2003) Root exudation and rhizosphere biology. Plant Physiol 132:44 LP–44 51. https://doi.org/10.1104/pp.102.019661

Yadav BK, Akhtar MS, Panwar J (2015) Rhizospheric plant-microbe interactions: key factors to soil fertility and plant nutrition BT. In: Arora NK (ed) Plant microbes Symbiosis: applied facets. Springer India, New Delhi, pp 127–145. https://doi.org/10.1007/978-81-322-2068-8_6

Yang J, Kloepper JW, Ryu CM (2009) Rhizosphere bacteria help plants tolerate abiotic stress. Trends Plant Sci 14:1–4. https://doi.org/10.1016/j.tplants.2008.10.004

Yang L, Danzberger J, Schöler A, Schröder P, Schloter M, Radl V (2017) Dominant groups of potentially active bacteria shared by barley seeds become less abundant in root associated microbiome. Front Plant Sci 8:1005. https://doi.org/10.3389/fpls.2017.01005

Yoshida K, Saitoh H, Fujisawa S, Kanzaki H, Matsumura H, Yoshida K, Tosa Y, Chuma I, Takano Y, Win J, Kamoun S, Terauchi R (2009) Association genetics reveals three novel Avirulence genes from the Rice blast fungal pathogen Magnaporthe oryzae. Plant Cell Online 21:1573–1591. https://doi.org/10.1105/tpc.109.066324

Yu C, Chen H, Tian F, Leach JE, He C (2014) Differentially-expressed genes in rice infected by Xanthomonas oryzae pv. Oryzae relative to a flagellin-deficient mutant reveal potential functions of flagellin in host-pathogen interactions. Rice (N Y) 7:20. https://doi.org/10.1186/s12284-014-0020-7

Yuan M, Chu Z, Li X, Xu C, Wang S (2010) The bacterial pathogen Xanthomonas oryzae overcomes rice defenses by regulating host copper redistribution. Plant Cell 22:3164–3176. https://doi.org/10.1105/tpc.110.078022

Zahedi H, Abbasi S (2015) Effect of plant growth promoting rhizobacteria (PGPR) and water stress on phytohormones and polyamines of soybean. Indian J Agric Res 49:427–431. https://doi.org/10.18805/ijare.v49i5.5805

The manufacturer's authorised representative in the EU is Springer Nature Customer Service Centre GmbH, Europaplatz 3, 69115 Heidelberg, Germany. If you have any concerns regarding our products, please contact ProductSafety@springernature.com

Printed and bound by CPI Group (UK) Ltd, Croydon, CR0 4YY

30/09/2026

02225557-0013